Energy Policies, Politics and Prices Series

The Struggle for Energy Independence in America: Regulations, Overreach, and Security
Victoria Whitehead (Editor)
ISBN: 979-8-89530-558-4
eBook ISBN: 979-8-89530-585-0

Pipeline Safety: Background, Practices and Regulatory Changes
Michael E. Massaro (Editor)
ISBN: 979-8-89530-058-9
eBook ISBN: 979-8-89530-066-4

Sustainable and Green Energy Policies
Kürşad Emrah Yıldırım, PhD,
and Eser Gemici, PhD
ISBN: 979-8-89113-773-8
eBook ISBN: 979-8-89113-852-0

The Development of Crude Oil and Natural Gas Prices and the Energy Security of the European Union
Matej Korček, PhD,
and Saleh Mothana Obadi, PhD
ISBN: 979-8-89113-357-0
eBook ISBN: 979-8-89113-591-8

The Future of Energy: Challenges, Perspectives, and Solutions
Thomas Valone, PhD (Editor)
ISBN: 978-1-53618-186-9
eBook ISBN: 978-1-53618-348-1

A Strategic Evaluation of Energy Security in the Eastern Mediterranean
Michail Chalaris, PhD
ISBN: 978-1-53619-273-5
eBook ISBN: 978-1-53619-346-6

Caspian: Status, Challenges, and Prospects. An Analysis into the Legal Classification, Security and Environmental Concerns, Geopolitics and Energy Flow Impact of the Caspian Plateau
Anis H. Bajrektarevic, PhD,
Mak Bajrekataravic,
and Tanvi Kamlesh Chauhan
ISBN: 978-1-53618-956-8
eBook ISBN: 978-1-53619-001-4

International Policies and Issues of Nuclear Safety and Management
Matthew D. Leeland (Editor)
ISBN: 978-1-61761-596-2
eBook ISBN: 978-1-61122-087-2

More information about this series can be found at
https://novapublishers.com/product-category/series/energy-policies-politics-and-prices/

Energy Policy and National Securit

Strengthening America's Economic and Environmental Futu

snova
New York

Kuzman Sukanya
Editor

Energy Policy and National Security

Strengthening America's Economic and Environmental Future

snova
New York

Library of Congress Cataloging-in-Publication Data

ISBN: 979-8-89530-879-0 (Hardcover)
ISBN: 979-8-89530-996-4 (eBook)

Published by Nova Science Publishers, Inc. † New York

Contents

Preface

This book explores the critical role of energy in shaping the nation's security, economy, and environmental landscape. It examines legislative strategies for ensuring affordable and reliable energy access, fortifying supply chains, and addressing climate change. Through analyses from congressional committees, the book provides insights into the expansion of American energy production, the implications of depleting the Strategic Petroleum Reserve, and the broader impact of energy policy decisions on economic stability and national security.

Chapter 1

American Energy Expansion: Strengthening Economic, Environmental, and National Security*

Committee on Energy and Commerce

Tuesday, January 31, 2023
House of Representatives,
Committee on Energy and Commerce,
Washington, DC.

The committee met, pursuant to call, at 11:02 a.m., in the John D. Dingell Room 2123, Rayburn House Office Building, Hon. Cathy McMorris Rodgers (chair of the committee) presiding.

Members present: Representatives Rodgers, Burgess, Latta, Guthrie, Griffith, Bilirakis, Johnson, Bucshon, Hudson, Walberg, Carter, Duncan, Palmer, Dunn, Curtis, Lesko, Crenshaw, Joyce, Armstrong, Weber, Allen, Baldersen, Fulcher, Pfluger, Harshbarger, Miller-Meeks, Cammack, Obernolte, Pallone (ranking member), Eshoo, DeGette, Schakowsky, Matsui, Castor, Sarbanes, Tonko, Clarke, Cárdenas, Ruiz, Peters, Dingell, Veasey, Kuster, Kelly, Barraga´n, Blunt Rochester, Soto, Craig, Schrier, Trahan, and Fletcher.

Staff present: Kate Arey, Content Manager and Digital Assistant; Jolie Brochin, Clerk, Health; Sarah Burke, Deputy Staff Director; Michael Cameron, Professional Staff Member, Innovations, Data, and Commerce; Lauren Eriksen, Clerk, Oversight and Investigations; Theresa Gambo, Financial and Office Administrator; Jessica Herron, Clerk, Innovations, Data, and Commerce; Nate Hodson, Staff Director; Tara Hupman, Chief Counsel; Noah Jackson, Clerk, Communications and Technology; Sean Kelly, Press Secretary; Peter Kielty, General Counsel; Emily King, Member Services Director; Elise Krekorian, Professional Staff Member, Energy; Giulia Leganski, Professional Staff Member, Communications and Technology; Mary Martin, Chief Counsel, Energy and Environment; Brandon Mooney, Deputy Chief Counsel, Energy; Clare Paoletta, Professional Staff Member, Health; Kaitlyn Peterson, Clerk, Energy and Environment; Peter Spencer, Senior Professional Staff Member, Energy; Michael Taggart, Policy Director; Hannah Anton, Minority Staff Assistant; Waverly Gordon, Minority Deputy

* This is an edited, reformatted and augmented version of Hearing before the Committee on Energy and Commerce, House of Representatives, One Hundred Eighteenth Congress, First Session, Serial No. 118–1, dated January 31, 2023.

In: Energy Policy and National Security
Editor: Kuzman Sukanya
ISBN: 979-8-89530-879-0

Staff Director and General Counsel; Tiffany Guarascio, Minority Staff Director; Anthony Gutierrez, Minority Professional Staff Member; Caitlin Haberman, Minority Staff Director, Environment, Manufacturing, and Critical Minerals; Perry Hamilton, Minority Member Services and Outreach Manager; Kris Pittard, Minority Professional Staff Member; Greg Pugh, Minority Staff Assistant; Caroline Rinker, Minority Press Assistant; Kylea Rogers, Minority Policy Analyst; Andrew Souvall, Minority Director of Communications, Outreach, and Member Services; Medha Surampudy, Minority Professional Staff Member; Rebecca Tomilchik, Minority Junior Professional Staff Member; Isaac Velez, Minority Intern; Tuley Wright, Minority Staff Director, Energy, Climate, and Grid Security; and C.J. Young, Minority Deputy Communications Director.

Mrs. RODGERS. The committee will come to order.

The Chair recognizes herself for 5 minutes for an opening statement.

Opening Statement of Hon. Cathy Mcmorris Rodgers, a Representative in Congress from the State of Washington

Welcome to our first Energy and Commerce Committee meeting hearing of the 118th Congress, especially to our new subcommittee chairs: Jeff Duncan, who is our new chair of Energy, Climate, and Grid Security Subcommittee, and Bill Johnson, who will chair the Environment, Manufacturing, and Critical Minerals Subcommittee. Energy is foundational to every aspect of our lives. Whether it is making energy more affordable and reliable, securing our supply chains, beating China, protecting the environment, addressing climate change, or putting energy security back at the center of policy making, these should be bipartisan goals, and we have already proven they can be over the last few weeks.

One of our actions on the floor was passing H.R. 22, the Protecting America's Strategic Petroleum Reserve from China Act. And it passed with 113 Democrats voting for it. And then just last week H.R. 21, the Strategic Production Response Act, in another bipartisan vote to help ensure a vital energy asset isn't drained for non-emergency political purposes.

These are just the beginning of a robust agenda to restore American energy dominance. We need to be doing more to secure and unleash American energy. Rush-to-green policies have had a devastating effect in Europe. Countries are rationing energy, switching back to coal-fired plants, and people are dying due to a lack of heat and electricity. America doesn't have to follow Europe down this path. Addressing emissions and unleashing abundant, affordable, and reliable energy aren't mutually exclusive.

In the last decade we have led the world in emissions reduction while increasing the availability of affordable energy through innovation. We did this while maintaining some of the highest environmental and labor standards in the world, and we must continue innovating and taking advantage of our abundant natural resources and reducing emissions.

Natural gas is one of our greatest economic and strategic resources. We cannot afford to shut it down. Instead, we must expand production and continue building pipelines, the safest, most reliable infrastructure to move it around the country, and increase our capacity to export LNG to the rest of the world, to reduce carbon emissions, and combat Russia and China's use of energy to expand their authoritarian influence.

We also have an opportunity to achieve a new renaissance in American nuclear technology and once again lead the world in its development and deployment. But to be successful, we need to rethink our restrictive regulatory approach and combat the antiscience opposition to expanding nuclear energy. We need to review the Nuclear Regulatory Commission's procedures and practices to be sure it can efficiently license new and advanced reactors.

Thankfully, we are close to finishing the Vogtle plant in Georgia, but one new nuclear plant isn't enough. Ensuring safe nuclear power is essential for ensuring American energy and also for protecting the environment and addressing climate change. And we can do this without inefficient, burdensome regulations. Safety cannot be an excuse for doing nothing.

To reestablish American nuclear energy globally, we need to approve and build new advanced reactors, establish a permanent spent fuel repository, while exploring the potential for fuel recycling, and establish a reliable domestic fuel industry that supplies and enriches uranium and reduces our reliance on Russia.

Across the board we need to update the regulatory landscape to pave the way for energy sources and technologies of all kinds: hydropower, nuclear, natural gas, and oil, hydrogen, along with wind, solar, and batteries. Republicans support all of these technologies, but we do not support picking winners and losers through massive subsidies and rigging regulations to favor certain industries.

Our energy solutions are climate solutions. We have a responsibility to our generation and future generations for cleaner water, air, and reduced emissions. Every energy technology and source has a role to play, but we need to be able to ensure that we are not becoming entirely dependent on unreliable, intermittent energy sources that rely on Chinese supply chains.

It is time to flip the switch and unleash American energy, and this committee is at the very center of securing our global leadership and making people's lives better.

I want to work with everyone, Republicans and Democrats, to achieve these goals. These cannot continue to be partisan issues. And my door is open to every member of this committee who wants to lead on unleashing American energy.

[The prepared statement of Mrs. Rodgers follows:]

Chair Cathy McMorris Rodgers Opening Remarks, Full Committee Hearing, American Energy Expansion: Strengthening Economic, Environmental, and National Security, January 31, 2023, as Prepared for Delivery

Welcome to our first Energy and Commerce Committee hearing of the 118th Congress...

… especially to our new subcommittee chairs... Jeff Duncan, who's our new Chair of the Energy, Climate, and Grid Security Subcommittee...

...and Bill Johnson, who'll Chair the Environment, Manufacturing, & Critical Minerals Subcommittee.

Energy is foundational to every aspect of Americans' lives.

Whether it's making energy more affordable and reliable, securing our supply chains, beating China, protecting the environment, addressing climate change, or putting energy security back at the center of policymaking.

These should be bipartisan goals, and we've already proven they can be over the last few weeks.

One of our actions on the floor was passing H.R. 22, the Protecting America's Strategic Petroleum Reserve from China Act…

...which passed with 113 Democrats voting for it.

And last week, we passed H.R. 21, the Strategic Production Response Act, in another bipartisan vote...

...to help ensure a vital energy asset isn't drained for non-emergency, political purposes. These are just the beginning of a robust agenda to restore American energy dominance. We need to be doing more to secure and unleash American energy.

'Rush to green' policies have had a devasting effect in Europe...

...countries are rationing energy, switching back to coal fired plants, and people are dying due to lack of heat and electricity.

America doesn't have to follow Europe down this path.

Addressing emissions and unleashing abundant, affordable, and reliable energy aren't mutually exclusive.

In the last decade, we've led the world in emissions reductions while increasing the availability of affordable energy through innovation.

We did this while maintaining some of the highest environmental and labor standards in the world.

We must continue innovating, taking advantage of our abundant natural resources, and reducing emissions.

Natural gas is one of our greatest economic and strategic resources... we cannot afford to shut it down.

Instead, we must expand production and continue building pipelines…

… the safest, most reliable infrastructure to move it around the country…

...and increase our capacity to export LNG to the rest of the world to reduce carbon emissions and combat Russia and China's use of energy to expand their authoritarian influence.

We also have an opportunity to achieve a new renaissance in American nuclear technology and once again lead the world in its development and deployment…

… but to be successful we need to rethink our restrictive regulatory approach and combat the anti-science opposition to expanding nuclear energy.

We need to review the Nuclear Regulatory Commission's procedures and practices, to be sure it can efficiently license new and advanced reactors.

Thankfully we are close to finishing the Vogtle Plant in Georgia, but one new nuclear plant isn't enough.

Ensuring safe nuclear power is essential for securing American energy, and also for protecting the environment and addressing climate change...

...and we can do this without inefficient, burdensome regulations. Safety cannot be an excuse for doing nothing.

To reestablish American nuclear leadership globally, we need to approve and build new, advanced reactors…

… establish a permanent spent fuel repository, while exploring the potential for fuel recycling…

… and establish a reliable domestic fuel industry that supplies and enriches uranium and reduces our reliance on Russia.

Across the board, we need to update the regulatory landscape to pave the way for energy sources and technologies of all kinds…

Hydropower, nuclear, Natural gas and oil, hydrogen…

… along with wind, solar, and batteries.

Republicans support all of these technologies, but we do not support picking winners and losers through massive subsidies and rigging regulations to only favor certain industries.

Our energy solutions are climate solutions. We have a responsibility to our generation and future generations for cleaner water, air, and reduced emissions.

Every energy technology and source has a role to play...

...but we need to be able to ensure we are not becoming entirely dependent on unreliable, intermittent energy sources that rely on Chinese supply chains.

It's time to flip the switch and unleash American energy...

...and this committee is at the very center of securing our global leadership and making people's lives better.

I want to work with everyone, Republicans and Democrats, to achieve these goals.

These can't continue to be partisan issues, and my door is open to every Member of this Committee who wants to lead on unleashing American energy.

I'll now recognize our full committee Ranking Member, Mr. Pallone, for five minutes for his opening statement.

* * *

Mrs. RODGERS. And now I will recognize our full committee ranking member, Mr. Pallone, for 5 minutes for his opening statement.

Opening Statement of Hon. Frank Pallone, Jr., a Representative in Congress from the State of New Jersey

Mr. PALLONE. Thank you, Chairwoman Rodgers.

And we just had our organizational meeting, and for those in the public, we are very proud of the fact that we now have the first woman chairperson of the Energy and Commerce Committee in its long history that goes back to almost the first days of the Republic. So let me just make that announcement. I am sure all of you recognize it, but I want to mention it again.

Now, I want to say that we are meeting for the committee's first hearing of the 118th Congress. The first few weeks of any Congress are an opportunity to lay out our priorities for the next 2 years. But, unfortunately, the energy bills that Republicans have brought to the floor this month have only demonstrated just how misguided and misplaced their priorities are.

Both bills that were mentioned by the chairwoman—the one that dealt with the Strategic Petroleum Reserve in China and the one we just dealt with last week that dealt with the Strategic Petroleum Reserve and said that we couldn't use it unless we opened up more public lands on the same percentage basis—they are both misguided because they limited this administration—or it was an attempt to limit this administration's ability to bring down gasoline prices at the pump. And instead of moving towards clean energy, which is the future and the only way that we are going to have less dependence on dictators and have less volatility in the market, it was an attempt to try to give more opportunities to Big Oil, which, frankly, they don't even want, right.

I mean, we know that during this gasoline crisis that we have had over the past year or so, this idea that somehow Big Oil wanted to pump more, they wanted to pump the leases they already had, was simply not the case. They wanted to keep the price artificially high. That is another reason why they opposed using the Strategic Petroleum Reserve: because it brought prices down.

So all I am basically saying is, as we move towards an energy transition, let's keep in mind that encouraging renewables—as we did with the Inflation Reduction Act, as we did with the Bipartisan Infrastructure Law—this is the way to go in the future.

I am certainly one of the key people that believes that we have to use all sources of energy, whether it is nuclear, hydropower, the other things that the chairwoman mentioned, certainly all of the above. But the bottom line is that we are going to be left behind if we don't move towards renewables and encourage them the way we did with the Inflation Reduction Act.

Now, the chairwoman mentioned China. Now, make no mistake that the rest of the world, including China, are already embarked in a major transmission to clean energy. In fact, China's investments in clean energy are so vast that the International Energy Agency forecasts that over the next several years 40 percent of solar and wind energy growth will come from China alone. And yet Republicans consistently oppose our efforts to catch up. If we don't move towards clean energy, if we don't use the proceeds of and the credits and the other incentives from the Inflation Reduction Act, we are going to be left behind. And that is going to mean left behind in terms of innovation, left behind in terms of the number of jobs that are created. And the irony of it is that although most, if not all, Republicans in the House voted against the Inflation Reduction Act, that many of those credits are going to the very red States and the industries in the very red States that want to take advantage of it.

So this is not a Republican or Democratic issue. This is a bipartisan issue. We need to move towards clean energy.

Madam Chair, I just wanted to mention that one of our witnesses today is Dr. Unruh Cohen. I want to thank her for all of her work as staff director for the Select Committee on the Climate Crisis. I look forward to her testimony and the testimony of all of the panel. But I just wanted to take a moment to thank my long-time—or our long-time—staff director of our Energy and Environment policy team, Rick Kessler, who is retiring this week. After 25 years of dedicated service to Congress, of which 12 years was spent with this committee, Rick has more than earned his retirement.

Earlier in his career Rick also worked in my personal office on energy and environmental issues. And last Congress, Rick was instrumental in our work on the Inflation Reduction Act, fighting to make sure meaningful climate investments, including many provisions from our own Clean Future Act, were included.

And during his tenure he was also instrumental in helping us pass the Frank Lautenberg Chemical Safety Act, the Pipeline Safety Act, legislation phasing down the use of dangerous HFCs, and many, many more energy and environmental laws.

So he has been a devoted member of my team and our team, always guiding us by a sincere desire to improve people's lives. And I am personally grateful for his many years of service, wish him nothing but the best in the future.

And with that, I yield back the balance of my time. [The prepared statement of Mr. Pallone follows:]

Committee on Energy and Commerce, Opening Statement as Prepared for Delivery of Ranking Member Frank Pallone, Jr., Hearing on "American Energy Expansion: Strengthening Economic, Environmental, and National Security," January 31, 2023

Thank you, Chair Rodgers. We are very proud of the fact that we now have the first woman to chair of the Energy and Commerce Committee and its long history that goes back to almost the first days of the Republic.

I want to say that we are meeting for the Committee's first hearing of the 118th Congress. The first few weeks of any Congress are an opportunity to lay out priorities for the next two years. Unfortunately, the energy bills that Republicans have brought to the floor this month have only demonstrated just how misguided and misplaced their priorities are.

Two bills were mentioned by the Chair, one that bans the sale of oil from the Strategic Petroleum Reserve (SPR) to China and one that prevents the use of the SPR unless we opened up more public lands for drilling on the same percentage basis. They're both misguided because they're an attempt to limit this Administration's ability to bring down gasoline prices at the pump.

Moving towards clean energy is the future and the only way we're going to have less dependence on dictators and less volatility in the market. Both these bills were an attempt to try to give more opportunities to Big Oil.

Republicans have pushed this idea that somehow Big Oil wanted to pump more but couldn't. In reality, they wanted to keep the price artificially high.

That's another reason why they opposed using the Strategic Petroleum Reserve because it brought prices down, so all I'm basically saying is as we move towards an energy transition let's keep in mind that encouraging renewables as we did with the Inflation Reduction Act and Bipartisan Infrastructure Law.

This is the way to go in the future. I'm someone who believes that we have to use all sources of energy like nuclear and hydropower.

The bottom line is we're going to be left behind if we don't move towards renewables. The Chair also mentioned China. Make no mistake: the rest of the world, particularly China, has already embarked on a major transition to clean energy. In fact, China's investments in clean energy are so vast, the International Energy Agency forecasts that over the next several years, 40 percent of solar and wind energy growth will come from China alone. Yet Republicans consistently oppose our efforts to catch up.

If we don't move towards clean energy and if we don't use the incentives from the Inflation Reduction Act, we're going to be left behind—left behind in innovation and left behind in number of jobs created. The irony of it is, most if not all, Republicans in the House voted against the Inflation Reduction Act, yet many of those credits are going to Red States and industries in the Red States that want to take advantage of those credits.

This is not a Republican or Democratic issue. This is a bipartisan issue. We need to move towards clean energy.

Madam Chair, I just wanted to mention that one of our witnesses today is Dr. Unruh Cohen. I want to thank her for all of her work as Staff Director for the Select Committee on the Climate Crisis. I look forward to her testimony and the testimony of all the panel.

I just want to take a moment to thank our longtime staff director of our energy and environment policy team, Rick Kessler, who is retiring this week. After 25 years of dedicated service to Congress – of which 12 years were spent with this Committee – Rick has more than earned his retirement. Earlier in his career, Rick also worked in my personal office on energy and environmental issues. Last Congress, Rick was instrumental in our work on the Inflation Reduction Act, fighting to make sure meaningful climate investments – including many provisions from our own CLEAN Future Act – were included. During his tenure, he was also instrumental in helping us pass the Frank R. Lautenberg Chemical Safety Act, the Pipeline Safety Act, legislation phasing down the use of dangerous HFCs, and many, many more energy and environmental laws.

He has been a devoted member of our team, always guided by a sincere desire to improve peoples' lives, and I am personally grateful for his many years of service. I wish him nothing but the best in the future.

* * *

Mrs. RODGERS. All the best, Rick.

The Chair recognizes Energy, Climate, and Grid Security Subcommittee Chair Duncan for 5 minutes.

Opening Statement of Hon. Jeff Duncan, a Representative in Congress from the State of South Carolina

Mr. DUNCAN. Thanks for recognizing me.

I first want to start by saying congratulations on holding your first hearing as chair of the Energy and Commerce Committee. You have been a champion of expanding all forms of American energy, and I looked forward to working with you and the rest of the committee to deliver affordable and reliable energy to all Americans. I look forward to also getting to know my colleagues on the subcommittee on both sides of the aisle and trying to find common ground that we can work on to move America forward.

I also want to thank all the Democrat Members that voted for the SPR last week. It was a true bipartisan bill that voted out of the floor, and thank you for that.

I want to thank the four witnesses that have attended the hearing today. I look forward to your insights on the state of American energy.

America became the global leader in crude oil and natural gas production while also leading the world in emission reductions. We did this by encouraging the entrepreneurial spirit in the private sector. The shale evolution is a perfect example of this. It is estimated that shale evolution saved U.S. consumers $203 billion annually. That breaks down to about $2,500 per family of four. It also lowered energy-related greenhouse gas emissions by 527 million metric tons per year, the most in the world. American energy production and emissions reduction are not mutually exclusive.

Unfortunately, Democrats and President Biden's regressive energy agenda takes a whole-of-government approach to phase out the oil and gas industry, and that is just wrong. Their agenda discourages investment, and the top-down aggressive regulatory approach seeks to create government-controlled economy and energy sectors.

Republicans on Energy and Commerce have solutions to reverse the Democrats' regressive energy agenda. It starts with unleashing all forms of American energy by creating a regulatory structure that encourages investment and innovation to bring all forms of energy online.

Energy is the foundation of our economy, and it impacts every aspect of American life. High energy prices hit low-income and middle-class Americans the hardest. According to the EIA, one-third of American households struggle to pay their energy bills. One in five households have to forego basic human necessities like food and medicine in order to keep the lights on and heat their homes. The Democrats rush-to-green policies are making energy unaffordable for American families. High energy increases the price of everything and contributes to rising inflation.

Our policies put the American consumer first so they don't have to decide between putting food on the table or keeping the lights on.

Ensuring we not only have affordable energy but also reliable energy is critical. There is no question a secure and resilient power grid is necessary for national security and a strong economy.

For these reasons, our energy grid is an attractive target for our adversaries. Pipelines, refineries, and other energy infrastructure are also extremely vulnerable to attack. Our policies must prioritize strengthening and hardening our grid and energy infrastructure. Energy security is national security, and our ability to export plays an important role in this.

Since Congress lifted the crude oil export ban in 2015, we have seen the benefits of free trade and open markets in the energy sector. When America is the world leader in energy production, the world is a safer place. There is no doubt about that.

We can and should export to energy-dependent countries who rely on corrupt nations. We have seen how Putin has used energy as a political weapon to keep this stranglehold over Europe. The best way to decrease the leverage of OPEC and Russia and the best weapon against Putin is a more aggressive U.S. energy production policy.

Nuclear energy also plays a critical role in energy security. Exporting our nuclear technologies gives us the ability to set global nuclear norms. Many of our adversaries recognize this and are prioritizing building up their nuclear sector. It is time the United States takes a hard look at the regulatory obstacles that stand in the way of nuclear advancement in this country.

Republicans have solutions to bring more nuclear on line by addressing inefficiencies in the permitting process while also maintaining our global gold standard in nuclear safety.

I look forward to working with my colleagues on both sides of the aisle to bring more nuclear on line so that we can once again be a global leader.

The ranking member mentioned China. There is no doubt that China is adding wind and solar and other renewables to their energy matrix. They are also building a heck of a lot of coal-fired power plants, fossil fuel generation, mining rare earths that they need for all the technology, which is very detrimental to the environment.

So we have got to address those issues. China will be a big part of that addressing.

And I want to thank Chair Rodgers for holding this timely hearing. It is time to flip the switch, unleash American energy production, and achieve American energy dominance. I look forward to the hearing and from our witnesses.

And with that, I yield back.

[The prepared statement of Mr. Duncan follows:]

Jeff Duncan (SC-03) Opening Remarks, Full Committee Hearing, American Energy Expansion: Strengthening Economic, Environmental, and National Security, January 31, 2023, as Prepared for Delivery

Thank you for recognizing me.

I want to start by congratulating Chair Rodgers on holding her first hearing as Chair of the Energy and Commerce Committee. She has been a champion of expanding all forms of American energy and I look forward to working with her and the rest of this committee to deliver affordable and reliable energy to all Americans.

I also want to thank our four witnesses for being here today, I look forward to your insights on the state of American energy.

America became the global leader in crude oil and natural gas production while also leading the world in emissions reductions. We did this by encouraging the entrepreneurial spirit in the private sector. The shale revolution is a perfect example of this. It is estimated that the shale revolution saved US consumers $203 billion dollars annually, breaking down to $2,500 dollars for a family of four. It also lowered energy-related greenhouse gas emissions by 527 million metric tons per year.

American energy production and emissions reductions are not mutually exclusive.

Unfortunately, the Democrats and President Biden's regressive energy agenda takes a whole of government approach to phase the oil and gas industry out of existence.

Their agenda discourages investment, and the top-down aggressive regulatory approach seeks to create a government-controlled economy and energy sector.

Republicans on Energy and Commerce have solutions to reverse the Democrats' regressive energy agenda. It starts with unleashing all forms of American energy by creating a regulatory structure that encourages investment and innovation to bring all forms of energy online.

Energy is the foundation of our economy and impacts every aspect of American life. High energy prices hit low-income and middle-class Americans the hardest. According to the EIA, one-third of American households struggle to pay their energy bills. One in five households have to forgo basic human necessities like food and medicine in order to keep the lights on and heat their homes. The Democrat's "rush to green polices" are making energy unaffordable.

High energy increases the price of everything and contributes to rising inflation. Our policies put the American consumer first, so they don't have to decide between putting food on the table or keeping the lights on.

Ensuring we not only have affordable energy, but also reliable energy is critical. There is no question – a secure and resilient power grid is necessary for national security and a strong economy. For these reasons, our energy grid is an attractive target for our adversaries. Pipelines, refineries, and other energy infrastructure are also extremely vulnerable to attack. Our policies must prioritize strengthening and hardening our grid and energy infrastructure.

Energy security is national security and our ability to export plays an important role in this. Since Congress lifted the crude oil export ban in 2015, we have seen the benefits of free trade and open markets in the energy sector. When America is a world leader in energy production, the world is a safer place. We can and should export to energy dependent countries who rely on corrupt nations. We have seen how Putin has used energy as a political weapon to keep his stranglehold over Europe. The best way to decrease the leverage of OPEC and Russia and the best weapon against Putin is more U.S. energy production.

Nuclear energy also plays a critical role in energy security. Exporting our nuclear technologies gives us the ability to set global nuclear norms. Many of our adversaries recognize this and are prioritizing building up their nuclear sector. It is time the United States takes a hard look at the regulatory obstacles that stand in the way of nuclear advancement in this country. Republicans have solutions to bring more nuclear online by addressing inefficiencies in the permitting process while also maintaining our global gold standard in nuclear safety. I look forward to working with my colleagues on both sides of the aisle to bring more nuclear online so we can once again be a global leader.

Thank you again Chair Rodgers for holding this hearing. It is time we Flip the Switch, Unleash American Energy Production, and achieve energy dominance.

I look forward to hearing from our witnesses and I yield back.

* * *

Mrs. RODGERS. The Chair recognizes Ranking Member DeGette for 5 minutes.

Opening Statement of Hon. Diana Degette, a Representative in Congress from the State of Colorado

Ms. DEGETTE. Thank you so much, Madam Chair. It is really fun to say that.

And I want to say one of the things that makes the Energy and Commerce Committee so unique is not just our vast jurisdiction but the willingness of Members to work together in a bipartisan way to get things done for the American people.

I have had the honor of serving on this storied committee since I was a freshman Member of Congress. As the chair of the Oversight and Investigations Subcommittee for the past 4 years, I am proud of the work that we did to hold the oil and gas industries accountable for the high gas prices last year and to help move our country forward.

In just the past 4 years, the full committee has passed legislation to strengthen our energy infrastructure, accelerate the deployment of new clean energy technologies, and help reduce the harmful emissions that are driving the climate crisis.

Thanks to those efforts, we have put our Nation on course to becoming truly energy independent and no longer dependent on the global oil market, which is the goal. And so I am hoping this hearing doesn't indicate that the majority is looking for ways to turn back the clock on that progress.

I was disturbed that the majority memo that I read didn't mention renewable energy sources one single time and instead doubled down on, frankly, outdated policies advocating increased fossil fuel production which will impede U.S. energy independence and make us more—ironically make us more dependent on an unstable international market. And that is because oil and gas is an international market, and so if we produce for here, it just goes into the international market.

We have all heard the slogans "drill, baby, drill," "energy dominance," and now "energy expansion," but don't be fooled. These policies will not expand our potential for new renewable energy sources, such as wind and solar, and that will only increase our dependence on oil and gas. So that will weaken our economic, environmental, and national security in the process. They are nothing more than a giveaway to the oil industry to further pad their profits.

Now, like Ranking Member Pallone, I was dismayed by the early energy bills considered on the floor. Just last week Congress considered H.R. 21, a bill that would open up our public lands to oil and gas exploitation and limit the President's ability to respond to future energy emergencies through the SPRO. We have seen this all-out push to expand oil and gas drilling before, and it just doesn't work.

Last February the United States produced more than 11 million barrels of oil per day, one of the highest levels of crude oil production in U.S. history. But when Vladimir Putin launched his war of aggression in Ukraine, the price of gasoline skyrocketed. And we as a nation were immediately reminded of how reliant we continue to be on a global commodity that is subject to disruptions of international markets.

The only way to secure our economic prosperity, our environment, and our national security is to break the addiction to oil as soon as possible. And so to do that, we must transition to clean energy technologies.

Now, I think that we could work across the aisle on this. I think that we could work together on supporting the research that we need to advance new clean energy technologies that will both make us energy independent and help us reach our Nation's climate goals.

I think—I am sorry that my colleagues didn't support the CHIPS and Science Act that has made investments in our energy and science agencies. But I will also say I do think—and I am looking forward to working with Chair Duncan and with our friends on the other side of the aisle. We have already talked. And I do think that we can find bipartisan ways to both reduce our dependence on the international oil market and also to develop clean energy that will address the climate crises.

But what we have to do, Madam Chair, in my opinion, we have to stop retreating to our corners and seeing everything as a zero sum game for whatever side. The markets and the utility companies and others have realized this. They have been moving to clean energy sources as fast as they can, and they need our help to be able to have the research and development to be able to make this secure, to make the grid secure, and to make the sources so that we can transition to a clean energy economy, be independent, and also address the climate crisis.

So I look forward to working with you and with this committee. I am excited about it. But we have to really hold out the branch and not talking about just drilling.

With that, Madam Chair, I yield back. [The prepared statement of Ms. DeGette follows:]

Committee on Energy and Commerce, Opening Statement as Prepared for Delivery of Subcommittee on Energy, Climate, and Grid Security Ranking Member Diana DeGette, Hearing on "American Energy Expansion: Strengthening Economic, Environmental, and National Security," January 31, 2023

Thank you so much, Madam Chair—it's really fun to say that.

I want to say one of the things that makes the Energy and Commerce Committee so unique is not just our vast jurisdiction, but the willingness of members to work together in a bipartisan way to get things done for the American people.

I've had the honor of serving on this storied committee since I was a freshman member of Congress. As the Chair of the Oversight and Investigation subcommittee for the past four years, I'm proud of the work that we did to hold the oil and gas industries accountable for the high gas prices last year and to help move our country forward.

In just the past four years, the full Committee has passed legislation to strengthen our energy infrastructure, accelerate the deployment of new clean energy technologies, and help reduce the harmful emissions that are driving the climate crisis.

Thanks to those efforts, we've put our nation on course to becoming truly energy independent, and no longer dependent on the global oil market, which is the goal. I'm hoping this hearing doesn't indicate that the Majority is looking for ways to turn back the clock on that progress.

I was disturbed that the Majority memo that I read didn't mention renewable energy sources one single time and, instead, doubled down on frankly outdated policies advocating for increased fossil fuel production, which will impede U.S. energy independence, and, ironically, make us more dependent on an unstable international market.

That's because, oil and gas is an international market—if we produce more here, it just goes into the international market.

We've all heard the slogans, "drill, baby drill," "energy dominance," and now, "energy expansion." But don't be fooled; these policies will not expand our potential for new renewable energy sources, such as wind and solar. They will only increase our dependence on oil and gas, and that will weaken our economic, environmental, and national security in the process.

They're nothing more than a giveaway to the oil industry to further pad their profits.

Like Ranking Member Pallone, I was dismayed by the early energy bills considered on the floor.

Just last week, Congress considered H.R. 21, a bill that would open our public lands to oil and gas exploitation and limit the President's ability to respond to future energy emergency through the SPR.

We've seen this all-out push to expand oil and gas drilling before, and it just doesn't work. Last February, the United States produced more than 11 million barrels of oil per day— one of the highest levels of crude oil production in US history. But when Vladimir Putin launched his war of aggression in Ukraine, the price of gasoline skyrocketed. We as a nation were immediately reminded of how reliant we continue to be on a global commodity that is subject to disruptions of international markets.

The only way to secure our economic prosperity, our environment and our national security is to break the addiction to oil as soon as possible. To do that, we must transition to clean energy technologies. I think that we could work across the aisle on this. I think that we could work together on supporting the research that we need to advance new clean energy technologies that will both make us energy independent and help us reach our nation's climate goals.

I'm sorry that my colleagues didn't support the CHIPS and Science Act that made investments in our energy and science agencies. I'm looking forward to working with Chair Duncan and with our friends on the other side of the aisle; we've already talked and I do think that we can find bipartisan ways to both reduce our dependence on the international oil market and also to develop clean energy that will address the climate crisis.

But what we have to do, Madam Chair, in my opinion, we have to stop retreating to our corners, seeing everything as a zero-sum game for whatever side. The markets, the utility companies, and others have realized this. They have been moving to clean energy sources as fast as they can. They need our help to be able to have the research and development to be able to make this secure, to make the grid secure, and to make the sources so that we can transition to a clean energy economy, be independent, and address the climate crisis.

I look forward to working with you and with this committee–I'm excited about it. We have to really hold out the branch and not talk about just drilling.

With that, Madam Chair, I yield back.

* * *

Mrs. RODGERS. The Chair recognizes subcommittee chair for Environment, Manufacturing, and Critical Minerals, Subcommittee Chair Johnson, for 5 minutes.

Opening Statement of Hon. Bill Johnson, a Representative in Congress from the State of Ohio

Mr. JOHNSON. Thank you, Madam Chair.

I too want to congratulate you for being selected as the chair of the best committee in the House of Representatives, the House Energy and Commerce Committee, and for holding your first full committee hearing as chair and making it about such a vital issue, "American Energy Expansion: Strengthening Economic, Environmental, and National Security," and I too look forward to working with you and our Members on both sides of the aisle to do the right things.

I agree that it is time for all members of our committee to get to work to come up with solutions to the major problems driving inflation in our country, a staggering inflation which threatens to lower all Americans' standard of living and make us less secure.

Just over 2 years ago, America was energy dominant for the first time in 70 years. We were the largest energy producer in the world, and fuel for our homes and automobiles was affordable for working Americans. Unfortunately, the President and the Democrat majorities in the last Congress decided to wage an all-out legislative regulatory and public relations campaign to suppress that energy dominance, leading to skyrocketing and unaffordable gasoline, natural gas, and electricity prices for many Americans. Not just that, but rather than relying on superior American innovation and ingenuity to solve emission issues—which we have demonstrated, by the way, we are clearly able to do—this campaign against American energy and energy workers is attempting to cancel proven, increasingly cleaner forms of existing energy resources in favor of only a few forms of weather-dependent energy sources. And that has very serious negative economic and national security implications.

Put simply, energy security is national security. We have heard it said over and over again, and it is true. It takes energy to produce raw materials, to manufacture products, to produce farm products, like produce, eggs and poultry, and pork and beef, to get those products to market.

In fact, energy is at the center of everything. Americans have seen firsthand—they are living it—that limiting domestic energy production and cheapening the value of the dollar through reckless Federal spending is a textbook way to drive inflation higher. We can do better. We must do better.

Now, I am not reflexively opposed to some of the Democrats' solutions being proposed, like, for example, greater renewal deployment in our country, but where I struggle is with the practical implementation of these policies and plans, the cost and potentially negative impacts on the quality of life for American consumers.

We can't just take what sounds good in townhalls and wealthier coastal or metropolitan areas and suggest that folks in less wealthy, more rural areas just jump in line and do it as well. If we want viable, sustainable energy solutions that make all Americans secure, that appreciates the diversity of our landscapes and doesn't rely on premature aspirational technologies or the availability of mineral supplies from foreign countries being available, then it is time to turn the page on the current national energy strategy, or a lack thereof, particularly since it is one that makes us more dependent on nations that hate us.

To illustrate the point, I ask my colleagues to consider a recent report in the Washington Post. It suggested senior U.S. military officials are predicting and preparing for war within 1 to 5 years with China, the very country we depend on for the critical minerals required for many of the rush-to-green solutions. This is why we need to maximize the production of clean, reliable, affordable American energy and the associated critical minerals right here at home, along with improving the permitting process so investors aren't deterred by government-imposed barriers and uncertainty for new energy infrastructure projects.

And, friends, it is time to get serious about making rolling blackouts on our energy grid a thing of the past and lowering gasoline prices with increased American domestic production, rather than being dependent on artificial market interventions by the White House.

More domestic supply is necessary to meet these challenges, and we shouldn't fool ourselves into thinking that draining our Strategic Petroleum Reserve assets is a recipe for success. We produce energy better, safer, and cleaner than any other place in the world. These facts used to be a source of national strength and pride, not to mention American international diplomacy.

Let me conclude with a hearty welcome to our witnesses today. I appreciate that you are here, and I look forward to you sharing your experiences with us.

With that, Madam Chair, thank you for yielding to me. I yield back the balance—well, I have exceeded my time.

[The prepared statement of Mr. Johnson follows:]

Bill Johnson (OH-06) Opening Remarks, Full Committee Hearing, American Energy Expansion: Strengthening Economic, Environmental, and National Security, January 31, 2023, as Prepared for Delivery

Thank you, Madame Chair.

I want to congratulate you for being selected as Chair of the best committee in the House of Representatives…the House Energy and Commerce Committee…and for holding your *first* full committee hearing as Chair and making it about such a vital issue - *American Energy Expansion: Strengthening Economic, Environmental, and National Security.*

I agree that it's time for *all* members of our committee to get to work to come up with solutions to the major problems driving inflation in our country…a staggering inflation which threatens to lower all American's standard of living and make us less secure.

Just a little over 2 years ago, America was energy dominant…for the first time in 70 years. We were the largest energy producer in the world, and fuel for our homes and automobiles was affordable for working Americans.

Unfortunately, the President and the Democratic majorities in the last Congress decided to wage an all-out legislative, regulatory, and public relations campaign to suppress that energy dominance, leading to skyrocketing and unaffordable gasoline, natural gas, and electricity prices for many Americans.

Not just that…but, rather than relying on superior American innovation and ingenuity to solve emissions issues… which we've demonstrated clearly we can do… this campaign against American energy and energy workers is attempting to cancel proven, increasingly *cleaner* forms of existing energy resources in favor of *only a few* forms of weather-dependent energy sources. And that has serious negative economic and national security implications.

Put simply, *energy* security *is national* security. It takes energy to produce raw materials, to manufacture products, to produce farm products – like produce, eggs and poultry and pork and beef, to get those products to market. In fact, *energy* is at the center of everything.

Americans have seen firsthand…they're living it…that limiting domestic energy production and cheapening the value of the dollar through reckless Federal spending is a textbook way to drive inflation higher.

We can do *better*…we *must* do better.

Now, I'm not reflexively opposed to some of the Democrats' solutions being proposed…like, for example, greater renewables deployment in our country. But, where I struggle is with the practical implementation of these policies and plans…the costs and the potential negative impacts on the quality of life for American consumers.

We can't just take what *sounds* good in town halls in wealthier coastal or metropolitan areas and insist that folks in less wealthy, more rural areas just jump in line and do it as well.

If we want viable, sustainable energy solutions that make *all* Americans secure, that appreciates the diversity of our landscapes, and doesn't rely on premature, aspirational technologies or the availability of mineral supplies from foreign countries being available; then it's time to turn the page on the current, national energy strategy, or lack thereof. Particularly since it's one that makes us more dependent on nations that hate us.

To illustrate the point, I ask my colleagues to consider a recent article in the Washington Post. It suggested senior U.S. military officials are predicting, and preparing for, war within one to five years with China – the very country we depend on for the critical minerals required for many of the "rush to green" solutions.

This is why we need to maximize the production of clean, reliable, affordable American energy and the associated critical materials right here at home…along with improving the permitting process so investors aren't deterred by government imposed barriers and uncertainty for new energy infrastructure projects.

Friends, it's time to get serious about making rolling blackouts on our energy grid a thing of the past…

and lowering gasoline prices with increased American domestic production, rather than being dependent on artificial market interventions by the White House.

More domestic supply is necessary to meet these challenges, and we shouldn't fool ourselves into thinking that draining our strategic petroleum reserve assets is a recipe for success.

We produce energy better, safer, and cleaner than any other place in the world. These facts used to be a source of national strength and pride – not to mention American *international* diplomacy.

We should embrace what we can do and what we are capable of, not demonize our proven successes. And we should encourage free enterprise and free-market-driven innovations to bring out the best, most affordable, and most practical solutions for our economy, our environment, and our national security.

Let me conclude with a hearty welcome to our witnesses today. I appreciate that you're here…and I look forward to you sharing your experience and perspectives with us.

With that, Madame Chair, thank you again for yielding to me. I yield back the balance of any remaining time I have.

* * *

Mrs. RODGERS. The Chair recognizes Ranking Member Tonko for 5 minutes.

Opening Statement of Hon. Paul Tonko, a Representative in Congress from the State of New York

Mr. TONKO. Thank you, Madam Chair.

And first let me congratulate you on your great achievement and Chair Johnson for his taking on leadership of the Environment Subcommittee.

I have had the privilege of working with the two previous Republican leaders of the subcommittee, Representatives John Shimkus and David McKinley. During that time we certainly had our disagreements, but we also found issues where we could go beyond partisanship and work together for the benefit of our constituents' public health and indeed the environment.

So whether it is drinking water, brownfields, harmful air pollution, or the many other issues under the subcommittee's jurisdiction, I do hope we will be able to find policies where we can work together. And as we begin to consider how our Nation can improve our energy security, foster innovation, and become the global leader in the energy economy of the future, I believe there have been great strides made by Congress in the past several years.

In the 116th and 117th Congresses, this committee played an instrumental role in the development and enactment of several historic energy and environmental laws, including the Energy Act of 2020, the AIM Act to phase down hydrofluorocarbons, the Infrastructure Investment and Jobs Act, and the Inflation Reduction Act.

The IIJA included $62 billion for DOE clean energy investments, including programs that will make our electric grid smarter and more resilient, build a domestic battery manufacturing and recycling supply chain, and support the development of cutting-edge industries through hydrogen and DAC hubs.

The IRA built upon that down payment with nearly $317 billion in funding for DOE, EPA clean energy programs, and expansion and extension of a wide range of tax incentives and other climate and clean energy investments. These efforts took our long-term national energy security seriously, positioning the United States to become the leader in many of the energy technologies and supply chains that will come to dominate the global energy system over the next several decades.

And we are already starting to witness the benefits of this long-term commitment to a clean energy future. More and more private-sector companies are making plans to manufacture and deploy clean-energy technologies across our country. Electric vehicles, semiconductors, solar components, batteries, offshore wind turbines, and electrolyzers are all part of this growing effort to create new industries right here at home.

It has been estimated that the IRA could result in 9 million jobs over the next decade, and many investments will support the deployment of affordable clean-energy solutions resulting in major savings for American consumers.

These economic and job-creation benefits are, of course, only part of the story. There will also be a significant reduction of pollution, which will mean fewer hospital visits and missed days of school and work. Reducing climate and traditional pollution will make us a healthier and more productive society as we continue to fight to avoid the worst potential consequences of global climate change.

But we do know that, even after the IRA and the IIJA, there is still much more work to be done. So, as we consider how to best build upon these historic bills, I would caution that creating loopholes in our Nation's environmental laws for the benefit of polluting industries will not improve our long-term national energy security.

On the contrary, our environmental laws can be a driver of innovation, complementing the investments we have already made to support the development and deployment of next-generation technologies.

I do want to welcome a great public servant to the committee, Dr. Ana Unruh Cohen, who can advise us on how to build upon the progress of the past 4 years. Dr. Unruh Cohen and former Select Committee Chair Kathy Castor played a critical role in the climate and clean-energy successes of the 117th Congress. I can think of no one better to help us understand where we have come from, what we have achieved, and what is left to do to transition our Nation and our planet to a more just and sustainable energy system.

Thank you, Dr. Unruh Cohen, for joining us today.

And, Madam Chair, I want to thank you for making this our first hearing of the new Congress. The American people should know more about the work that has been done to make our country less reliant on economically volatile and polluting industries. And soon we will be able to provide more information to our constituents and local businesses about how to take advantage of the generous tax credits and rebates in the IRA.

So I look forward to having many more opportunities to share that information with the public in the months ahead.

Earlier our past chair, Frank Pallone, acknowledged Rick Kessler for the great work he has done with this committee and certainly with the subcommittee that I chaired.

Rick, I thank you for an outstanding bit of professionalism that you brought to this committee. I thank you for the hard work, the passion, and the success that we have all enjoyed. God bless you as you go forward with your new career opportunities.

And with that, Madam Chair, I yield back.

[The prepared statement of Mr. Tonko follows:]

Committee on Energy and Commerce, Opening Statement as Prepared for Delivery of Subcommittee on Environment, Manufacturing, and Critical Minerals Ranking Member Paul D. Tonko, Hearing on "American Energy Expansion: Strengthening Economic, Environmental, and National Security," January 31, 2023

Thank you, Madam Chair.

First, I want to congratulate you, as well as Chairman Johnson for taking on leadership of the Environment Subcommittee.

I have had the privilege of working with the two previous Republican leaders of the Subcommittee, John Shimkus and David McKinley.

During that time, we certainly had our disagreements.

But we also found issues where we could go beyond partisanship and work together for the benefit of our constituents' public health and the environment.

So, whether it is drinking water, brownfields, harmful air pollution, or the many other issues under the Subcommittee's jurisdiction, I hope we will be able to find policies where we can work together.

And as we begin to consider how our nation can improve our energy security, foster innovation, and become the global leader in the energy economy of the future, I believe there have been great strides made by Congress in the past several years.

In the 116th and 117th Congresses, this Committee played an instrumental role in the development and enactment of several historic energy and environmental laws, including the Energy Act of 2020, the AIM Act to phasedown hydrofluorocarbons, the Infrastructure Investment and Jobs Act, and the Inflation Reduction Act.

The IIJA included $62 billion for DOE clean energy investments, including programs that will make our electric grid smarter and more resilient, build a domestic battery manufacturing and recycling supply chain, and support the development of cutting-edge industries through hydrogen and DAC hubs.

The IRA built upon that down payment with nearly $370 billion in funding for DOE, EPA clean air programs, an expansion and extension of a wide range of tax incentives, and other climate and clean energy investments.

These efforts took our long-term national energy security seriously, positioning the United States to become the leader in many of the energy technologies and supply chains that will come to dominate the global energy system over the next several decades.

And we are already starting to witness the benefits of this long-term commitment to a clean energy future.

More and more private sector companies are making plans to manufacture and deploy clean energy technologies across the country.

Electric vehicles, semiconductors, solar components, batteries, offshore wind turbines, and electrolyzers are all part of this growing effort to create new industries here at home.

It has been estimated that the IRA could result in 9 million jobs over the next decade. And many investments will support the deployment of affordable clean energy solutions, resulting in major savings for American consumers.

These economic and job creation benefits are of course only part of the story.

There will also be a significant reduction of pollution, which will mean fewer hospital visits and missed days of school and work.

Reducing climate and traditional pollution will make us a healthier and more productive society as we continue to fight to avoid the worst potential consequences of global climate change.

But we know even after the IRA and IIJA there is still much more work to be done.

So, as we consider how to best build upon these historic bills, I would caution that creating loopholes in our nation's environmental laws for the benefit of polluting industries will not improve our long-term national energy security.

On the contrary, our environmental laws can be a driver of innovation, complementing the investments we have already made to support the development and deployment of next-generation technologies.

I want to welcome a great public servant to the Committee, Dr. Ana Unruh Cohen, who can advise us on how to build upon the progress of the past four years.

Dr. Unruh Cohen and former Select Committee Chairwoman Kathy Castor played a critical role in the climate and clean energy successes of the 117th Congress.

I can think of no one better to help us understand where we have come from, what we have achieved, and what is left to do to transition our nation— and our planet— to a more just and sustainable energy system.

Thank you, Dr. Unruh Cohen for joining us today.

And Madam Chair, I want to thank you for making this our first hearing of the new Congress.

The American people should know more about the work that has been done to make our country less reliant on economically volatile and polluting industries.

And soon we will be able to provide more information to our constituents and local businesses about how to take advantage of the generous tax credits and rebates in the IRA.

I look forward to having many more opportunities to share that information with the public in the months ahead. With that, I yield back.

* * *

Mrs. RODGERS. Our first witness is the Honorable Paul Dabbar, former—well, he was Under Secretary of the Department of Energy. And you are recognized for 5 minutes.

Statements of Paul M. Dabbar, Former Under Secretary of Energy; Donna Jackson, Director of Membership Development, Project 21, National Center for Public Policy Research; Ana Unruh Cohen, Ph.D., Former Majority Staff Director, House Select Committee on the Climate Crisis; and Robert Mcnally, President, Rapidan Energy Group

Statement of Paul M. Dabbar

Mr. DABBAR. Chair McMorris Rodgers, Ranking Member Pallone, and members of the committee, I am honored to be again before this committee. I was with the chair last in

Washington State a couple of years ago as we were making good environmental progress adjacent to her district.

Not so long ago, under the leadership of Secretaries Perry, Brouillette, the U.S. became the undisputed dominant country in energy. When I was Under Secretary in 2017, crude oil exports jumped from zero to 1 million barrels a day. We were amazed by the jump.

By 2019, exports had jumped to 3 million barrels a day. That year, for the first time since the 1950s, we went from the largest energy importer in the world to a net energy exporter.

The Jimmy Carter founding mission for DOE had finally been achieved, and the U.S. became the undisputed dominant country in energy. We became the number-one producer of crude oil in the world, increasing from a low of 4.8 million barrels a day to 12.9, a monumental jump. We became the number-one global producer of natural gas. We became the marginal producer and the global price setter of crude oil, taking the crown from OPEC.

We extended our lead as a top country in the discovery of new energy technologies, including solar, lithium ion chemistry, and new drilling technologies. We were a solid manufacturer, including for turbines, nuclear, electric vehicles, and drilling.

We allowed the country to build, whether it was wind, batteries, LNG, or pipelines. This resulted in energy price deflation, and the U.S. led the world in the reduction of tons of emissions. No other country came close to that combination—production, technology, and manufacturing—not China, not Russia, and not Saudi Arabia. This dramatic turn in energy posture not only created economic growth and security, it produced a new foreign policy playing field, and we were no longer at the bidding of a cartel.

This allowed us to take the lead in stabilizing the global energy markets when COVID hit. We were so strong that the 2020 OPEC production agreement was announced from the White House. That was energy dominance.

We did this by concurrently focusing on all-the-above energy policies, increasing all-the-above energy supply, strengthening energy national security and diplomatic posture, producing energy price deflation, and reducing emissions.

This shift happened for three reasons: U.S. energy technology innovation, government policies that encouraged all-the-above energy supply, and investment by the private markets.

The trigger was a vast amount of energy innovation technologies. Much of which is now being deployed did not exist 20 years ago. Then this committee, in a bipartisan and after many years of negotiations, passed several acts that made a big difference. Those E&C compromises I can summarize as, "If you give me my pro supply policies, I will give you your pro supply policies." But this balanced approach has taken a detour, and we are now back to asking Riyadh, Caracas, and Tehran for their help.

Here are the steps the committee can take to establish an energy dominance: Require agencies such as Interior, EPA, and FERC to permit, license, and approve all-the-above energy types. FERC needs significant legislative reform, including a reform of the Federal Power Act and the Natural Gas Act, to make them do their statutory obligation to ensure that there is enough energy supply. They need to be required to approve transmission projects for all types of energy, and they need to radically overall ISO rules that encourage baseload that is being shut down faster than intermittent is being built.

Capital providers needs to be pushed so that investment criteria is not all about ESG and that they should also care about prices, reliability, national security, and supply chain versus supporting energy production moving overseas, the autocratic regimes, including one using slave labor. That does not seem very ESG to me.

And we need to continue to support discovery, science, and innovation, a leading strength of America.

Our moment of energy dominance was a culmination of many factors: innovation, all-the-above pro-supply set of policies, markets willing to invest, and strong energy diplomacy.

As we confront the challenges of the energy markets today, we do well to remember the ingredients that made us so successful not so long ago.

[The prepared statement of Mr. Dabbar follows:]

Statement by the Honorable Paul M. Dabbar, Former under Secretary, U.S. Department of Energy CEO, Bohr Quantum Technology, Distinguished Visiting Fellow, Columbia University Contributor, Wall Street Journal, before the House Energy and Commerce Committee, on American Energy Expansion: Stregthening Economic, Environmental, and National Security, January 31, 2023

Chair McMorris Rodgers, Ranking Member Pallone, and Members of the Committee, I am honored to again be before this Committee. I was with the chair last in Washington state a couple of years ago where we were making strong environmental progress near your district.

Not so long ago, under the leadership of Secretaries Perry and Brouillette, the U.S. became the undisputed dominate country in energy. When I was Under Secretary in 2017, US crude exports jumped from zero to 1mm bbl/d. We were amazed by the leap. By 2019, exports jumped to 3mm bbl/d. And that year, for the first time since the 1950's, we went from the largest energy importer, to a net energy exporter. The Jimmy Carter founding mission for DOE had finally been achieved, and the U.S. became the undisputed dominate country in energy.

We became the #1 producer of oil as crude production increased from a low of 4.8 to 12.9 mm bbl/d, a monumental increase. We also became the #1 global producer of natural gas. We became the marginal producer and global price setter of crude oil, taking the crown from OPEC. We extended our lead as the top country in discovery of new energy technologies, including solar, lithium-ion chemistry, and new drilling technologies. We were a solid manufacturer including for turbines, nuclear, EV's and drilling. We allowed the country to build, whether it was wind, batteries, LNG, or pipelines. This resulted in energy price deflation. And the U.S. led the world in the reduction of tons of emissions.

No other country came close to that combination: production, technology and manufacturing, not China, Russia or Saudi Arabia. This dramatic turn in energy posture not only created significant economic growth and security, it produced a new foreign-policy playing field and we were no longer at the bidding of a cartel.

This allowed us to take the lead on stabilizing global energy markets when Covid hit. We were so strong, that the 2020 OPEC production agreement was announced from the White House. That was energy dominance.

We did this by concurrently focusing on all the "All the above energy policies": increasing all the above energy supply, strengthening energy national security and diplomatic posture, producing energy price deflation, and reducing emissions.

This shift happened for three reasons: U.S. energy-technology innovation, government policies that encouraged all-the-above energy supply, and investment by private markets.

The trigger was a vast amount of innovation of new technologies, and much of what is now being deployed did not exist in commercial form twenty years ago. Then this committee, bi-partisan, and after many years of negotiation, passed several Acts that made a big difference. Those E&C compromises can be summarized: If you give me my pro-supply policies, I'll give you your pro supply policies.

But this balanced approach has taken a detour. And we are now back to asking Riyad, Caracas and Tehran for their help.

Here are steps this committee can take to re-establish energy dominance:

- Require agencies such as Interior, EPA and FERC to permit, license, and approve all the above energy types.
- FERC needs significant legislative reform to make them do their statutory obligation to ensure there is enough energy supply of all types. They need to be required approve transmission projects for all types of energy. And they need to radically overhaul ISO rules to encourage baseload power, that is being shut down faster than new intermittent plants are being built.
- Capital providers need to be pushed so that investing criteria is not all about ESG, that it includes caring about prices, reliability, and national security & supply chain risks. Vs supporting energy production moving overseas to autocratic regimes including one using slave labor camps. That doesn't seem very ESG to me.
- And continue to support discovery science and innovation, a leading strength of America.

Our moment of energy dominance was the culmination of many factors: innovation, an all-the-above pro-supply set of policies, markets willing to invest and strong energy diplomacy. As we confront the challenges in energy markets today, we'd do well to re-implement the ingredients that made us so successful not so long ago.

* * *

Mrs. RODGERS. Thank you.

Thank you for your testimony.

Our next witness is Ms. Donna Jackson with the National Center for Public Policy Research, Project 21.

You are recognized for 5 minutes.

Statement of Donna Jackson

Ms. JACKSON. Chair Rodgers, Ranking Member Pallone, and members of this committee, I thank you for allowing me to testify today. My name is Donna Jackson. I am the director of membership development for Project 21, a Black leadership network of the National Center for Public Policy Research.

Project 21 is one of the oldest and largest Black conservative think tanks in the country. Our hundreds of members come from all walks of life, from small business owners, law enforcement, healthcare workers, teachers, energy industry workers, clergy, healthcare workers, both professional and nonprofessional. Most of us are not career activists, lawyers, or lobbyists, and more than a few of us actually live in the communities that we hope to improve. I applaud this committee for kicking off a new session of Congress by looking at increasing domestic energy supply and reducing energy costs. Needless to say, these costs are tough enough on the middle class, but they are even tougher on folks struggling to reach the middle class. In fact, they are actually harder than you may think.

The U.S. Energy and Information Administration had some very useful statistics. It talked about the problems that Americans have in paying their energy bills. One-third struggle to pay their energy bills. A fifth of them have to forego necessities. But, of course, these numbers are double, nearly double for minority communities. And this report was actually a couple of years old. So added inflation makes it even that much harder.

But perhaps even more damaging is their immediate difficulty of paying for sky-high energy bills, making it from paycheck to pay-check, and the long-term barriers to upward mobility and achieving a piece of the American dream. In so many ways, it is painful and unaffordable. Domestic energy is part of the ticket out of poverty. And we have so much of it, so much of it.

For example, we see the entrepreneur spirit of the Black community. In many Black-owned businesses, small-owned businesses, these businesses are struggling and failing under the weight of high energy costs. And they also have regulatory compliances that they barely can afford. And it is disproportionately impacting small businesses that really want to make it in this country.

It all adds up to a regressive tax on those seeking to better themselves, their families, and their communities by owning and operating their own businesses.

Unaffordable energy means less—means the individuals have less to spend on necessities. The industrial employment has historically led to the emergence of a vibrant Black middle class. But what we are seeing is these energy costs are destroying these high-paying gateway jobs for low-income and minority communities. And we have very few options to earn money to be able to make it into the middle class.

The energy industry itself represents a very reliable source of well-paying jobs for people that don't have the options of being able to participate in the keyboard industry. And, yet, we are destroying those options for these individuals who have the American spirit, who want to do better, want to be self-reliant and not dependent on government assistance. We should not be making minority and low-income individuals an afterthought in saving this planet. Their lives matter. Our lives matter.

And I applaud you that you are considering the fact of the hardships that unaffordable energy poses to people like me who have self-esteem, self-respect, and we want to be a part of this great American experience.

Thank you.

[The prepared statement of Ms. Jackson follows:]

House Committee on Energy and Commerce, Hearing on "American Energy Expansion: Strengthening Economic, Environmental, and National Security," January 31, 2023, Remarks of Donna Jackson, Director of Membership Development, Project 21

Chair Rodgers, ranking member Pallone and members of this Committee, thank you for allowing me to testify today. My name is Donna Jackson, and I am the Director of Membership Development for Project 21 – the black leadership network of the National Center for Public Policy Research. Project 21 is one of the oldest and largest black conservative think tanks in the country. Our hundreds of members come from all walks of life, from small business owners to law enforcement to teachers to energy industry workers to clergy to health care workers – both professionals and non-professionals. Most of us are not career activists, lawyers, or lobbyists and more than a few of us actually live in the communities we hope to improve.

I applaud this Committee for kicking off the new session of Congress by looking at increasing domestic energy supplies and reducing energy costs. Needless to say, these costs are tough enough on the middle class, but they are even harder on those still struggling to reach the middle class. In fact, it may be worse than you think.

The U.S. Energy Information Administration has some very useful statistics on the problem. For example, it found that nearly one-third of households struggle to pay their energy bills, and that one in five report having to reduce or forego necessities like food or medicine in order to pay an energy bill. And of course, these numbers are nearly double for low-income and minority households. Also note that this study is several years old and does not reflect the much higher energy prices we have experienced during the Biden Administration.

But perhaps even more damaging than the immediate difficulties of paying sky-high energy bills and making it to the next paycheck are the long-term barriers to upward mobility and achieving a piece of the American dream. In so many ways, plentiful and affordable domestic energy is part of the ticket out of poverty and dependence.

For example, we see the entrepreneurial spirit of the black community in many black-owned small businesses, but those businesses struggle and sometimes fail under the weight of high energy costs. And expensive energy comes on top of other regulatory compliance costs that the Small Business Administration has shown to have disproportionate impacts on small businesses. It all adds up to a regressive tax on those seeking to better themselves, their families, and their communities by owning and operating their own businesses.

Unaffordable energy also means less of the industrial employment that has historically led to the emergence of a vibrant black middle class. Without these high paying blue-collar gateway jobs, low-income and minority communities have fewer options to earn what is needed for home ownership and decreasing the wealth gap.

In addition, the energy industry itself is a source of many well-paying jobs, and in fact Bobby Rush and others on this Committee helped enact job training programs so that more minorities could take advantage of the opportunities created by the shale revolution. But there will be far fewer such opportunities if the President makes good on his promises to kill off the American oil and natural gas sector.

And perhaps most importantly, every extra penny struggling households must spend on expensive energy is a penny we can't save for home ownership or our children's education or retirement savings as well as other things that build family wealth and help lift up the next generation.

Let me conclude by telling you what the black community does not want – and that's climate change policies and environmental justice initiatives that take us in exactly the wrong direction by blocking domestic sources of energy, raising costs, destroying jobs and stifling our pursuit of the American dream. I believe that energy affordability should always come first, and that's a message the black community wants to hear from Congress. Thank you.

* * *

Mrs. RODGERS. Thank you for your testimony.

Next we will hear from Dr. Ana Unruh Cohen, former majority staff director of the U.S. House Select Committee on Climate.

You are recognized for 5 minutes.

Statement of Ana Unruh Cohen, Ph.D.

Dr. UNRUH COHEN. Good morning, Chair Rodgers, Ranking Member Pallone, and the members of the committee.

I am Ana Unruh Cohen, and I most recently served as a majority staff director for the Select Committee on the Climate Crisis.

For a climate scientist turned congressional staffer, it will be hard to stop serving as staff director for the select committee during the last 4 years under the leadership of Representative Kathy Castor and with other members of this committee while Congress passed the most consequential climate and clean energy laws in U.S. history.

The last 4 years have underscored how energy, climate, and economic security are tied together. Between the economic impact of prevaccine COVID lockdowns, ongoing supply-chain disruptions, Russia's invasion of Ukraine, and extreme weather events, it has become clearer than ever that the United States needs to invest in building an energy economy that meets the needs of the 21st century. To thrive we will need affordable energy that addresses the climate crisis, improves public health, and reduces our reliance on volatile global powers and their resources.

The legislation enacted in the 116th and 117th Congresses, especially the Infrastructure Investment and Jobs Act, the Inflation Reduction Act, and the CHIPS and Science Act, make critical down payments on achieving those goals. We are already seeing the benefits of these laws accrue across America.

Since the enactment of the IRA in August of 2022, almost $90 billion in clean energy projects have been announced, initiating new economic development opportunities across the country, and we will see even more in the coming years.

The need for the clean energy transition is clear and growing in urgency. The United States faced $35 billion extreme weather in climate-related disasters in 2021 and 2022, with a cumulative price tag of more than $180 billion in direct economic losses alone.

Other countries around the world have faced devastating heatwaves, droughts, and floods in recent years. Without comprehensive climate action, the science points toward continued disasters with regional, national, and global consequences.

The climate crisis and need for energy security will require governments at all levels to increase their engagement on energy and climate policies, to realize the potential created by the news laws, and to use the available tools to ease impacts on family when new climate and energy challenges arise.

Congress, and especially this committee, will need to understand and respond to the increasingly dynamic global energy landscape to maintain U.S. global leadership and competitiveness.

Successful energy and climate legislation must meet the test of cutting climate pollution guided by science, reducing energy costs on families, advancing equity and justice, and creating good, family-sustaining jobs. As this committee takes up energy and climate legislation, Members should evaluate the bills with these metrics in mind, just as we did at the select committee and working with the standing committees in the 116th and 117th Congress. My written testimony summarizes many of the accomplishments of those laws.

My testimony also summarizes the current global energy investment landscape as countries ramp up climate action. But climate considerations are not the only driving force behind increased interests in clean energy. Just yesterday BP's chief global economist pointed to the Russian invasion as a reason that countries are seeking to increase access to domestically produced energy, most of which he said will likely come from renewables and other nonfossil fuels.

As we have heard, China is leading that clean energy race, and we must catch up. They may be out front now, but the United States can close the gap to power America in ways that improve our global competitiveness and help solve the climate crisis.

American ingenuity is our superpower. Giving up on a technological competition is not the American way. Instead, we need energy and climate solutions that meet the needs of the 21st century, and we need them now.

I will close on a personal note. Many of you who know me from DC may not know that I was born and raised in Corpus Christi, Texas, or that my dad spent his career in the chemical industry, and my brother currently works in safety operations for oil and gas.

I am well acquainted with the challenges that communities face in balancing energy production and public health, and I know we can do it in a way that—find a way forward for a clean energy future that works for everyone and doesn't pit public health against economic security.

Thank you.

[The prepared statement of Dr. Unruh Cohen follows:]

Testimony before the U.S. House Energy and Commerce Committee, "American Energy Expansion: Strengthening Economic, Environmental, and National Security" January 31, 2023, Ana Unruh Cohen, PhD

Good morning, Chair Rodgers, Ranking Member Pallone and the Members of the committee. I am Ana Unruh Cohen and I most recently served as the Majority staff director for the Select

Committee on the Climate Crisis. I came to DC in August 2001 with a freshly minted doctorate in Earth Science. I was eager to explore how a climate scientist could contribute to national policy and have subsequently worked in some form on every major energy and climate legislation that Congress has considered during that time. The highlight of my career was serving as staff director for the Select Committee during the last four years under the leadership of Rep. Kathy Castor and with other Members of this committee.

The last four years were also tumultuous for global energy markets and consequential for climate policy. Between the economic impact of pre-vaccine Covid lockdowns, ongoing supply chain disruptions, Russia's invasion of Ukraine, and extreme weather events, it's become clearer than ever that the United States needs to invest in building an energy economy that meets the needs of the 21st century. The tools we have for meeting these challenges have also changed during the last decade. The cost of solar, wind, batteries and other renewable technologies have plummeted even as U.S. shale development has influenced global oil and methane markets. To thrive we will need affordable energy that address the climate crisis, improves public health, and reduces our reliance on volatile global powers and their resources.

The legislation enacted in the 116th and 117th Congresses, especially the Infrastructure Investment and Jobs Act (IIJA), the Inflation Reduction Act (IRA), and the CHIPS and Science Act, make critical down payments on achieving those goals. We are already seeing the benefits of these laws accrue across America and we will see even more in the coming years as we accelerate our transition to a net-zero emission, clean energy economy.

The need for the clean energy transition is clear and growing in urgency. The United States faced 35 "billion-dollar" extreme weather and climate-related disaster events in 2021 and 2022, with a cumulative price tag of more than $180 billion in direct economic losses alone.[1] Other countries around the world have also faced devasting heatwaves, droughts and floods in recent years. Without action, the science points toward continued global catastrophe. In fact, the United Nations Secretary General Antonio Guterres referred to the IPCC Sixth Assessment Report as a "code red for humanity," pointing to its dire findings on some of the irreversible trends set in motion by climate change.[2] However, the IPCC also concluded that it is still technologically possible to halve global climate pollution by 2030 and warned that global climate pollution must peak by 2025 in order to avoid the worst devastation.[3]

The transition to a net-zero emission economy will require governments at all levels to increase their engagement on energy and climate policies to realize the potential created by recent legislation and to use the available tools to ease impacts on families when new climate and energy challenges arise. Congress, and especially this committee, will need to understand and respond to the increasingly dynamic global energy landscape to maintain U.S. global leadership and competitiveness.

---

[1] National Centers for Environmental Information, "Summary Stats | Billion-Dollar Weather and Climate Disasters," *National Oceanic and Atmospheric Administration*, 2022.

[2] United Nations, "IPCC report: 'Code red' for human driven global heating, warns UN chief," *UN News,* August 9, 2021.

[3] IPCC, "Climate Change 2022: Mitigation of Climate Change," 2022.

Energy and Climate Are Inextricably Linked

Every energy decision has implications for the climate system. Every energy decision impacts the amount of heat-trapping pollution in the atmosphere. We are already experiencing the consequences of the ~1°C increase in global temperatures –largely from burning fossil fuels— and the increasing risks to health, the economy and national security. Based on the extensive collection of scientific and economic research, the Council of Economic Advisors noted that:

> A growing literature suggests that as temperatures and sea levels rise, and extreme weather becomes more common, the physical damages that stem from the warming of the planet will have substantial, adverse effects on macroeconomic outcomes at the local, national, and international levels. Though not all caused by climate change, across the United States, estimated damages from storms, floods, wildfires, and other extreme weather events have grown to about $120 billion a year from 2016-2020. Climate-driven extreme events can also result in cascading damages to critical and interconnected systems such as energy, public health, ecosystems, water, and food. These damages have a variety of effects on the economy, including but not limited to straining government budgets, changing asset values and insurance costs, and shifting migration patterns and labor supply. The economic effects vary across U.S. regions and industries and will likely disproportionately harm disadvantaged communities.[4]

The energy sector has also experienced the impacts of extreme weather and other consequences of the climate crisis. In recent years, electricity failures have occurred from extreme temperatures – both heat and cold – and from destruction from wildfires and storms. The water cycle is incredibly sensitive to climate change leading to challenges for water used in energy production, cooling of thermal electricity generation and hydropower. Rising sea levels puts coastal energy facilities at risk to flooding even as increasing temperatures contribute to the severity of storms. Extreme rainfall events increase the risk of flooding and failures at energy waste facilities across America. As in the rest of society and the economy, the climate crisis is exposing and exacerbating vulnerabilities in the energy sector and underscoring the need for change.

Government and private investments are also responding to the climate crisis and the need for climate pollution to peak in the next few years and then reduce to achieve a net-zero emissions by mid-century, if not sooner. For the first time in 2022, investments in the clean energy transition reached the same level as fossil fuel investments at $1.1 trillion.[5] Despite reaching this level, the clean energy transition needs three times this level of investment for the rest of the decade to achieve transition to a net-zero economy according to analysis by Bloomberg NEF. BP's annual energy outlook also found that governments and industries are behind in the transition to net-zero emissions by 2050.[6]

Climate considerations are not the only driving force behind increased interest in clean energy. The Russian invasion of Ukraine has refocused attention on energy security, especially ways to meet their energy needs with domestic resources. According to BP's chief economist, "The increased focus on energy security as a result of the Russia-Ukraine war has the potential

---

[4] Council of Economic Advisers, "Climate-related Macroeconomic Risks and Opportunities," April 4, 2022, pg 3-4.

[5] BloombergNEF, "Energy Transition Investment Trends 2023," January 2023.

[6] Tom Wilson, "BP cuts long term forecast for oil and gas demand," *The Financial Times*, January 30, 2023.

to accelerate the energy transition as countries seek to increase access to domestically produced energy, much of which is likely to come from renewables and other non-fossil fuels."[7] The importance of leadership in clean energy technologies and investments will only grow in the coming years.

China already has a head start on the rest of the world in clean energy investments. Just less than half of the global energy transition funding in 2022 comes from China's $546 billion spending. The United States trails in a very distant second place with $141 billion.[8] U.S. investments will ramp up in the coming years as recently enacted laws unleash new federal funding that will leverage additional private capital. While China may be out front now, the United States can, and must, close that gap to power America in ways that improve our global competitiveness, help solve the climate crisis, and strengthen our partnerships with other countries to do the same.

In making the clean energy transition, we also have the opportunity to overcome environmental harms of the past. While the consequences of the climate crisis affect us all, it does not affect us all equally. Increasingly environmental justice is a cornerstone of environmental and climate policies to integrated equity and environmental justice into building a cleaner and more resilient economy. Legislation enacted in the 117th Congress reflected this growing emphasis. In addition, the Biden-Harris Administration is prioritizing environmental justice through the Justice40 Initiative and is taking steps to "narrow the racial wealth gap" by supporting underserved entrepreneurs and small businesses in federal procurement.[9]

Successful energy and climate legislations must meet the tests of cutting climate pollution guided by science, reducing energy costs on families, advancing equity and justice, and creating good, family-sustaining jobs. As this committee takes up energy and climate legislation, Members should evaluate the bills with these metrics in mind as we did at the Select Committee in working with the standing committees in the 116th and 117th Congresses.

Climate and Clean Energy Accomplishments in the 116th and 117th Congresses

During the last four years, Congress has produced legislation to set the United States on the path towards solving the climate crisis by passing policies to significantly cut climate pollution, help our communities adapt to its effects, and make them more resilient to its consequences. These climate and clean energy investments will also advance cleaner, cheaper energy generation that enhances America's energy security and global competitiveness. While the 116th Congress enacted some important legislation, especially the energy-related provisions of the 2020 year-end omnibus legislation, the 117th Congress seized the opportunity for transformational action and passed three pieces of legislation that will guide climate action for the next decade: the Infrastructure Investment and Jobs Act (IIJA), the Inflation Reduction Act (IRA), and the CHIPS and Science Act. All told, these policies will reduce heat-trapping pollution in a way that creates good-paying American jobs, bolsters domestic manufacturing of clean technologies, reduces energy costs for families and businesses, invests in historically

---

[7] Ron Bousso, "Ukraine war to accelerate shift to clean energy, BP says," *Reuters*, January 30, 2023.
[8] BloombergNEF, "Energy Transition Investment Trends 2023," January 2023.
[9] The White House, "FACT SHEET: Biden-Harris Administration Advances Equity And Economic Opportunity Through Federal Procurement And State And Local Infrastructure Contracting," July 6, 2022.

disadvantaged communities, and firmly positions the United States to remain the global leader in clean technologies of the 21st century.

The majority staff of the Select Committee on the Climate Crisis summarized the climate and clean energy legislative accomplishments of the 116th and 117th Congresses in the December 2022 majority staff report "Solving the Climate Crisis 2022: Key Accomplishments and Additional Opportunities."[10] Below are brief summaries of four of the bills:

2020 Omnibus

The 2020 year-end omnibus legislation helped advance climate solutions and clean energy research, invested in our nation's resilience, and used America's technological leadership to expand opportunities for all Americans. The legislation included support for energy and environmental innovation, clean energy tax extensions, pipeline safety, and water infrastructure. Notably the legislation directed the Environmental Protection Agency (EPA) to phase down the production and consumption of heat-trapping hydrofluorocarbons (HFCs) which, along with the Senate's ratification of the Kigali Amendment to the Montreal Protocol, will enable the United States to lead global efforts to avoid increasing global temperatures by up to 0.5°C.

Infrastructure Investment and Jobs Act

The 2021 Infrastructure Investment and Jobs Act (IIJA) included historic investments to strengthen our nation's core infrastructure for transportation, water supply, and electric grid resilience and modernization. It also provided major investments in energy efficiency, advanced environmental justice by remediating legacy pollution, and prepared communities for costly weather disasters. These investments include the largest investment ever in public transit, broadband to expand access for low-income American families, clean electric school and transit buses and a nationwide electric vehicle charging network. The IIJA also supported projects to reduce flood and wildfire threats to communities, including strengthening housing, public buildings, and infrastructure against climate-fueled extreme weather.

The CHIPs and Science Act

The CHIPS and Science Act authorized critical investments in American innovation and workforce development by advancing research to expand clean energy, modernize the grid, develop carbon removal and clean industrial technologies, and advance the fields of climate science, clean water systems, and critical minerals. The CHIPS and Science Act also benefits environmental justice communities, with investments to build regional technology hubs, expand grants to economically-distressed communities and labor markets, and diversify the STEM workforce. The investments included for semiconductor chips will support the American-led development of increasingly essential components of electric vehicles, building electrification, renewable energy, and electric transmission technologies.

The Inflation Reduction Act

The Inflation Reduction Act (IRA) is the largest climate and clean energy investment in U.S. history, including roughly $370 billion to expand access to cleaner, cheaper energy and cost-

[10] Select Committee on the Climate Crisis majority staff, "Solving the Climate Crisis 2022: Key Accomplishments and Additional Opportunities," December 2022.

saving technologies that will help reduce pollution, boost resilience across America, ensure environmental justice for vulnerable communities, address the national security threats posed by extreme weather, and meet our climate goals. In addition to lowering energy costs and creating jobs, the IRA is expected to help put the nation on a path to reducing harmful carbon pollution by as much as 42% by 2030. The IRA will:

- Deploy cheaper, clean energy faster and support energy efficiency upgrades and clean electrification to American families and businesses to help save money on their energy bills;
- Help deploy cheaper, cleaner vehicles through tax credits for electric vehicles, and enhance transportation resilience, equity, and sustainability;
- Create millions of good-paying jobs, develop the workforce, and promote economic development, along with continued development of strong domestic supply chains for clean energy technologies;
- Reduce climate pollution from our industrial sector and tackle releases of potent heat-trapping pollutants like methane;
- Focus investments into disadvantaged communities to combat environmental injustice and empower communities across the United States to ensure that they share in the benefits of the transition to a clean economy;
- Provide billions of dollars for conservation, restoration, and resilience projects on public and private lands, including working lands, to promote healthy environments that provide clean air and water, food, jobs, and store climate pollutants;
- Strengthen community and infrastructure resilience, including investments to help rural communities and tribes build resilience and adapt to unavoidable climate impacts; increase the resilience and reliability of the electricity grid; improve energy and water efficiency and resilience for affordable housing; and increase the resilience and capacity of coastal and marine habitats to withstand weather events;
- Advance renewable energy development and ensure oil and gas companies pay their fair share for extraction activities on public lands and waters; and
- Bolster advancements in climate and weather research and clean technology innovation.

In addition to the laws summarized above, the 116th and 117th Congresses also took other important conservation, environmental justice, climate and clean energy actions through the enactment of additional legislation, including the Great American Outdoors Act, the American Rescue Plan, and the annual National Defense Authorization Acts and appropriations acts.

The Benefits of These Climate and Clean Energy Investments Are Arriving Now with More Anticipated in the Future

By the second anniversary of President Biden's inauguration, companies had announced $300 billion investments in innovative technologies from biomanufacturing to batteries and semiconductors to solar.[11] Since the enactment of the IRA in August 2022, $89.5 billion in clean energy projects alone have been announced, initiating new economic development opportunities across the country.[12] Companies are already responding to the changing U.S. clean energy investment and demand landscape especially for globally competitive technologies like solar, batteries and electric vehicles.[13]

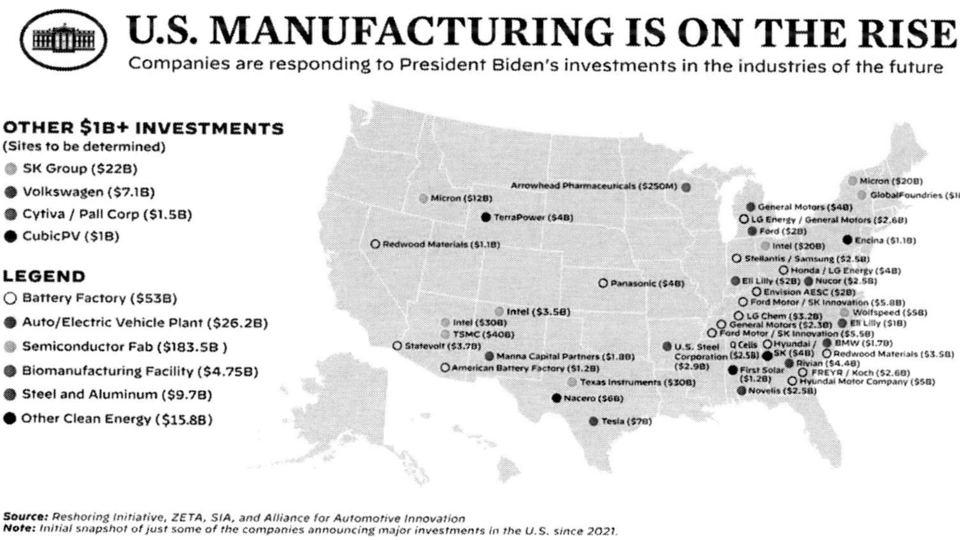

As the initiatives from the IIJA, IRA and CHIPs and Science Act ramp up, these investments and subsequent economic development will continue to grow. Various analyses suggest that additional jobs, health benefits and cost savings will come from the IRA, including the following:

- The BlueGreen Alliance and the University of Massachusetts Amherst Political Economy Research Institute found that the IRA would create more than 9 million good jobs over the next decade in clean energy, manufacturing, and transportation as well as building efficiency, environmental justice and natural infrastructure.[14]
- According to Energy Innovation, the IRA "could lead to between 3,700 to 3,900 avoided deaths in 2030, in addition to 99,000 to 100,000 avoided asthma attacks, and 405,000 to 417,000 avoided lost workdays … As a percentage decrease, avoided deaths are concentrated in communities of color … and on balance, the bill's provisions reduce health burdens more in communities of color."[15]
- The Princeton University-led REPEAT Project found that the IRA would lower annual U.S. energy expenditures overall by at least 4% in 2030, equal to nearly $50 billion

---

[11] Brian Deese, "The Biden Economic Agenda, Two Years In," January 20, 2023.
[12] Personal calculation based on Clean Energy Project Tracker, accessed on January 29, 2023.
[13] Kelsey Tamborrino and Josh Siegel, "Big winners from Biden's climate law: Republicans who voted against it," *Politico*, January 23, 2023.
[14] BlueGreen Alliance, 9 million Jobs from Climate Action: The Inflation Reduction Act (Aug. 2022).
[15] Energy Innovation, "Modeling the Inflation Reduction Act Using the Energy Policy Simulator," August 2022.

per year in savings for households, businesses, and industry.[16] U.S. households in particular would save hundreds of dollars annually on energy costs.

- The Rhodium Group estimated that due to lower electricity rates, lower fuel costs, and greater use of energy efficiency, household energy costs will decline by $717 - $1,146 in 2030, including electricity bills, home heating fuel bills, and transportation fuel costs.[17]
- Resources for the Future (RFF) estimated that the IRA will help reduce retail costs of electricity by 5.2-6.7%, generating aggregate savings of $209 - $278 billion, and helping the average U.S. household save roughly $170 - $220 each year from lower electric bills and lower costs for goods and services over the next decade.[18] RFF expects that the IRA's deflationary impacts on electricity prices will begin quickly – as soon as 2023. RFF also estimated that the IRA will help protect ratepayers from volatility in natural gas prices.
- Rewiring America also published an analysis of the community and household benefits available by congressional district.[19] All 120.7 million households in the United States are eligible for an average of $5,739 in tax credits and an average of $367 in rebates. Since they have triple the energy burden (the portion of their income spent on home energy) as other households, the 53 million low- and moderate-income households in the United States will greatly benefit from being eligible for an average of $10,377 in rebates through the IRA.

As encouraging as these analyses already are, the ultimate benefits may be even greater as the IRA works in conjunction with the IIJA and the CHIPs and Science Act. The history of past technical change suggests that to enable successful, ongoing innovation, governments must invest in R&D to push development of new technologies and couple those with demand policies to pull them to deployment.[20] The IIJA, the IRA and the CHIPs and Science Act provides this push-pull combination for clean energy as illustrated in this graphic by Dr. Constantine Samaras and America will reap the benefits for years to come.[21]

---

[16] Jesse Jenkins, Erin Mayfield, Jamil Farbes, Ryan Jones, Neha Patankar, Qingyu Xu, and Greg Schively, Preliminary Report: The Climate and Energy Impacts of the Inflation Reduction Act of 2022, REPEAT Project, (Aug. 4, 2022).

[17] John Larsen, Ben King, Hannah Kolus, Naveen Dasari, Galen Hiltbrand, and Whitney Herndon, A Turning Point for U.S. Climate Progress: Assessing the Climate and Clean Energy Provisions in the Inflation Reduction Act (Rhodium Group, Aug. 12, 2022).

[18] Nicholas Roy, Dallas Burtraw, and Kevin Rennert, Retail Electricity Rates Under the Inflation Reduction Act Of 2022 (Resources for the Future, Aug. 2022).

[19] Rewiring America, The IRA will deliver huge savings: If we maximize its electric potential for our communities and households, September 13, 2022.

[20] Vicki Norberg-Bohm, "Creating Incentives for Environmentally Enhancing Technological Change: Lessons From 30 Years of U.S. Energy Technology Policy," *Technological Forecasting and Social Change*, 2000.

[21] Costa Samaras, available at https://twitter.com/CostaSamaras46/status/1616631930274865152.

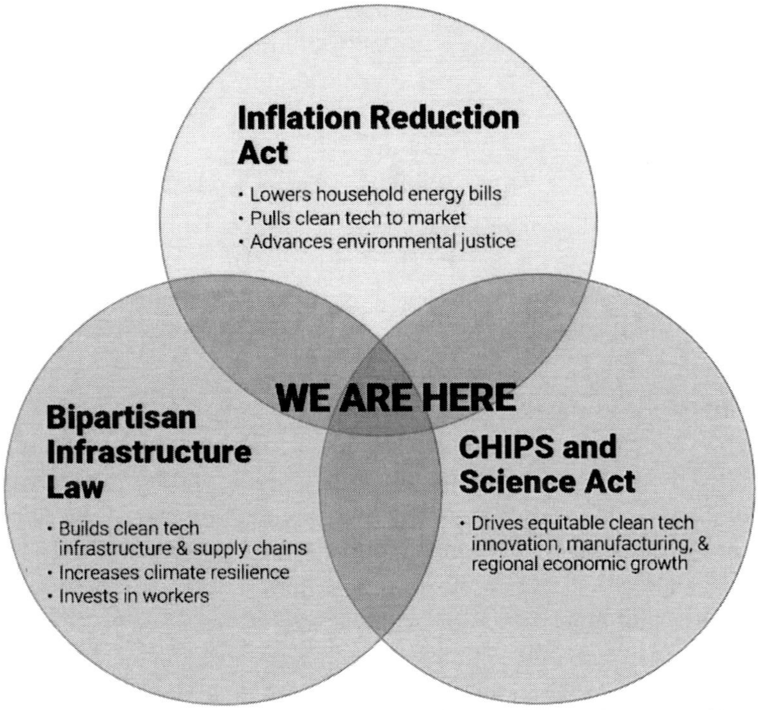

Solving the Climate Crisis

Building on the achievements of the 116th and 117th Congresses, there is still more to do to fully address the climate crisis. Foremost is the need for implementation of the legislation that was enacted. This requires concerted effort from the Biden-Harris administration in partnership with state, local, and tribal governments as well as businesses, investors, civil society and individuals taking advantage of the tax credits and programs that will support energy efficiency and clean energy deployment. In addition, the Biden-Harris administration is developing a number of standards and safeguards that will work in tandem with the recent laws to achieve further climate pollution reductions. These are crucial to meeting the climate and clean energy goals that President Biden has set.

Congress should also consider further appropriations and legislation to drive the transition to a clean energy economy that supports a healthy, resilient and just America. The 2022 Select Committee majority staff report highlighted additional opportunities for future Congresses to address the climate crisis,[22] including the need for:

- A comprehensive transmission strategy to meet the increased electric load from electrification of vehicles, buildings, and industrial processes;
- A Clean Electricity Standard, Zero Emission Vehicle sales standards, and other sector-specific standards to provide certainty for investments in pollution reduction;

[22] Select Committee on the Climate Crisis, "Select Committee Democrats Release Final Report on Key Accomplishments, Additional Opportunities," December 14, 2022.

- A comprehensive approach to critical minerals sourcing and recycling, including updating outdated mining laws to ensure critical minerals are secured in an environmentally, economically, and socially responsible way;
- Continued investments in research for hard-to-decarbonize sectors like off-road transportation and industry, and for carbon removal;
- Improving community engagement in the permitting process, addressing the cumulative impacts of plastic production and disposal in fenceline communities, and supporting efforts to strengthen the environmental justice focus of agencies;
- Increasing support for workforce development and communities experiencing energy transitions, including through registered apprenticeships and a reimagined Civilian Conservation Corps, and ensuring all workers are protected by securing strong labor standards, especially during any extreme weather conditions or events;
- A National Adaptation and Resilience Plan, including strategies to advance climate science and tools, expand technical assistance to improve planning and access to federal programs, and prioritize investments in environmental justice communities;
- Advancing resilience-based codes and standards against rising flood, wildfire, and extreme weather risks, and accelerating disaster recovery and bridging the resilience and protection gap for communities at greatest risk;
- Implementing nature-based solutions on public, private, and working lands and waters, enshrining climate mitigation and adaptation in federal natural resource and land management, and protecting and conserving at least 30% of lands and waters;
- Increasing research on climate and public health impacts, prioritizing health equity in federal planning for climate impacts on the healthcare sector and public health, recognizing the disproportionate burden on disadvantaged communities; and ensuring all health sector infrastructure is resilient to the impacts of climate change;
- Fulfilling our commitments to mobilize climate finance for developing countries, and expanding diplomatic, humanitarian, and military capacity to address climate issues; and
- Embedding climate considerations into all facets of national security and defense policy, including acquisitions, vehicles, technologies, construction, and other approaches.

* * *

Mrs. RODGERS. My favorite line is “American ingenuity is our superpower.”

Next let’s turn to our final witness, Mr. Robert McNally, president of Rapidan Energy Group.

You are recognized for 5 minutes.

Statement of Robert Mcnally

Mr. MCNALLY. Chair Rodgers, Ranking Member Pallone, distinguished members of the committee, I am Bob McNally, president of Rapidan Energy Group. With the exception of 21⁄2 years in the Peace Corps and 2 years working for President Bush, I've been an analyst, historian, and a student of energy markets policy and geopolitics.

I would like to briefly summarize my view on where we are headed in oil prices and touch on some energy prices.

Let me start with the good news. For the best part of the last 20 years, OPEC is no longer controlling the global oil market. Let me tell you the bad news. For the best part of the last 20 years, OPEC is no longer controlling the global oil market. That is why we have seen the return of Space Mountain price volatility. It is paradoxical, but true: When the oil market does not have a successful, durable manager like OPEC or, before it, the Texas Railroad Commission, oil prices wildly gyrate in boom-and-bust cycles that hurt our economy, national security, and environment.

What to expect in the future as you do your work: More of what we have seen in the past. Oil prices are going to gyrate between 5- to 7-year boom phases when they go up and bust phases when they go down. In the late summer of 2021, we exited a bust phase, lower oil prices, and we are in the foothills of multiyear boom cycle. President Putin's decision to invade Ukraine hit fast forward on this boom cycle last year, as we saw with $5 gasoline, but oil prices quickly retreated when all the Russian supply we thought we were going to lose did not go offline.

Now, the shale boom is a huge economic and security benefit, but it will not insulate our consumers from oil price volatility. Therefore, we cannot retreat into energy isolationism. We must work with allies to support domestic energy production and support infrastructure, bolster free trade, maintain ample stockpiles, and prevent adversaries from holding the world hostage to economy-reckoning disruptions.

As we enter this new multiyear boom phase, it is all the more important that we resist or correct policy mistakes. These include old ones, like 1970-era ideas of windfall profits taxes or restricting exports. Look, on the windfall profits taxes, they are going to—profits are going to go up during booms and down during busts. Seizure of them by the Government will only worsen supply and increase prices later.

Restricting energy exports will not lower pump prices at all—I mean, for any meaningful period of time, if at all, for our consumers. Longer term it will raise them. Restrict export, you get less supply. Less supply, you get higher prices. Moreover, restricting the exports now would help Putin, help Iran, and hurt our allies.

Newer errors to resist or correct include an overall shift unfortunate from all-the-above to keep it in the ground. As we enter a boom cycle and face geopolitical risk, it is critical the capital flow back to fossil fuel production, refining, and infrastructure. Fossil fuel is 82 percent of our primary energy supply; oil alone, 31 percent; renewables—they are great—7 percent.

Policies that delay or impede the return of capital expenditure to fossil energy will only tighten supply/demand balances further and punish consumers with even higher energy prices.

So the Strategic Petroleum Reserve cries out for course correction, Members of Congress. At 372 million barrels, the reserve is down by about a half, the lowest in 40 years, due to recent emergency and nonemergency sales.

Now, the energy gyrations after Russia's invasion of Ukraine illustrate the folly of frittering away our emergency supplies. To their credit, President Trump—excuse me—President Biden and the Congress agreed to cancel 140 million barrels of planned energy sales, and it signaled an interest in refilling. And, Chair Rodgers, you shepherded a bill that will also impose restrictions.

Finally—and this is often overlooked but very important as you begin your work—I would like to highlight for the committee's attention an overlooked but crucial area for improvement. A large gap has opened between increasingly unrealistic energy transition targets on the one hand and objective and timely analyses to require to assess their costs and likelihood on the other.

President Biden and some Members of Congress have called for outlawing the use of natural gas and coal and electricity by 2035. That is 60 percent of our generation. It is astounding there has been no cost estimate by EIA of this proposal.

Meanwhile, under political pressure, the EIA has ceased providing a policy-neutral reference case scenario and adopted a peak rosy demand scenario. On that basis they have called for no new investment in oil and gas in the world. I can't think of anything more catastrophic for energy than that.

Now, to conclude, our country is blessed with enormous energy resources. The sweats and the smarts exhibited every day by our intrepid energy-sector workers underwrite our standard of living. Perilous market and geopolitical conditions may loom, and policy risks certainly abound. But if you leverage and if we leverage realism, pragmatism, and innovation, we can protect our economy, security, and environment.

I wish you and your committee success.

[The prepared statement of Mr. McNally follows:]

Testimony by Robert McNally at the House Energy and Commerce Committee Hearing on Tuesday, January 31, 2023, Entitled "American Energy Expansion: Strengthening Economic, Environmental, and National Security"

Summary

Outlook for Oil Prices

Since 2008, the oil market has not had an effective, consistent supply manager and therefore extreme oil price volatility has returned. Oil prices have fluctuated between highs that hammer consumers and impede economic growth and lows that force producers to shut in wells. The world is back to riding "Space Mountain" oil prices for the first time since the interwar period.

Where are we now amidst these alternating booms and bust phases? In late 2021 we began transitioning from a bust phase to a boom one. *Consequently, expect structurally higher prices as recovering demand slams into inadequate supply.*

The Shale Oil Boom Is a Huge Blessing but Will Not Insulate Us from Price Volatility

The shale boom confers huge economic and security benefits but won't insulate us from global crude oil price volatility. Therefore, we must continue working with allies to support domestic energy production and infrastructure, bolster free trade in energy, maintain ample emergency stockpiles, and prevent adversaries from holding the world hostage to economy-wrecking supply disruptions in key regions around the world.

Resisting or Correcting Policy Mistakes

Successful management of looming challenges includes resisting or reversing erroneous policy trends and proposals that have unfortunately cropped up recently. They include reviving 1970s-era mistakes such as windfall profits taxes and restricting or banning energy exports.

Newer errors to resist or correct include an overall policy paradigm shift from "all of the above" to "keep it in the ground," draining the Strategic Petroleum Reserve, and closing the gap between wildly unrealistic transition targets on the one hand and objective data and analyses of their real-world impacts on the other.

The Geopolitical Outlook Features Important Risks

Near term, potential supply losses from Russia, Iran, and Libya would far exceed OPEC+ spare production capacity. Foreign and domestic actors have begun to attack our vital domestic energy infrastructure.

Longer term, China is striving to secure oil, gas, and critical minerals supply around the world while defending its dominance of renewable energy supply chains.

FAL Policy Papers

Given the breadth of the topic and space restrictions, the author would request submitting six detailed policy papers recently published by the Forum for American Leadership, whose energy security working group he chairs. They cover subjects ranging from climate policy, confronting China, and making the US an arsenal of energy.

Testimony by Robert McNally at the House Energy and Commerce Committee Hearing on Tuesday, January 31, 2023, Entitled "American Energy Expansion: Strengthening Economic, Environmental, and National Security"

Chair Rodgers, Ranking Member Pallone, distinguished Members of the House Energy and Commerce Committee, my name is Robert (Bob) McNally, and I am founder and president of Rapidan Energy Group, an independent Washington, DC-based energy advisory firm. I am honored that you have invited me to contribute to your important hearing today. This testimony reflects my views and not those of Rapidan Energy Group.

I have worked for 32 years at the intersection of energy markets, policy, and international politics. Except for service in the Peace Corps and as President George W. Bush's energy advisor from 2021-2003, my professional responsibilities entail providing clients with analysis and forecasts of energy market, policy, and geopolitical trends and events. Neither Rapidan nor I represent or lobby for any person, group, or company.

I would like to review the main market, policy, and geopolitical energy terrain you will encounter as you formulate and implement energy policies to bolster our economy, national security, and environment. I will focus mainly on the oil market, which has been my principal area of professional activity, and respectfully suggest some policies for your consideration.

I would also like to request to submit six energy policy papers appended at the end of this note and published by the Forum for American Leadership, a non-profit foreign policy and national security advisory group whose Energy Security working group I chair. These papers include specific recommendations to improve energy security by making the US an arsenal of energy, counter China's attempt to dominate critical minerals and renewable energy supply chains, and set a course for a sound and serious climate policy.

"Space Mountain" Oil Prices – Buckle up for a Multi-Year Boom

I need not explain to Members of Congress or any elected officials that gasoline and diesel pump prices matter for our fellow citizens' sense of happiness and well-being, as well as for economic growth.

Last year's sudden spike in gasoline prices to record highs around $5 per gallon and then abrupt reversal were largely caused by shifting perceptions of crude and product supply losses due to Russia's invasion of Ukraine. Looking ahead to this spring, oil prices are in a tug of war between bullish factors such as EU sanctions on Russia's exports and China's economic emergence from COVID and bearish drivers such as inflation and central bank tightening.

While I am happy to discuss during the hearing the drivers for recent and near-term oil price volatility, please allow me to step back and make a broader point about where oil prices are headed in the coming years as your Committee conducts its work. To cut to the chase, members should anticipate steady and sharp increases in oil prices this decade, particularly if the US and global economies recover from COVID and inflationary risks to resume a healthy growth trajectory. This section explains why.

Domestic pump prices are driven mainly by global crude oil prices. Crude oil is largely fungible, widely traded, and globally priced. While countries and states may impose taxes or subsidies on refined products such gasoline or diesel fuel, they all face the same price for the basic input – crude oil.

Crude oil prices have always been volatile, but volatility has been unusually high over the last 20 years compared with the previous 90. From 1932 through the early 2000s oil prices usually spiked only during Middle East wars and crashed during recessions. Otherwise, they tended to be relatively stable.

But over the last two decades, oil prices have soared without wars in the Arabian Gulf and crashed without recessions. It is important to understand why extreme oil price volatility has returned, how long it will last, and what we can expect later this decade.[23]

Since the beginning of the modern oil market in 1859, oil prices naturally tended toward *extremely high* volatility. Oil's unusual volatility stems from the fact that both demand and supply are insensitive or inelastic to price changes. In a bust supply exceeds demand, the resulting surpluses push oil prices down to levels that force drillers to stop drilling. In a boom demand exceeds supply, forcing prices up to levels that force consumers to reduce demand, usually triggering or causing a recession.

Everyone is hurt by continual boom and bust price cycles. The resulting uncertainty discourages investment both in the oil industry and broader industrial sectors, hammers consumption, and complicates policymaking from the Federal Reserve to the Environmental Protection Agency.

---

[23] For historical and economic background on crude oil price volatility, see McNally, Robert. Crude Volatility: The History and the Future of Boom-Bust Oil Prices. Columbia University Press, 2017. See also 2012 testimony by Robert McNally before the House Committee on Small Business entitled "'Space Mountain' Pump Prices." https://smallbusiness.house.gov/uploadedfiles/mcnally_testimony.pdf.

Therefore, oil policy from the earliest days can be described as a quest for price stability. The only proven cure to ruinous oil price volatility is having some producers act as swing producers, sometimes also called supply managers. Supply managers attempted to tame oil's natural price volatility by adjusting supply to prevent large inventory imbalances and thus the need for huge price swings.

There have been three successful long term supply managers throughout history: John D. Rockefeller's Standard Oil in the late 1880s, the Texas Railroad Commission from 1932-1972, and OPEC which reigned from 1973 until 2008.

Since 2008, the oil market has not had an effective, consistent supply manager and therefore extreme oil price volatility has returned. Oil prices have fluctuated between highs that hammer consumers and impede economic growth and lows that force producers to shut in wells. The world is back to riding "Space Mountain" oil prices for the first time since the interwar period.

Where are we now amidst these alternating booms and bust phases? In late 2021 we began transitioning from a bust phase to a boom one. *Consequently, expect structurally higher prices in coming years as recovering demand slams into inadequate supply.*

I wish I had better news on that front.

Isn't OPEC+ the New Supply Manager That Will Prevent a Ruinous Oil Price Spike This Decade?

Yes and no. Throughout history, whenever extreme oil price volatility erupts, oil producers attempt to cooperate to tame it. Sometimes they succeed, often they fail. Since 2016, a new group of oil producers including all 13 OPEC countries and ten producers outside OPEC have attempted to play the role of the supply manager. Called "OPEC+" and led by Saudi Arabia and Russia, this group has enjoyed both success and setbacks at stabilizing oil prices.[24]

But looking forward, OPEC+ will be unable to prevent soaring oil prices later this decade as global demand outstrips supply outside OPEC+ because its members do not hold enough spare production capacity. OPEC+ is likely to lose control the same way its predecessors the Texas Railroad Commission and OPEC did – by running out of producible oil supply in a boom cycle amidst rising prices.

The Shale Oil Boom Is a Huge Blessing but Won't Insulate Us from Price Volatility

The shale boom confers huge economic and security benefits but won't insulate us from global crude oil price volatility. The shale oil boom emerged around 2010 and has turned the US into the world's largest oil producer and, starting in 2020, a net oil exporter for the first time in 70 years. It's important to understand that while we are a net oil exporter, we still rely heavily on imports of crude and products.

Our refiners, for example, import large amounts of foreign crude due to crude quality and refinery configurations. And the east coast relies heavily on gasoline and diesel imports.

---

[24] As a historical matter, it is too soon to say whether OPEC+ will join the ranks of Standard Oil, the Texas Railroad Commission, and OPEC as successful *multi-decade* supply managers. Russia's refusal in November 2014 to contribute to supply management inaugurated the bust phase. After enduring sharp crude price drops in 2015, Moscow agreed to become a supply manager. OPEC+ has operated since 2016 mostly with success, excepting March and April of 2020, when Russia's refusal to support supply control triggered another oil price crash. With the urging of President Trump, OPEC+ resumed supply management and has operated successfully since the summer of 2020.

The economic and national security benefits of being a net oil exporter are clear as they are massive, including employment and international competitiveness gains and acting as an arsenal of energy for our allies facing energy supply interruptions. But as we have seen many times in the past few years, the fact that we export more oil than we import doesn't protect our economy from price shocks caused by turbulence in the global oil market. A disruption anywhere causes a price shock everywhere, including here.

Finally, the outlook for future shale oil production growth is challenging due to price collapses during the bust phase, preference by investors for cash instead of new investment, and high costs for material inputs like steel and labor. Moreover, to the degree governments adhere to a "keep it in the ground" policy stance it will dampen enthusiasm by investors and shale executives to ramp up activity.

Therefore, despite being a net oil exporter, we must continue working with allies to support domestic energy production and infrastructure, bolster free trade in energy, maintain ample emergency stockpiles, and prevent adversaries from holding the world hostage to economy-wrecking supply disruptions in key regions around the world.

Energy Policy – Old and New Risks to Economic Growth, National Security, and the Environment

The looming shift from structurally loose to tight oil markets will pose economic and national security challenges in the coming years. Being a net oil exporter confers some resilience to our economy as a whole when oil prices rise. But our consumers and businesses will suffer. Meanwhile, our principal geopolitical adversaries like Russia and Iran will reap a windfall from higher revenues from their oil exports.

Successful management of looming challenges includes resisting or reversing erroneous policy trends and proposals that have unfortunately cropped up recently.

They include reviving 1970s-era mistakes such as windfall profits taxes and restricting or banning energy exports.

In boom-bust eras oil industry profits will rise and fall with prices. During the recent bust phase, profits fell. As prices start to boom, they will rise. Government seizure of oil industry profits will only exacerbate underinvestment and prolong and worsen the boom.

Restricting oil exports will not lower domestic pump prices for very long, if at all, and in the longer term will raise them. Export restrictions would quickly lead to less investment and therefore higher prices.

Moreover, banning or restricting energy exports would hurt our allies and help our adversaries like Russia and Iran. Maintaining the bipartisan consensus to support energy exports, exemplified in the bipartisan decision in 2015 to end the crude oil export ban, is therefore essential for our economy and national security.

Newer errors to resist or correct include an overall policy paradigm shift from "all of the above" to "keep it in the ground," draining the Strategic Petroleum Reserve, and insisting public energy forecasting agencies provide timely and objective analyses you and your colleagues require to evaluate policy options.

As signaled by President Biden's deeply unfortunate decision to cancel the Keystone XL pipeline project, federal energy policy has shifted from years of bipartisan, balanced agreement to support all energy sources and technologies to an across-the-board, immediate attack on fossil fuel exploration, production, and development. Examples include proposed ban on drilling on federal lands and waters and moving to use National Energy Policy Act, Securities and Exchange Commission, and other permitting authorities to block, delay, and raise the cost of capital for energy infrastructure projects.

As we enter a boom cycle, it's critical that capital flows back to fossil fuel production, refining and infrastructure. Fossil fuels supply 82% of global primary energy and oil alone 31%. Renewables supply 7%.[25] Shifting from the latter to the former at large scale will take many decades. Therefore we literally cannot afford to artificially ration or restrict investment in fossil energy this decade. Policies that ban or delay the needed capital expenditure will only tighten supply-demand balances further and punish consumers with even higher energy prices.

Alarmed by rising energy prices starting in late 2021, the Biden administration has tentatively begun to walk back, at least verbally, its across-the-board anti-oil and gas stance. Congress can encourage the White House to return to a more balanced and realistic approach to energy policy that will support both fossil and non-fossil fuels and technologies.

Nowhere does federal policy require immediate course reversal than with the Strategic Petroleum Reserve (SPR). At 372 million barrels (mb), the reserve is at the lowest level in 40 years. The reduction is due primarily to the emergency sale of 180 mb since last March. However, non-emergency sales such as congressionally mandated drawdowns since 2017 and the November 2021 non-emergency exchange have contributed to this record low level of protection against severe supply interruption.

If Russia's invasion of Ukraine has a positive aspect, it has alerted officials to the folly of frittering away our emergency energy supplies. To their credit, late last year President Biden and the Congress agreed to cancel 140 mb of planned mandatory sales and signaled an interest in refilling the reserve.

The House under Chair Rodgers' leadership recently passed a bill restricting future non-emergency sales.

Reasonable minds can disagree on the size and composition of the SPR. The subject deserves careful study and debate. But meanwhile, Congress and the President should resist non-emergency sales to raise revenue or in a fruitless attempt to control oil prices. I would respectfully suggest Congress cancel the remaining 125.6 mb of planned mandatory, non-emergency sales.

Space does not permit the appropriate elaboration that these and other policy questions deserve. The appended FAL papers address many topics, from climate to China. However, I would like to highlight for the Committee's attention an overlooked but crucial area for improvement. A large gap has opened between increasingly ambitious climate policy proposals and objective analyses required to assess their costs and likelihood of success. *It is critical that Congress close this gap and insist on receiving expert and unbiased information and analyses from public agencies responsible for producing them.*

The United States leads the world in providing comprehensive, timely, and accurate data on our energy markets. Ably led astute new Administrator Joseph DeCarolis, EIA's data are critical for industry, consumers, policymakers, and analysts and it deserves full funding and support.

However, as described below in this citation from FAL Paper "Congress is Key to Restoring Realism in U.S. Energy Policy," the EIA is being underutilized while its influential

---

[25] BP, World Energy Outlook, 2022.

international counterpart, the International Energy Agency, has allowed its forecasting to become politicized. Timely, sound and objective data and analysis are foundational for policy evaluation and therefore I urge the Committee to insist it receive the support it requires and deserves from these tax-payer financed agencies.

Those advocating for unrealistic energy transitions have so far not been held accountable by having official energy agencies analyze their unrealistic, extreme proposals. Specifically, the International Energy Agency (IEA) has recently skewed its forecasts to please climate extremists while depriving policy makers of the ability to evaluate costs and benefits of energy and climate proposals. And the U.S. Energy Information Agency (EIA) has failed to provide Congress with feasibility and cost-benefit analyses of President Biden's executive orders and legislative proposals that would mandate abrupt, massive bans on hydrocarbon energy.

To help right the ship and avoid costly and dangerous policy errors, Congress should reassert its role in setting U.S. energy policy. As a starting point, Congress should insist that taxpayer-funded agencies provide unbiased forecasts of energy markets, as well as objective evaluations of proposed energy and climate policies. Toward this end, the Forum for American Leadership Energy Working Group calls for more congressional oversight of the International Energy Agency (IEA) and U.S. Energy Information Administration (EIA) and recommends the establishment of a National Commission on Energy Transition Realism, an expert, non-partisan commission of renowned energy experts to advise government officials and evaluate policy options for energy transitions.

The Paris-based International Energy Agency is composed of 31 member countries (the U.S. and primarily European nations) and nearly a dozen association countries, including China, India, Brazil, and Argentina. Established in the wake of the 1973 oil crisis, IEA is a forum under the Organization for Economic Cooperation and Development with a mandate to respond to disruptions in the global oil supply and provide policy recommendations, as well as data analysis, 2 on global oil and energy supplies to bolster global energy security.

In recent years, however, the IEA has strayed from its assigned role as a watchdog for energy security and instead has transformed into a lap dog for climate zealots advocating for unrealistic energy transition targets. Congress should steer U.S. policy to walk IEA back to its security mission. Specifically, in recent years the IEA's analyses have veered from security and unbiased analysis to feeding the newly fashionable myth that the United States, and the world, can afford to immediately ban investment in fuels that compromise over 80% of global energy.

Moreover, IEA began skewing its energy forecasts to hide the costs of extreme climate policies while depriving elected officials of the ability to make informed cost-benefit assessments of energy and climate proposals. For example, the IEA 2020 World Energy Outlook abolished its "business-as-usual" (BAU, formally known as Current Policies Scenario) reference case scenario, an unwarranted break with decades of forecasting convention that makes it impossible to evaluate the costs and benefits of climate proposals.

Meanwhile, the non-partisan EIA is being underutilized as an agency that was created to deliver accurate energy data and transparent, objective forecasts and analyses to policymakers and the American people. For example, the EIA has so far not provided Congress with analysis of the energy security and economic implications of recent presidential executive orders and legislative climate proposals. The list includes President Biden's executive order calling for the federal government to reduce its emissions by 65%

by 2030 and reach net-zero emissions by 2050, as well as a bill (CLEAN Future Act) introduced by Democrats, mirroring the President's campaign proposal, to ban the use of fossil fuels—which currently account for 60% of U.S. electricity generation—in U.S. power plants by 2035.

We recommend Congress seize an opportunity to use hearings, letters, legislation, and other oversight tools to restore realism in U.S. energy policy by focusing on U.S. policy at the IEA and the EIA. Furthermore, the creation of an expert, non-partisan National Commission on Energy Transition Realism will further enable Congress to access fact-based data, analysis, and counsel from energy sector experts.

Geopolitical Risks Abound

The geopolitical outlook features important risks. Near term, potential supply losses from Russia, Iran, and Libya would far exceed OPEC+ spare production capacity. Foreign and domestic actors have begun to attack our vital domestic energy infrastructure. Longer term, China is striving to secure oil, gas, and critical minerals supply around the world while defending its dominance of renewable energy supply chains (please see FAL papers for background and proposals pertaining to China).

The oil market's reaction to Russia's invasion of Ukraine illustrates the fact that prices reflect expectations of the future as well as perceptions of the present. Crude oil prices closed at $96 per barrel the day before the invasion and within 7 trading days it hit almost $140 on an intraday basis. The main reason oil prices soar was widespread expectations and government warnings that sanctions on Russia would immediately disrupt 3 mb/d of oil supply, a large amount and roughly equal to spare capacity held outside of Russia.

Stresses in the refining sector also showed up last year: Refined product prices rose more sharply than crude last spring, because Russia is a major product exporters and China, which has also become a large product exporter, was restricting exports due to COVID restrictions.

Oil prices reversed and collapsed in the summer mainly because the feared loss of Russian supply did not materialize. During the second half of last year, oil prices were generally softer mainly on concerns about overall economic growth, especially due to China's bumpy exit from COVID restrictions and the return of China's product exports.

Looking ahead, geopolitical risk is likely to continue to put upward pressure on oil prices. OPEC+ spare production capacity remains relatively low at around 2.8 mb/d and is concentrated in the Arabian Gulf. As the economy recovers and production growth stagnates, OPEC+ spare capacity will shrink.

Geopolitical risks include principally the EU's ban on Russian crude imports that took effect and the ban on refined product imports on February 5. Iran continues to pose a threat to Arabian Gulf oil and gas infrastructure and choke points.

Finally, increasingly brazen and direct attacks on our vital domestic energy infrastructure threaten our energy security. The May 7, 2021, Russian-backed Colonial Pipeline cyberattack resulted in by far the biggest loss of domestic energy supplies due to hostile foreign action against the U.S. homeland.

- The Colonial pipeline supplies 45-50% of East Coast liquid fuel supplies, 90 military bases and installations, and seven major airports.

- The attack also highlighted the importance of engaging in strategic deterrence against future, potentially catastrophic, attacks on our critical energy infrastructure and exposed significant national security gaps that require timely legislative and executive branch remedies.
- Congress must work with the Executive Branch to take robust steps to deter and punish cyberattacks on critical energy infrastructure while preparing the country to manage future attacks better than it did in May 2021.

A spate of recent attacks on electricity infrastructure across the United States has also underscored the importance of improving our grid security.

Conclusion

To conclude on a positive note: Our country is blessed with enormous energy resources. The sweat and smarts exhibited every day by our intrepid energy-sector workers underwrite our high standard of living. Our vibrant, innovating energy sector confers solid benefits for our economy, national security and environment and can do so in the future. Perilous market and geopolitical conditions may loom and policy mistakes certainly abound. But if we leverage realism, pragmatism, and innovation we can protect our economy, security, and environment. I hope you will find these views and suggestions to be helpful and wish your Committee success.

Forum For American Leadership Papers

Eight Necessary Steps to defend U.S. Critical Energy Infrastructure from Cyber Attacks, October 2021

Creating an Arsenal of Energy, April 2022

Blueprint for a Serious and Sound Climate Policy, April 2022

Congress is Key to Restoring Realism in U.S. Energy Policy, September 2022

Setting U.S. Climate Policy Straight: Recommendations for the 118th Congress, December 2022

Restoring U.S. Energy Security: Recommendations for the 118th Congress, December 2022

* * *

Mrs. RODGERS. Thank you for your testimony.

I will begin the questioning and recognize myself for 5 minutes. Restoring American energy dominance and putting security at the center of our energy policy should be the top priority, and we should be sure that our policies don't put Americans and our allies at strategic disadvantage to geopolitical adversaries like China.

The Biden administration's national security strategy released in October mentions climate more than 60 times. It talks about, quote, "an urgent need to accelerate the transition away from fossil fuels." By contrast, the Chinese President's annual report to the National Congress of the CCP a few weeks later only mentions climate twice and warns a gradual withdrawal from traditional energy must be based upon safe, reliable replacements. He talks about less restrictions on fossil fuels. There seems to be a mismatch of priorities here.

Mr. Dabbar, your testimony makes the point that as a dominant energy producer with an all-of-the-above mindset, America strengthened its national security, lowered prices, and helped drive down carbon emissions. The developing world will continue to demand affordable traditional energy for the next several decades.

So does it make sense for the United States to accede its energy dominance in traditional energy to China?

Mr. DABBAR. Thank you, Chairwoman, for the question.

I would like to make the point that China is absolutely not the benchmark for clean energy. Some people like to point out about how much they manufacture. They produce more emissions than the whole OECD combined, and we lead the world in the reduction of tons of emissions, and China is increasing by 3 times per year than we are cutting. These are just numbers; right. These are just facts.

And by far—I wrote a paper at Columbia I published last week. Fifty-seven percent of all the Btu's in China from all energy uses is from coal, and it is going up. It is going up. And so China is not the benchmark, right? They do manufacture some things well, but they are not the benchmark. I wish everyone kind of just knew— knew kind of the fact pattern on that.

U.S. leadership for energy allows us flexibility around prices, as we were hearing earlier. And as many people here who are veterans, there are less people in the way than when I was younger when I was in the military and allows flexibility that did not exist, and it allows us, from a security point of view, to support our allies, which if we hadn't been building what we have been building in Texas and elsewhere, we would not be in the place today to help our friends.

Mrs. RODGERS. Would you speak briefly about how we make sure that we can export our energy and technological innovations like advanced nuclear to developing countries rather than having China and Russia increase their influence?

Mr. DABBAR. So, Chair, as Congressman Tonko pointed out, the U.S. leads the way in energy innovation. We need to do better at manufacturing. We are not bad at it, but we can do better. But almost everything that is out there in the world was invented in America. The lithium ion battery chemistry was funded by DOE and the National Science Foundation. I was there when we won the Nobel Prize for the lithium ion battery. Drilling technologies, we lead the world in drilling technologies. Solar, thanks to Secretary Chu and DOE, drove down those costs.

So the important part is about how do we take—we need to continue to support innovation. We lead the world on that, and we need to figure out how to take some of those and continue to build on our leadership to export.

Mrs. RODGERS. Thank you.

I would like to turn to Ms. Jackson.

And you make a very compelling point about energy affordability. You know, just a recent review of the Energy Information Administration data shows that residential electricity price is up 24 percent since January 2021, industrial rates up over 30 percent.

Would you just talk briefly about why lowering energy prices is so important to the communities that you work with?

Ms. JACKSON. You know, minority communities are already at the low end of the spectrum in terms of resources and finances. The average income for Black Americans is

$48,000 a year. They are basically living on about $3,700 a year. They don't have that much extra money to be able to spend on unaffordable energy. It disproportionately impacts their standard of living.

We are experiencing the largest decrease in the standard of living that we have ever experienced in my lifetime. We are struggling to be able to pay our bills, and be able to go to work. We are struggling to be able to buy food. We are struggling to buy shoes for our kids, clothing. We don't have any extra money left over for even necessities—not luxury items—necessities.

And we want home ownership, but we can't save any money. We want to be able to start our own businesses, but we can't save any money.

Mrs. RODGERS. Thank you.

Ms. JACKSON. Every week we are in a negative deficit.

Mrs. RODGERS. Thank you. Thank you.

Pleased to yield to the ranking member of the full committee, Mr. Pallone, 5 minutes.

Mr. PALLONE. Thank you, Chairwoman Rodgers. I appreciate the opportunity to speak about energy expansion and security today. However, as I alluded to in my opening statement, I reject the premise that fossil fuels are the key to expanding our energy security here in the 21st century.

So Dr. Unruh Cohen, in your testimony you highlighted some of the actions that Congress took during the 116th, 117th Congress and that President Biden took last year to enhance our energy security, particularly through the Energy Act of 2020, the Infrastructure Investment and Jobs Act, the CHIPS and Science Act, and of course the Inflation Reduction Act, all of which this committee played important roles in developing.

So let me ask you, could you speak about how the investments we made in our electric grid in the Bipartisan Infrastructure Law and the hundreds of billions of dollars of investments we made in the Inflation Reduction Act will enhance our energy security by speeding the energy transition?

Dr. UNRUH COHEN. Thank you for the question. Enhancing and expanding the grid is one of the most critical aspects to transitioning to a clean energy economy, reducing pollution from the electricity sector, which then has knock-on effects for other sectors. And so, in both the infrastructure bill and in Inflation Reduction Act, we had critical programs to invest in the resilience of the grid and the expansion of the grid, and I think we have it—those will come to fruition soon.

We are already seeing some good signs on building more transmission as well as using the existing footprint to replace, to upgrade with the technology so that we can carry more electricity and electrons on the given footprint.

Mr. PALLONE. Well, thank you. Now, I also suspect we will hear quite a bit today about the so-called benefits of dirty sources of energy, but even from a pure energy security perspective, fossil fuels come with plenty of costs. So would you talk about how some of the vulnerabilities of the fossil fuel-based economy, both as it relates to our dependence on petro dictators as well as closed to consumers, and could you talk about the risks we would face to our energy security if some of the critical investments included in the IRA were rolled back?

Dr. UNRUH COHEN. That is a big question, but yes. As oil and methane are now international markets, so as we stay dependent on them here in the United States, the demand and use of them around the world as well as either geopolitical attacks or disasters that impact the delivery have repercussions on both the price and, therefore, our U.S. economy.

In the United States, we have a few States that are oil and gas producers, and they see that benefit in their economy. Most of the States are actually consumers, and so, as we see volatility in price, that has a negative consequence on everybody's economic outlook.

So we have to keep that connection in mind, and as I said in my oral testimony and in my written testimony, the future is moving to reducing the use of fossil fuels, ramping up clean energy fuels. That enhances our domestic energy security, it helps improve public health, and as we are seeing, it means more economic development and jobs in the United States.

And all of that movement is underscored by especially the infrastructure bill and the Inflation Reduction Act.

Mr. PALLONE. I know you have only a minute left, but I wanted you to—you mentioned with regard to the Select Committee on Climate, and, you know, actions that Congress could take to build on the work of the BIF and the IRA towards a more clean energy economy. What are some of the things this committee should be focusing on in this Congress in that regard? You have got 45 seconds.

Dr. UNRUH COHEN. OK. Well, I have three that I will try and get through. One, I think there does need to be more focus on grid enhancement. Actually, Senator Manchin's proposal had some language on the grid that I think Energy and Commerce staff also liked, to help deploy more things, deal with cost allocations. So I think that is something the committee should look at.

Workforce issues are incredibly important, and I know in the last Congress, this committee, along with Education and Labor, worked on some very important legislation going towards the energy workforce. I think that needs to be a focus.

And finally, the consequences of the climate crisis are here. Adaptation and resilience has to be a focus of every committee that has some jurisdiction. EPA has some important programs there that fall under the jurisdiction of the Energy and Commerce Committee.

Mr. PALLONE. Thank you. Madam Chair, that sounds like some bipartisan things we could work on.

Mrs. RODGERS. For sure.

Mr. PALLONE. Thank you.

Mrs. RODGERS. The Chair recognizes the gentleman from Ohio, Mr. Latta.

Mr. LATTA. Well, thank you, Madam Chair, for holding today's hearing, and also congratulations to you on your first hearing today, and look forward to working with you to get energy to the forefront here in the United States.

I also want to thank our witnesses for being with us today, and your testimony. It is always important to have you here so we can get your views out.

But if I could start with my questions. Secretary Dabbar, in your testimony, on page 3 you said about reestablishing energy dominance here in the United States, but I want to point and to hone in on one of your sentences. You state, "And they need to radically overhaul ISO rules to encourage baseload power that is being shut down faster than new intermittent plants are being built."

I come from Ohio. You know, what we do, we make things. We have major manufacturing in my district and across the State. That means, when I am talking about steel, when I am talking about float glass, and I am talking about our Central Foundry over in Defiance, and I think about all the different things that we have, we have to have massive amounts of power.

In 2014, we had a polar vortex that went across the Midwest, and every power plant in the State was up and running, that we did not have one blackout or brownout because every one of those power stations was up.

I think it is important to point out about baseload power. Can we get baseload power—and I am—we are for all-of-the-above energy strategies we fought for as Republicans in 2008. But my question is, when it comes to baseload and peaking—and I think that is a problem that a lot of people don't understand—but where is that limit where we have to be at to make sure that we can make sure that these lights come on every day and those plants go on every day with baseload capacity?

Mr. DABBAR. So I am a big fan of wind and solar, and they are great at low costs on kind of a spot basis. They are the worst technologies when it comes to availability and intermittency. OK. So these are just kind of technology kind of fact patterns. And what has happened with ISOs is that they have become kind of Rube Goldberg kind of contraptions that they used to be when many people here were part of initial setting up ISOs about free markets, about having market clearing prices, and have transparency.

In reality, what has happened, RPS standards, production tax credits, all these different things have made these constructs incredibly fragile and encouraged baseload from being shut down in New York. They shut down good, well-running nuclear power plants to replace it with intermittent.

And many areas of this country, more power plants are being shut down than are being built, and that is just a fact. That is just kind of numbers.

And so at the end of the day, the ISOs, under the Federal Power Act, under the authority of FERC, are allowing an increased vulnerability in the power markets year after year, and I strongly think that FERC needs some more guidance from this committee on increasing the reliability before it tends in the wrong direction.

Mr. LATTA. Let me go on here. You talk about nuclear. I think it is also important to bring this up, because I know that I am working with Senator Barrasso over in the Senate to advocate for policies that build our domestic needs through industry because it is so important. And I think it is important, one of the questions I have got is, how critical is it the United States builds up its nuclear fuels infrastructure in this country?

Mr. DABBAR. So, Congressman, I wrote a paper about this and an article in The Hill. There's portions of nuclear fuel fabrication that are fine, such as making fuel rods, but I think, as many people here know, the U.S. is significantly exposed to Russia in terms of uranium enrichment.

Forty-seven percent of all the uranium enriched services, that are a key component to making nuclear fuel, is from Russia. It makes OPEC seeming like a minor component of crude.

And I know there has been a lot of discussion about how to incentivize enrichment being rebuilt in this country when DOE and Sentra shut down the last enrichment plant in Kentucky. The U.S. has had no ability. And as a side point, the U.S. has lost its ability to make high enriched uranium for the nuclear weapons program. Most people have no idea that we have lost the capability of making new nuclear material for weapons. And that is all related to this topic.

I think this Congress should take a look at how to incentivize bringing it back from Russia.

Mr. LATTA. Well, and again, because again, when you think about with Russia, what is going on today and our dependence on them, it is critical for this country to get out there and make sure that we got a nuclear fuel security program that we can rely on.

But I want to thank our witnesses, and Mr. Chairman, I yield back the balance of my time.

Mr. DUNCAN [presiding]. I thank the gentleman. The Chair will now go to Ms. Eshoo.

Ms. ESHOO. Thank you, Mr. Chairman. I want to thank each one of the witnesses. I have to tell you that I am having a little bit of a tough time figuring out what you are for, really what you are for. I hear a mix of, you know, "Woe is us. We were in great shape, we are not anymore." Some are for a mixed portfolio of energy sources.

If you were to sum up in a sentence, starting with—I don't have my glasses on—Mr. Dabbar, what is it that you want the Congress to do relative to energy? Is it displacement of other sources of energy? Are you saying that we should, you know, pursue more oil?

One sentence, what is your top line to the Congress of the United States today?

Mr. DABBAR. I think in one sentence, Congresswoman, I think properly incentivizing and authorizing all forms of energy that support producing emissions concurrently with that is the right general set of policies to do.

Ms. ESHOO. What does that mean, though? What does that mean? Are you—I mean, if you are promoting oil and that there be more oil exploration, just say it so that, you know, it is clear, and then it is up to Members to decide how that—I think that we have to have a mixed portfolio. I think that countries around the world, including the United States, are in big trouble, given what emissions have caused us.

We know that we have to have energy to operate, but it is a matter of, you know, what it is. So if it is oil, say so, so that it is clear, because otherwise it sounds like, to me, kind of a convention of just—generally speaking, a convention of energy suppliers having a discussion.

Do you want to redo your one sentence, or are you sticking with it?

Mr. DABBAR. I would make one comment, Congresswoman, that a colleague of mine at Columbia made a point that there's a difference between supply and demand. And as long as there's a demand for an energy type, so as long as we are still having internal combustion engines for a period of time—maybe long, maybe short, you can have that debate—it is better for America to produce that oil. Saudi Arabia right now intends——

Ms. ESHOO. Well, your colleague over here said the good news is that OPEC is not in charge of the bubble or whatever anymore. So you don't see—I don't think—well, whatever. It is a difference of opinion of witnesses here.

How about the gentleman at the end, what is your one sentence?

Mr. MCNALLY. Congressman Eshoo, so my one sentence would be: Go back to all of the above. Federal policy in the last 2 years has shifted from bipartisan all of the above to anti-fossil fuels. Keystone Pipeline decision, call to end Federal leasing, the ban on gas and coal electricity——

Ms. ESHOO. That is a long sentence. OK. Because my time is running out.

Doctor?

Dr. UNRUH COHEN. Congresswoman, I would say what Congress needs to evaluate their actions on is getting America and the world to a net-zero-carbon-pollution economy. So you need to evaluate everything that is coming through here of whether that will help advance that goal in the United States or us working with partners. It may not mean the total end of fossil fuel, but it will be a managed transition to achieve that goal.

Ms. ESHOO. Ms. Jackson?

Ms. JACKSON. Yes. I am for letting the markets decide what energy sources we should use. If you prop up an energy industry——

Ms. ESHOO. OK, one sentence, that is one sentence. Thank you.

Ms. JACKSON. Well, if you subsidize something, then of course they will produce more of that.

Ms. ESHOO. Thank you. I yield back. Thank you, everyone.

Mr. DUNCAN. The gentlelady yields back. The Chair will now go to Mr. Guthrie from Kentucky.

Mr. GUTHRIE. Thank you very much. I think someone said more power plants are being shut down now than being built. You know, the concern with that is, on December 23rd, when we were all wondering if we were going to get home for Christmas because of all the snow is coming in and the ice coming in, the storms coming in, are planes going to fly, you know, all that tension, and are we going to pass a bill or not. And then you get a phone call from my wife saying, guess what, we were out of power.

So in Kentucky, where I represent, I represent the Paradise plant that John Prine's parents are from, that area, that finally shut down the coal plant after that song has been around for a long time.

But we are having rolling blackouts in Kentucky, December 23rd and December 24th. So my constituents experienced that.

So, Mr. McNally, according to the Energy Information Administration, carbon emissions from coal from the United States is 2.5 percent of global carbon emissions. And yet, we are shutting down coal plants, where people are not able to have heat on the coldest days of the year.

So my question for you: Can you talk more about implications for policymakers of, like, the International Energy Agency? They provide forecasts that do not include accurate baseload power needs. And also let me just go to my second question for you. What information are expert statistical and energy information agencies like EIA and IEA not providing us, and do you think this—these forecasts gloss over the factors that should be considered, like grid reliability?

Mr. MCNALLY. Thank you, Congressman Guthrie, for that question. And if I could have added a comma and a sentence or a phrase to my sentence, it would have been: "and get objective and timely data so you can do your work."

Indeed, one of the main reasons we need to have IEA, especially, return to what we call business as usual or a reference case scenario, is, you can't do a cost-benefit analysis without one.

When folks refuse to put a forecast out saying, Look, here's the future if we just freeze policies in place—they don't do that—then you cannot assess the cost and benefits of different policies to maintain adequate electricity supply for your constituents while we are trying to decarbonize, which is important.

So we have to insist that the International Energy Agency go back to providing the tool you need to evaluate policy.

On the EIA side, they are ably led by Joe DeCarolis, and they are a wonderful resource for this country. However, they have been AWOL in terms of providing you with objective, honest forecasts of the implications of the policies that are being pursued by this administration, or proposed. That is simply unacceptable.

We ought not be afraid of science and analysis and forecast, but you are literally flying blind at this time.

Mr. GUTHRIE. OK, thanks.

And, Mr. Dabbar, a lot of times when I hear people back home, these questions they will ask me, let's say we do what—I think California has electric-only vehicles by 2035. If that was to come to pass for the country, something to that nature, you are dismissing—you have some environmental issues you are trying to address, but you create so many other implications and issues.

People ask me, what am I going to do with the batteries once they are used? The rare earth minerals, I think we left 3 trillion in Afghanistan when we pulled out in that chaotic decision the Biden administration did. I think the Chinese were waiting to move in. I mean, so what are the implications of—I know we are trying to decarbonize, but what are the other implications of some of the decisions coming from things like the Green New Deal, the IRA, other things such as that?

Mr. DABBAR. Yes. So one of the things I always find very problematic is arbitrarily picking technologies by legislators. The technology market moves much faster than the legislative process. I will give you an example. I just wrote a paper in The Hill in which I pointed out that California has looked at banning all internal combustion engines.

But in Congresswoman Kelly's district, they figured out how to manufacture negative carbon gasoline, that when you burn it, it is net zero, OK, so a technology-neutral option, and something interesting coming out of Illinois. Rather than saying everything has to be electric vehicles, why not let innovation drive the—you know, the opportunities that we have in front of us had not been invented.

And so, I think there's a lot of great examples that we should have a technology-neutral policy and let us kind of drive to the future based on that.

Mr. GUTHRIE. Thank you. And just about a half-minute that I have left, what are some lessons learned from Europe and their decisions in their carbon fight? What's some good lessons? Maybe you or Mr. McNally can answer.

Mr. MCNALLY. Don't become dependent on Russia or any other adversary for your critical supplies of energy, a lesson we ought to take heart, as we have heard with regard to China.

Mr. GUTHRIE. Thank you. And you hear people say that is the reason we need to do all the wind and solar because—we can be independent ourselves and still have fossil fuels in this country. We lost that independence.

So my time is expired, and I yield back.

Mr. DUNCAN. I thank the gentleman.

The Chair will now recognize the ranking member of the Energy, Climate, and Grid Security Subcommittee, someone I look forward to working with in this Congress, Ms. DeGette, for 5 minutes.

Ms. DEGETTE. Thank you very much, Mr. Chairman. Mr. Chairman, I ask unanimous consent to submit an article from Energy Wire entitled "Ukraine War to Accelerate Shift Away from Fossil Fuels."

Mr. DUNCAN. Without objection.

[The information appears at the conclusion of the hearing.]

Ms. DEGETTE. Thank you. This article reports that BP's own annual energy outlook acknowledges that, quote, "The increased importance placed on energy security as a result of the Russia-Ukraine war leads over time to a shift away from fossil fuels toward locally produced nonfossil fuels, accelerating the energy transition," end quote.

And so this hearing is about discussing the strength of our economic, environmental, and national security. In other words, its energy independence.

Now, as I mentioned in my opening statement, my colleagues have put together—put forward policies and ideas that suggest that they think energy independence hinges on increased domestic development of oil and gas resources. But as I said, energy independence can only happen with a rapid transition to clean energy.

And so, Dr. Unruh Cohen, first of all, I want to thank you for your work on the select committee. And I also want to thank my colleague Kathy Castor for her wonderful work on that committee too.

In your expert opinion, will increasing oil and gas production help protect the American public from the volatility of the global oil market, and if not, why not?

Dr. UNRUH COHEN. Increased U.S. production will not necessarily protect American consumers from volatile energy prices because we are still connected to the international market, and so if something a world away causes production to go out in Saudi Arabia and we have a price spike, then that passes through our entire economy.

Ms. DEGETTE. Even if the oil and gas is produced here?

Dr. UNRUH COHEN. That is correct.

Ms. DEGETTE. OK. Now, will increasing oil and gas production reduce the cost of extreme weather and climate-related disasters?

Dr. UNRUH COHEN. It will not.

Ms. DEGETTE. And will increasing oil and gas production ensure that our economy, the U.S. economy, is thriving?

Dr. UNRUH COHEN. It will not.

Ms. DEGETTE. Will increasing oil and gas production mean that the United States no longer imports crude oil from countries like Saudi Arabia?

Dr. UNRUH COHEN. Most likely not.

Ms. DEGETTE. OK. So if increased oil and gas production won't protect consumers for price shocks, from the increasing cost of natural disasters, from the boom-and-bust oil economy, and it won't make us energy independent, what do we need to do, very briefly, to achieve energy independence, and how do we need to get there?

Dr. UNRUH COHEN. I think our goal should be energy security, which means relying on clean energy produced here in the United States and working with our allies around the world. And there is a whole host of technologies that will help us do that, and we will see an acceleration in those based on the laws that have recently been signed.

Ms. DEGETTE. Now, you are not suggesting that tomorrow we would have to stop using oil and gas, but you are saying we would need to——

Dr. UNRUH COHEN. Correct.

Ms. DEGETTE [continuing]. Have a transition? Is that right?

Dr. UNRUH COHEN. We need a managed transition. That is what we need for our economy, that is what we need for our energy workers, that is what we need for our communities that depend on energy for their tax base. That is why we had so many things in the Inflation Reduction Act to help those communities and those workers transition.

Ms. DEGETTE. Great. Thank you.

And speaking of those communities, Ms. Jackson, I wanted to just talk to you for a minute because in my congressional district, I have a lot of low-income communities that are also environmentally very vulnerable.

And I have a neighborhood, Globeville-Elyria-Swansea, which is just downwind from an oil—from a gas plant, from a gas refinery. And they have had terrible, terrible environmental problems there: lead in their soil, terrible asthma increases, all kinds of other problems.

So I wanted to ask you the question. I know you advocate for low energy costs, and I agree with that, because those people have a hard time paying for their energy in the winter. I agree with that, but wouldn't you love to see an economy where we could both reduce those environmental impacts on low-income communities and also have low energy prices? That would be a yes or no answer.

Ms. JACKSON. That would be a yes with stipulations.

Ms. DEGETTE. OK. Thank you. I yield back.

Mr. DUNCAN. I thank the gentlelady, and I need to apologize to the gentleman from Kentucky for not recognizing him as the chairman of the Health Subcommittee, so welcome to that and good luck.

I now want to go to the chair of the Oversight and Investigations Subcommittee, Mr. Griffith, for 5 minutes.

Mr. GRIFFITH. Thank you very much, Mr. Chairman.

Mr. Dabbar, Inflation Reduction Act, we have heard about that, and we have heard about trying to, you know, do all this stuff in America, and one of the stipulations in that Act, when they were doing all the Green New Deal-type stuff inside of what was supposed to be the Inflation Reduction Act, was that we would do it in America, we are going to do it in the United States.

So I have got a couple of questions in that regard, and that is, one, recently the DOE awarded a $200 million grant to battery maker Microvast even after the company told the Securities and Exchange Commission that they were heavily dependent upon policies made by the Chinese Communist Party's apparatus in China. Do our DOE officials do a decent job of vetting these, or do they just plain just fill them out, and if they have got an address in Texas, that is good enough?

Mr. DABBAR. Yes, Congressman, I can't exactly define what DOE did on the Microvast award, but as you correctly said, the company itself told the Securities and Exchange Commission it was a Chinese-controlled entity and that it did not meet American accounting standards.

And the Securities and Exchange Commission was looking at delisting them from the United States. Even though technically it was legally incorporated here, the vast majority of the operations were in China, by their own admission, and with the SEC.

It is—I think it was the intent of the Inflation Reduction Act and the Infrastructure Act to strengthen American companies, and to— for those battery technologies, amongst others, get us away from China, and that is a bit of a question on why they decided to see that.

Mr. GRIFFITH. So you would think that was a bad decision by DOE, to give them $200 million?

Mr. DABBAR. I think that there are plenty of other companies— there are plenty of battery companies, and I am not certain why awarding it to someone who self-admits that they are a Chinese-controlled entity, they could probably—there's probably other ones in the United States to support.

Mr. GRIFFITH. Let me ask you about Contemporary Amperex Technology Company Limited. Now, this is a Chinese battery manufacturer and also attempting to get both Virginia money and Federal money, made a deal with Ford, and in that deal, Ford would own the physical plant but the Chinese company would have all the technology. They would control all the technology and all the actual manufacturing. Is that really a way to bring American, green technology into the forefront, or are all we are doing is becoming a colony of China when it comes to battery technology, when we take our taxpayer dollars and give it to the Chinese, even if they have a Ford nameplate on the outside of the factory?

Mr. DABBAR. Yes, yes, Congressman. It appears that Ford and the largest Chinese battery manufacturer, the largest one in the world, CATL, decided to structure legally to get around the Inflation Reduction Act American requirements, by having Ford front them, but it was really going to be a CATL plant.

I applaud the Governor of Virginia to turn that down. But whether it is this committee or Ways and Means, I would hope that someone would write a letter to Treasury as they look at writing the letter—sorry—the Tax Code underneath the Inflation Reduction Act and make certain that lawyers—no disrespect to lawyers, I am a nuclear engineer—don't come up with all these options that legally get around the Treasury rules but allow Chinese companies to get a greater foothold in the United States. Hopefully, that is quite bipartisan.

Mr. GRIFFITH. Well, I hope they will be bipartisan, and I can assure you that we will continue to do oversight from this committee. Hopefully other committees will do as well.

With the remaining time that I have left, I would be remiss if I didn't give Ms. Jackson some time. If you could expound on how late and missed electric bill payments, in light of the fact that the rates are going up for everybody in the country, including poor folks, how the late and missed electric bill payments compound the negativity on families living paycheck to paycheck?

Ms. JACKSON. Well, we are not living paycheck to paycheck. We are living paycheck to Wednesday, and then we are borrowing money for people to be able to last us to the end of the week. So the impact is devastating.

This is the deal. Here is the situation. We should be making sure—our community shouldn't be an afterthought. And so, when you are sitting down there thinking about how you are going to work with energy, we should also be considering the cost.

Mr. GRIFFITH. And the cost for the poor families in particular?

Ms. JACKSON. Yes, yes.

Mr. GRIFFITH. Thank you very much. My time is up. I yield back.

Mr. DUNCAN. I thank the gentleman. The Chair will now go to Ms. Matsui for 5 minutes.

Ms. MATSUI. Thank you, Mr. Chairman. I want to thank the witnesses for being here today.

Transportation is one of the most polluting sectors of our economy. Emissions from gas and diesel engines are directly responsible for the premature deaths of 17,000 to 20,000 Americans per year, and that doesn't even take into account the much larger impact of vehicle emissions as one of the leading causes of climate change.

And that is why I have been a long-standing champion of stronger vehicle emissions rules and stronger fuel economy standards, and why I lead yearly appropriations efforts supporting clean transportation across different agencies.

Dr. Unruh Cohen, what are the health benefits of clean transportation, especially with regard to frontline communities?

Dr. UNRUH COHEN. Thank you for the question, Congresswoman. The public health benefits of switching to cleaner transportation, transit, other ways to get around the country and our communities, is huge.

So much—so many communities are impacted by NOx, smog, particulate matters, black carbon, that come from both on-road and off-road fuel use that we can—significantly decreasing those leads to immediate health benefits, both for asthma in kids and elderly. I mean, we are even at the point where we have identified some of those pollutants can cross the placental boundary in women, and so, you know, babies still in their mother's womb are being impacted by air pollution.

Ms. MATSUI. Thank you. In 2021, I co-led the Medium and Heavy Duty Electric Vehicle Infrastructure Act, which would fund EV charging equipment for medium- and heavy-duty vehicle fleets. This would help State and local governments and municipal service providers and trucking companies electrify their fleets.

The Inflation Reduction Act expands the 30(c) Alternative Fuel Refueling Property Tax Credit to help governments, businesses, and individuals buy and install EV charging stations.

Dr. Unruh Cohen, what will the EV charging infrastructure look like in America in 5 years, given the incentives in the Inflation Reduction Act and the Bipartisan Infrastructure Law?

Dr. UNRUH COHEN. Well, as Yogi Berra is said to have said, it is hard to make predictions, especially about the future, but what I am confident in saying is, things will look totally different. The investments from both the infrastructure bill and the Inflation Reduction Act are really going to mean a huge expansion and improvement in our charging structures across the country.

You know, we are going to see a huge increase in electric school buses, which is going to be incredibly beneficial to the budgets as well as the health of school districts all across the country.

And, you know, I think we will be in a position where people will feel very comfortable traveling, you know, going on the great American road trip in their electric vehicle.

Ms. MATSUI. OK. Thank you. California leads the country in roof-top solar capacity with more than 8 million California homes benefiting from the cheapest source of energy known to man.

I am a strong supporter of rooftop solar and proud to champion expanding the 25D Residential Clean Energy Tax Credit in the Inflation Reduction Act. When we talk about grid resilience, it is important to remember that many homes in Florida and Puerto Rico with rooftop solar did not lose power even after they were battered by powerful hurricanes.

Dr. Unruh Cohen, what role does rooftop solar have in the clean energy transition, and how can it help make our electric grid more resilient?

Dr. UNRUH COHEN. Thank you for that really important question. Rooftop solar has incredible benefits, both to save consumers money on their electric bills and build back neighborhood resilience. I think, you know, increasingly, we will see neighborhoods with rooftop solars developing microgrids, so that when those—and additional placement of power, of PV and batteries, other things that critical schools, fire stations, those types of places in our community, so that when disaster does happen, there is still a power source, there is a place where people can come to get electricity and do the things they need in the aftermath of disaster.

Ms. MATSUI. OK. Thank you. I am running out of time, so thank you very much for your testimony. I yield back.

Mr. DUNCAN. I thank the gentlelady and now will go to the subcommittee chair for Innovation, Data, and Commerce, Mr. Bilirakis, for 5 minutes.

Mr. BILIRAKIS. Thank you. Thank you, Mr. Chairman. I appreciate it. Congratulations on your subcommittee chairmanship as well.

Since the Biden administration came into office, Americans have been faced with a persisting energy crisis. We are in the midst of unprecedented increases in costs of living, and I continue to hear from my constituents on a daily basis regarding how difficult it is to make ends meet. I want to highlight one in particular, if I may. Patricia is an 84-year-old constituent who lives in Hudson, Florida, in Pasco County, with her 62-year-old disabled daughter. They live on an extremely strict, fixed income. Their electric bill has gone up an average of $50 per month in the past year despite the fact that they do not use central heat or air conditioning in their home—and you know what it is like not to have air conditioning, particularly in the State of Florida.

So, when coupled with other inflationary increases in the price of food and medicine, this mother-daughter senior duo has been pushed to the brink of financial hardship.

Sadly, many of my constituents have similar stories to Patricia, and very clearly, from Mrs. Jackson's testimony, the most vulnerable Americans are disproportionately affected.

The United States needs energy policies that make energy more affordable for Americans, not more expensive. That is the bottom line.

Particularly, I want to focus on improving the permitting process for pipelines and expanding pipeline infrastructure.

I mentioned that this stat, I mentioned it earlier in the roundtable that we had last week, but according to the American Petroleum Institute, Florida will face a 42 percent market—higher market-forward prices for natural gas in the average this calendar year.

One of the main factors contributing to these higher costs is a lack of adequate pipeline infrastructure. Mr. Dabbar, can you please talk a bit about how delays in pipeline permitting, like for the Mountain Valley Pipeline and the canceled Atlantic Coast Pipeline, impact the ability for these projects to be completed or even deter the undertaking of these projects in the first place?

And, again, what recommendations do you have to help expedite pipeline infrastructure for the American Southeast?

Mr. DABBAR. Yes, Congressman. So when a company comes along and wants to build a pipeline or a transmission line, they estimate how much it is going to cost, and how long it is going to take to start building, and how long it is going to take to build. There's two things that have really exploded the challenges around that.

One is that the extended approval processes—when people come up with a project cost and the approvals take a very long time, those project costs, because of inflation, become uneconomic.

And we are seeing that, for example, in Massachusetts, actually, with offshore wind right now. They proposed a contract, the approvals have taken so long, and then the inflation topic is on top of it, all of a sudden the offshore wind projects, they withdraw them because they don't make economic sense anymore.

So the combination of inflation, plus very, very long approval processes are really impeding any sort of construction in this country today.

Mr. BILIRAKIS. Thank you.

Another question for Mr. Dabbar: In addition to addressing delays on the agency's side, is there anything we can do in Congress to reduce the delays brought about by the courts?

Mr. DABBAR. Congressman, I am not—never written a law from scratch like everyone here has, but I do think that the Federal Power Act and the Natural Gas Act are not definitive enough to make FERC approve and meet their statutory obligation to deliver energy, of any type. Everything is getting held back.

So I do—I would recommend that this committee take a look at how to facilitate, and maybe things such as time stops, time clocks, the maximum amount of time to review. That is not to say yes or no. It is just, put a timeline on, so it doesn't take 10 years for a 10-mile road.

Mr. BILIRAKIS. Very good. Thank you. I yield back, Mr. Chair. Appreciate it.

Mr. DUNCAN. Yes, I thank the gentleman. The Chair will now go to Ms. Castor for 5 minutes.

Ms. CASTOR. Thank you, Mr. Chairman.

Dr. Unruh Cohen, thank you for returning so soon to help us focus on policies that will help America build greater energy security and policies to help solve the urgent climate crisis.

While it is disappointing that Republicans in Congress have dismantled the Climate Committee, I have incredible pride in the progress that our committee and the Democratic-led Congress made for America. Talking about lower-cost clean energy, creating good-paying jobs, and building healthier, more resilient communities. We could not have accomplished all that we accomplished without your leadership and expertise, so thank you very much.

You know, at the outset of the creation of the Climate Committee, we convened and did broad outreach to people all across the country: business, labor, faith leaders, scientists, farmers, entrepreneurs, and Members on a bipartisan basis.

It culminated in this "Solving the Climate Crisis Action Plan" that made 715 policy recommendations to help solve the climate crisis. And I am so proud to report that, as of today, 438 of our policy recommendations were passed in the House of Representatives, and 314 were passed into law.

The report was described as the most detailed and well-thoughtout plan for addressing the climate crisis that has ever been part of U.S. politics, an extraordinary synthesis of expertise from social and scientific fields.

What that did is, it really culminated in a lot of the new law, the Inflation Reduction Act, the largest clean energy and climate investment in U.S. history, the Bipartisan Infrastructure Law that will help make communities more resilient and electrify transportation, the CHIPS and Science Act that will empower industries to lead the clean energy transition.

And I love the stat that you cited in your testimony. In the 9 months since the Inflation Reduction Act was signed, companies have announced more than $90 billion into clean energy investment.

So these are magnets now for capital, investing in our workers, in our communities, and in the clean energy future.

And though the Climate Committee ended with the last Congress, our fight has to continue. And I want to highlight to everyone—I am going to send this around to their offices—we did a year-end report that highlights key accomplishments and then additional opportunities, because there are so many more opportunities that we have got to focus on.

Meanwhile, the last 8 years have been the hottest on the planet. The Intergovernmental Panel on Climate Change, in their recent assessment, made clear that the next few years are critical to limit warming. Thankfully, progress begets progress, and with our accomplishments, we are poised now to move now to significant implementation.

You answered a question from—to Rep. Eshoo about the most important is to focus on net zero, getting to net zero, as soon as possible. So go into a little more detail on the most important steps of implementation on how we get there to reach our climate goals and reduce climate pollution.

Dr. UNRUH COHEN. Thanks for the question. And, you know, one of the good news, just going back to some of the permitting discussion in the Inflation Reduction Act, we also had nearly $1 billion set aside to agencies to help them do these analyses and get the projects permitted that we need to get permitted.

So I think that is one of the first things that we need to see. We need to see that implementation on the ground. We need to see— we need to build the EV infrastructure. We need to upgrade and expand our transmission infrastructure and using the programs that we passed in the infrastructure and the Inflation Reduction Act.

We also need to see the partnership with States and local communities and business organizations, because much of what we passed in the Inflation Reduction Act are—there's tax credits, there's other programs and initiatives that are going to help States do even more.

So I think it will be really important for Congress to work with the Biden administration to see those implementation—the implementation of those two laws in particular come to fruition.

Ms. CASTOR. We really are at an exciting pivot point, and thank you again, and I yield back.

Mr. DUNCAN. The gentlelady yields back.

The Chair will now go to the chairman of the Subcommittee on Environment, Manufacturing, and Critical Minerals, my friend from Ohio, Mr. Johnson, for 5 minutes.

Mr. JOHNSON. Thank you. You know, today it is clear from our testimony that America needs a course correction on our national energy strategy, because the current strategy simply is not working.

In fact, it is increasingly evident that we don't really have a national energy strategy at all. A change in direction must be made. Republicans on this committee have the plans and the legislative proposals to do just that.

It has already been said, energy security is national security. That goes for us here in America and for the rest of the world.

Other countries are thinking about this. Look at China. Do you think they cornered the market on rare earths and critical minerals by accident? Do you think it never occurred to them that in a time of great power competition, this would give them the upper hand?

I will get back to that in a moment, but right now America is in a similar spot with natural gas. Global demand is booming, and demand will remain high for decades to come. We have plenty for use here at home and to export abroad. Natural gas can and should be our global power differentiator. We should want the world to rely more on us for natural gas, not the other way around.

So first, Mr. McNally, I appreciated your recommendations in your testimony making changes to the Natural Gas Act, to cut the red tape surrounding LNG exports. I have got legislation that would do that.

Can you explain to our committee, when we are leading in exporting energy resources around the globe, and even though it certainly supports jobs and investment here at home, how is it that that also improves America's geopolitical position abroad?

Mr. MCNALLY. Thank you for that question, Congressman Johnson.

Nothing less than making the United States an arsenal of energy, as we were an arsenal of democracy under President Roosevelt during World War II. We are the world's largest natural gas producer. This year we are going to be the biggest exporter, and two-thirds of our exports have been flowing to our beleaguered allies in Europe facing the largest war since World War II on the continent there.

And we are beating the Qataris, the Norwegians, the Azerbaijanis put together. And, you know, Winston Churchill said, you know, security and energy comes from diversity, diversity alone. And thank God our allies, both in Europe and in Asia, in Japan and South Korea, they know they can rely on the United States as an arsenal of energy, an ample source of energy imports, especially when they face our geopolitical rivals.

So it is hard to overstate when you look at history, when you look at our current predicament and the outlook, how important being an arsenal of energy, keeping domestic production strong, and policies favoring exports alive.

Mr. JOHNSON. Well, continuing with you, Mr. McNally, back to critical minerals and rare earths, I mean, it is hard to believe our military relies on China for this stuff. We need to scale up domestic production rapidly, and if you read the reports, I do believe America can do this safer and cleaner than anywhere else that it is currently being done.

Can you go a bit further into how environmental regulations could be improved or changed to allow this to happen here in the United States?

Mr. MCNALLY. Thank you, Congressman. Congress might want to consider establishing a National Stockpile for critical minerals, and tax credits and other incentives to produce it here at home. And we might want to think about restricting the dependence, at least at some point in the future. It may be hard to go cold turkey, but in the coming years, to require that we get off of Chinese sources for our critical minerals and so forth.

So I think there is a variety of things I think you and others have been thinking of doing. Frankly, it is really just common sense, and I think you should pursue that course and will. Thank you.

Mr. JOHNSON. In a final kind of statement here, you know, I have heard it mentioned several times, you know, to get to these bipartisan solutions, one of the things we have to do is not go into our respective corners on this important issue.

And words matter. And I think one of the things that if we could just come to an agreement, the word "transition" doesn't mean what a lot of people out in America thinks that it does, right?

The Biden administration uses the word "transition" to talk about getting rid of fossil fuels. That is a negative way to use that term. I mean, we can walk and chew gum at the same time. We can increase market-driven innovation on renewables without throwing out the bath water and destroying the three-legged triad of oil and gas, coal, and nuclear that provide the resilience, the reliability, the affordability, and the sustainability to our electric grid.

With that, I yield back.

Mr. DUNCAN. The Chair now recognizes Mr. Sarbanes for 5 minutes.

Mr. SARBANES. Thanks very much, Mr. Chair.

Ms. Unruh Cohen, I want to talk a little bit about the profits of the oil and gas companies, which, to be very candid, drive me insane. So we have seen, through 2021 and 2022, ExxonMobil, Chevron, BP, Shell, made a record-breaking combined profit that is anticipated to be well over $200 billion.

They were also getting taxpayer subsidies during this time. I will come back to that in a moment.

We also have information today that Chevron announced $36 billion of profit, and Exxon this morning announced a profit of $56 billion.

Some of our colleagues on the other side suggest that, you know, this is a necessary evil, it is what has to be done to fund research and production and so forth on the part of that industry.

But then you look at how they are using these profits, and we see that Chevron announced that it would be spending $75 billion to buy back its own shares, and only investing $12 million into its business to increase production.

And, of course, these profits we have seen over the last few years have come at the expense of ordinary citizens out there being gouged at the pump. So it is clear that, at a minimum, taxpayer subsidies to that industry don't make any sense given the profits they are posting. Would you agree with that?

Dr. UNRUH COHEN. Yes, I would, Mr. Sarbanes.

Mr. SARBANES. And to the extent the Government is going to provide investments, aren't the investments that are reflected in the Inflation Reduction Act, the Infrastructure Investment and Jobs Act, in clean energy, in building out our future there—aren't those the kinds of investments that make a lot more sense?

Dr. UNRUH COHEN. Yes, Mr. Sarbanes, that is correct. And in our work in the select committee, you know, we argued that we should be aligning our tax code and our investments to drive the U.S. economy to this net-zero-carbon pollution.

Mr. SARBANES. I mean, look, it makes—there is something to be gained by going back and looking historically. The United States wanted to establish its fossil fuel industry, at one point, as premier in the world, and so the Government stepped up as a partner to try to drive that kind of innovation, and it was a very successful partnership.

Now we have the opportunity to go create a clean energy future, again, lead globally. That is what we are trying to do with the pieces of legislation that we passed.

But the model is really no different. It is just this mature, successful, very profitable fossil-fuel industry is no longer in need of support. If anything, we should be looking at ways to suss out what about their profit-making enterprise is fair to the American consumer and what is unfair, and how we can look to different kinds of investments to stand up this other set of opportunities.

But before we run out of time, just speculate with me on why the industry can't seem to free itself of this alternative reality. I would think that with the profits they are raking in, they would want to step into that clean energy future, using all of their assets and leverage and exercise leadership globally, establish themselves as real players in that arena.

But for some reason their reflex, their instinct, is to keep doubling down on the old way of doing things. What is the problem there? What is the difficulty with the frame of thinking?

Dr. UNRUH COHEN. Well, I think as you look at oil companies, you know, you really need to examine whether they want to be oil companies or whether they want to be energy companies, because we need energy companies in the 21st century, and that may be a whole range of things. And even if they are saying they want to be energy companies and that may mean they are diversifying their portfolio, then next you have to look, as you mentioned, at where they are putting their money. Is it just a little token piece of investment in other renewables or other technologies, or, you know, are they making a significant shift.

Mr. SARBANES. Thanks very much. Yield back.

Mr. DUNCAN. I thank the gentleman.

And the Chair will now recognize the gentleman from the crossroads of America, the great State of Indiana, Mr. Bucshon.

Dr. BUCSHON.

Mr. BUCSHON. Thank you, Mr. Chairman.

My constituents in southwest Indiana need access to reliable, affordable, environmentally sustainable energy. Look, all of us want to breathe clean air, we want clean water to drink, and we want clean land to utilize and enjoy in the future. That is kind of universal, right?

You know, but when out visiting energy stakeholders recently, someone explained the current idealogically driven efforts to replace all of the fossil fuel generation with wind and solar in this way—and I am paraphrasing—basically said, "This is a battle between politics and science and physics, and I can tell you which one will win out." And I think what he meant by that is, we need to start looking at the facts and get away from the idealogy.

And in that vein, I mean, Dr. Cohen, I just want to ask you a yes-or-no question quickly: Does the U.S. Government own any gasoline stations in America?

Dr. UNRUH COHEN. Did you say gasoline, gas stations?

Mr. BUCSHON. Yes. Like you go to pump your gas, does the U.S. Government own any of them?

Dr. UNRUH COHEN. Well, I assume DOE has some fueling——

Mr. BUCSHON. Did we build them? The answer to that question is no. And so I just don't see why the U.S. taxpayer should be funding EV charging stations across the country. Look, I support EVs. But why should we be doing that? Because—we are doing that because the market won't bear it. The free market at this point in our country won't bear it. If it in the future does, I am all in. Let the private sector——

Dr. UNRUH COHEN. Well, the U.S. Government has certainly engaged——

Mr. BUCSHON. I didn't ask you a question. I didn't ask you a question.

So every point you make on energy issues in this hearing depends on massive Federal Government subsidies and not the free market. So I just wanted to point that out. I mean, we have a disagreement, I think, between our two sides here: Believers in the free market and let, you know, technology, wherever it goes, all the above, or massively subsidize green energy at the expense of energy security and national security and cost.

So, Mr. Dabbar, I will ask you this question: Do you believe that current investors in the energy sector are primarily motivated—by what? Politics, physics, science? I mean, if they are looking to invest, where—and what are the pressures being put on them by politics?

Mr. DABBAR. So, Congressman, they are still primarily focused on profit. That is excellent for America.

Mr. BUCSHON. Of course.

Mr. DABBAR. But there is pressure from investors around ESG because of, you know, people who feel like that they need to do a certain shift, and they do certainly get that pressure.

Mr. BUCSHON. Of course they do.

And what do you think—same question, Mr. McNally. Do you have any comments on that?

Mr. MCNALLY. I would agree with my colleague. It is profit return on capital.

Mr. BUCSHON. Right.

Mr. MCNALLY. And I would just point out——

Mr. BUCSHON. Yes. I mean—but I will just give you my view. But we are using idealogical politics to try to divide—to direct investment in a free market capitalist system, and that is a huge mistake. Let the system work.

Mr. Dabbar, again, you mentioned in your testimony that FERC needs to radically overhaul ISO rules to facilitate increased base-load power.

What should we do?

Mr. DABBAR. The markets that right now try to support baseload, the capacity markets, don't work, and more power plants are being shut down than are being built. I think that FERC needs to come up with new rules that specifically require either the ISOs or maybe even going back to the utilities, like in the old days, where they have an obligation to serve. Right now there is no obligation to serve, and that is the challenge in the markets right now. The individual power plant owners don't have an obligation to serve, and the wires companies don't have an obligation to serve other than keeping their wires up. But supply itself, where it used to be the individual utility had a, quote, "obligation to serve," meaning to actually make the power, have power plants, that has been pushed out to the ISOs, and they are not incentivizing power to actually be built.

Mr. BUCSHON. Thank you for that answer.

I just want to point out someone—one of you mentioned we shouldn't be using China as a benchmark, and let me just partially tell you why. I mean, since 2021, they are building 33 gigawatts of coal-fired power generation, 3 times more than the rest of the world combined. They are building 14 new power plants since 2021 in their own country, and here—listen to this—but they will finish another 27 abroad. They are driving this worldwide. Of course, it is true they are trying to expand the renewables also, no doubt.

Currently there's about 1,118 coal-fired power plants in China, 225 in the United States. So I just want to concur with that testimony that they are not our benchmark. And, first of all—the other thing is we are our benchmark. We should be the benchmark driving the global energy economy.

I yield back.

Mr. DUNCAN. I thank the gentleman.

The Chair will now go to Mr. Tonko for 5 minutes.

Mr. TONKO. Thank you, Mr. Chair.

And before I get into my questions, Dr. Unruh Cohen, I believe you wanted to respond to the last exchange that you had.

Dr. UNRUH COHEN. Just—thank you for that, Mr. Tonko. Just a couple of points.

You know, the physics that Democrats have been concerned about are the physics of climate change, and we have known since 1856, in experiments that happened in Seneca Falls, New York, by Eunice Foote, that putting carbon dioxide in the atmosphere warms things up.

Secondly, the Government is totally involved in many ways in subsidizing transportation, whether it is the hundreds—you know, the centuries-long oil subsidies, biofuel subsidies, you know, we are involved, the U.S. Government, for many years building highways. We are involved in transportation. This is nothing new.

Mr. TONKO. OK. Thank you.

And, Ms. Jackson, I agree totally that we need to be focused on energy affordability and advancing solutions to reduce low-income Americans' energy burdens. It is a must.

In 2015, though, I would acknowledge that one in three households reported experiencing energy insecurity, and one in five reported reducing or foregoing necessities, such as food or medicine, to pay an energy bill. Eleven percent of households reported keeping their home at an unhealthy or unsafe temperature. So this is not a new challenge.

We have and have had real problems with energy affordability in our country, but we also have very important programs, programs that can help alleviate these burdens, like LIHEAP and DOE's Weatherization Assistance Program.

So, Dr. Unruh Cohen, the select committee's report recommended significant support for DOE's Weatherization Assistance Program. Can you explain why we should be providing upfront funding to low-income families to support cost-effective energy retrofits of their homes?

Dr. UNRUH COHEN. Yes, absolutely. Thank you for that question. Weatherization is one of the most important things we can help low-income families deal with. The incentives for renters sometimes precludes, you know, upgrading on energy efficiency. And so instead of seeing their energy bills skyrocket, with insulation, with that weatherization, they can stay warmer in the winter and cooler in the summer and reduce their energy use and, thus, their energy bills and insulate them also from the volatility of electricity markets.

On a personal note, my mother benefited from this program in Colorado, and it has meant a warmer house for her as she has seen some deadly cold weather there in Colorado Springs lately.

Mr. TONKO. Thank you.

And in the 117th Congress the Bipartisan Infrastructure Law provided some $3.5 billion for the Weatherization Assistance Program. This funding will do a lot of good. Households receiving weatherization assistance save on average of $372 each year on their energy bills. These are real savings and make a huge difference for families struggling to pay their utility bills. And there are other modest changes we can make to the program to improve it further, like allowing more funding to go to each household to account for the increased cost of construction materials and to allow service providers to train and retrain a qualified workforce. Dr. Unruh Cohen, I mentioned the Weatherization Assistance Program funding in the IIJA, but are there ways that the Inflation Reduction Act is expected to reduce energy costs for Americans?

Dr. UNRUH COHEN. Yes, there is a whole host of additional things in the Inflation Reduction Act that will help homeowners be able to purchase more efficient appliances. We will see more money flowing to State programs that help provide rebates and upfront— you know, cutting the costs upfront as people are purchasing those appliances, as well as just the overall advancement of adding more clean electricity. People will see just a general reduction in electricity costs.

My written testimony quotes some of the analyses that have been done on the Inflation Reduction Act that gives us some sense of the benefits to come.

Mr. TONKO. Thank you.

And, Mr. Dabbar, based on your testimony, you seem very proud of the work you did at the Department of Energy to support the development of innovative energy technologies. You mentioned breakthroughs in battery chemistry, solar technologies, and nuclear energy.

Can you explain what you see as the role of the Federal agencies like DOE to partner with the private sector and academia to support emerging energy industries?

Mr. DABBAR. Yes, Congressman. As we have talked in the past about Brookhaven and elsewhere around the country, new energy technologies, we lead the world, in New York and elsewhere. And one thing I want to point out is that the beyond lithium ion chemistries—lithium ion is great, great American invention, Nobel Prize, say it again—but there are many other technologies in the battery space that can outperform potentially lithium ion, and that will reduce some critical mineral requirements and imports from elsewhere around the world and actually have them perform even better. And the support by this Congress, many labs, DOE is leading the way to even more innovation on that front.

Mr. TONKO. Thank you.

And with that, I yield back, Mr. Chair.

Mr. DUNCAN. I thank the gentleman from New York.

And now I will go to the Tar Heel State, North Carolina, Mr. Hudson for 5 minutes.

Mr. HUDSON. Thank you, Mr. Chairman.

Let me just say, you are doing an excellent job. You look good with that gavel in your hands.

It is really an honor for me to serve on this committee again this Congress. You know, our committee has a long history of coming together to produce bipartisan solutions, and I look forward to continuing to work with all of my colleagues on both sides of the aisle in the coming Congress.

And I want to thank our witnesses for your time today and for the excellent testimony you have provided.

For the last 2 years, the Biden administration has failed to take meaningful steps to address energy production. Energy prices have surged to historic highs because of this, and all of our constituents have been paying the price. This winter has been extremely difficult for my constituents, particularly those on fixed income. Bad energy policy and the rising energy prices and the disruptions of supply that result from that disproportionately impact working families and lower-income communities.

And, Ms. Jackson, I appreciate your testimony to that effect. And I would like to extend you an opportunity to answer the question earlier about LIHEAP and affordability for lower income, if you would like to expand on your answer.

Ms. JACKSON. Yes. I wanted to piggyback on her.

You were talking about weatherization programs. The problem with minority communities is that they don't own their homes. What they want is homeownership, and we have climate policies that restrict that because it makes the building materials and the cost of building new homes so expensive that the price creates artificial scarcity and pushes them out of the marketplace.

The other thing is when you talk about electric vehicles, when you create a mandate on electric vehicles, basically what you are doing is excluding Black people from being able to own a private vehicle, which is very much needed so they can be able to get jobs that are outside of their communities, because most of those jobs don't exist where they are.

So weatherization benefits, of course, the landlord who owns the home. Subsidies for solar panels benefits who? The landlords who own those homes. We are renters, the majority of us, and creating higher energy costs is increasingly keeping us out of homeownership.

Mr. HUDSON. Thank you for that answer.

Changing the direction a little bit, last month two power substations in Moore County, North Carolina, were targeted with vandalism, causing a multiday loss of power for nearly 45,000 residents. Besides not having heat during the coldest part of the year, residents of Southern Pines and the surrounding communities lost thousands of dollars as a result of spoiled food and medicine, lost wages for workers and income for business owners, disrupted transportation, and created a lack of accessibility and essential services.

Now, I am proud of my neighbors for the way our community rose up and took care of one another, but this was a tough time. And just 2 weeks ago another substation was intentionally vandalized in neighboring Randolph County.

Disruptions to our energy supply are dangerous. They are costly, and they are unacceptable, especially when more than one-third of U.S. households reported skipping or reducing expenses like medicine or food in order to pay their energy bill in the first place.

I am looking forward to working with my colleagues to find solutions for these problems and hear from stakeholders on how we can address grid security.

I will start with you, Mr. Dabbar. How do energy disruptions caused by a vulnerable grid impact our national security?

Mr. DABBAR. So, Congressman, it is a high degree of worry. When we were at the department instead of the office, the CESER office—there is a significant amount of equipment that is produced overseas, including we found from China that they could easily have backdoors. Because as Mr. McNally has pointed out, energy is just in time. There is very little storage in electricity. There's some, but it is incredibly minor compared to demand. It is fragile. The system is very fragile because it is very just in time.

And so small disruptions of the electrical grid can have a massive impact. And the importance of building software, hard support, it is an amazing vulnerability for the country.

Mr. HUDSON. Well, thank you.

Mr. McNally, we have got 30 seconds left. Do you want to chime in there?

Mr. MCNALLY. Congressman, as a new resident of Southern Pines, I was directly impacted by that outage. I was also directly impacted by the brazen Russian-based attack on the Colonial Pipeline. It is astounding we have not spent more attention on that. That was the largest foreign attack on our energy sector ever. The closest we came is when the Japanese forgot to bomb the tanks at Pearl Harbor, would have set back World War II a few years. That was the closest risk. They hit us. They cut that vital artery that serves airports, major airports, and military bases. We are not deterring sufficiently. And I think we have some recommendations or we have put in the testimony how to improve—how we manage the next attack. I don't think it was well managed at all, sir.

Thank you.

Mr. HUDSON. Thank you.

Mr. Chairman, I see I am out of time, but with your indulgence I will submit in writing questions for our witnesses on sort of what they see as the role of the Department of Energy and the Congress in working towards this goal of grid security.

Thank you.

Mr. DUNCAN. Members are reminded they can submit questions to witnesses in writing.

The Chair will now go to Ms. Clarke for 5 minutes.

Ms. CLARKE. Thank you very much, Mr. Chairman, and to our ranking member, Mr. Pallone, for holding this important hearing today.

Let me also add my voice in congratulating Chairwoman Rodgers on being appointed as chair of this historic committee.

That said, I am deeply disappointed with the majority's failure to explicitly include any reference to environmental justice, racial or economic disparities, or inequality in their document organizing the jurisdiction of this committee.

I remain steadfast in my commitment to environmental justice and protecting the most vulnerable of our citizens. The exclusion of a clear commitment to environmental justice demonstrates a complete disregard for the historic marginalization and unjust systems exacerbated by our country's dependence on fossil fuels.

The impacts of the climate crisis, including climate disasters like historic heatwaves, hurricanes, and flooding, disproportionately hurt the most vulnerable among us, historically marginalized communities of color, those living with medical conditions that require electricity, older people, communities where there are language barriers, pregnant women, and children.

Far too often outdated energy sources, like dirty peaker plants— power plants which spew twice as much carbon dioxide and 20 times as much nitrogen oxides than regular power plants—are most often located in low-income and communities of color. Democrats understand that our country cannot continue to invest in sources of energy that exacerbate environmental injustices, pollute our communities, and rely on marks dominated by foreign dictators.

That is why we passed the Inflation Reduction Act which, invested a historic $369 billion in clean energy made in America and bringing down greenhouse gas emissions. This bill, which not a single Republican in the House voted for, included provisions to lower energy costs, electrify homes, and so every American, not just the wealthiest, can be part of the transition to the clean energy community.

This transition must be both just and inclusive, which is why we have to look back as well as forward to remediate past injustices.

I want to thank all of our witnesses for joining us today.

My first question is to Dr. Unruh Cohen. Can you expand on the intersections between climate change and racial injustice?

Dr. UNRUH COHEN. Thank you, Congresswoman Clarke, for that important question.

You know, unfortunately, the climate crisis hits the most vulnerable, and that tends to be poorer communities, communities of color, because, you know, they have had to develop in that lowerlying area, places that are flood prone on the coast, not to mention our Tribal communities in particular.

So time after time we see those communities and people who are least able to adapt really being hit. I mean, what we saw in the U.S. Virgin Islands and in Puerto Rico now on multiple times, I mean, it just illustrates within our own country the challenges that we face.

And then when you look globally, you know, you see countries that have contributed just a minimal amount to the warming that we have experienced feeling the huge impact from the, you know, warming that we have already seen.

Ms. CLARKE. So as natural disasters caused by climate change continue to increase in frequency and severity, we must invest in resilience in low-income communities and communities of color so that our homes and critical infrastructure are prepared for when the next superstorm, wildfire, flood, or heatwave strikes.

That is why I have partnered with my colleague Congresswoman Barraga´n to introduce the Energy Resilient Communities Act in the last Congress.

So, Dr. Unruh Cohen, could you explain the role of distributed energy sources, like microgrids powered by renewable energy, in building out our Nation's resilience to climate change disasters?

Dr. UNRUH COHEN. Yes. As I said earlier that, you know, rooftop solar and other—in microgrids in a normal day contribute, help lower costs for the electricity. When disaster strikes, they can be isolated from the grid and continue to be able to provide power and electricity to communities as they are confronting, you know, some terrible times.

Ms. CLARKE. Thank you.

Mr. Chairman, I just wanted to ask for unanimous consent to submit this report from the American Public Health Association entitled "Energy Justice and Climate Change: Key Concepts for Public Health" into the record.

Mr. DUNCAN. Without objection.

[The information appears at the conclusion of the hearing.]

Ms. CLARKE. I yield back, Mr. Chairman.

Mr. DUNCAN. I thank the gentlelady.

The Chair will now go to Michigan. It must be cold up there. It is also known as the Mitten State. Mr. Wahlberg for 5 minutes.

Mr. WAHLBERG. Thank you, Mr. Chairman. Three degrees this morning.

I appreciate the panel being here today. I wish we had more time to talk about some reality issues as well, and I appreciate the reality that has been brought.

Recently utilities in my State have announced—and in my district—that they will be closing some of our coal power plants even faster than initially planned. While I support cleaning up our environment and diversifying our power supply, we need to ensure that it is done intelligently, that it does not come at the expense of reliability, cost, or jobs. And did I say reliability?

In some areas the coal plants are being replaced by clean natural gas, which has cleaned up the world, or at least cleaned up our country, and could clean up the world if we allowed it to take place and pushed it. In others, wind and solar will take up the mantle. This summer there were already capacity shortfall concerns in the region. I don't know if you have ever been to Michigan in January, but I worry that solar and wind just won't cut it.

Mr. McNally, companies seem to be in a race to transition to green energy. Meanwhile, demands for energy is growing, especially with the Biden administration's push for EV adoption and electrification. And I am not sure these companies even believe in this race, but they lack the courage to stand up to the—I guess my issue—the green religion.

How is this transition impacting reliability across the country? And should we expect concerns about the brownouts and blackouts to grow?

Mr. MCNALLY. Thank you, Congressman.

May I suggest there is an absence of just core knowledge. If you think from Moses until Harriet Tubman, five millennia, the fastest a human being could travel on land was a horse. And the reason we transitioned from horses to cars and planes was not a PCORS policy. It wasn't central planning. It was a human innovation to solve kind of cravings we had. Actually, illumination is where oil started.

And so we have to be very wary when we try to think that government central planning can effectuate a rapid and quick energy transition. The truth is energy transitions happen very rarely, very slowly, and due to innovation and so forth. And we see government get involved, you have unintended consequences.

The earliest version of the corporate average fuel economy programs, the National Academy of Sciences determined about 20 years ago it was inadvertently killing people by forcing automakers to downsize and downward deplete. We had an inadvertent, kind of a human cost. We were trading blood for oil out of a well-intentioned policy to conserve oil.

So my recommendation, sir, would be that everyone—all of us, me included—study the oil history and energy history, have humility, learn the facts, learn about energy transitions, and be very careful when calling for sweeping government intervention in energy markets that sustain our quality of life.

Mr. WAHLBERG. "I am from the Government. I am here to help you," right?

Fermi and Cook nuclear plants are in my district, one on Lake Michigan, one on Lake Erie. The last nuclear reactor we built in the U.S. came on line in 2016, and it took 40 years to get it from conception to operation. Meanwhile, China is now the world's fastest expanding nuclear power producer. We need to reevaluate our country's approach to this technology.

Mr. Dabbar, the cost and timeline to jump through all of the regulatory hoops seems to discourage investment in new traditional nuclear reactors. I get that. How can small modular reactors reshape U.S. investment in nuclear power, or can they?

Mr. DABBAR. So, Congressman, in theory, small nuclear SMRs are similar to the Navy reactors. I ran a Navy reactor. They are built on time and on budget. It is much more likely that, if something is contained, can be produced in a central factory, it is going to be on time and on budget.

That is not good enough. I ran the environmental program at DOE. I ran a very large nuclear business. It was not on time and on budget for a long time. When we left, it was back on time and on budget. People didn't believe that could happen.

And so part of it is good technology, but part of it is culture, transparency, and running projects well. And that is a—I am going to do a forum on that at Columbia on how to make certain we are able to do that in nuclear going forward.

Mr. WAHLBERG. I wish you good success. And with that, I yield back.

Mr. DUNCAN. I thank the gentleman.

Members are advised votes are going to be called about 1:30. We will try to get to two more, one on each side.

Mr. Cárdenas from California is recognized for 5 minutes.

Mr. CÁRDENAS. Thank you very much, Mr. Chairman.

I grew up with freeways and power plants surrounding my community. As a child I remember not being able to go outside and play because of smog alerts. That was normal for us back then. It has improved because of government responsibility, both State, local, Federal, et cetera. That is the responsible way that government should be involved in trying to manage what is good for the American people, not just consider it to be a laissez-faire, let the markets take care of it themselves.

By the way, the oil industry in America is very subsidized, so it is not necessarily a free market system, hasn't been for a long time and isn't today.

Since I was a kid, science has shown that realities of climate change have worsened. The fact is that communities like mine have felt the brunt of climate change intimately because we have been on the frontlines unjustly paying the price for the—and also for the power plants on all sides of my—excuse me—on my side of town, but not on the other side of town, while certain companies and oil companies get richer.

Unfortunately, my Republican colleagues are beholden to the special interests of the fossil fuel industry and continue to fight for the expansion of oil, gas, and coal. Expanding our Nation's dependency on fossil fuels doesn't strengthen our economy or national security or our response to climate change. It hurts our economy and burdens the American people unjustly.

Democrats understand that, when we don't prioritize our climate and health, it hurts the American people and leads to higher prices in energy, at the gas pump. It makes our national situation vulnerable to our adversaries and, once more, it has devastating longterm effects.

That is why both the Bipartisan Infrastructure Law and the Inflation Reduction Act included critical investments in clean energy infrastructure and grid resilience.

In your testimony, Dr. Cohen, you referred to the IPCC's sixth assessment report being a Code Red for humanity. And I note that the damages that stem from the climate crisis will disproportionately harm disadvantaged communities.

Can you elaborate on why it is necessary to prioritize equity and justice if we truly want to void the worst devastation?

Dr. UNRUH COHEN. Thank you for that question, Congressman. And, you know, I think it goes back to your childhood experience. Until we had government standards and safeguards for public health, you know, communities like you, kids like you were growing up, you know, surrounded by smog and other things.

Mr. McNally made a point about horsepower, and I think it would just be important for this committee to remember that the U.S. automobile industry was still using whale oil as a lubricant until we banned the import of it in 1971.

So there are important reasons for the Government to intervene, and equity and addressing environmental injustice is a criminal one right now. We spent a lot of time on the select committee thinking and recommending we integrate that. I think we are in a place where we have turned that corner, and every climate and energy bill really needs to have that environmental justice and equity lens.

The Biden administration is moving forward on that with their Justice40, and I think, you know, Congress needs to continue to support those type of programs.

Mr. CÁRDENAS. When was the EPA created? Wasn't it early 1970s?

Dr. UNRUH COHEN. It was, yes, in the 1970s. Yes.

Mr. CÁRDENAS. And has the EPA been a good impact on improvement of our environment in our country, or is it something that maybe should have never come about?

Dr. UNRUH COHEN. I think the EPA and its mission to protect public health and welfare has been incredibly critical to our economic development. We have seen the U.S. have phenomenal economic growth since the 1970s, as well as cuts in air and water pollution. We need to do more as we learn more, but we can absolutely protect every American's health even while we grow our economy.

Mr. CÁRDENAS. One of my colleagues talked about foreign attacks, but yet at the same time, I think they called it the Freeze that happened in Texas. And that turned out to be not only deadly but also, beyond that, long-term shocking for families that were paying hundreds of dollars a month for their energy bills and then found themselves at $5,000 and $10,000 or more in energy bills.

Dr. UNRUH COHEN. Yes.

Mr. CÁRDENAS. Was that an attack from another country, or was that Mother Nature?

Dr. UNRUH COHEN. That was extreme cold and the natural gas industry in Texas not being prepared for it.

Mr. CÁRDENAS. Or not being regulated very well either?

Dr. UNRUH COHEN. I think they could do better, speaking as a native Texan.

Mr. CÁRDENAS. OK. Thank you.

Thank you, Mr. Chair. I yield back.

Mr. DUNCAN. Thank you to the gentleman.

The Chair will now go to the Peach State, even though we grow more peaches than you do in my district, Mr. Carter from Georgia, 5 minutes.

Mr. CARTER. Thank you, Mr. Chairman.

And I want to take this opportunity to begin with to congratulate the Chairwoman, and I am excited about what is ahead for us on this committee in the next 2 years, in this session, and I want to congratulate her.

I also want to congratulate her on the choice for the first full hearing that we have. What better bipartisan subject could we have than American energy? And I think, you know, we all agree that we want America to be a positive energy leader in the world. So kudos on choosing this, and it is extremely important.

Mr. Dabbar, you mentioned the fact that we want to return to American energy dominance and that it is very important. And this would require allowing America to unleash all of our energy potential, and we are certainly not doing that at this point.

When we talk about energy security globally, we have got to be talking about reliability. It is paramount. We understand that. And I think that is probably why we haven't heard more about global power relying on solar or wind, because they are just not reliable at this point. But nuclear is, because it is baseload. It is 24/7. It is clean. It is everything we want.

And, you know, it is exciting to think about the future as far as the innovations go, the advancements that we have seen in fusion, permitting SMRs. We just talked about SMRs. I am excited about that, and I think people should be excited about it.

But let me ask you this: In the context of global energy security, when we are talking about global energy security, how important is it for us to have a domestic—a robust domestic nuclear energy to compete with adversaries like China and Russia around the world?

Mr. DABBAR. So our energy costs have been the lowest, amongst the lowest in the world, including in electricity. That is a great sucking sound of industry, and we are seeing the opposite of that in Germany at the moment, right: high-risk energy policies, high prices. Their industry is moving out, and it is great. They are moving to America. It is actually a big strength of what we have as a result of all of that.

And so that is—energy security drives the economy, drives us as a nation. It is also about us in the world. Once again, I am making reference to all the veterans here. We fought a lot of wars. A lot of wars were around energy. The whole Kuwait War was, frankly, about them invading and about energy.

And our ability—and with all due respect, when we move from 4.8 million barrels a day of crude oil to 12.9, we came off—if you just look at the numbers, we came off of energy risk from OPEC. That is just numbers. We imported less from OPEC. We are less exposed to OPEC. And the energy prices were set in the Permian Basin for the world, and that is actually something driving innovation, driving costs down. That allowed us to have flexibility that is incredible.

Mr. CARTER. Good. And thank you for that.

And I read with interest your—the article that you coauthored in The Hill about what is at risk due to Russia's nuclear power dominance. And, you know, it was interesting. It was also alarming, alarming that, you know, we have got to address this with our allies, and it is something we have got to work together with our allies.

You also mentioned about Europe just then. I had the opportunity to go to Europe with the Conservative Climate Caucus, and what we witnessed there is a situation where they have allowed their policies to get ahead of their innovation. They are shutting down nuclear plants and relying on wind and solar when they are not ready for it yet, when—we all want to go to renewables. We all want to do that, but we have got to do it in such a way that we don't allow our policies to get ahead of our innovation, and that is exactly what they have done.

But, again, the points that you bring up in this article that you coauthored, I am—they are alarming, and I appreciate you bringing it too.

I want to go quickly to Ms. Jackson.

Ms. Jackson, you bring up some great points, and I had the opportunity to read an article that you wrote as well about the minority exploitation game called environmental justice. And you mentioned in there about Ford Motor Company and about how they have $60 billion for—in the IRA, there was $60 billion for environmental justice programs about the size of Ford Motors who employs over 200,000 people.

You know, all of this money that we put into the IRA—or was put into the IRA—I am not saying we, but was put in the IRA, it could have been used for better ways. Can you give us an example?

Ms. JACKSON. Yes, absolutely. You know. The biggest injustice of environmental justice is poverty and the fact that we are using environmental justice money to—that is going to lobbyists and never going into the community to create economic stability. It is not creating any business districts, not industrial parks, nothing that is going to improve the lives of those individuals.

That money was equivalent to 2,000—the Tulsa Black Wall Streets, we could have built 2,000 Black Wall Streets for the equivalent of that environmental justice money. In other words, we wouldn't be having a conversation about poverty right now because all of our Black communities would be thriving. But, instead, we are going to get a few retrofits, a few windows, and nothing is going to change about the economic condition.

Mr. CARTER. Thank you. Thank you for that. And I yield back, Mr. Chairman. Thank you.

Mr. DUNCAN. Thank you.

I apologize for the Members and the witnesses, but votes have been called on the floor, so we are going to recess until 10 minutes after the close of the second vote. There are two votes. The witnesses can take care of business, and we will be back 10 minutes after the close of the second vote.

And the committee stands in recess. [Recess.]

Mr. DUNCAN. All right. The subcommittee will come back to order.

In the order of things, Dr. Ruiz is next from California, 5 minutes.

Dr. RUIZ. Thank you, Mr. Chairman.

I want to focus today on the question of reliability. As an emergency medicine physician, I know how important reliable electricity is for powering critical medical equipment and preventing lifesaving medicine from spoiling. And one of the best ways to increase reliability, expand our Nation's power grid, and move towards a carbon-free energy system is to dramatically increase the battery storage capacity on our Nation's grids.

Lithium batteries are a critical component of electrical vehicles, but they are also an obvious solution to securing our clean energy future, while also providing the grid reliability that we need. In fact, batteries are already being used to increase reliability on the power grid.

In my district, California's 25th District, the Crimson Energy Storage in Blythe is currently the second largest energy storage project in the world, constructed by union labor, including the International Brotherhood of Electrical Workers, I believe also the Teamsters, powering vital medical equipment, electric vehicles, and so much more.

Our country needs more batteries and more batteries storage projects like this. Batteries need lithium, and lithium is found at the Salton Sea in Imperial Valley, California. In my district the Salton Sea has a massive supply of raw materials that can power a clean energy future. The Salton Sea has the fifth largest deposit of lithium in the world, and it has the potential to supply the lithium needed not only for electric vehicles but also the batteries that can make our electric grids resilient.

Lithium Valley, as we like to call it back home, is the key to unlock our clean energy future, cement the U.S.'s leadership, secure our energy independence, strengthen our battery supply chain, and protect our national security.

Dr. Unruh Cohen, in your testimony you mention that China has a head start on the rest of the world in clean energy. I am sure you know that the head start includes both the chemical supply of lithium and the manufacturing of lithium batteries. In fact, according to the World Economic Forum, the United States is eighth in lithium production in the world, and China has 60 percent of the lithium refining capacity in the world.

Can you talk about the effects of the United States lagging in producing a lithium supply and in manufacturing batteries?

Dr. UNRUH COHEN. Thank you for the question, Dr. Ruiz.

And let me just underscore the importance of reliability first and say, you know, wind and solar availability is predictable. And so when you couple that with battery resources, we start to have a full spectrum 24/7, our opportunity to provide electricity to the grid from renewables.

To do that, you are right, we need more lithium. And I know, you know, there are projects in Salton Sea and around the country that are, you know, getting their permits, moving forward, getting investments. And I think, you know, the investments that we see in the Infrastructure Act and in the Inflation Reduction Act are going to drive that further.

The additional good news when it comes to the grid is we actually can take—when an EV has used their batteries and they are no longer great for that, you can put them in ways and use them on the grid.

Dr. RUIZ. Yes.

Dr. UNRUH COHEN. So there is a lot of energy that——

Dr. RUIZ. So in terms of China—us importing the majority of batteries, et cetera, lithium from other countries, we all know too well that it is dangerous when our Nation is dependent on foreign supply chains, because if a foreign country decides to block exports or favor its own domestic companies for whatever geopolitical reasons, our clean energy revolution is stopped dead in its track.

That means we need strong domestic supply chains that start in communities like in the Imperial Valley. That means making sure companies don't simply come in, extract community resources, and turn it into batteries elsewhere. And that means that those jobs are filled by local workers, and those benefits come home to our communities.

All of this is more important than ever to ensure that the benefits of this lithium goes back to the Salton Sea, which could become a worse environmental disaster with reduced waters flowing from the Colorado River and the exposed playa putting fine particular dust in the air that can affect people's lungs, blood, et cetera.

So the Inflation Reduction Act will help us build a strong supply chain for a clean energy future here at home.

In what ways do you expect recently passed legislation like the IRA to strengthen our domestic supply chains and encourage domestic manufacturing of batteries in the U.S.?

Dr. UNRUH COHEN. I think we are already seeing the results of that. I mean, of that $90 billion in investment that I mentioned in my testimony, a lot of that is electric vehicles. I think we just had one today that maybe I didn't get to incorporate. So almost every day there is a new announcement.

Mr. DUNCAN. The gentleman's time has expired. I now recognize myself for 5 minutes.

I want to thank the panelists for being here. It has been very informative.

I approach energy in two broad categories: transportation fuels, which is fossil fuels and electricity for EVs, and then baseload power generation. And in baseload power generation, you have nuclear power, which can't be ramped up and ramped down very quickly or easily, so it remains constant.

And then you have renewables as part of that energy metrics as they mix everything for transmission. Renewables come on line when the sun comes up or the wind is blowing. It could be a climatic event that makes solar go offline or they can see it starting to climb, or it might be the sun setting.

And so when that happens, they have got to fill the gap to make sure that the baseload power is constant for our consumers. That is our constituents' homes, that is manufacturing, that is municipalities.

And so what works in that gap-filling area is called hydro, if it is available, and natural gas. And so we have got to make sure that natural gas is a big part of that. And what I hear from a lot of my utilities is that we have a pipeline issue in this country, that our pipelines are about at capacity for the amount of gas that is drawn off of it. And if we don't meet that future demand, the gas isn't going to be available.

So I wish the infrastructure bill would have included more gas pipeline permitting and projects, but unfortunately it didn't. It focused a lot on EVs and charging stations and all that.

And look, as Mr. Dabbar said, I like wind and solar. I think it is groovy, and it ought to be part of the matrix, absolutely. But I also know about what works to provide that baseload power that our manufacturers need: always on, always ready, always available 24/7, 365 baseload of power.

And right at Christmas Eve in South Carolina, even we just about had a Texas event because of some of those situations where wind and solar wasn't possible. Coal-fired power plants take 3 days to crank up to thermal capacity to meet energy needs. Natural gas is 30 minutes. So it has got to be a big part of that.

Pipelines is the new word from "The Graduate." It was plastics in the movie "The Graduate." It is pipelines today.

So I want to focus on nuclear. Nuclear energy must be a part of our energy matrix moving forward. Unfortunately, other countries are outpacing us. Adversaries like Russia and China have continued to develop and invest in their nuclear industries. In just the past few years, China has brought 21 reactors online—many of them are advanced reactor designs—and 31 plants are under construction.

In this country we have got to focus on advanced nuclear or small modular reactors, thorium reactors, things that work, things that could be walkaway safe.

Unfortunately, lack of development in U.S. industry is in large part due to regulatory and licensing structure that disincentivizes private-sector investment. I have got a piece of legislation we filed last Congress we hope to reintroduce, if we haven't already, that expedite the permitting process for nuclear power plants. If you have got a proven design that works, we can replicate that, do the geology, make sure that there ares no things there with the earth and earthquakes and other things, that it can be sited there, the water source there for cooling, but replicate that in an expedited permitting process. Let's gets nuclear power online as we continue to look at SMRs and other advanced nuclear reactors.

I was glad to see an article this week that the first SMR was actually permitted to come on line and start producing energy as a test case. Once we see that it works—heck, it has worked in the United States Navy for, what, 50, 60, 70 years—it works. Modular reactors work.

So those are some of the things that I believe in.

Mr. Dabbar, we, of course, don't have a government-controlled energy sector. We need to incentivize private-sector investment. What are some of the high-level suggestions to modernize the nuclear regulations of this country to invite innovation, to invite growth, to encourage exports of nuclear technology and lead, once again, this globe in nuclear energy?

Mr. DABBAR. So, as you pointed out, Congressman, the Nuclear Regulatory Commission is extremely conservative, as you know. You interview regulators all the time. Over the course of time, they get more and more conservative because they don't want anything overturned in the courts. And as a result, it is drug out, you know, tremendously as a result of that kind of increased view of, like, let's review it 10 times to Sunday.

So I think any sort of, you know, legislation that might allow them to facilitate that, to make it—to get them focused, I think that is important, because that's where we have ended up.

The other thing that I think we should look at doing is taking a look at some of the Federal power entities like WAPA, like Bonneville, and so on, and take a look at them willing to contract as like a first customer for some of these new reactors. So I think any authorization for DOE or TVA or others to be a first customer of those plants, I think that will provide a big jolt for the industry. Mr. DUNCAN. Yes, that is spot on. This committee will have an RCN to talk about those things, ask some of these same questions, also look at closing the fuel cycle and possibly recycling of spent fuel at commercial reactors, because there is a uranium concern and a fuel concern.

So my time has expired, so I will now go to Mr. Peters from California.

Mr. DABBAR. Congressman, 5 seconds. Mr. DUNCAN. Five seconds.

Mr. DABBAR. The Savannah River National Lab proposed to me a recycling plant when I was Under Secretary, and I think you should be talking to them about their idea, because it is exactly what you just said.

Mr. DUNCAN. We are. And there are great sites like Hanford, Idaho Flats, and others across the country that can play a part in that.

Mr. Peters from California is recognized for 5.

Mr. PETERS. Thank you, Mr. Chairman.

Our path to true energy security is not to double down in oil and gas. The oil embargo in the 1970s, the Gulf War in the 1990s, Hurricane Katrina in 2005, the 2008 financial crisis, COVID, Putin's war on Ukraine all contributed to a roller coaster of harmful oil price volatility under both Republican and Democratic Presidential leadership. It is essentially what Mr. McNally testified to.

And despite the pain of these price fluctuations, Republicans continue to propose the same false solution—more oil and gas—and that will only lead to more price uncertainty and pain as we have seen at the gas pump recently and as San Diegoans are seeing on their natural gas bills today.

The solution long term is clean innovative energy technologies, which are becoming cheaper than fossil fuels and produce energy domestically without reliance on foreign adversaries. It is not radical. It is smart. Even Texas takes 40 percent of its power from noncarbon sources, and it is not because Texas is against oil and gas. It is because it is the sensible thing to do.

I am proud that we provided significant funding in the 117th Congress to build this clean and more secure energy future, but I want to talk today about the 118th Congress. Something that we haven't talked about much today is speed, because we can have all the money in the world, but we will still fail if we don't act faster. Look at high-voltage electric transmission lines. According to research from Princeton, 80 percent of the projected emissions reductions from the Inflation Reduction Act depend on building transmission faster.

Berkeley Lab found that there are currently enough wind, solar, and storage projects in the pipeline to power nearly 85 percent of our economy, but 80 percent of those projects could be canceled due to insufficient transmission.

According to Jesse Jenkins at Princeton, the current power grid took 150 years to build. To get to net-zero emissions by 2050, we have to triple in size in the next 30 years. That means 200,000 miles of new transmission lines by the 2030s, 200,000. Over the last decade, we have built just 1,800 miles per year because each one takes more than 10 years to complete—and 7 of those 10 years, seven of those 10 years—are just for planning and permitting.

Other countries are doing it just fine. According to the Americans for a Clean Energy Grid, North America has built just 7 gigawatts interregional transmission. Less than half of that is the U.S. So let's say we are about 3 or 4. South America is 22, Europe is 44, China is 260.

In the 1970s, our environmental priority was to stop dirty, destructive projects. The National Environmental Policy Act, or NEPA, was designed to require public input and to ensure Federal agencies assess the environmental impacts of proposed Federal projects and their alternatives before they made decisions. And we have NEPA to thank for a great deal of environmental presentation, but its implementation is inevitably slow, with documents that are thousands of pages long, reviews that take more than up to 4 years to complete, and NEPA is also the most litigated environmental statute, with lawsuits dragging on for more than a decade.

And the simple threat of litigation can prevent new projects, including the climate action projects I have named. We must build to beat our climate goals. Climate action is about building things, not stopping things. To save the planet, we must build transmission, utility-scale solar power, passenger rail, hydrogen pipelines, direct-air carbon capture, bike lanes, tons of in-fill housing. And, ironically, many of the laws intended 50 years ago to protect the environment could undermine our climate action.

Some claim that NEPA isn't the problem or that it can't be touched. But as someone who has practiced law in this field, I must say I don't believe we can sustain project-by-project litigation on climate initiatives and still achieve climate action with the time and the money that we have.

These excessive process requirements effectively inflict a punitive tax on clean energy. And fixing laws to serve the public good, that is our job. We can achieve high environmental standards with less time.

NEPA was signed into law in 1970. One hundred sixty-five of our congressional colleagues were not even yet born. We are as far in time from 1970 as 1970 was from 1917. This is an old law, folks. We are charged to update it for our times, and that is OK, because it didn't come from Moses on tablets. It came from people just like us who had IBM Selectric typewriters.

Excessive process is not the only climate challenge, but it is one undeniably. And those of us committed to climate action who say that it is the greatest threat we face, we are called on to rethink cumbersome requirements that are decelerating clean energy products.

To unlock a future that doesn't depend on greenhouse gas energy, we have to update our environmental laws to make it easier, not harder, to build. I encourage all of us to engage in a constructive process with the White House, with the communities of interest to come up with a real bipartisan process to enact sensible permitting of forms, deliver energy security and environmental protection for the American people.

I look forward to working with all of you on that, and I yield back.

Mr. JOHNSON [presiding]. I thank the gentleman for yielding back.

The Chair now recognizes Mr. Palmer from the great State of Alabama for 5 minutes.

Mr. PALMER. Thank you, Mr. Chairman.

I want to start off by talking about how these Green New Deal policies impact people. Right now there are about 2 billion people around the world who have little to no access to reliable energy. And it reminds me of a quote from Thomas Hobbes' book "The Leviathan" when he described life at that time as solitary, poor, nasty, brutish, and short.

Just to give you an idea of what having access to reliable energy means, the World Bank reported that since 1980, energy consumption has doubled. It was half what it is today in 1980, and extreme poverty was 4 times higher then than it is today.

Life expectancy because of China's expansion of their energy grid—and, obviously, it is mostly coal-fired power plants, but their life expectancy has increased by 10 years. This is not just—I mean, I worked for two international engineering companies. You are going to hear me repeat that multiple times. I have a pretty good idea of what it takes to generate power. I worked in everything from refuse energy to aerospace, different projects. And I will just be honest with you—and it is not just my opinion. There was a report from the Electric Power and Research Institute that basically says that this study shows clean electricity plus direct electrification and efficiency are not sufficient themselves to choose net-zero economywide emissions.

In other words, no amount of wind turbines, solar panels, hydropower, nuclear power, battery power, electrification of fossil technologies, any of that is going to get us to net zero by 2050. That is a Wall Street Journal article describing that policy—that research paper.

Ms. Jackson, you and I have had this discussion before. I grew up dirt poor. My dad had an eighth-grade education. He built a house that I grew up in. He only finished what he could pay for. Unlike the Federal Government, he didn't spend more than he had. But we heated the house with a coal heater that sat in the kitchen, and in the wintertime we slept under about a foot of blankets that my mom and grandmother and great aunt quilted.

What does it mean to people when they see their power bills going through the roof and they are having to make decisions on how much food they can afford to buy and still be able to adequately heat or cool their home?

Ms. JACKSON. You know, one of the things that we always advocate for, because I hear a lot of people talking about climate change policies, but I have never heard anybody talking about actually how many degrees it is going to save.

If climate change policies are more harmful than what we are trying to save, then we shouldn't be enacting it. One of the things we advocate for is that you do an economic impact study on the lives of the people that it is going to affect, because the climate change policies right now are more harmful for low-income and minority families than the climate change—than what climate change is supposed to be doing.

Mr. PALMER. Well, we are hurting people.

Ms. JACKSON. We are hurting real people. I get calls every day from people that don't have my political lean, asking for help, Black people, NAACP people, Urban League people, saying, "We need help." They can't afford to pay these bills, but nobody is listening.

Mr. PALMER. What gets me is, we keep hearing report after report from Europe, from United Kingdom, about the number of people dying as a result of excess winter deaths because they can't adequately heat their homes in the wintertime.

We see that here as well. And people can't afford to heat their homes adequately and still afford their medicine. It really impacts people with respiratory disease, cardiovascular, that sort of thing, and it is almost like my colleagues across the aisle think this is collateral damage and it is acceptable to achieve what clearly they are not going to achieve by 2050.

Ms. JACKSON. Yes.

Mr. PALMER. So, Mr. McNally, I want to ask you, I want to pivot now, China is building coal-fired plants all over the world. There was a report came out last year, last fall, that they had built 14 coal-fired plants. They are building out infrastructure around the world, developing allies, and yet we are going to make ourselves dependent on China for our energy production. Does that make any sense to you?

Mr. MCNALLY. Congressman Palmer, if there is one area, perhaps, of bipartisan agreement, I think we are all clear-eyed—or hopefully are—that China views energy and the transition and dominance of both the power and the transportation aspects of energy as a way to replace the United States as the world's pre-eminent power. It is a militarized, superpower strategy, and we ought to see it as such and build our own clean energy here.

Mr. PALMER. Mr. Chairman, that time went by quick, but my time is expired, so I yield back.

Mr. JOHNSON [presiding]. The gentleman yields back.

The Chair now recognizes the gentlewoman from New Hampshire, Ms. Kuster, for 5 minutes.

Ms. KUSTER. Thank you, Mr. Chairman and Ranking Member Pallone, and our Energy and Environment subcommittee leaders for holding this important hearing. I want to start by taking a quick moment to congratulate Chairwoman Cathy McMorris Rodgers on making history as the first woman to chair this great committee.

While we may represent districts on opposite sides of the country, our constituents rely on similar energy resources, including hydropower, and that is why I look forward to working with all of you to unlock the full potential of hydropower to provide affordable, reliable, and clean energy to communities from the Pacific Northwest to the Northeast.

It is important today to remind ourselves that Democrats and Republicans share many of the same values when it comes to our Nation's energy resources. We believe that all Americans should have access to low-cost, reliable energy that does not depend on foreign resources or foreign technologies.

While we may have our policy disagreements, if we focus on these shared goals, I believe we can find enough common ground to deliver on.

Last Congress, we passed historic legislation to invest in our country's clean energy future and put the U.S. back on track to lead the clean energy revolution.

The Infrastructure Investment and Jobs Act, IIJA, included approximately $800 million in investments in our Nation's dams, including to upgrade our electrical grid for hydropower energy.

In his testimony, Mr. Dabbar spoke about the importance of FERC prioritizing baseload energy, and I agree: Hydropower is the baseload power that we need.

With black start capabilities, and the potential for additional energy reserves through pumped hydropower storage technologies, hydropower can provide additional resilience to our energy grid.

I wanted to ask you, Dr. Unruh Cohen, can you speak to the ability of clean energy resources and technologies such as hydropower and hydropower storage, to bolster the resilience of our grid as we transition to clean energy?

Dr. UNRUH COHEN. Thank you for the question, Congresswoman, and hydropower has been an incredibly important power source for the history of our country. I mean, it was the first power source for our industrial work up in the New England and the Northeast. And it continues to be critical.

And I think the important thing to consider for this committee, and especially looking at the investments that we have made in dams over the past few Congresses, is the water cycle is very sensitive to climatic change, and so that is—we are going to see challenges to hydropower.

Unfortunately, our friends in the West are really seeing that challenge now with the Colorado River level, but it impacts in the Northeast too.

And so, in order to continue to provide that incredibly important clean energy from hydropower and from pump storage, we really need to look at the investments and make sure we are maximizing the clean power that we are getting from our dams around the country.

Ms. KUSTER. Yes. And I just would add, this pumped hydro storage, I was out in Mr. Curtis' district in Utah and meeting with a company there that is interested in pumped hydro storage, and how that can help with baseload and with peak and how we manage that.

So my time is limited, but I will just say it is important that this committee does not overlook the potential of our Nation's hydropower resources to deliver abundant, reliable, clean energy. It exists. We have 90,000 dams. Some of them can be retrofitted with turbines.

So I remain committed to pursuing holistic and comprehensive policies to bolster our hydropower-generating capacity and to ensure our Nation's energy independence, and I hope that will be on a bipartisan basis with our committee.

I stand ready and willing to work with each of my colleagues here today to make good on these promises for the American people, and, Mr. Chairman, I yield back.

Mr. JOHNSON. The gentlewoman yields back, and I thank you.

The Chair now recognizes the gentleman from Florida, Dr. Dunn for 5 minutes.

Mr. DUNN. Thank you, Mr. Chair. I saw an interesting fact yesterday. The EIA projects a 50 percent increase in world energy use by 2050. The U.S. should be the global leader in innovative technologies to maximize efficient energy production, be at the forefront of filling that demand.

Entrepreneurial spirit drives innovation, but overly proscriptive policies from the Biden White House are drowning American innovators with burdensome regulation. This puts our economy and our national security at risk, as Mr. McNally pointed out a few minutes ago.

This is why I will be focusing on combating the administration's radical energy policies that destroy American leadership in the energy space.

I look forward to working with my colleagues on policies that will unleash American innovation and return America to the forefront of energy production and unlock resources for future generations. I have to pause for a moment to say, Ms. Jackson, it is great to see you again. We have talked a lot just recently. I hope to see you in Tallahassee before too long, or your team at least.

But a first question, Mr. McNally, I am going to give you this question, and Mr. Dabbar, I would like you to answer after him. American innovation is driven first and foremost by ideas. Those innovative ideas then must be supported by efficient allocation of capital investment. This administration's policies and the private-sector ESG ideology are stymieing rational capital investments in the energy sector, specifically oil and gas but also in renewables and in the nuclear field.

Capital allocators direct their investments to renewable technologies because they are supported by favorable government subsidies. Mr. McNally, do you feel that the current administration's policies and the ESG investing are obstacles to American innovation and development in energy?

Mr. MCNALLY. I do. I think that the administration is signaling that it doesn't want to see capital flow into oil and gas production at a time, as I mentioned earlier, when we are in the foothills of what I believe will be a multiyear boon cycle where we are going to want every drop we can.

They are doing that on based on the idea that we need to enforce disclosure, or raise the cost of capital based on carbon and climate change and so forth. As I stated in some of the papers I will be submitting with my testimony, we need a sound and serious climate policy starting with depoliticized science, cost-benefit analyses, legislation, not regulatory rules and courts.

Well, we should do that first, and then, if as a result of that, we need to start telling capital where to go and where to deploy, maybe we consider it down the road. But that is——

Mr. DUNN. You got to be careful. You cautioned us on that earlier, be careful how you ready-shoot-aim, right, thing.

Mr. Dabbar?

Mr. DABBAR. Dr. Dunn, I have always had a view that the best place for Federal investment is in discovery and new innovation that is too far away from cash flow for the venture capital community and the corporate community to invest.

The private sector, luckily, in this country is still vastly larger than the Federal Government, and the amount of capital is widely available for things that can be applied. And the better use for Federal money is where the private sector doesn't see anything quite yet to invest in, and therefore discovery, science, and innovation in new things is more efficient for the Federal spending.

Mr. DUNN. OK. Very good. And on that subject, Mr. Dabbar, I am going to ask you to spend the rest of this time opining on what we can do to make America a global leader in the nuclear energy technology. And this is a pointed question because we have a large gathering of the nuclear industry this evening, so take it away.

Mr. DABBAR. I think the advanced development reactor program that was passed here a few years ago has really triggered a lot. There is a lot of excitement within the industry right now, and I think that is moving along very well. So I think—I think Congress should be—has done a lot with that. I think——

Mr. DUNN. Is there more we should do?

Mr. DABBAR. I think that trying to find—that the U.S. system, trying to be maybe a first customer of some of that power, and once again, I said a little bit earlier, maybe TVA or WAPA or Bonneville should maybe look at being a first customer for some of the power coming off of those plants.

Mr. DUNN. True. And we have talked about it, using DoD a number of times.

Mr. DABBAR. Or DoD certainly.

Mr. DUNN. Yes. So I like these ideas. Thank you all, all the panelists, for coming.

Mr. Chair, I yield back.

Mr. JOHNSON. The gentleman yields back.

The Chair now recognizes the gentlewoman from Delaware, Ms. Blunt Rochester, for 5 minutes.

Ms. BLUNT ROCHESTER. Thank you, Mr. Chairman. Congratulations to all of the new Members and congratulations to Chairwoman McMorris Rodgers, and thank you to all of our witnesses. I have been proud to support and vote for the Bipartisan Infrastructure Law and the Inflation Reduction Act, which made historic investments in our clean energy future.

These investments not only help to expand domestic clean energy, they also help to diversify our renewable energy sources, which is a key to building a resilient and sustainable clean energy system.

Before I start with my questions, I would also like to note that I was pleased to see in the testimony of Ms. Jackson and also hear from Dr. Unruh in her testimony a positive recitation about former Energy Subcommittee Bobby Rush's, his work on increasing diversity in the energy industry. This is really important to me.

And while I was disappointed that we could not find bipartisan support and agreement in the last Congress, I really would love to see a compromise, and I am hopeful that in the 118th Congress we will focus on workforce and jobs.

As a former secretary of labor in Delaware and the former CEO of the Urban League in Delaware, this issue is an incredibly important one, and during this Congress I will continue to work on that issue, starting on creating programs within the Department of Energy to meet the workforce needs of the energy sector, and also to further train and get more people in the underrepresented communities into the energy workforce.

It will lower the cost for those individuals that are in communities that are underrepresented and communities of color, and it will also increase opportunities and salaries. These are jobs that are good-paying jobs.

I also want to address my questions mainly to you, Dr. Unruh Cohen. We have heard a lot today about energy production, but I also want to emphasize the importance of energy efficiency, again, back to the lowering costs.

The global energy demand is only going to grow, and we need to focus on initiatives that optimize energy. Can you discuss the importance of energy efficiency when we think of things like national security, our economy, and also public health? Can you talk to us a little bit about the benefits?

Dr. UNRUH COHEN. Thank you, Congresswoman, for that question. You know, the most important piece of energy is that what we don't use. So energy efficiency is actually our biggest energy resource if you look at our history since, you know, since the 1970s.

And it—energy efficiency is crucial for taking the edge off of those bills, for making sure that households are not making those hard decisions about paying their medical bills versus their energy bills. It reduces the use of energies all over, and so, you know, where we have pollution concerns, that is going to lower that, and just the less energy we use, the less dependent we then may be on international providers of that energy.

Ms. BLUNT ROCHESTER. Thank you. Transitioning to clean energy is not only necessary to protect human health and the environment, but it is also an enormous opportunity to create a more equitable economy.

Mr. Chairman, I ask unanimous consent to submit this report titled "Improving Equity Outcomes for New Federal Investments in Clean Energy Infrastructure" from the Bipartisan Policy Center into the record.

Mr. JOHNSON. Without objection, so ordered.

[The information appears at the conclusion of the hearing.]

Ms. BLUNT ROCHESTER. Thank you, Mr. Chairman.

And, Dr. Unruh Cohen, can you discuss how clean energy investments from the Bipartisan Infrastructure Law and the Inflation Reduction Act will help build a more equitable economy?

Dr. UNRUH COHEN. Yes. In both of those bills and through the Biden administration's initiatives, you know, there is a huge focus on making sure that communities that are building out the clean energy, that those jobs are flowing there.

Earlier, we were talking about the weatherization program, and one of the benefits there is not just to the people living in the house being more comfortable, but also the job training and getting, you know, skills—doing skilled training for those workers.

I mentioned my mom benefited from that, and she loved talking to the young men who were there doing the work and learning about, you know, what—the skills they were building up in their hopes for the jobs that they would have in the future.

Ms. BLUNT ROCHESTER. Mr. Chairman, I thank you for the time. I just want to add that I think this is an area where I have seen young people, older people, people of color, different communities come together and recognize that we have got a lot of opportunity here, and so I hope as a committee we take up this opportunity. Thank you, Mr. Chairman, I yield back.

Mr. JOHNSON. The gentlewoman yields back. The Chair now recognizes the gentleman from Utah, Mr. Curtis, for 5 minutes.

Mr. CURTIS. Thank you, Mr. Chairman. It is a delight to be here. A little warning to my staff: I am not delivering the remarks that we worked so hard to prepare, but rather my impressions of being here for several hours with all of you today.

I couldn't help but think if an alien was listening to us today, they might conclude that those on my left hated fossil fuels and would stop at nothing to stomp them out. Likewise, they might surmise that those of us on the right want to go crazy with fossil fuels and do nothing else.

To any alien friends that are listening, I have some good news and some bad news. First the good news. Despite the dialogue, there are many areas of agreement between those of us on the right and the left.

Representative Peters has pointed out the need to deal with NEPA reform and permitting reform. Representative Ruiz has pointed out the rich natural resources—I have heard him talk many times—in his district that we will desperately need.

Representative Duncan has addressed nuclear. Many of us agree that nuclear is part of our future and that we can't have a reliable, affordable, clean future without it.

We all like emerging technologies like hydrogen fusion and better battery storage. We all agree, actually, that wind and solar are important. We may not agree on the mix, but we all agree that they are important. We all agree we don't want to lose to China.

And, Ms. Cohen, we all agree energy efficiency is important.

And, Ms. Jackson, I don't know anybody that would listen to you and would disagree that we have to take into account those who are least able to afford this.

And rather than ask you all, I will just take your nodding heads in agreement that we are on the same page. This was a list quickly put together by me. I am sure there is much more.

But now the bad news. We spend too much of our time in the areas where we disagree. And for those of us on the right, we feel there is too much misrepresentation of our position, and I have no doubt my friends on the left would feel the same way.

I don't speak for either group, but let me say this: I don't know anyone in my circle—and I represent oil and gas and coal, I lead a group of almost 80 Republicans that talk about climate—I don't know anyone in my circle who doesn't want to leave a clean Earth better than we found it, who thinks it is OK to leave something for our grandchildren not as good as we found it. I don't know anyone who thinks that more pollution is better than less pollution.

At the same time, those I hang around with think that it is wrong to demonize fossil fuels and those that produce them. As far as I know, all of us in this room are highly dependent on fossil fuels.

I don't know any energy expert anywhere, right or left, who won't tell you that we will be using fossil fuels in the year 2050. I think this is the problem: We too often mistake fossil fuels with emissions. We need to be clear. Do we hate fossil fuels, or do we hate emissions?

I challenge my friends on the left to substitute their anger with fossil fuels with that of emissions, and then I ask this question: If fossil fuels can compete with other energy sources in cleanliness, why do we insist that they die? Why do we demonize the very people who have produced these for decades and decades? Why can't they be viewed as part of the solution and not the problem?

On my side, we see hypocrisy like posing in an electric Hummer and bragging about it. The reality of it is, a gas Chevy Malibu produces as much greenhouse gas emissions, lifecycle, as an electric Chevy Hummer.

We see hypocrisy of shutting down Federal lands. My district, seven of my counties are 90 percent federally owned, and yet the President then goes to other countries, in many cases our enemies, and asks them to produce more.

To my colleagues who express they don't understand our approach, let me be clear: We believe we have been falsely told that we must sacrifice affordability, reliability, and national security so that we can be clean.

I believe we can have it all. I believe we can be energy independent. I believe we can be reliable, affordable, and clean.

Now, if we can get together and talk, I actually believe my colleagues on the left believe the same thing. And in the tiny 30 seconds I have left, I don't know if any four of you want to weigh in on that.

Ms. Cohen, you are the witness for the other side. Would you like to take my last 20 seconds?

Dr. UNRUH COHEN. I will just add, Mr. Curtis, that in the select committee, across the aisle, we actually would often agree on the challenges that the country faced, and sometimes we agreed, as you said, on the opportunities to solve those. And other times, we had disagreements. And I encourage this committee to, as you said, focus on the things maybe you agree.

Mr. CURTIS. Thank you. Regrettably, I am out of time. I would love to have you all share your thoughts, but I yield my time.

Mr. JOHNSON. The gentleman yields back. The Chair now recognizes Mr. Veasey from the State of Texas for 5 minutes.

Mr. VEASEY. Mr. Chairman, thank you very much, and so glad that we are holding this important hearing today, particularly on energy security. I think that energy security, our environment, the air that we breathe is something that we can never, quite frankly, talk enough about.

And one of the things that have really been worrying me lately is China. We know that China has been under very strict COVID lockdown. They are probably going to, quote/unquote, "reopen here" sometime soon. And I was hoping that maybe Mr. McNally could tell me what he thinks China reopening is going to do as far as prices are concerned, what sort of upward surge in prices might that impose on the American public?

Mr. MCNALLY. Yes, sir. Congressman Veasey, thank you for the question. That is my bread and butter. So oil prices, or pump prices, have been in a tug of war between this Russian disruption risk, which makes them go up, and this macroeconomic weakness risk, which makes them go down.

China was on the macro weakness side until the end of last year because, as you noted, they were in severe lockdowns, their demand was depressed, and very importantly, China is a big exporter of refined products like diesel and gasoline. That helped gasoline prices get to $5 a barrel.

However, sir, to your question, as we come into this year, President Xi has decided to let COVID run rampant, burn through, so that by the second quarter we and most analysts think that China will be recovering back to its pre-COVID demand level. That means close to about 16 million barrels a day.

The big chunk that has to come back is jet fuel if they let their citizens start to fly. So that would, all else equal, put upward pressure on prices. You have seen prices started to rise.

Now we have to go back and look and see what happens with Russia because, like the boy who cried wolf, we didn't see the disruption last summer, but we might going forward.

So China's return has sort of switched sides in the tug of war from being a downward price factor to being an upward price factor, sir.

Mr. VEASEY. Yes. And some reports are even saying that we could get to about 101 million barrels a day, which would be a record for the world.

And so one of the things that I have heard a lot from my friends on the other side, on the majority side, is that we need to unleash all of this energy that we have, and, you know, I am proud to say that as a Texan, you know, we do a pretty good job of unleashing. Not only have we revolutionized things in the fracking area, but we have also been one of the world leaders when it comes to renewable energy. About 25 percent or so of ERCOT's grid is renewable energy. So I think that we are doing it right there.

But when I hear about unleashing American energy and being energy independent, one area that continues to really hurt the U.S. Congress and we haven't been able to find any sort of compromise on is the issue of immigration.

And if you talk to people in the Permian Basin and you talk to people that run the oil and gas companies, a lot of people don't realize this, but the permits that ranchers use, agriculture, the permits that hospitality use, those same immigration permits are not the ones that are needed for people to have temporary work visas in the oil and gas sector.

And so, when you talk to people in the oil and gas field, you know, people that say, "Hey, we need more swabbing units, we need more rigs," guess what? You ain't unleashing a thing unless you do something about immigration reform.

And so, when you hear people talk about being able to unleash, you can't unleash anything if the dog don't have a handler. So, please, can you please tell me, how are we going to unleash all of this energy potential if the other side is not willing to work with us on immigration reform and will continue to use this as sort of a wedge between the American people and offer absolutely no sort of solution?

Mr. MCNALLY. Congressman, I will step around the land mine of immigration and the border and so forth and just concur, though, that when I hear the same thing from my clients drilling in your State and others, that its input costs are really high, and part of that is finding good people.We are scouring Sri Lanka now. All over the world they are trying to bring workers in, and steel tariffs have raised the cost of steel and casing and so forth.

So there are real supply-side pressures that your constituents are dealing with, my clients are dealing with, and Congress ought to look across the board at how to alleviate those. But I am going to steer clear of immigration, sir, if you don't mind.

Mr. VEASEY. Yes. And before I get ready to turn back over my time, Mr. Chairman, I will just say that, if we really want to address this issue and we really want to unleash, we better do something about immigration reform because just saying "unleash" and not addressing immigration reform means absolutely nothing.

Mr. JOHNSON. The gentleman yields back.

The Chair now recognizes the gentlewoman from Arizona, Mrs. Lesko, for 5 minutes.

Mrs. LESKO. Well, thank you, Mr. Chair. I am really excited to be on the Energy and Commerce Committee and on the Energy Subcommittee and talking about energy, because it is vital to every aspect of our lives.

I represent the Phoenix, Arizona, area and its suburbs. And right outside of my district is the Palo Verde Nuclear Plant. It is the largest power producer in the Nation for almost the last 30 years. Recently last year, Palo Verde Nuclear Plant in Arizona was awarded funds from the Department of Energy to increase its flexibility by creating hydrogen when selling to the grid isn't economical.

My question, Mr. Dabbar: What more needs to happen to increase the deployment of energy technology like hydrogen?

Mr. DABBAR. So we awarded the money around increasing hydrogen production across the nuclear facilities. Hydrogen is a storage vehicle. So if you think about it, the comparison, you produce electricity and you can store it in chemical form, or you can store it in gaseous form, in hydrogen. Or you can do it in liquids, as I was talking about earlier. And so, hydrogen has a great opportunity for an additional storage.

The other thing that is really exciting about hydrogen is that some industrial uses cannot really use electricity, OK? So the likelihood of electrifying certain industries is quite poor, right, around certain things like steel and concrete.

So if you can take that electricity and put it into something that can produce a much higher heat rate for that manufacturing, it is going to create an opportunity for those industrial to, quote, electrify, but it is electrified via converting it to hydrogen.

Mrs. LESKO. Exciting. That is the one thing I like about energy. A lot of new technology going on, exciting that I think can solve a lot of our energy needs in America.

Mr. McNally, in your written and verbal testimony, you recommended the establishment of a national commission on energy transition realism, an expert nonpartisan commission of renowned energy experts to advise government officials and evaluate policy options for energy transmissions.

I love it. I think it is common sense. Sounds like a great idea to me.

The United Nations Intergovernmental Panel on Climate Change, the UNIPCC, in a report issued last year pointed out that collectively the G20 members are not on track to meet their goals under the Paris Climate Agreement.

The UNIPCC also states in its latest assessment that global warming at the end of the century is estimated at 2.7 degrees Celsius, not even close to the Paris Climate Agreement of 1.5 degrees. The Biden administration is spending trillions of dollars on solutions that are not working, according to the UNIPCC. Don't you think it is time the administration has an honest conversation on spending trillions of dollars on ideas that according to the UNIPCC aren't working?

Mr. MCNALLY. Thank you, Congressman Lesko. I certainly do. I see a sound and serious climate policy as having a foundation, like a house has a foundation, and the quality of the foundation will determine whether the house is safe to build in and live in and so forth.

And the foundation ought to be depoliticized transmission of science to nonexperts like all of us. And the concern with the IPCC reports, actually, and the summary for policymakers, the SPM, is the folks who are decoding the complicated climate science and explaining it to all of us are government officials, and they don't have the rigorous peer review and requirements to be transparent like you do in actual climate science.

So I would respectfully suggest that perhaps your side of the aisle—and make it very clear you embrace climate science. The problem is not climate science, it is how it is transmitted to the rest of us, because it is pretty complicated, in these IPCC reports. And let's have reforms to make sure it is honest, peer-reviewed, and a good foundation upon which for us to debate sound policies to address the problem.

Mrs. LESKO. Fantastic. And in the 13 seconds I have left, thank you for being here once again and telling us—and representing low-income communities and minority communities and speaking up. Thank you. And I yield back.

Mr. JOHNSON. The gentlewoman yields back.

The Chair now recognizes Mr. Soto from Florida for 5 minutes. Mr. SOTO. Thank you, Chairman. There are a few elephants in the room worth mentioning. The Washington Post reports today, "Oil companies post record-smashing profits as gas prices creep up." Record-smashing profits.

ExxonMobil: $55.7 billion in 2022, a record. Chevron: $36.5 billion, 2022, another record.

Inflation, corporate greed, record profit, share buybacks—something this committee needs to keep in mind as we are navigating this.

Elephant number 2: Climate change demands our attention. Intensifying hurricanes, massive floods, prolonged droughts, rising seas, extreme heat, extreme cold. Climate is changing, and it takes more than just saying the word "climate" or delay, delay, delay, delay, delay to actually solve it.

The Inflation Reduction Act is now law. This committee should join the President and the Senate in shepherding the fair implementation of this landmark law, pursuing diversifying our energy production with clean renewable energy.

Number 3: exporting oil abroad. There used to be a ban on that, and now oil is being exported and Americans are paying high prices at the pump. This isn't an America-first policy. This is the exact opposite.

If you want to kick-start American energy dominance, we should start by looking at the exports that just started just a few years ago.

Under Secretary Dabbar, what do you think we should do about record oil profits at the expense of the American people?

Mr. DABBAR. Yes, Congressman, it is almost ironic that when— a few years ago, when we were the energy dominant side on the oil markets and the gas markets, that the oil and gas companies were actually making very low profits, and the biggest problems when that was going on, when we had so much production, was that they were—the prices had been dropped so much for the consumer.

And it is a funny dialogue, right, so that when we have been supportive of export terminals and pipelines and so on, that actually the oil and gas companies make less money and consumers actually pay less money.

But when we put lots of restrictions on them and there's lots of turmoil, prices go up and then emissions go up, right? Emissions are because of coal plants coming online, especially in Germany. So it is a funny dynamic around those different issues.

Mr. SOTO. Well, thanks. I appreciate you describing the issue, but it would be helpful for some solutions on it.

Dr. Cohen, a key role for this committee is implementing the $369 billion in incentives to boost renewable energy, boost conservation. What are some of the things you think this committee can do to help with that fair implementation?

Dr. UNRUH COHEN. Thank you for the question. We have talked a lot about the importance of the grid, and both the infrastructure bill and the Inflation Reduction Act have important provisions to help improve the resilience of the grid and expand it. I think that is an area that this committee can give some really close attention to.

In addition, there are a number of other—we have also talked about nuclear power, and both the Infrastructure Act and the Inflation Reduction Act provide ways to keep our current nuclear power—to support currently operating nuclear power plants that are safe to operate. I think taking a look at that will be important, so we keep that carbon-free emission coming.

And then, as we talked about, just the deployment, making sure the deployment of wind, solar, the electric vehicle infrastructure is happening.

Mr. SOTO. Thank you so much, Dr. Cohen. I, like many on this committee, agree it should be an all-of-the-above solution and we should look at different sources of energy.

I hope the mistakes of the past with the Affordable Care Act of trying to eliminate it for many, many years, only for it to continue to be the law of the land—over 3 million Floridians now get their care from there—that we learn from those mistakes when we look at the Inflation Reduction Act knowing that it is the law and will be the law for the foreseeable future. And rather than trying to undermine it, we should work together on maximizing it.

And under the Inflation Reduction Act, we do have incentives for things like modular nuclear and carbon capture in there. So for colleagues who are talking about it as if it is all just renewable energy, that actually is false. There is a great opportunity for us to work together to pursue these in this committee, and I yield back.

Mr. JOHNSON. The gentlemen yields back.

The Chair now recognizes Mr. Crenshaw from Texas for 5 minutes.

Mr. CRENSHAW. Thank you, Mr. Chair. Thank you for hosting this hearing. Thank you all for being here. You know, we heard earlier that Republican priorities were misguided, and that was a little shocking because I think our priorities are guided quite well and are very clear, and they are this: Provide energy to the American people that fits three criteria: reliable, affordable, and clean. I don't think that is very controversial. Reliability is number one, of course, it has to be, because energy production is pointless if it isn't reliable. A society cannot function if it cannot keep the power on.

Affordability is second. Americans need to be able to afford reliable energy without massive government subsidies so they can drive to work, heat their homes, and benefit from a thriving economy.

Third, Republicans want our energy to be clean, and yes, we can have energy that is reliable, affordable, and clean. It is possible.

Also worth noting that America's natural gas revolution is the single largest factor in reducing America's carbon emissions. Switching from coal to gas accounted for 61 percent of emission reductions in the U.S. since 2005.

If we cared about global emission reduction, we might note that the U.S. natural gas is 42 percent cleaner than Russian natural gas, which is why Republicans advocate for more natural gas exports to displace foreign coal, which, by the way, accounts for about 50 percent of total global power emissions. Seems that would be some pretty low-hanging fruit if carbon emissions was actually the goal.

Also worth noting that Republicans are by far the strongest supporters of nuclear energy, which is both reliable and 100 percent clean. Maybe our colleagues would join us in fixing the outdated permitting process that makes a 4-year project last for 15 years.

My Democrat colleagues mostly just want to talk about wind and solar, which isn't reliable and only affordable if you subsidize it. They say we can just build a bunch of batteries to deal with the intermittency problem of renewables. They say we can clear hundreds of square miles of land for solar and wind farms and pay China to mine and process all of our critical minerals so that we can pretend that we aren't actually responsible for the environmental devastation of Chinese mining practices and the enormous amount of emissions that result from processing it all.

So my question to my colleagues is simple: If the goal is actually reducing global emissions, then will you work with us to improve our absurd environmental permitting regulations that are choking off our ability to not only build pipelines, but also build the solar and wind farms and battery backup systems that you claim to care so much about?

I know some of them are, because my friend Scott Peters was just talking about it. And surely my colleagues would find it troubling that something like the Ten West Transmission Line, whose groundbreaking ceremony was attended by Vice President Harris last year, won't actually be online until 2025. The planning actually started in 2016, just for a simple 125-mile transmission line on public land. So that is nearly 10 years from start to finish, nearly 4 years of which was just to get the environmental impact statement approved.

We put a man on the moon in less time than that. Now, does that seem like a healthy permitting and regulatory system to anyone? Surely not.

It is not just transmission lines, it is critical minerals too. In Nevada we couldn't build a lithium mine because of some useless plant called Tiehm's buckwheat. I am not joking. You can look it up.

In Oregon we can't mine for lithium because of a sage grouse, which is basically just a fancy version of a chicken.

In Minnesota this administration halted the Twin Metals Project over vague environmental concerns. Now, this mine would have produced taconite iron ore, copper, nickel, cobalt, all of which are needed for any renewable energy project.

My point is this: The false narrative that we can transition smoothly to a wind-and-solar-only future is not based on anything that resembles reality. It is a fantasy, and it is a dangerous one that will quickly take us down the failed energy scenario that we now see in Europe.

Wind and solar certainly have their place, but when energy demand will increase by 50 percent over the next 50 years, intermittent renewables will never, ever, ever be enough, and it is time to let that fantasy go.

So am I promoting oil and gas? Yes, yes, I am. In fact the quickest way to reduce global emissions would be to ensure that our cleaner natural gas is displacing foreign, coal-fired power plants. That single feat would have a larger impact on global emissions than any other solution offered.

In fact, the industry thinks we could quadruple gas exports in the next 10 years, if we let them. If that gas displaced foreign coal, it would reduce emissions more than the combined impact of doubling our wind capacity, installing solar panels on every home, and electrifying every vehicle in the country.

So you want solutions for reducing global emissions, build pipelines, build export terminals, lease the land for drilling, and send some special trade reps to countries like India and Indonesia to make a deal. That is a realistic solution, and it is actually doable. Reality has to guide our solutions for the future. We cannot sacrifice energy reliability for radicalism. We must be rational environmentalists, not radical environmentalists, and we have to remember that the prosperity of the American people depends upon reliable and affordable energy. And I yield back.

Mr. JOHNSON. The gentleman yields back.

The Chair now recognizes the gentlewoman from Minnesota, Ms. Craig, for 5 minutes.

Ms. CRAIG. Thank you so much, Mr. Chairman.

I am going to focus my attention here this afternoon on our Nation's biofuel sector and its role in strengthening economic, environmental, and national security, and I am actually going to ask you a question, so get ready.

As many of my colleagues know, I have been a champion on the Energy and Commerce Committee for expanding and enhancing this vital segment of our economy. For example, I was the first Member of Congress to pass year-round E15 through the House in the Lower Food and Fuel Cost Act last Congress, and I am eager to work with Chair Rodgers to pass an E15 fix again this year.

We should take up and pass the Consumer and Fuel Retailer Choice Act, which is bipartisan, bicameral legislation, that would allow the year-round nationwide sale of ethanol blends higher than 10 percent, helping to lower fuel prices and improve stability and certainty in the U.S. fuel market.

This bill was supported by the largest unified group of farming, biofuels, and oil companies to date, and I look forward to reintroducing the proposal in this Congress again.

E15 creates opportunities for our family farmers, supports economic growth in rural America, and lowers prices at the pump for Minnesotans. With this sentiment in mind, I want to direct a couple of questions to the panel for your thoughts and observations on biofuels policy and where we go from here.

Mr. McNally, you recently spoke, I believe, at the National Ethanol Conference about the future of liquid fuels and told the group to keep the faith, because there is no evidence that consumers or governments are on course to decarbonize as rapidly as the consensus expects.

In a minute or so, I am hoping you can talk more about those comments and why you predict American drivers will still be filling up with home-grown biofuels for many years to come. So why don't we go ahead and ask you to comment on that.

Mr. MCNALLY. I will have to talk very quickly, Congresswoman. Thank you very much for the question. The consensus has decided to believe that liquids fuel demand—so gasoline and ethanol, biofuels—is going to peak globally in about 10 years. This is a very controversial, in my view unjustified if perhaps attractive, vision. If you believe that, that means demand for biofuels and oil, because they go together, is going to plateau and go down. In my view, that consensus is wishful thinking, and that demand for energy is going to grow much stronger than that, including liquid fuels, which means there will be a bigger pie for gasoline and diesel, which biofuels complement. Ms. CRAIG. Thank you so much.

And now, Dr. Cohen, the Select Committee on Climate Crisis recommended that Congress and specifically this committee develop a low-carbon fuel standard to build on the renewable fuel standard. I have been a cosponsor on the Next Generation Fuels Act, which would gradually ramp up the use of home-grown ethanol at gas stations across the country, making Americans less reliant on foreign oil and less vulnerable to the anticompetitive tactics of OPEC. I am wondering if you can speak a little more about the design of a low-carbon fuel standard and why the time is right to begin this important work here on the Energy and Commerce Committee.

Dr. UNRUH COHEN. Thank you for the question, Congresswoman.

I actually was able to work on the 2007 energy bill where we did the renewable fuel standard, which is now in need of reauthorization. So the time is right for this committee to take a hard look at our biofuels policy.

We recommended doing a low-carbon fuel standard. We have seen that work for transportation fuels, the agriculture community out in California. It would be a way to take the holistic view about our liquid fuels and provide that signal and that standard to move towards lower carbon fuels.

I agree with Mr. McNally, we are going to be using liquid fuels. We invested in sustainable aviation fuels in the recent laws, and so we need to provide the signals to producers so that we are getting the type of transportation fuels we need to meet the climate pollution reduction goals that we know we need to achieve.

Ms. CRAIG. Well, I would just like to end by saying thank you to all the panelists, I know it has been a long day for you. And with that, Mr. Chair, I yield back.

Mr. JOHNSON. The gentlelady yields back.

The Chair now recognizes the vice chair of the Environment, Manufacturing, and Critical Minerals Subcommittee, Dr. Joyce from Pennsylvania, for 5 minutes.

Mr. JOYCE. Thank you, Mr. Chairman. I thank all of the witnesses for being here today. Realize that Energy and Commerce, the first formal hearing that we are holding, is directed to energy and how that unleashing of American energy is so important.

As we begin the 118th Congress, this hearing is the first chance for Energy and Commerce Republicans to begin tackling the issues that are facing all American people. It is great to be back together with my colleagues in one room to do the job that our constituents sent us to Washington to do.

Our new Republican majority is ready to enact the commitment to America. A core piece of that plan is to ensure that our Nation has a robust and a reliable energy supply. Let's be clear—and you have heard us say it repeatedly—energy security is national security. Our physical and economic well-being is tied to maintaining energy.

Ms. Jackson, your words resonated with me. You messaged to us how important that energy sources at affordable prices are to all Americans and how the high cost of energy is certainly having more impact on lower socioeconomic Americans and how that impact affects each individual American every day of the week.

American policymakers in front of you have recognized that reality, and it is why the Department of Energy was established in 1977, in order to decrease our reliance on foreign adversaries.

After years of energy development under both Republican and Democrat Presidents, our Nation finally had achieved that goal. But under the last administration, energy had superceded that goal. We were an energy exporter. We were energy dominant—energy dominant—supplying those necessary energy forms to our friends and to our allies.

You know what, sadly, it is no longer true. The Biden administration has waged a war on American energy industry by creating restrictive and burdensome regulations that have left us less secure and more exposed to bad actors.

It has focused on poor alternatives, like wind and solar, instead of baseload power capacity, has made our grid less reliable and less resilient. This becomes clear in my district in Pennsylvania when I heard from constituents that they were asked by their utility companies over the Christmas holidays to conserve energy or risk outages.

I will tell you, it is shocking. It is shocking in the State of Pennsylvania, where we have strong energy portfolios—we have coal, we have natural gas, we have nuclear, and they all play critical parts in the Commonwealth's electricity supply.

Generations of Pennsylvanians have gone into coal mines to power America, and new drilling technology has enabled an explosion of natural gas production from the Marcellus Shale industry. It is unacceptable that in a State and in a nation as blessed as we are with natural resources for our citizens to be at risk of black-outs because of bad government policies.

Now is the time to abandon these failed policies and unleash the reserves, the reserves that are under the feet of my constituents in Pennsylvania.

With our new House majority, Republicans are ready to begin implementing policies that will allow new leases for oil and gas production, reform the permitting process for energy infrastructure, and prevent burdensome government regulation to reclaim American energy dominance.

My first question is for you, Mr. Dabbar. Can you speak on how critical it is for grid reliability to have baseload power capacity, and how does the closing of coal and natural gas power plants in favor of renewables affect that grid resilience?

Mr. DABBAR. Yes, Congressman. Until, you know, relative recent past, during the last kind of 20 years, almost all the power plants in the country were dispatchable, whether it was hydro or gas or coal or nuclear. And the advent of wind and solar, which are great at emissions, has increased instability in the grid.

Lithium-ion is great, but it is only good at the edges on the grid. It doesn't come close to having other peaking sources.

Mr. JOYCE. And how would—I am going to interrupt if I may— how would government policies forcing electric vehicle adoption further strain the grid?

Mr. DABBAR. Yes, Congressman. So, obviously, electrification overall, and EVs are certainly a big part of it, are adding more demand. And what we are seeing is that, on the supply side, more power plants are being shut down than are being built, and the ones that are getting built are less available.

So that's increasing risks in our system and as a result is actually increasing prices. In New England the electricity prices were a hundred percent on the energy side above where it was a year ago.

Mr. JOYCE. Thank you. I appreciate all of you being here, and I yield.

Mr. JOHNSON. The gentleman yields back.

The Chair now recognizes the gentlelady from Washington, Ms. Schrier, for 5 minutes.

Ms. SCHRIER. Thank you, Mr. Chairman. I am delighted to be back on this committee, and let me just also say that it is a special privilege to work with my fellow Member from Washington State, our new chairwoman of the committee.

You know, I agree with my Republican colleagues here that high and volatile energy prices are a real problem. They are a problem for my constituents, who, by the way, are still paying in many areas over $4.50 a gallon for gas. Prices are coming down, but they are not down enough.

My colleague, Mr. Soto, just talked about, he called it "record-smashing profits" once again by the oil and gas industry, and so I would just like to continue to call out what sure feels like price gouging. And that is why last Congress we passed my bill, the Consumer Fuel Price Gouging Prevention Act, through the House, and I hope we will take that up again.

I wanted to pivot to national security. I think Russia's war on Ukraine has really highlighted and refocused our attention on how tightly tied our energy independence is to our national security, as we are watching what is happening in Europe.

National security is not just geopolitical, though. It is also security from the fire and the weather disasters that we are experiencing more and more frequently all across this country, and with that in mind, and also sort of conjuring up what Mr. Curtis said, like, I think we can all agree that we need to be bringing down emissions and we need to be transitioning away from fossil fuels toward cleaner sources of energy.

And I would add, Mr. McNally, you mentioned that even with so much domestic production of gas and oil, we still have extreme price volatility. And so, there's many reasons—energy independence, where we need to transition, stability in energy prices, pollution prevention, climate action, national security. That all should make us want to redouble our efforts here to move away from 20th century energy sources to the energy sources of the future, and the U.S. needs to lead here, not China.

We took tremendous action last Congress in really putting in incentives, economic incentives, to boost innovation here, hydrogen hubs, battery manufacturing. It really supercharged research.

And by the way, I will also note that some of the things that were in that bill, Ms. Jackson, would help individual customers afford heat pumps and other innovations that will help with energy efficiency and to bring down those monthly bills.

I also note it is because of clean hydropower that Chairwoman McMorris Rodgers and I pay some of the lowest electricity costs in the Nation.

So I wanted to just ask, Dr. Cohen, in your testimony, back to national security, you noted that China leads the globe right now in clean energy investments. I think we all know that this is not because they are great environmentalists. They are doing it because that is where the future is, that is where the money is.

Can you talk about why the United States needs to catch up, surpass, and lead in this area?

Dr. UNRUH COHEN. Thank you for the question, Congresswoman. And just going back to your point on the consequences of the climate change, I mean, the western wildfires, they don't just stay in the West. Everybody, you know, has the health impacts that come from that, from the burning of those forests.

But to your question about China, in 2022, you know, China invested about $546 billion in clean energy transmission, and the U.S. was a distant second at just $141 billion. That was pre—that doesn't count what we just passed, so we expect that to change very quickly.

But as a number of people have mentioned, increasingly every country is looking for energy security, and so they are looking at what can they do domestically, which means they start looking at what can they produce from a renewable energy. And we—I quoted it, Congresswoman DeGette quoted it. BP's economist has said, you know, we are seeing that. And we have seen a huge change in Europe in the face of them realizing they can no longer rely on Russia for energy.

Ms. SCHRIER. Thank you.

And that is why we need to be doing the innovation that this committee will do. We are at the crossroads right now where we can't depend on China for those resources. We need to do the mining, the recycling, and everything else that it will take to invest in ourselves here at home.

Thank you. And I yield back.

Mr. CURTIS [presiding]. The gentlewoman yields.

I recognize the gentleman, my colleague from North Dakota, Mr. Armstrong.

Mr. ARMSTRONG. Thank you, Mr. Chair.

We need to level set and be honest about the use of energy on the planet and its importance for economic growth and national security. Next year, regardless of any policies pushed by the Biden administration, the world will use more carbon energy than it did this year, and the year after that and the year after that.

In the United States this administration has taken a whole-of-government approach to dissuade development. But in Europe we are seeing a realignment accompanied by substantial long-term investment in carbon energy as countries try to make up for Russian supply chains. The Norwegian Energy Ministry has proposed putting 92 new offshore oil and gas blocks on offer in its next licensing route. And Spain has laid the groundwork to become EU's national gas hub. They have the capacity to onboard it. They just need to move it. Germany is building an infrastructure to support up to 56 billion cubic meters of LNG import capacity, roughly the same amount imported by pipe from Russia in 2021.

Given the right regulatory environment, these investments are for the long term. Despite past rhetoric, there is an awakening in Europe that energy security is essential to national security and economic growth.

And there are lessons to be learned from Europe, and if we don't think the challenges there are relatable to the United States, we only have to look to the Northeast, where natural gas availability is threatened by an artificial supply crunch. Like Continental Europe, natural gas is shipped in from producing areas.

However, unlike countries such as Spain and Germany, Northeastern States have yet to recognize the threat posed by unreliable access to energy, despite warnings from grid operators, industrial users, and utilities.

At nearly every opportunity, the permitting and certification process for carbon energy infrastructure have been mired in legal and regulatory entanglements. And worse yet for my friends who support green energy, we have taught them how to stop those projects as well.

You don't have to look any farther than the Northern Pass Pipeline, which is trying to bring hydropower from Montreal to Boston, and it cannot get built in any State on the Northeast corridor. Do you know who shut it down? Indigenous Tribes in Canada and the Sierra Club.

From the defeat of the Constitution Pipeline to the cancellation of the Northeast Supply Enhancement Project, years of coordinated activists and environmental opposition have prevented the construction of 1,000 miles of interstate pipeline across the Northeast. Instead of expediting deployment of the necessary infrastructure to move abundant energy resources, Northeastern States, supported by an antagonist Federal regulatory environment, have done everything in their power to actually halt that development. Since day one, the Biden administration has taken that playbook and applied it on a national scale.

Last month, through an overly vague and expansive definition of waters of the United States, the Biden administration further empowered activist environmental entities within States to abuse the Section 401 certification process to stifle pipeline development for political reasons.

Now, only a few weeks later, a new guidance from the Council on Environmental Quality will further muddy the waters as agencies evaluate greenhouse gas emissions and climate change when considering proposed actions under NEPA. Under this latest proposal the Council on Environmental Quality is pushing for agencies to use the NEPA process to link decisions with the national climate change agenda, connecting NEPA reviews to environmental justice, pushing arbitrary alternatives, and providing a new pathway to consider both upstream and downstream emissions.

While it remains to be seen how this new guidance will play out, it almost certainly will further slow the development of much-needed energy infrastructure projects and drive capital away from the carbon energy resources.

Under Secretary Dabbar—and I appreciate what you said about the lithium ion battery, because when we figure out a better way to store electric energy, it will have as great of an impact on our economy as the steam engine or the microchip. The problem is that the lithium ion battery ain't it.

But in your testimony you talked about the need for reform of FERC as part of the permitting and approval process. Can you walk through how the activist capital, combined with FERC straying from its statutory charge, has distorted the marketplace for carbon energy infrastructure?

Mr. DABBAR. Yes, Congressman. So, as you laid out, the very, very long approval processes are effectively a tactic to never build, right, for the people who are building. That is what it is.

And, you know, in the last administration, actually, you could see if you put a time stop, if you may remember. It was, like, it is not a yes or a no, but it is you got to get it done it in a certain period.

Mr. ARMSTRONG. I called it an unreasonable amount of time.

Mr. DABBAR. And so I think some sort of—and Congressman Peters talked about NEPA reform. I think one of the key things that's holding up things from getting built is the incredible long time period from the time that you—that someone, a company tries to lay it out to the approval, plus the inflation, right, the inflation makes projects with a long delay undoable economically at the end of the day.

So I think some sort of time stop on a NEPA review, you know, of the reform. That doesn't say it has to be built or it is going to be approved or not approved. It is just do something reasonable from a time period, I think, is good for everything.

Mr. ARMSTRONG. Thank you. And I yield back.

Mr. CURTIS. The gentleman yields.

The Chair recognizes the gentlewoman from Massachusetts, Mrs. Trahan.

Mrs. TRAHAN. Thank you, Mr. Chair.

So I don't think we can have a hearing about unleashing our domestic energy supply and not mention the record profits the big oil companies are about to announce from last year. Five massive oil corporations are slated to earn a combined profit of almost $200 billion.

These same companies sat here before this committee last year, and they groveled for more drilling permits so that they could lower prices. Apparently, they couldn't use any of those profits to start drilling on the thousands of permits they already have, and they couldn't allocate another dime to restart the refineries that they shut down. But you bet they used those products for stock buybacks, to inflate their share price and make sure they got their multimillion-dollar bonuses.

Mr. Chair, I agree with you and members of your party that we need to achieve energy independence. We need to stop relying on the whims of OPEC, but we can't swap foreign oil oligarchs for domestic oil barons. We can't trade Vladimir Putin for ExxonMobil if they are going to keep doing the same thing, influence the market to pad padding their profits while working families pay the price.

Mr. Chair, the American people want lower energy costs and a planet that their kids can inherit. To achieve that goal, me must focus on sustainable alternatives, and we are close to having one, in particular, that will change energy as we know it.

Recent breakthroughs, including most recently at Lawrence Livermore National Laboratory, have finally put fusion energy within reach. Unlocking this virtually unlimited source of clean energy will drive down costs for families struggling with gas and oil prices. It will reduce our carbon footprint and help ensure a healthy future of our planet for generations to come. But it won't happen without building on the public- and private-sector investments in research and development that have gotten us to this point.

So, Dr. Unruh Cohen, first, thank you for your incredible work over the last two Congresses.

In your testimony you mention that U.S. public-sector investment in clean energy trails other countries, including China. The Department of Energy is considering applications from private fusion companies for $50 million in public/private partnerships and a new milestone-based funding program that would support building fusion pilot plants. But according to the Fusion Industry Association, the funding opportunity announcement was significantly oversubscribed with applications requesting close to three times as much funding as was allocated.

Can you just tell us why public investment in fusion energy is so critical for accelerating the impact of existing private investment in the United States?

Dr. UNRUH COHEN. Yes. Thank you for the question.

And I spent a lot of my career working for Ed Markey, both in the House and the Senate. So I am very familiar with the great clean energy technology in Massachusetts and the companies that are spinning out from that, and fusion is one of those. Obviously, you have got a great one there. And, you know, the promise of fusion is amazing, and we have spent a lot of important research, Federal research dollars going into that. We have had these really exciting breakthroughs, and I know, you know, we are on the cusp of being able to understand the potential there more and, hopefully, move forward to commercialization. I hadn't realized it had been oversubscribed. I hope that is something that this committee and the Appropriations Committee can work on to see if there are more resources available for that.

Mrs. TRAHAN. Same—did you want to add to that?

Mr. DABBAR. So, Congresswoman, probably one of the leading fusion companies in the country or world is in or around your district in Devens. I think that plant is going to—when it comes on line in 2025 will well outperform what happened at Livermore. They don't like to talk about how much, but I have a pretty good idea. I ran a fusion energy program for the country. We decided when we were there to engage with the private sector where DOE had not engaged with the private sector before. We had to, you know, break a little glass on that.

I think the milestone program that we started, that we started moving along needs to be expanded. And I know that some of the proposals in Build Back Better had made that larger. I would highly encourage that we take the momentum on technology, innovation in particular, in your State and your district but elsewhere, and I would recommend that everyone here look at expanding that.

Mrs. TRAHAN. And I look forward to bipartisan work on that. I mean, I will leave the committee with this. The Nuclear Regulatory Commission recently released their options for licensing and regulating fusion energy systems which highlights the important safety and security benefits of fusion energy: No high-level nuclear waste, no chance of a meltdown, no special nuclear material like plutonium or uranium. But the paper also leaves some regulatory ambiguity suggesting that future fusion power plants may be regulated like fission, which is a very different energy process.

So I look forward to making sure that we put the right regulatory in place, environment in place so that we don't stifle that innovation as it is coming to bear.

Thank you. I yield back.

Mr. CURTIS. The gentlewoman yields.

The Chair recognizes the gentleman from Texas, Mr. Weber.

Mr. WEBER. Thank you, Mr. Chairman.

First I want to correct something that was said earlier by Scott Peters. He said in Texas 40 percent of our power is renewable. That is just not accurate. You go to EIA.gov, Energy Information Administration, look at that, and Texas is 50 percent natural gas. It is 18 percent coal. Nuclear is 8 percent. And that is 76 percent. I am not good in math, but that leaves about 24 percent: 20 of it is wind, and 4 percent is solar.

So we appreciate renewables in Texas, we do. But the real important fact is that the renewables cannot be the leading actor. Renewables can be a supporting actor, but renewables cannot be the leading actor. We found that out in Winter Storm Uri.

You all talked about pipelines, Mr. Dabbar. I think it was you and one of our Members. The Keystone Pipeline would have come into my district over by Beaumont, Texas. It carries 830,000 barrels of oil a day. The Colonial Pipeline system that you talked about carries about

3.1 million barrels of product a day. The Key-stone Pipeline is literally one-fourth of the output what feeds the entire Southeastern part of the United States. Why didn't the President come to Texas and beg us to drill for more oil? It has been a real mystery to us.

Also, in my gulf coast district, I have Strategic Petroleum Reserve. We house also about 59 percent, not quite 60 percent of the SPR in my district. I have been watching it for a long time. I served 4 years in the Texas House. I was on the Environmental Energy Subcommittee, and I will tell you that—or Environmental Reg Committee, and I will tell you that we paid a lot of attention. Texas has 225,000 miles of pipeline. The pipeline industry has a 99.925 percent safety rating. We can move product safer than anybody else. We can store it in the SPR. We can have it ready for emergency, not because the President wants to bring down gas prices in an election year. I mean, he is trying to help. Maybe he is storing classified documents in his garage to keep from paying storage fees. I don't know. He is just trying to help.

But I just want to go to you, if I can, Mr. McNally. You made an interesting comment. You said an arsenal of energy is what we need. That is a great term. You talked about the things that we had—you talked about World War II attack, if I remember correctly. Americans need to know what is important and why we need to be energy dominant, energy independent. It is energy. It is absolutely energy safety. It is domestic safety. It is political safety. It is military safety. You just—economic safety. You just can't express how important it is.

For the President to draw down the SPR in an election year is totally uncalled for. I am told you are an expert on the SPR, and you sound like you are pretty knowledgeable about it. So my question to you is, in your opinion—and you have probably gone back and looked. I think the SPR capacity—it depends on who you talk to—is about 714 million barrels, maybe 730, depends on who you are reading.

In your opinion, what is the proper use of the SPR? And what is the history of it, if you know that far back?

Mr. MCNALLY. Thank you, Congressman Weber.

Yes, I think we topped out at about 725 million barrels. It is now down to about half that level. We are at a 40-year low, so it hasn't been since 1983 that we have seen it this low.

And it is really unfortunate, because in a way, in my view, it is a bipartisan mistake to start to sell off the SPR just to pay sort of regular expenses in 2017. We did the same thing in the mid-1990s, and I worked for President Bush after 9/11 when we re-stocked the SPR at higher prices.

So we kind of went around the circle once before. I was hoping we wouldn't have to do it again.

But, in addition to the mandatory nonemergency sales, which I think are very unfortunate but starting to reverse, President Biden, especially in November of 2021 before the Russian invasion of Ukraine, used it purely for political reasons. I mean, there was no disruption. The IEA wouldn't go along, and he had to beg China, of all people, of all countries, to go along with us.

And that is up there in the halls of infamy with the decision by President Clinton September 2020, 2 months before the election, to invite Al Gore to announce a release at that time too.

So with the emergency release in March of last year, at least we thought we were going to lose Russia. When it started out, we had a real emergency, we thought. But within a couple of months, we realized Russian supply wasn't going off. They should have suspended the sales at that point.

So we have a mixed history with the SPR. Fortunately, some Presidents have used it for purely political price control, very limited, by my count twice. We have had, no kidding, emergency releases. And I would again say the administration probably took the right decision initially, because we thought in March of 2022, the IEA said we are going to lose 3 million barrels a day of Russian supply in April, like 2 weeks later. Again, that wolf did not come into the village. We should have suspended those sales, but we did not.

Thank you for the question.

Mr. WEBER. I thank you for that. And I yield back.

Mr. CURTIS. The gentleman from Texas yields.

The Chair recognizes the gentlewoman from Texas, Mrs. Fletcher.

Mrs. FLETCHER. Thank you, Mr. Chairman.

And I join in welcoming our new Members and congratulating Chairwoman McMorris Rodgers on her historic chairmanship.

And I am thanking our witnesses for being here today. It has been a long day, but I have been here and appreciated your testimony.

As we have heard throughout the day, our country leads in energy production and innovation, and we lead in these areas because of the work that is done in my district and surrounding areas in Houston, ones where we touch every single segment of the energy industry, from exploration and production, transportation, transmission, marketing, technology, both traditional and renewable.

And I am going to resist the congressional urge to spend 5 minutes talking about how great my district is, although it is, but I want to share these data points because I think it really helps underscore both the importance of the issues to me and also the depth and breadth of knowledge of my constituents that I bring to the work of this committee.

And it is with that in mind that I am glad to say I have heard a lot of things I agree with today from our witnesses, from my colleagues on both sides of the aisle, but I have also heard a few things I disagree with. And I can't go through all of them here, but I am disappointed to hear some of the attacks on the 117th Congress' energy policies rather than more ideas about how we can work together to build on the historic investments that we made in the last Congress.

And, you know, we have talked about some today especially the focus on permitting issues that affects all sectors and which I very much look forward to working on. But I do think it is important to underscore that the Infrastructure Investment and Jobs Act and the Inflation Reduction Act that we passed last Congress are historic pieces of legislation that make energy and infrastructure investments that further strengthen American energy security and drive innovation for our energy future.

At this moment we have an energy sector that is looking stronger than ever, with the EIA expecting U.S. domestic oil production to set a new record high this year of 12.4 million barrels a day, which would surpass the previous record set in 2019. The red count is back up. It was 771 last week. Net exports are also expected to rise, further strengthening our energy security and, importantly, that of our allies. And this is happening while we are working to meet growing global demand for energy and reducing admissions. The bottom line is that we need to do all of the things we are doing and more, and we need a holistic approach. So I really look forward to talking about what that is and how we come together to do that on this committee this Congress.

But for today I want to focus my questions on one of the programs that we passed in the Bipartisan Infrastructure Law: funding for the implementation and design of regional hydrogen hubs. And through this program, Congress authorized the development of multiple hydrogen hubs to advance the country's clean hydrogen sector. And the Department of Energy will select 6 to 10 regions to establish these hydrogen hubs.

Mr. Crenshaw and I recently wrote a letter to Secretary Granholm making the case, not surprisingly, for the Department to select Houston as a location for a hydrogen hub under the program. So I want to direct my question first to Dr. Unruh Cohen. In the interests of time, I am going to ask this, and I have a quick follow-up if we can get to it. If not, I will submit it to you for the record. But can you just talk a little bit about how this program will enhance hydrogen technology deployment and why Federal support is essential in emerging technological sectors like hydrogen?

Dr. UNRUH COHEN. Yes. And I will sing Houston's praises in the process because they are a great example of where industry, academia, the local government have come together to identify we need to take our strengths and talents in the industry and expand that.

So I think you will be a good candidate for the hydrogen hub. The program will drive these types of partnerships between industry and the Federal Government to develop the new technology. And hydrogen is critical for also all of the petrochemical work that happens in the Houston area, because we are going to need clean green hydrogen to help decarbonize some of those processes so that we can, you know, benefit and have clean chemicals coming out of our domestic plants.

Mrs. FLETCHER. Well, thank you for that.

And with the time I have left, I do want—I may have to get your question for the record, and if anyone else wants to submit a response, I would appreciate that, because I think while this investment is essential to jump starting the technology, I worry there's still a lot of hurdles that Congress needs to address before we can see widespread adoption.

DOE's 2020 Hydrogen Program Plan identified rights-of-way and permitting issues for hydrogen pipelines as challenges for hydrogen delivery infrastructure. And I think there's still a lot of unresolved questions regarding siting, political Federal/State jurisdictional conflicts, and the regulation of pipeline rates and terms of service that need to be resolved.

So with the 12 seconds I have left—oh, have I gone over?

Mr. CURTIS. You are going the wrong direction.

Mrs. FLETCHER. I am going the wrong direction. So with that, that is my question. If you could respond to that for the record, I would appreciate it.

And I will yield back the balance of my time that I have gone over. Thank you very much, Mr. Chairman, for indulging me.

Mr. CURTIS. Yes. All right. Thank you. The gentlewoman from Texas yields.

The Chair recognizes the gentleman from Georgia, Mr. Allen.

Mr. ALLEN. Thank you, Mr. Chairman.

And, you know, energy security is national security. We have heard that over and over again. And when we were in the last— the last time we were in the majority, we passed the Congressional Review Act. And what it did was, you know, it revised a lot of the regulations, updated a lot of regulations around the oil and gas industry. And what we saw we couldn't believe. We became energy dominant, and we had the power to control the cost of a barrel of oil. That to me is maybe the greatest power that you can experience. In fact, we drove it—you know, market price was $30, $40 a barrel. We created an oil war with $7 a barrel. And now we are, what, $90 to $100? It has been over $100. And, Mr. McNally, you said it was going to fluctuate even more than that.

But, in fact, we did this throughout the economy. And, Ms. Jackson, everybody benefited. It was the greatest economy I've ever seen in my lifetime. So there is no secret that President Biden and his administration have declared a war on fossil fuels. I mean, he said, "There will be no more drilling." I think that was a quote.

And so we see what has happened now. In fact, back in the greatest economy, we had 1.26 percent inflation.

And so, you know, Mr. McNally, how could we unleash the oil and gas industry and become energy dominant again? Is there a secret sauce?

Mr. MCNALLY. Did you mean me or Mr. Dabbar?

Mr. ALLEN. I mean to both, yes.

Mr. MCNALLY. OK. Well, again, if I could rephrase my attempt to answer Congressman Eshoo's very good question: Get back to all of the above and get honest data and analysis. If we can just do that, we will be in a much better place.

To his credit, President Obama—to his credit—helped get rid of the crude oil export ban, which directly threatened the shale oil boom, and recognized not only was that good for our energy production, but helped us offset the loss of Iran, which we were sanctioning and so forth. He understood that transitions are multidecade affairs, and he, to his credit, thought that gas was part of the clean energy future.

But as you have pointed out, we have moved to keep it in the ground, to a war on fossil fuels, and that must end. And I think, honestly, if we could just get back to where we were in the last few Presidents, including President Trump, where it was all of the above, we can unleash, you know, our energy potential.

Mr. ALLEN. Mr. Dabbar?

Mr. DABBAR. So, Congressman, it started with great innovation by the Member of Texas. The Permian Basin cost year by year was dropping because of innovation. That allowed us to drive prices down globally as a result of that innovation. Then the Federal system and the State systems allowed things to get built to move that energy to where it needs to go——

Mr. ALLEN. So let me ask you this: Would it be fair to assume that free market drives prices down, government intervention drives prices up?

Mr. DABBAR. That is the irony of what we are seeing today.

Mr. ALLEN. OK. So we have got—so what we have done—I was on the House Energy Action Team, and I met with a lot of—I was a small business owner. I met with a lot of small business owners who were in the business of drilling and refining. They are out of business. They were driven out of business.

You know, if you have more demand than you have supply, guess what? The price is going to go up. Now, obviously, if you are one of the few companies remaining, you are going to benefit from that. Now, how do we reverse that? We have got to open up the free markets, and then you will drive down prices. I mean, you know, there is no secret to the way our economy works.

As far as—also, I wanted to ask you a quick question—I have got about 45 seconds—about Russia being a leader in exporting enriched fuel. And, of course, the strategy was to urge for U.S. production of our own nuclear fuel. I have got Vogtle 3 and 4 hopefully coming on line. We have got plenty of electricity in Georgia. In fact, you are probably going to have to bring your car to Georgia to charge it at some point in time.

But can you give me a little background on that?

Mr. DABBAR. So Russia has almost half of the global enrichment market. The biggest exposed company to that is the United States. If Russia tomorrow decided to stop exporting their enriched uranium to us over the course of a couple refueling cycles, we might lose half of the fuel needed to run the nuclear power plants. That is 20 percent of the country. That is 10 percent of the power plants would be at risk of not having fuel.

Mr. ALLEN. And I would hope that this Congress will do something about that.

Thank you.

Thank you. And I yield back.

Mr. CURTIS. The gentleman yields.

The Chair recognizes the gentlewoman from California, Ms. Barragán.

Ms. BARRAGÁN. Thank you, Mr. Chair.

When I received the Republican committee memo for today's hearing, what stood out to me was the complete absence of the need for U.S. energy policy to address climate change or environmental justice. Not a word.

An effective U.S. energy policy must keep costs down, create the jobs of the future, and reduce the fossil fuel pollution that warms our planet and harms the public health of many communities, including my own Latino community and communities of color in my district.

Dr. Unruh Cohen, let me thank you. Thank you for your tireless work in the last Congress with the select committee to make sure that we were doing everything we could to save the planet, addressing climate change, and doing the hard work of the select committee.

Now, one important program from the Inflation Reduction Act the Democrats passed is $3 billion for climate and environmental justice block grants based off of my bill, the Climate Justice Grants Act.

Can you tell us a little bit about how can this program help to reduce energy costs and reduce pollution in communities of color and other communities across the country?

Dr. UNRUH COHEN. Thank you for the question, Congresswoman. That is one of the most exciting programs in the Inflation Reduction Act. That is going to provide funding to empower communities, to look at the challenges that they are facing when it comes to energy costs, climate costs.

You know, as we have heard from Ms. Jackson and others, you know, there are challenges that constituents are facing. High prices are, you know, problems for everybody.

And so this program is going to really empower communities to figure out the solutions that work the best for them, to help bring them affordable clean energy and respond to the climate crisis consequences that they are seeing already in their communities.

Ms. BARRAGÁN. Great. Thank you.

My next question is also for you. I want to talk a little bit about geothermal, something that the chair and I are working on together, and how we invest in geothermal. The question is related to the Climate Crisis Action Plan that the Select Committee on the Climate Crisis worked on, highlighted the development of more geothermal energy as a building block of growing American clean energy production.

Could you describe how increased development of geothermal energy in California and elsewhere would enhance U.S. energy security with 24/7 clean power?

Dr. UNRUH COHEN. Yes. Geothermal is tapping into the power of the earth and what is all around us. California has been a leader because of the particular geologic benefits and profile in California. But as the technology has improved from investment in DOE over the years, we are now at a place where we are close to commercialization of energy supply in areas that don't have quite the great resources that California and other parts of the West have.

And so it will also—it will add to that need to provide dispatchable power to have—to fill in at times that we need it. And I think it is actually one of the most exciting opportunities coming our way.

In addition, quickly, it also will use the skill set and training of many members who are working—or workers in the oil and gas industry right now. And so it provides an additional opportunity for them to take their skills and help to continue to provide this country with energy.

Ms. BARRAGÁN. OK. Thank you.

Mr. Chair, I would like to enter into the record a January 30 Climatewire article titled "China Invests $546 Billion in Clean Energy, Far Surpassing the U.S." This is China's 2022 investment.

Dr. Unruh Cohen, the Inflation Reduction Act passed by Democrats includes $369 billion in clean energy and climate programs. It is important to fight climate change and to compete with China for the jobs and industries of the future. More must be done to support clean energy.

What are the most important steps that the U.S. can take in the next few years to keep pace with China on clean energy?

Dr. UNRUH COHEN. I think we are seeing it already. You know, just since the passage of the Inflation Reduction Act, we have seen hundreds of announcements about bringing clean technology manufacturing to the U.S. and expanding what we have already. I had in my testimony, you know, nearly $90 billion in projects, and that is private capital coming in, you know, ready to partner and have that synergy with the Federal Government investments.

Ms. BARRAGÁN. Thank you. Thank you for your responses.

And I yield back.

Mr. CURTIS. The gentlewoman from California has asked for a document to be entered into the record. We are waiting for that document. Peter, we will reserve that until the end of the meeting.

Ms. BARRAGÁN. Thank you.

Mr. CURTIS. Yes.

The Chair recognizes the gentleman from Ohio, Mr. Balderson.

Mr. BALDERSON. Thank you, Mr. Chairman.

Thank you all for being here today. I know it has been a long day.

Mr. Dabbar, my first question is for you. It is no secret that the shale revolution in the United States has brought immense economic benefits. In Ohio, natural gas and oil development contributes more than $50 billion to the State's GDP and supports hundreds of thousands of jobs.

In addition to the economic benefits, we have also seen clean environmental benefits from increased use of natural gas. According to the Department of Energy, use of natural gas for electric power production has led to a 57 percent reduction in domestic emissions of airborne particles such as soot. This has resulted in an estimated $17 billion in annual health benefits.

And in 2021, U.S. natural gas exports were the highest on record, and the United States has been an annual net exporter of natural gas since 2017.

Can you discuss the public health and environmental benefits America has seen as a result of the shale revolution?

Mr. DABBAR. So, as you pointed out, the big shift from coal to natural gas has had a big impact. Another thing that has been kind of underreported—and, actually, this is kind of efficiency when it comes to natural gas—the natural gas power plants that you produce in your State, the combined cycle gas plants, are about 50 percent more efficient than they were. And so they are running on natural gas, and they are 50 percent more efficient. That means it takes less BTUs, it takes less energy to make the same amount of electricity, and it produces 50 percent less emissions.

So the emissions have been driven down in large part because of your State, because of both the natural gas production, but the turbine improvements. The combination of those two has dramatically reduced emissions as a result.

Mr. BALDERSON. Thank you.

In followup to that, can you discuss the possible global environment benefits in exporting our cleaner natural gas to developing nations?

Mr. DABBAR. Yes. So the developing nations—if you go to any international conference, you go to especially the African nations, they are very upset today, OK, because they say that the IMF and the World Bank and saying, like, "You can't get any money to build energy. You guys did it, but we are not allowed."

And it is pretty stark when you go out to these international meetings and listen, in particular to the African nations. Right now they burn coal being made by Chinese companies, coal-fired power plants. They don't produce much natural gas in China. They have to import.

So our ability to export natural gas to those countries and do what we did in this country versus what the Chinese are doing to the world is quite—it is quite obvious.

Mr. BALDERSON. Thank you.

When green-at-all-costs advocates in Europe shut off nuclear and fossil fuel power generation years ago without concern for their own baseload needs, they found themselves beholden to the likes of Putin for oil and natural gas.

What has happened in Europe is certainly not a path my constituents want our country to follow.

Mr. Dabbar, you referenced the problems facing Germany earlier, and I would like to expand on that. Do you have concerns that this administration is putting the cart before the horse when it comes to the transition to renewables? That is the first question.

And instead of picking winners and losers, why is it so important this administration and Congress promote all forms of energy production?

Mr. DABBAR. If you want to see the worst-case energy policies is look at Germany. They decided to shut down nuclear when it was perfectly safe because of what happened in Japan. They decided to increase their risk with an autocrat in terms of their energy exposure, and they expanded renewables where Germany is not particularly sunny and it is hard to site because it is a relatively crowded country.

I think one of the starkest things you can see on the internet was when the former President was at the U.N. saying to the Germans that they were going to be increasing their risk of energy exposure to Russia and they better stop it. And there was a picture of the Germans at the U.N. laughing at him. I think we know where things actually turned out.

Mr. BALDERSON. Agree.

Mr. McNally, I would also like to hear your thoughts on how we can ensure energy policies don't lead us into the same situation much of Europe has found itself in.

You have 30 seconds. Thank you.

Mr. MCNALLY. Yes. So, again, do not let China dominate the supply lines for renewable power and electric vehicles should those really take off in scale into the future. Maintain our strategic stockpiles. Remain to become an arsenal of energy, remain that. Stay open to exports. Keep diversity and global energy and be at the center of it.

Mr. BALDERSON. Thank you very much. Mr. Chairman, I yield back.

Mr. CURTIS. Thank you. The gentleman from Ohio yields.

The Chair recognizes the very patient gentlewoman from Illinois, Mrs. Schakowsky.

Mrs. SCHAKOWSKY. Thank you, Mr. Chairman.

It is my own fault. You know, the rule is, if you miss the gavel, then you go to the end of the line. I didn't realize that. That wasn't true when we were in the pandemic.

Anyway, I am happy to be on the committee and happy to be here today, and I am really looking forward to the rollout of the Inflation Reduction Act and the Infrastructure and Jobs Act because I think it is really going to make a difference. It is going to make a difference to communities all over this country, and it is going to make an unprecedented investment into America's clean energy future, creating millions of good, clean jobs and lowering the prices of people at home on their energy, on their energy bills, and finally, really addressing the climate crisis.

And I agree with a previous Democrat here saying that I don't hear enough about that. You know, we have all—we should put in the same level all of the energy options. This is an existential issue right now dealing with the climate, and I am so happy that we are going to be addressing that as well as maintaining and I think even enhancing the economy of our country.

One of the things that was in the bills that we passed would be an investment in a historic $15 billion for the removal of lead service lines, a problem in my community and in so many communities where we are not drinking clean water because of these lead service lines, but also billions of dollars that are going to help communities that are most affected: low-income communities, communities often most populated by people of color.

But I also want to mention that over almost 100 years Americans have been subsidizing the fossil fuel industry to the tune, even today, of billions of dollars. And I think it is important just to mention that, that somehow spending the amount of money that we are on addressing the climate crisis is certainly as important.

I wanted to—meant to ask Dr. Unruh Cohen the question similar to the one that was asked earlier. We haven't rolled out the kind of programs that are going to deal with environmental justice communities, communities that are suffering right now.

What are some of the important things that we are going to be doing that will serve those communities and address the absolute relevant issues that Ms. Jackson raised?

Dr. UNRUH COHEN. Thank you for the question.

And I think, you know, we are in a really exciting time. So for, you know, the first—these were the first laws that we really have been able to put money forward in specifically environmental justice programs. The Biden administration is focusing in its investments and has committed to fulfilling the Justice40 so that we see these benefits flow to these communities that have been underinvested in, that have experienced, you know, the impact of pollution that comes from fossil fuel development and refining.

Mrs. SCHAKOWSKY. Aren't there also workforce opportunity benefits that might accrue?

Dr. UNRUH COHEN. Yes, absolutely. And I think we will see as these clean energy programs roll out, there will be a focus of developing jobs for people in the community and getting them the training. You know, we talked about earlier workforce is a place that I think we need to have more focus from the Congress and this committee. And so I hope, while we will see some improvement, that this committee can work on some of those issues going forward.

Mrs. SCHAKOWSKY. Let me ask you this. Instead, it seems to me in working to pass what I think is an effective legislation that had to do—that was passed that had to do with—am I over? I think I am.

Mr. CURTIS. Yes, a minute over.

Mrs. SCHAKOWSKY. OK. I will write that and send it to you. Thank you.

Dr. UNRUH COHEN. I look forward to it. Thank you.

Mrs. SCHAKOWSKY. I appreciate it.

And I yield back.

Mr. CURTIS. I don't mean to be aggressive——

Mrs. SCHAKOWSKY. No, you're right.

Mr. CURTIS [continuing]. But we are going to have to get everybody in before votes.

So the gentlewoman yields.

The Chair calls on the gentleman, Mr. Fulcher, from Idaho.

Mr. FULCHER. Thank you, Mr. Chairman.

I just need to respond to a comment a few moments ago from my friend from California that indicated that the United States somehow needs to follow China's lead on clean energy investment. And I just want to state for the record that, if anyone believes that that is a model that we want to follow, pay China a visit, and I think they may change their mind.

But on to my specific topic. This is going to be a question for Mr. Dabbar. But it is in regard to geothermal energy. And in my State, Idaho, we have been a little bit of a pioneer on that front. There is—the first district heating system in the country was in Boise, Idaho, in 1892, and we have got a very good resource there. It is carbon free. It is baseload. And we are a little bit familiar with it there, but it doesn't seem to get a lot of attention.

And you made a comment, Mr. Dabbar, earlier today that caught my ear that these—you need to be technology neutral, I think was the term you used, when making decisions on these sources and not some political idea and try to drive it with subsidies.

And so, with that comment, I wanted to just get your opinion. I have got a—I have had a bill that I have run for several years called the Enhancing Geothermal Production on Federal Lands Act that basically allows geothermal exploration and production on Federal lands where there's already existing leases, oil and gas leases.

It hasn't made it very far, but I wanted to just get your opinion. Does this satisfy the tech neutral argument that you laid out before? What are the opportunities and challenges for geothermal in this country?

Mr. DABBAR. Congressman, I think expanding geothermal is absolutely something that we should be looking at. And not knowing every detail of the bill, I think facilitating lands that are already being used for energy production that have the geology already mapped out is going to drive down costs and increase the likelihood of finding the right formation. And further investment in driving down costs of geothermal equipment, right, and the thermal efficiency, it gives great baseload generation, right. It is great baseload generation wherever it is at. And I would think any place that it is available, we should be aggressively attacking that.

Mr. FULCHER. Thank you for that.

And I need to pivot and just utilize my time as best as possible. I am going to shift, and I have a question for Mr. McNally here. It does go to the nuclear arena. And also in my State we have the Idaho National Lab, and there is a lot of research there with small modular reactors.

And one of the arguments that we hear is that, in addition to the efficiency of the energy production, there could also be some upside when it comes to grid security, specifically because we are threatened with sabotage, we are threatened with cyber attacks on our energy systems so much now.

By having an energy source that you can isolate—for example, one reactor could potentially power the City of Boise, Idaho. By isolating that and staying off ultimate connectivity of an overall grid, that could help with a cybersecurity or cyber attack threats.

Your comments on that? Is that true or is it false? Is it benefited—is it a benefit to be isolated off the grid, or is it a negative to do that?

Mr. MCNALLY. Congressman Fulcher, on that question, that is outside my area of expertise. I would defer to my panel mate, Mr. Dabbar, if you might have a view on that. But that is outside my area of expertise, I am afraid.

Mr. FULCHER. OK. Mr. Dabbar?

Mr. DABBAR. Yes. So I think nuclear power is absolutely supportive of grid security. First of all, it is really hard to penetrate a nuclear reactor in terms of security, and I think that has a lot of value. And then, at the end of the day, the availability of it is unparalleled in this system.

Mr. FULCHER. But the isolation of staying off of a greater grid, do you think that is a good thing or a bad thing?

Mr. DABBAR. The ability to separate in times of any challenges in the grade, including an attack, has great value.

Mr. FULCHER. OK. Thank you, Mr. Chairman. I yield back.

Mr. CURTIS. The gentleman yields.

The Chair recognizes the gentleman from Texas, Mr. Pfluger.

Mr. PFLUGER. Thank you, Mr. Chairman.

I would like to thank the witnesses as well. Most of the good questions have been asked today. Thank you all for being here.

I represent the Permian Basin that has been mentioned several times. This is literally the heartbeat of America's energy. It is the heartbeat of the shale revolution. I am so proud of the men and women in the Permian Basin, Midland and Odessa and throughout that part of the country, that have innovated to a point where we have literally helped raise a billion people out of poverty throughout the world. We have literally lowered the cost of living for every American family to the tune of $2,500 per year. We have allowed this economy, prior to January of 2021, to soar, to absolutely take off.

And I am trying to figure out right now, the discussion about the climate crisis, what—who is saying that this is the greatest threat we face with any sort of facts?

And, Mr. McNally, to your point, I want to see the analysis here. I want to see the data. You all have made some incredible points on that.

I go back to some of the campaign promises that the President made. And I quote: "We are going to get rid of fossil fuels. There will be no more coal plants. We are going to phase out fossil fuels." Those were three quotes that he made. I got to hand it to him, he has done everything he possibly can with every tool to assault the 2 million Texans who are in this industry who are, like Dr. Cohen's family in Corpus Christi, who are doing the things cleaner and more efficiently and better than anywhere else on the planet. He has assaulted this industry in a way that is directly attacking— and thank you, Ms. Jackson, for your testimony today— the most vulnerable in our country.

I appreciate the regulatory and the permitting discussions.

Mr. Dabbar, if we continue with the policies that this administration is pushing, what is our country going to look like? What is our economy going to look like? What is our—you know, we are expecting, what, 45 percent increase in demand over the next 30 years in electricity. What are we going to look like?

Mr. DABBAR. I think the key word you just mentioned there, Congressman, is demand. So I think, no matter how you approach this topic, as long as there is a demand for a product, then it has to be supplied from someplace.

And taking aside all the other debates, if we are still going to be needing petrochemicals to make this pen or to drive some cars because they are not all electric vehicles, someone has to provide it. I think we can all agree on that from both sides.

And so, if it is not produced in Midland or Odessa, it is going to be produced in Caracas or outside Tehran. So that is it. That is it. So if we restrict it because of ESG, if we restrict it because of siting in this country, it will be produced overseas. It will emit more. It will be more environmentally hurtful for the world. It will be less jobs. It will be jobs in Tehran, not jobs in Odessa. I think it is relatively straightforward.

So even if you have this debate on this topic, you know, about demand, you know, as long as the demand is there, you know, for national security and the economy and the environment, it is better to be produced in America than it is in Russia or someplace else.

Mr. PFLUGER. Yes. Right now in San Angelo, Texas, my hometown, which is about 120 miles to the east of Midland/Odessa, it is 27 degrees. It is snowing. There is no wind at all, and the sun is not going to shine until Friday morning. And it will be 27 degrees on Friday morning. I just looked at the weather forecast.

I asked Mr. Kerry, the climate—I don't know what his official position is. They call him the climate czar. But I asked him if renewables provide baseload capacity. And I will ask, Dr. Cohen, the same to you. We have a ton of wind energy. In fact, we are so proud of the wind energy we have, which is more than the State of California in my district.

Do renewables, whether it is solar or wind or another form I am not familiar with, provide baseload capacity for my family right now?

Dr. UNRUH COHEN. Renewables provide predictable electricity for your family and families all across the country.

Mr. PFLUGER. And predictable is, I think, not always reliable.

Dr. UNRUH COHEN. Predictable means predictable. We know when the sun will shine. We know when the wind will blow. Smart grid operators can then provide—make the energy decisions they need to keep the lights on.

I just happened to look at the Texas electricity map yesterday, and I noticed there is a lot of stranded electricity down in my part of the world. So if Texas could continue its leadership in building transmission, they actually would be able to free up some of that renewables that's blowing between Corpus Christi and San Antonio and get it up to other parts of the State.

Mr. PFLUGER. It takes every form of energy, every amount of energy, and every bit of energy to service the demands. I could talk for another 10 minutes, but I don't have that much time.

I yield back.

Dr. UNRUH COHEN. Well, maybe we can talk again another time.

Mr. PFLUGER. But they won't let you. Thank you.

Mr. CURTIS. The gentleman from Texas yields.

The Chair calls on the gentlewoman from Tennessee, Mrs. Harshbarger.

Mrs. HARSHBARGER. Thank you, Mr. Chairman.

And thank you all for being here. You are tired. I am tired. So I am going to make a couple of comments, and then I will go to my question for Mr. Dabbar.

Ms. Jackson, I want you to know that I represent a rural district with two distressed counties. The average median income is 49,000, so I understand exactly. I get calls every day from constituents— who I call my friends and family, by the way—and they tell me, "If I pay my power bill, I am going to have to make decisions about, do I buy groceries or do I buy my medication?"

So I want you to know I totally understand; OK.

And, Mr. McNally, your statement about foreign and domestic actors beginning to attack our domestic energy infrastructure has never been more evident than with the Colonial Pipeline attack. I was on Homeland Security last Congress on the Cybersecurity and Critical Infrastructure Subcommittee, and it was very frightening to hear how that attack occurred and that it could happen again. And it hit me because I drive from my district in East Tennessee all the way to Washington, DC, every weekend, and I had to stop at every gas station along the way to make sure that I had fuel just to come here and do my job. So that hit home.

And you saying—I agree with you that Congress needs to be better prepared for more cybersecurity attacks, and we need to hold those bad actors accountable. That is a big deal.

And, Mr. Dabbar, your comments about Russia having the largest overall HEU stockpile in the world and your statement that the U.S. needs to work on our uranium enrichment capabilities is on point.

And just so you know, I have nuclear fuel services in my district, and I have talked to many of those people there, and they absolutely understand the importance of this very issue, why we need to do enriched uranium.

So with that said—and I will continue to talk to them, so stay tuned—I agree that we absolutely need a higher baseload of power because we have seen how fragile we are to these disruptions.

And my question is this. I am from East Tennessee, the 1st District, and just before Christmas, East Tennesseeans experienced rolling blackouts after the Tennessee Valley Authority was unable to meet those energy demands required to heat your homes during Storm Elliott. We know that natural gas facilities can come fully on line around 30 minutes and our best option for wrapping up energy production in a pinch.

My question is this: How much additional investment in natural gas will we need to combat these unexpected increases in energy demands in the future?

Mr. DABBAR. So, Congresswoman, you know, TVA shut down baseload just like other places have.

Mrs. HARSHBARGER. Yes, they did.

Mr. DABBAR. And that is the reason why I think you ran into those problems, was that TVA was shutting down more plants. They are coal plants. You can kind of understand it, but you are going to need to replace them with something. You need to replace them with something that can be dispatched.

Mrs. HARSHBARGER. Yes.

Mr. DABBAR. And natural gas is the obvious place to go, subject to some of your nuclear power plants, which would be wonderful for East Tennessee.

Mrs. HARSHBARGER. Yes.

Mr. DABBAR. So the gas pipelines in general are not there to replace plants that used to be coal and rail and so on. So things are getting more unreliable in the whole TVA area, including in your region. That is the reality.

And if gas pipelines are not built, you are going to run into that problem more and more. As your great economy grows—I know East Tennessee well from being at DOE——

Mrs. HARSHBARGER. Oh, yes.

Mr. DABBAR [continuing]. You are going to have additional problems because your economic growth is so strong.

Mrs. HARSHBARGER. You know, the scary thing was that they didn't give notice—TVA didn't—to emergency management services. The emergency broadcast systems went down because of the rolling blackouts. There was equipment damage. We are trying to track that too as we go along, to see how long they were down, how much notice did they get, how much damage was done monetarily, because there was one industry that lost $3.5 million just from the shutdown and had to work through the weekend.

And I have a question, and anybody can answer this in the amount of time I have left. You know, the American people tell us, "Come up here and work, do your job, and get results." But, like so many other projects, whether it be energy production or highway improvements, they get stalled in the NEPA process, and I guess my question is, What would a rerun of NEPA look like, and how that would allow for us to build natural gas production so that environmentalists can't stall those projects into nonexistence? Anybody?

Mr. MCNALLY. How about a bipartisan agreement to have legally enforceable deadlines for NEPA decisions? Clean energy, conventional energy, everything, just say you got a certain amount of time, you got to get it done here, and you cut back on the litigation risk that we have heard today, and it is for all energies, all businesses and so forth. I sense maybe there is bipartisan agreement behind that.

And then we might want to think about critical national security infrastructure, where there you supersize it, you say, "NEPA," you know, "you got to go even faster." But I would hope there is—I am sensing bipartisan agreement, we got to improve permitting for all energy here.

Mrs. HARSHBARGER. Fantastic.

Dr. UNRUH COHEN. Chairman, I would just underscore that the Inflation Reduction Act had nearly $1 billion for agencies to fund their permitting work.

Mr. CURTIS [presiding]. Thank you. The gentlewoman yields. Take us home, the gentlewoman from Iowa, Dr. Meeks.

Mrs. MILLER-MEEKS. Thank you very much, Chair Curtis, and thank you to all of our witnesses. Iowa is an energy State, so that may surprise people. Usually I don't need this, but Iowa is an energy State, and let me also say, unequivocally, that all of us, I think, bipartisan agree that we want a cleaner, healthier planet for our children and our grandchildren.

But we also want to be able to have an economy that can compete in a global economic environment. Energy demand that I learned at COP26 and COP27, to my surprise—I thought energy demand was increasing—I was surprised to hear at COP26 and 27 that they thought energy demand is increasing. What I have yet seen from this administration is a way to transition with reducing emissions.

So let me tell you a little bit about Iowa. Iowa has 50 percent of its energy from renewables, so that is wind, solar, biodiesel, ethanol, biomass, bio char, manure. We have this entire slew, and up until 2 years ago, we also had nuclear. We have an entire slew. Fifty-eight percent of our electricity is generated by wind.

And we were told this year that we would potentially have rolling blackouts. Why? Because we are an energy exporter. So why should a State that generates massive amounts of renewable energy be subject to rolling blackouts? And it is because there is not enough energy production.

On average, it takes 6.5 years to prevent transmission projects in this country. There are some examples of projects taking over 10 years, and they are still not fully transmitted—or permitted.

If we are serious about improving our grid security, modernizing our grid infrastructure, and diversifying our energy mix, we need permitting reform.

Last Congress, I introduced the Stay Off My Line Act, which seeks to address some of these permitting challenges. And just very briefly, if you can, because I have got a couple of questions, what other steps can we take to improve our Nation's permitting process when it comes to transmission?

So if you would, Mr. Dabbar, if you could take a swipe at that. Mr. DABBAR. Yes, Congresswoman. I keep coming back to the time clock. People don't want to have things built and use the time clock——

Mr. CURTIS. Excuse me. Your microphone, please.

Mr. DABBAR. The time clock of approvals under NEPA is basically preventing things from getting built. I think that is the biggest thing that could get fixed.

Mrs. MILLER-MEEKS. Thank you very much.

Dr. UNRUH COHEN. I would add that, in Senator Manchin's permitting legislation, the language that dealt with transmission, I think, is a good place for this committee to look about going forward.

Mrs. MILLER-MEEKS. Thank you, Ms. Unruh Cohen, and you had mentioned that earlier as well.

Iowa is also the Nation's largest fuel ethanol producer, and it accounts for about one-fourth of U.S. fuel ethanol production capacity. I have heard a lot from this administration and from people about electrifying our transportation sector.

So I want to bring up a little different source of energy than what we have talked about throughout this 4 hours. I am very supportive of ethanol-based aviation fuels or sustainable aviation fuels. According to the Department of Energy, replacing existing jet fuels with sustainable aviation fuels has been recognized as an effective strategy to help the aviation industry reduce greenhouse gas emissions, diversify fuel supply, and enhance energy security.

The technology needed for SAF production already exists, including ethanol-to-jet-conversion technologies. Compared to petroleum jet fuels, sustainable aviation fuel produced from today's corn ethanol offers a 15 percent lower carbon intensity and, as we have heard, can even be carbon negative.

This is, in part, because the technology to produce ethanol from corn is improved. In fact, lifecycle emissions of corn ethanol have decreased by roughly a quarter in the past 15 years.

What sort of R&D incentives and coordinated efforts would be needed to speed up the deployment of SAF in commercial aviation? Mr. McNally?

Mr. MCNALLY. I hate to—I learned a long time ago working for President Bush, you either know the answer to the question or you say, "I don't know and I can find out." So I am going to pass to any other of my colleagues who might have views on SAF. It is not something in research and development.

Mr. DABBAR. We funded quite a bit of this at DOE, and as you mentioned, it is a big area of focus. The conversion of wind energy into sustainable aviation fuel, through a series of chemistry steps, is completely doable. You can turn wind into aviation fuel. And so—it takes a few steps of energy, and some efficiency needs to be improved. But Iowa is in an excellent position to help drive that innovation.

Mrs. MILLER-MEEKS. Thank you very much. And I would agree with our, you know, Ames Laboratory, with what we are doing in the renewable space, and then also continued innovation in nuclear fusion, as we have just seen some landmark things occur, and then in hydrogen as well. I think we have a bright future if we focus on where we have agreement and that we all want a cleaner, healthier planet. Thank you.

Mr. CURTIS. The gentlewoman yields, and the Chair calls on the gentlewoman from Florida.

Mrs. CAMMACK. Thank you, Mr. Chairman. Thank you to our witnesses for being troupers. We are in hour 6 of this committee. So thank you for your diligence, persistence, and endurance.

But speaking of endurance, our economy needs some endurance, and that can only be provided through a reliable, domestic production of energy. So I am so excited this is our first hearing, our first topic.

It goes without saying, I think, regardless if you are Republican or Democrat, we are an energy economy. Everything begins and ends with energy. So I am excited for all the discussion that we have had today.

I am the author of the REINS Act, which seeks to rein in the regulatory environment, which costs our economy $2 trillion a year. You can look no further than the work that is being done at DOE or at FERC, or others, where this has been an ever-growing presence.

So I am going to go down the line. If you guys can keep them short because I do have a follow-up question. To all our witnesses, but I will start with you, Mr. McNally: Give me one regulation that we can take off the books that would help unleash domestic production of energy and bring down cost of energy in America. Just one.

Mr. MCNALLY. Take off a reform, it would be NEPA. NEPA is the taproot of all the problems. Fix NEPA and you fix a whole lot of things.

Mrs. CAMMACK. Dr. Cohen?

Dr. UNRUH COHEN. I think we have talked about unleashing some of the clean renewable energy that is tied up, and we need more transmission for that.

Mrs. CAMMACK. What regulation specifically?

Dr. UNRUH COHEN. That—I think FERC is doing some rulemaking right now that will look at their regional planning, and I think that improvement will bring a lot of new transmission building——

Mrs. CAMMACK. OK.

Dr. UNRUH COHEN [continuing]. To our country.

Mrs. CAMMACK. Thank you.

Ms. Jackson? Which, by the way, you have had remarkable quotes today. I have actually written a number of them down, so thank you for your testimony today.

Ms. JACKSON. Well, I am going to say, and I don't know if I have the regulation correct, but I would say, make the permitting reform so that we can have more energy production. It doesn't matter how many leases you have if you can't get the permits. It is useless.

Mrs. CAMMACK. Certainty. Certainty in government, it is a novel concept, I know.

Ms. JACKSON. Yes.

Mr. DABBAR. I will also go with NEPA. I think we need to clear out the litigation that has backed the standards that have been built up, and only you could clear out all that.

Mrs. CAMMACK. Perfect. Thank you. And it goes right back to you, Mr. Dabbar. We need to put energy security back at the center of our energy policy, both for our international security but also for our allies.

My question to you is an element of IP theft and China. You were at the Department of Energy during the Thousand Talents Program crackdown concerning Chinese researchers gaining access to intellectual property and other sensitive security information. And now you are also involved in a cutting-edge quantum technology exploration.

What specifically can we do, should we do, to protect our secrets and other sensitive information from the Chinese Communist Party, and how will this effort benefit our own technological development?

Mr. DABBAR. Well, with the spin out of Cal Tech, that a certain Member here, I think, also may have went to earlier, the reality is that the Chinese have a vast amount of effort for stealing technologies from National Labs at ICOT.

When I showed up at DOE, there was a significant amount of technology that was being appropriated legally because we had no regulation on the interaction, but also at universities.

And so when I was at DOE, we rolled out four orders to limit that, and I will give you one example of something that I think should be applied to other areas in addition to DOE, is that we banned grant money, the American taxpayer money, going to university researchers who were also, at the same time, talent program members for the Chinese Communist Party. We said the American taxpayer money should not go to those people who are also working for them.

Mrs. CAMMACK. Seems a little too commonsense, if you ask me.

Mr. DABBAR. Yes.

Mrs. CAMMACK. And I am sorry I cut you off, but to that point, I mean, you have the Thousand Talents Program, you have the Confucius Institutes on college campuses. I mean, is there something specifically a database that we are tracking, or is it simply just not on the books, or has it been done by Executive order on the connection between Thousand Talents and issues of national security programs that we are working on?

Mr. DABBAR. Yes, it has basically been a few agencies that have done it, and that is it. I would recommend that this Congress take a look at the best practices for what has been done at DOE and some others, at DoD, because at NSF, at NIST, the Federal Reserve has Chinese——

Mrs. CAMMACK. Wow. I appreciate your feedback——

Mr. CURTIS. The Chair greatly wants to excuse our witnesses. If you can endure it, we have got one more, and the Chair recognizes the gentleman from California.

Mr. OBERNOLTE. Thank you very much, Mr. Chairman. I want to thank all of our witnesses. You have done an amazing job. I know it has been a very long day. For you, it is almost over. I want to thank you for hanging in there with us.

Mr. Dabbar, you said something in your testimony that really resonated with me when you highlighted the fact that we are removing baseload generation from our grid at a much faster rate than we are adding intermittent power sources.

And as I am sure you know, that is a particularly acute problem in my home State of California. Because of that, we are having a situation where there are times when the sun is shining and the wind is blowing where we are having to pay adjacent States to take our excess energy because we have so much of it.

And then we have other times when we have brownouts, or we have to ask people to curtail demand because we don't have enough baseload. So the question for you is, What should be done about that? You know, what do you think we ought to be doing differently to solve that problem?

Mr. DABBAR. So, the only entity that is trying to pay attention to this, and I think they are doing a poor job, is the Cal ISO, and that is because FERC has allowed them to do what they are doing on what you just described.

I think that FERC should be under the—that you all should look at a reform of the Federal Power Act to guide FERC to actually do their job to order the California ISO to effectively set up efforts that would reinvigorate building of baseload.

Mr. OBERNOLTE. Right. I agree with you on the Federal side. On the State side, I actually put the blame more with the California State legislature, where I served for 6 years. I think the folks at Cal ISO do their best but sometimes are constrained by the State law that is imposed on them.

And that is one of the reasons why my constituents pay twice as much for residential electricity as neighboring States, three times as much for commercial electricity, four times as much for industrial. You know, it really puts a hard, heavy burden on the people in California.

Mr. DABBAR. Commissioner Bernie McNamee, former FERC Commissioner, has written a couple papers about how these layers of State, RPS standards, and tax incentives have turned what was an efficient market model that was anticipated under the Federal Power Act into very convoluted systems which does exactly as you described, sir.

Mr. OBERNOLTE. Right. So, you know, talking about baseload, I wonder if we could have a discussion about what a potential path forward would be. So, obviously, we are trying to rekindle some interest in fusion energy—I am sorry—fission energy, but there have been very promising developments in fusion energy. And I know we had a question on this from my cochair on the Fusion Caucus, Mrs. Trahan.

Could you talk about the future you see for fusion and maybe your level of optimism that that could be part of the solution to the problem?

Mr. DABBAR. So for a couple of decades, there was not much innovation in the material science and other areas around fusion. There was a lot of innovation in batteries and wind and solar.

But as of about 5 or 6 years ago, there were some big jumps in terms of innovation in fusion, in particular around material science that allowed for the magnetic fields to get stronger, that really makes the possibility of an add-out fusion possible.

The NIFT announcement at Livermore was great, but that was not made to be a power plant. That was dealing with the weapons program. So it wasn't made to be a power plant. But I think that there are a number of fusion companies, including in Southern California, including in Northern California, as well as Massachusetts, that are much farther down the road than even NIFT at Lawrence Livermore.

I would recommend that, given all—fusion has all the positives of all the other energy sources and literally almost none of the negatives. And so, given the advances in technology, I would recommend further, additional investment by the country into this now kind of beginning-to-break-through area.

Mr. OBERNOLTE. Right. Yes, I agree with you. I think it has the potential to really revolutionize this space and solve a lot of these big, hairy, societal problems that we have been grappling with. But I see I am out of time. I would ask you another follow-up question about this, but let it suffice to just say that I think it is going to require more than just investment. We are going to have to take a look at some of the regulatory framework that we have created that is appropriate for fission but might not be appropriate for fusion. I think we are going to have to amend the Atomic Energy Act to try and create a framework that works and that really catalyzes the growth of that industry.

But I want to thank you for your testimony. Thank you to all of our witnesses. I really enjoyed the hearing today. I yield back.

Mr. CURTIS. The gentleman yields back. The witnesses have made it, and I want you to know you were worth everything that you were paid to be here today. And I am going to talk to the committee about building in bathroom breaks in the future. Thank you so much to our witnesses and to all of our Members.

Without objection, I would like to request the following documents be entered into the record for today's hearings: a letter concerning the lack of natural gas infrastructure from Industrial Energy Consumers of America; a letter concerning energy regulations and productions from the American Exploration and Production Council; a statement concerning regulations and baseload generation retirements from America's Power; a paper entitled, "Creating an Arsenal of Energy" from Forum for American Leadership; a paper entitled "Blueprint for Serious and Sound Climate Policy" from the Forum for American Leadership; a paper entitled "Eight Necessary Steps to Defend U.S. Critical Energy Infrastructure from Cyber Attacks" from the Forum for American Leadership; a paper entitled "Congress is Key to Restoring Realism in U.S. Energy Policy" from the Forum for American Leadership; a paper entitled "Setting U.S. Climate Policy Straight: Recommendations for the 118th Congress" from the Forum of American Leadership; a paper entitled "Restoring U.S. Energy Security: Recommendations for the 118th Congress" from the Forum for American Leadership—by the way, the witnesses can leave—a report entitled "Restoring America's Competitive Nuclear Energy Advantage" from the Department of Energy; a report entitled "Reducing Russian Involvement in Western Nuclear Power Markets" from Columbia University Center on Global Energy Policy; an article entitled "Russia's State Nuclear Company Aids War Effort, Leading to Calls for Sanctions" from The Washington Post; an article entitled "Ukraine War to Accelerate Shift Away from Fossil Fuels" from E&E News; a report entitled "Energy, Justice, and Climate, Change: Key Concepts for Public Health" from the American Public Health Association; and a report entitled "Improving Equity Outcomes for New Federal Investments in Clean Energy Infrastructure" from the Bipartisan Policy Committee; and finally, an article from Climatewire on China clean energy investment.

[The information appears at the conclusion of the hearing.[26]]

Mr. CURTIS. I remind Members that they have 10 business days to submit questions for the record, and I ask the witnesses to respond to questions promptly. Members should submit their questions by the close of business date, February 14th.

Without objection, the subcommittee is adjourned. [Whereupon, at 4:48 p.m., the subcommittee was adjourned.] [Material submitted for inclusion in the record follows:]

---

[26] The Department of Energy report has been retained in committee files and is available at https://docs.house.gov/meetings/IF/IF00/20230131/115356/HHRG-118-IF00-20230131-SD014.pdf.

January 30, 2023
The Honorable Cathy McMorris Rodgers
Chair

House Committee on Energy and Commerce
2125 Rayburn House Office Building
Washington, DC 20515

The Honorable Frank Pallone, Jr.
Ranking Member
House Committee on Energy and Commerce
2125 Rayburn House Office Building
Washington, DC 20515

Re: Comments for the Record on the "American Energy Expansion: Strengthening Economic, Environmental, and National Security" Hearing

Dear Chair McMorris Rodgers and Ranking Member Pallone:

Thank you for the opportunity to provide comments for the record on the "American Energy Expansion: Strengthening Economic, Environmental, and National Security" hearing.

The Industrial Energy Consumers of America is a nonpartisan association of leading manufacturing companies with $1.1 trillion in annual sales, over 12,000 facilities nationwide, and with more than 1.8 million employees worldwide. IECA membership represents a diverse set of industries including: chemicals, plastics, steel, iron ore, aluminum, paper, food processing, fertilizer, insulation, glass, industrial gases, pharmaceutical, building products, automotive, independent oil refining, and cement.

I. Use the Defense Production Act (DPA) to build needed natural gas pipeline capacity for the manufacturing sector from Alabama to New York to boost American manufacturing investments, job creation and bolster national security.

Today, there is either none or very limited firm pipeline capacity available on the east coast and the firm capacity that is available is being taken by entities that can pass those scarcity prices to customers. This has put existing manufacturing operations at risk and threatens to halt investments in expansions or new facilities. The entire eastern seaboard is supplied by one major pipeline, the Transcontinental Gas Pipeline (Transco). Transco has plans to expand pipeline capacity, but the expansions are insufficient and will start operating in 2025 at the earliest. Also, it is our understanding that these pipeline expansions will not benefit Transco Zone 5 Virginia and North Carolina consumers. Without natural gas pipeline capacity, the U.S. has neither natural gas or electric reliability, which threatens national security, supply chains, investments, and job creation.

There is no good alternative fuel to replace natural gas in manufacturing processes. With capital investments, other fuels such as coal, fuel oil, or propane can replace natural gas, but are undesirable alternatives. Equipment using natural gas cannot be switched to electricity. Plus, a BTU of natural gas energy is much less expensive than a BTU of energy from electricity.

There are major economic and job implications. The manufacturing sector is a significant contributor to the Eastern seaboard (see Figure 1). In 2021, combined manufacturing sector contributions include 3,407,700 jobs, 30,461 facilities, and $638 billion in GDP, which is roughly 10 percent of combined total state GDP. Combined state natural gas consumption is as follows: manufacturing 1,340,881 million cubic feet (MMcf), electric power 4,700,586 MMcf, and residential 1,381,179 MMcf.

Because there is no new pipeline capacity available in the near term, the only solution is to use the Defense Production Act (DPA) to intervene at the state level to require electric utilities to temporarily delay the shutdown of coal-fired power generation and the building of new natural gas fired generation. And or, require companies that are holding firm pipeline transportation to release unused capacity for market use. Companies who hold this capacity are natural gas marketers, utilities, and LNG exporters. Finally, we urge use of the DPA to build more natural gas pipeline capacity.

There are 59 coal-fired generating units along the Transco pipeline that electric utilities plan to voluntarily shutdown and replace with natural gas-fired power. Replacing these facilities with natural gas generation would significantly increase demand by 10 billion cubic feet a day (Bcf/day) and directly impact manufacturing facilities by further reducing potential available pipeline capacity.[27]

On the East Coast, every year, manufacturing companies face growing natural gas scarcity due to the lack of interstate natural gas pipeline capacity. Inadequate pipeline capacity is negatively impacting our ability to operate existing facilities. During winter and summer peak natural gas demand, manufacturers are the first to be curtailed. When this happens, manufacturers must reduce, or stop operating their facilities with a significant economic impact that could cost millions of dollars per day. Last winter, for areas like Transco Zone 5, manufacturers paid over $20 per MMBtu for their natural gas. Until additional pipeline capacity or compression is added, our only hope to avoid more serious shortfalls of supply is to temporarily delay the shutdown of coal-fired generation.

Pipelines that would have provided needed supply like the Atlantic Coast Pipeline have been stopped by activists. Mountain Valley Pipeline (MVP) may provide limited relief to the area (its primary role is to supply LNG to the Gulf Coast). However, as you know, MVP has been stopped by activists.

Manufacturers will not expand in the region without a reliable supply of energy. Manufacturers will build new facilities in other states with robust supply.

In March 2022, the Energy Information Administration (EIA) released its report, "Exploration of the No Interstate Natural Gas Pipeline Builds Case," a study that explores the implications of not building interstate pipelines that adds insight into the seriousness of the problem.[28] The following is included in the study:

We find that not building interstate natural gas pipelines affects some regions significantly more than others. We found that East North Central, Middle Atlantic[29] and South Atlantic[30] regions were the most sensitive to changes in pipeline capacity because of limitations to growth in production in the Appalachia Basin (see page 2).

---

[27] S&P Global, March 3, 2022.

[28] "Exploration of the No Interstate Natural Gas Pipeline Builds Case," https://www.eia.gov/outlooks/aeo/IIF_pipeline/pdf/AEO2022_IIF_pipelines.pdf.

[29] Pennsylvania, New Jersey, New York.

[30] Delaware, District of Columbia, Florida, Georgia, Maryland, North Carolina, South Carolina, Virginia, and West Virginia.

These states and their supply chain will miss out on reshoring of manufacturing jobs. Because natural gas prices are lower in the U.S., European companies are investing in the U.S., but not in states without robust reliable supplies of natural gas. The Wall Street Journal stated that European foreign direct investment in the U.S. increased to about $3.2 trillion last year from a year earlier.[31]

States, 2021	Natural Gas Consumption (Million cubic feet)			# Industrial Consumers	# Manufacturing Employees (Thousands)	State GDP ($Millions)	State Manufacturing GDP ($Millions)
	Industrial	*Electric Power*	*Residential*				
Alabama	232,804	383,354	32,188	3,522	265.8	209,979.3	39,631.7
Georgia	162,005	418,660	126,670	2,772	405.2	575,292.2	64,944.7
Maryland	18,720	99,760	76,874	1,138	110.0	368,571.1	26,409.6
New Jersey	60,186	222,585	232,225	6,696	243.1	566,893.2	60,496.2
New York	89,731	448,621	445,564	6,044	415.6	1,514,779.2	75,235.9
North Carolina	122,230	355,937	72,067	2,672	465.9	541,933.8	102,344.7
Pennsylvania	240,907	859,281	226,453	4,482	545.4	710,973.1	101,947.2
South Carolina	99,792	178,223	33,205	1,429	254.0	221,045.0	39,264.7
Virginia	112,372	384,734	81,261	989	235.0	505,351.0	49,498.6
Delaware	32,826	21,920	11,524	130	25.1	64,404.7	5,600.1
Florida	131,299	1,307,497	19,330	496	396.6	1,029,575.6	64,480.3
West Virginia	38,009	20,014	23,818	91	46.0	71,343.2	8,488.8
Total	1,340,881	4,700,586	1,381,179	30,461	3,408	6,380,141	638,343

Sources: Natural Gas Annual, U.S. Energy Information Administration (EIA), https://www.eia.gov/naturalgas/annual/; State Employment, U.S. Bureau of Labor Statistics, www.bls.gov; and GPD by State, U.S. Bureau of Economic Analysis, www.bea.gov.

Figure 1. Statistics by State.

Finally, we note that the U.S. imports approximately $1 trillion of manufactured goods per year. These are products that could be manufactured in the U.S., which would substantially increase jobs, GDP, and reduce global GHG emissions. U.S. manufacturing GHG emissions per unit of production are only about one-third that of countries like China. But we cannot do that without increased pipeline capacity.

Thank you for the opportunity to provide input. We look forward to working with you.

Sincerely,
Paul N. Cicio
Paul N. Cicio
President & CEO
cc: House Committee on Energy and Commerce

The Industrial Energy Consumers of America is a nonpartisan association of leading manufacturing companies with $1.1 trillion in annual sales, over 12,000 facilities nationwide, and with more than 1.8 million employees worldwide. It is an organization created to promote

---

31 "U.S.-Europe Trade Booms as Old Allies Draw Closer," November 21, 2022, https://www.wsj.com/articles/u-s-europe-trade-booms-as-old-allies-draw-closer-11668914679?mod=Searchresults_pos1&page=1.

the interests of manufacturing companies through advocacy and collaboration for which the availability, use and cost of energy, power or feedstock play a significant role in their ability to compete in domestic and world markets. IECA membership represents a diverse set of industries including: chemicals, plastics, steel, iron ore, aluminum, paper, food processing, fertilizer, insulation, glass, industrial gases, pharmaceutical, building products, automotive, brewing, independent oil refining, and cement.

AXPC | AMERICAN EXPLORATION & PRODUCTION COUNCIL

January 31, 2023

The Honorable Cathy McMorris Rogers
Chair
Committee on Energy and Commerce
US House of Representatives
2125 Rayburn House Office Building
Washington, DC 20515

The Honorable Frank Pallone
Ranking Member
Committee on Energy and Commerce
US House of Representatives
2125 Rayburn House Office Building
Washington, DC 20515

RE: American Energy Expansion: Strengthening Economic, Environmental, and National Security

Dear Chair McMorris Rogers and Ranking Member Pallone:

The American Exploration and Production Council (AXPC) thanks you for holding today's important hearing "American Energy Expansion: Strengthening Economic, Environmental, and National Security."

The production of domestic oil and natural gas provides our nation with good-paying jobs, lower-cost energy, greater energy security, and a cleaner environment. America has achieved the greatest emissions reductions in the world because of our Shale Revolution, as natural gas displaced higher-emitting energy sources – without taxpayer funding and while simultaneously providing historically low energy costs to American families and businesses.

As you know, the world continues to face a global energy crisis and the need for lower-cost, responsibly sourced, reliable energy have never been more evident. The American independent oil and gas industry has a role in meeting this challenge both domestically and internationally. In addition to being the world's largest producer of both oil and natural gas, the United States has become a top exporter of crude and the top exporter of liquefied natural gas. American energy alleviates energy poverty around the world, helping other nations achieve greater economic opportunity and prosperity.

AXPC looks forward to working with this committee on the following policies that we believe can unleash American energy production and ensure the United States remains the world leader in energy production, energy exports, and climate leadership:

- **Strengthen our nation's energy infrastructure:** To ensure we can meet growing energy demands while continuing to reduce emissions, Congress should promote infrastructure development and reform the permitting process. Streamlining the National Environmental Policy Act (NEPA) process and ending obstruction of oil and natural gas projects in the permitting process will spur project investment in energy infrastructure. The Biden Administration's interim NEPA guidance on greenhouse gas (GHG) and climate change, however, threatens the permitting of necessary energy infrastructure, including emissions reduction technologies that are critical to a lower-emissions future such as Carbon Capture, Utilization, and Storage (CCUS) and hydrogen. This guidance creates even greater delays, uncertainty, and legal vulnerability – which could make energy more expensive and less reliable for the American people. It should be repealed. Additionally, as Congress considers permitting reform proposals, AXPC has outlined specific upstream permitting reform recommendations.

- **Allow access to produce on federal lands:** Amidst the global backdrop of high energy prices and skyrocketing inflation, Congress should ensure that the Biden Administration is supporting access to produce oil and natural gas on federal lands. The US Department of the Interior should hold regular lease sales, as statutorily required by the Mineral Leasing Act. Additionally, permits and Application for Permit to Drill (APD) should be approved in a timely manner. The Biden Administration's historically low federal acreage reduction in leasing,[i] and myriad new regulations, coupled with the omnipresent risk of litigation, create serious challenges for operators working to responsibly develop federal mineral resources.

- **Protect and promote energy exports:** The energy landscape is global and US energy exports are important to our country's role in the market, global security, and our climate leadership in the world. Congress should ensure that American energy producers can continue to export crude oil, refined products, and natural gas to the global markets – which helps keep Americans' energy prices lower and more stable and maintains geopolitical stability. In addition to supporting the necessary infrastructure here at home, policymakers should promote utilization of US LNG around the world as a cleaner, more responsible choice to alternative energy sources, and support policies that will encourage building additional import capacity around the globe. AXPC's LNG Export Agenda outlines policies that will support and encourage US LNG exports.

- **Increase regulatory certainty:** AXPC member companies often exceed federal regulatory standards as part of their efforts to operate safely and efficiently and decrease emissions from upstream operations. Regulatory certainty is important to our ability to grow production and to invest and deploy new technologies that further reduce emissions from our operations. We encourage Congress to engage federal agencies, including US Environmental Protection Agency, the US Department of the Interior, the US Department of Energy, and the Council on Environmental Quality, to ensure federal regulations are technology-neutral, flexible, cost-effective, and not duplicative of other federal and state regulations.

- **Encourage capital investments in American energy production:** Our industry is committed to transparency, collaboration, and engagement on climate solutions, but that does not mean that securities laws are the appropriate context for all climate-related disclosures. Congress should urge the Securities and Exchange Commission to reconsider its burdensome climate disclosure proposal, collaborate with our industry, and allow for voluntary efforts already underway to further consistency and comparability of climate-related reporting.

- **Relieve inflationary pressure and supply chain bottlenecks:** The complex and costly process for drilling and completing a new well in the United States takes time and significant resources, even without the challenges we face currently from supply chain disruptions and labor shortages. Inflationary costs, labor shortages, and supply chain disruptions are further hindering increased production. Inflationary pressures caused a 30 percent increase in drilling and completing a well from 2021-2022, and the price is expected to be an additional 12 percent higher in 2023. The average shale well cost just $7.3mn to drill in 2019, but will cost $9mn this year, according to independent research and business analyst Rystad.[ii]

America's oil and natural gas producers look forward to working with you and your colleagues to further policies and innovative, free-market solutions to reduce Americans' energy prices, support our economy, further our energy security, and reduce global emissions.

Sincerely,

Anne Bradbury
Anne Bradbury
CEO
American Exploration and Production Council

---

[i] "Federal Oil Leases Slow to a Trickle under Biden," by: Timothy Puko & Anthony DeBarros, The Wall Street Journal (September 2022). https://www.wsj.com/articles/federal-oil-leases-slow-to-a-trickle-under-biden-11662230816

[ii] "What the end of the US Shale Revolution would mean for the world," by: Derek Brower, Financial Times (January 2023). https://www.ft.com/content/60747b3b-e6ea-47c0-938d-af515816d0f1

* * *

January 30, 2023
The Honorable Cathy McMorris Rogers, Chair
The Honorable Frank Pallone, Ranking Member
Committee on Energy & Commerce
U.S. House of Representatives
Washington DC 20515

The Honorable Jeff Duncan, Chairman
The Honorable Paul Tonko, Ranking Member
Subcommittee on Energy, Climate, and Grid Security
U.S. House of Representatives
Washington DC 20515

The Honorable Bill Johnson, Chairman
The Honorable Diana DeGette, Ranking Member
Subcommittee on Environment, Manufacturing, and Critical Minerals
U.S. House of Representatives
Washington DC 20515

Dear Chair Rogers, Chairman Johnson, Chairman Duncan, Ranking Member Pallone, Ranking Member DeGette, and Ranking Member Tonko:

We commend the committee for holding the hearing tomorrow titled "American Energy Expansion: Strengthening Economic, Environmental, and National Security." To assist the committee, I am attaching a statement for the hearing record. My brief statement focuses primarily on coal retirements, EPA regulations, and the risk both pose for electric reliability. The statement also includes recommendations. We would be pleased to provide any additional information that could be helpful to the committee or its subcommittees.

Sincerely,
Michelle Bloodworth President & CEO

Attachment: "Statement of Michelle Bloodworth"

Energy and Commerce Committee Hearing: "American Energy Expansion: Strengthening Economic, Environmental, and National Security," Statement of Michelle Bloodworth, President and CEO, America's Power, January 31, 2023

Chair McMorris Rogers, Ranking Member Pallone, and members of the committee, my name is Michelle Bloodworth. I am president and CEO of America's Power which advocates on behalf of coal-fired electricity and its supply chain. We want to commend the committee for holding this hearing today on such a critically important topic and express our appreciation for the opportunity to submit this brief statement. My statement focuses primarily on the value of the coal fleet and the massive retirements of coal power plants which have become a major risk to the reliability of the electricity grid.

All of the Above

The war in Ukraine has taught us many lessons, especially the need for countries to be energy secure and the danger to national and economic security when countries fail to rely on a healthy mix of energy resources. The U.S. needs to keep these lessons in mind and develop a resilient all-of-the-above energy strategy that promotes economic growth and energy security. Unfortunately, our energy strategy is being biased by ideological preferences and unrealistic goals rather than by rational considerations. The U.S. needs to take advantage of fossil fuels, nuclear power, hydro, wind, solar, geothermal, battery storage, demand response, and likely others in the future. The combination of all (not just a few) of these resources can provide reliable, resilient, and affordable electricity.

The coal fleet is an essential part of an all-of-the-above strategy for the following reasons:

- Dependability
  Coal plants have a high accredited capacity value (90%) that helps prevent electricity shortfalls. Accredited capacity is a measure of how dependable a resource is when electricity demand peaks, such as during extreme weather. Based on capacity values, coal is six times more dependable than wind (16.7%) and at least twice as dependable as solar (50% but will decline to 20% in the future).[32] Battery storage will eventually improve the capacity values of wind and solar, and technology innovation is one of the reasons the grid transition needs to be gradual.
- Fuel security
  The coal fleet has maintained an average on-site coal stockpile equivalent to 76 days of coal burn during the past five years. Therefore, the coal fleet does not have to rely on weather conditions (wind and sun) or just-in-time fuel delivery (natural gas) to produce electricity. Based on preliminary data, coal was able to provide almost half (47%) of the additional electricity during the height of Storm Elliott in the PJM region, 35% in MISO, and almost 40% in SPP.[33] The coal fleet's on-site fuel gave coal plants immediate access to fuel when needed.
- Reliability
  Coal provides most of the attributes that are needed for grid reliability. These include essential reliability services (frequency support, voltage control, and ramping/balancing) and fuel security. The Midcontinent Independent System Operator (MISO) has identified six reliability attributes; coal provides five of the six.
- Optionality
  Because of its relatively stable and low price, coal is also a reliable option when other electricity resources are not available or are too expensive. For example, according to EIA, average delivered monthly coal prices over the past 15 years have ranged from \$1.88 to \$2.45/MMBtu, whereas natural gas prices have ranged from \$2.04 to \$15.73/MMBtu. During that period, coal prices averaged \$2.16/MMBtu and natural gas prices averaged \$4.39/MMBtu.

---

[32] See, for example, page 6 of PowerPoint presentation "MISO System Attributes Workshop," September 21, 2022. Wind's capacity credit is 16.7%, solar declines from 50% to 20% over time, hybrid declines from 60% to 30%, and battery declines from 100% to 75%. Coal, gas and nuclear are accredited between 90% and 100%.

[33] These percentages represent the contribution of resources that provided additional electricity due to Elliott.

Coal Retirements

Fifteen years ago, the generating capacity of the nation's coal fleet was more than 300,000 megawatts (MW). For a variety of reasons, the coal fleet has shrunk to less than 200,000 MW today but still provided 22% of U.S. electricity last year. By comparison, wind and solar provided 12%. Unfortunately, utilities have announced plans to retire more than half (106,000 MW) of the remaining coal fleet by 2030.

The chart below shows announced coal retirements (in MW) nationwide for each year during 2022-2030. These announced retirements do not take into account the impact of new EPA rules (see next section). The red bars show that announced retirements total 37,400 MW during 2026-2028. This three-year period is highlighted because we estimate that coal retirements will increase considerably during that time, given EPA's likely compliance schedule for six rules.

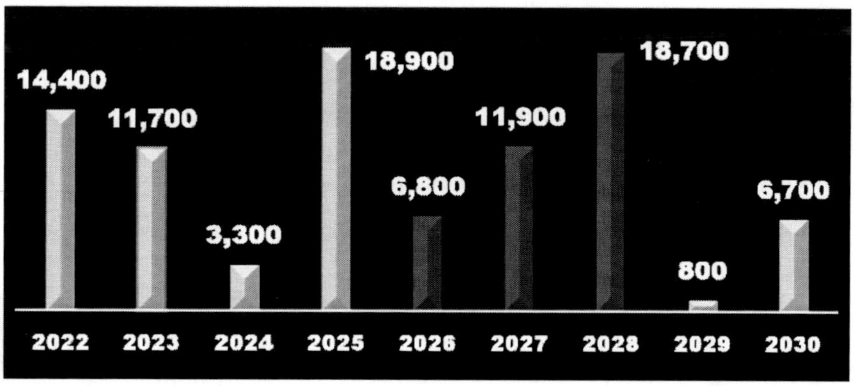

Coal retirements on such a massive scale and over a relatively short period of time will increase reliability risks significantly by depriving the grid of the reliability attributes of coal-fired generation.

EPA Rules

We estimate that at least six EPA rules will cause coal retirements to rise sharply during 2026-2028 and exacerbate the risk of grid reliability problems. The number of coal retirements will depend on the stringency of the rules.

For example, EPA estimates that its proposed Ozone Transport Rule (aka "Good Neighbor Rule") will cause the retirement of 23,000 MW of coal, more than 10% of the existing coal fleet, by 2025.[34] (Certain of these rules could also cause the premature retirement of gas- and oil-fired generation.) The other five are the Coal Combustion Residuals (CCR) Rule, Effluent Limitations Guidelines (ELG), Regional Haze Rule, a replacement for the Affordable Clean Energy Rule (carbon regulation), and revised Mercury and Air Toxics Standards (MATS).

---

[34] See Table 4-14, *Regulatory Impact Analysis for Proposed Federal Implementation Plan Addressing Regional Ozone Transport for the 2015 Ozone National Ambient Air Quality Standard*, EPA-452/D-22-001, February 2022. https://www3.epa.gov/ttn/ecas/docs/ria/transport_ria_final-csapr_2011-06.pdf.

EPA could use the Transport Rule, Regional Haze Rule, and a more stringent MATS to require the installation of the most expensive emission controls on most, if not all, coal capacity that does not already have advanced controls. We estimate that 92,000 MW of coal-fired generation, even though already well-controlled, lack the most expensive controls (e.g., selective catalytic reduction and/or flue gas desulfurization (FGD) systems). For example, an FGD system for a large coal unit can easily cost more than $300 million. It is almost certain that a significant number of coal plants would retire early rather than install these expensive controls. The retirement of coal due to EPA rules would add to the 106,000 MW of retirements already announced. Note that this at-risk coal generation does not include retirements (or idling) that are likely to result from the CCR or ELG rules.

Recommendations

We urge Congress to take steps, including the following, to promote grid reliability and minimize coal retirements:

- We commend the committee for holding this important hearing today and encourage the committee to continue its oversight by holding agencies accountable for identifying and properly addressing the many challenges facing the electric sector. One of these challenges is the impact of EPA rules on energy prices and grid reliability. Congress should ensure that EPA pays careful attention to the concerns of the Federal Energy Regulatory Commission (FERC), North American Electric Reliability Corporation (NERC), grid operators, utilities, and utility commissioners and design its rules so as to avoid causing reliability problems and minimize retirements of coal and other dispatchable resources. If EPA is willing, the agency could choose regulatory options that cause the fewest retirements; defer to states about how to implement certain rules; make regulations as flexible as possible, not prescriptive; and provide adequate time for retiring generating capacity to be replaced.
- Congress should pass legislation directing FERC to work with grid operators to identify specific attributes that are necessary for reliability and ensure that market rules provide just and reasonable compensation for those attributes. MISO, for example, has suggested a number of reliability attributes but, at the same time, indicated that "MISO, and the industry as a whole, are still defining attributes."[35] Legislation should include a deadline for attributes to be identified and a deadline for FERC to take action on changes to market rules to compensate those attributes.
- Congress should pass legislation requiring federal agencies to conduct formal reliability assessments for rules and policies that could adversely impact grid reliability. The reliability assessments should explain all risks to grid reliability that could result from the rules or policies and describe steps an agency has taken to avoid causing reliability problems. If a rule is projected to cause the retirement of dispatchable resources (coal, natural gas, or nuclear power), the reliability assessment should show that adequate accredited replacement capacity will be placed in service by the time retirements are projected to occur.

---

[35] "Mind the Gap," OMS Resource Adequacy Summit, August 8, 2022.

In closing, we appreciate the opportunity to submit these comments for the record and would be pleased to provide any additional information that might assist the committee.

Creating an Arsenal of Energy: How to Bolster U.S. Energy Security and Aid our Allies Confronting Authoritarian Aggression, April 1, 2022

Russia's invasion of Ukraine has roiled global energy markets, spawning oil and gas price spikes that threaten economic growth and geopolitical stability. Moscow's war of choice may instigate the most severe energy crisis since the 1970s.

If crises have silver linings, they force a hard re-examination of core national security and energy policies. The Biden administration and EU member states, mugged this winter by tight markets, soaring consumer prices, and President Putin's aggression, are beginning to rediscover the benefits of energy security, especially ample oil and gas production capacity and infrastructure.

To assist the White House and Congress in their newfound interest in bolstering domestic and international energy security, we suggest the following long-overdue policy changes for Congress and the Executive Branch that would increase the supply and transmission of all forms of energy. These recommendations would enable the United States to leverage its vast energy resources and technological prowess to protect our economy and become an Arsenal of Energy for allies threatened with authoritarian aggression.

Energy Recommendations

- Until the enactment of a serious and sound national and international climate policy, suspend executive branch regulatory initiatives, including by independent federal agencies, pertaining to greenhouse gas emissions emanating from the exploration, production, transmission, refining, and trade in all forms of energy.
- Abolish the requirement for a national interest permit for any cross-border energy infrastructure projects unless the President finds that it would gravely imperil the national security of the United States.
- Remove executive branch restrictions/bans on U.S. government support, technical assistance (such as at the Departments of Energy, Commerce, Treasury, and State), and financing (U.S. International Development Finance Corporation and Export-Import Bank) for fossil fuel projects and related infrastructure to enable the U.S. Government's ability to support energy security and development of partners and allies.
- Modernize and improve the leasing and management of the federal estate to ensure that taxpayers realize the full and complete benefit of the resource base of the United States.
- Amend the National Environmental Policy Act to enforce timelines and provide expedited permitting for critical national security energy projects.

Oil Recommendations

- Repeal non-emergency drawdowns from the Strategic Petroleum Reserve (SPR) and direct the Department of Energy to expand the SPR from its current 568 million barrels to one billion barrels and fill it.
- Direct the Department of State to ensure the International Energy Agency (IEA) will:
  - Correct the misunderstanding the IAE has created in the media that the IEA "calls for" a ban on new upstream oil and gas investment and clarify the agency does not oppose new investment.
  - Resume including a Current Policies Scenario in its forecasts to enable political leaders to evaluate the costs and benefits of energy security and climate policies.
  - Make all data and models available to the public at no charge.
- Approve all necessary permits for the Keystone XL pipeline.
- Reverse all Biden Administration policies restricting oil and gas development in Alaska, a critical region where Russia already excels, and establish a powerful U.S. Arctic presence founded on energy projects.
- Congress should reject the President's budget proposal to increase taxes on the oil and gas producers by $40 billion, which would reduce investment in supply and drive-up costs for consumers.

Natural Gas and LNG Recommendations

- Deem every LNG liquefaction project currently before the Department of Energy to be approved.
- Congress should amend the Natural Gas Act of 1938 to eliminate the public interest determination for LNG exports to non-Free Trade Agreement countries.
- Consider loan guarantees, co-funded with European countries, to expedite the construction of LNG liquefaction and regasification facilities.
- Direct the IEA to bolster the transparency of natural gas markets, especially in Europe.
- Require international lending institutions receiving federal government funding to prioritize natural gas infrastructure, including trade-related terminals, ships, and barges, decentralized production (e.g., Floating LNG), processing facilities, and pipelines.
- Facilitate government and the unlock of private sector financing to enable the U.S. Government's ability to support the energy security and development of partners and allies
  - Remove/overturn executive branch restrictions/bans on U.S. government technical & financing assistance for fossil fuel projects (including at the Departments of Energy, Commerce, Treasury, and State).
  - Create/reinstate programs financing tools and mechanisms to support fossil and infrastructure projects abroad (e.g., LNG regasification) via loan guarantees, risk insurance, and other financing supports (at the Department of State, U.S. International Development Finance Corporation, and Export-Import Bank).

Critical Minerals Recommendations

- Congress should legislate a streamlined permitting process for mineral extraction and processing in the United States to bolster our competitiveness against established producers in China.
- Congress should provide funds for bolstering critical mineral security through the Defense Production Act of 1950, a federal statute passed during the Korean War that gives the president broad authority to influence the domestic industry in the interest of national defense.
- Congressional leadership should make critical minerals legislation a higher priority. A bill introduced in January 2022 by Sen. Mark Kelly and Sen. Tom Cotton would establish a strategic rare earths reserve and restrict the use of rare earth elements from China in advanced defense technology in the United States.
    - Congress should enact legislation that would force defense contractors to stop buying rare earth-enabled products from China by 2026 and use the Pentagon's Defense Logistics Agency, to create a permanent stockpile of rare earth minerals. The U.S. has only one producing rare earths mine at Mountain Pass, California, run by MP Materials, and cannot currently process rare earth minerals downstream of the mine.
- Congress should consider establishing tax credits, modeled after wind and solar credits, for the domestic production, processing, and recycling of critical minerals.
- Prioritize domestic development and direct any foreign assistance for clean energy mining towards friendly and stable sources of supply, particularly U.S. allies such as Australia and Canada and partners in Latin America.
- Congress should mandate an update to the 2019 Critical Minerals Strategy, issued by the Department of Commerce, and require subsequent updates every two years.
- Establish a "one-stop-shop" consolidating and expediting the processes across the Department of the Interior, Department of Energy, and USDA for permitting, grants, loans, and other incentives to produce, process, and recycle critical minerals.
- Create an Export-Import Bank program that supports exportation of U.S.-origin rare earth mining equipment and services and debt and risk insurance for U.S. rare earth investments abroad.

Blueprint for a Serious and Sound Climate Policy, April 18, 2022

Maintaining U.S. living standards and national security depends on the reliable, ample, and affordable flow and conversion of hydrocarbon energy sources – oil, gas, and coal – that supply 80% of the world's growing energy supply. Hydrocarbons lifted humanity out of millennia of darkness and squalor and remain indispensable for all aspects of our daily life, including our domestic food and water supplies, our mobility, national security, and economic growth.

While massively beneficial, hydrocarbons also generate negative externalities, greenhouse gas emissions among them. The climate is warming and human activities, principally accumulating carbon dioxide from hydrocarbon combustion, exert a physically small but growing effect upon it. While the science is far from settled regarding how the climate will change under human influences and the net economic and environmental impacts of those changes, the issue requires a serious and sound policy response.

The prevailing narrative held by the current administration and many activists is ineffective, unscientific, and endangers America's economic growth and national security, as well as the environment it claims to protect. Moreover, it is damningly silent on how to provide reliable and affordable energy to the six billion people who need more energy to improve their lot.

This narrative starts by misrepresenting the science to alarm the public and then demands the expunging hydrocarbon energy from our energy options. The Green New Deal demands the solution of a permanent "wartime mobilization" on the nation and world. Its radical policy prescriptions include extensive government central planning, rationing of energy resources, individual compulsion, and punitive taxation. Their climate narrative defies physics, economic and political realities, human behavior and expectations, and rationality. It will therefore fail, as it should.

The United States and rest of the world need a new and improved blueprint for a sound and serious climate policy. Sound strategy must leverage practical but serious policies to address the real risks posed by human impacts on the climate while protecting free market capitalism, a healthy economy and environment, and national security and providing adequate energy to those who need it. Our blueprint for a sound and serious climate policy has three pillars:

1) Depoliticize and honestly explain to non-experts both the scientific certainties and uncertainties, as well as technical and-economic realities;
2) Apply sound economic and cost-benefit principles to all strategies and fuels; and
3) Legislate at home and lead abroad.

Depoliticize and Accurately Represent the Science and Technologies

The foundation of any sound and serious climate policy must be complete, transparent, and unbiased descriptions for non-experts of the scientific understanding of climate and human effects upon it. Such honest summaries are critical to assess the costs and benefits of policy options in the context of certainties and uncertainties in the science.

Climate science is complex, and most citizens and leaders are not climate scientists. The public's current understanding of "The Science" is distorted by the media and politicians who demand an almost religious conformity with their narrow viewpoint. They brand anyone – climate experts and non-experts alike - who objects to their alarmist, unscientific decrees to be a "denier." This approach is extremely unhelpful and has backfired, particularly among conservative and independent audiences.

Given the rampant bias and intimidation, it is not surprising that the popular perception of what the science says is quite different from what the actual science says. There is scientific consensus that the climate is warming. And human activities contribute a physically small, but growing, warming influence on the climate, principally by consuming hydrocarbon energy.

However, as the veteran climate scientist and President Obama's Undersecretary for Science at the Department of Energy Steven Koonin (among others) has noted, the science is far from settled about past human contributions to climate and is incapable of producing useful forecasts of future warming, much less human influences upon it. Historical data are often poor, natural influences on the environment are substantial, and the complexity of the physical and biological foundations of the Earth's climate is immense. Modeling only scratches the surface.

The problem is that while the scientific research is typically transparent, rigorous, and objective, the government summaries that inform non-experts in the media, government, and citizenry are not. The U.S. government should subject summaries of the science to the same objective, rigorous peer-review that the actual science enjoys.

FAL's Recommendations for Congress

- Sustain and enhance funding for scientific observations of the earth's climate system.
- Require that authors of Summaries for Policy Makers (SPMs) of UN IPCC reports be selected by independent, non-government experts and scientists without conflicts of interests.
- Require settlement of differences arising in the peer-review of assessment reports by an independent referee, as is the case for research papers, instead of allowing the assessment report's authors to discard criticisms without explanation, as is the case now.
- Require UN and U.S. assessment reports to undergo a formal review by a group of independent climate experts tasked with challenging the assumptions, conclusions, and presentation, probing for weak spots, distortions, and exaggerations. Require report authors to rebut any points raised.

Apply Sound Principles and Leverage All Strategies and Fuels

Consider All Climate Strategies

- There are three general climate strategies: (1) mitigating or reducing emissions; (2) geoengineering or enhancing the earth's reflectiveness and removing carbon from the atmosphere; and (3) adaption to live, if not thrive, within a future climate. A sound strategy would consider the cost-effectiveness and national security, economic and environmental impacts of each approach. Congress should require the Biden administration to provide a thorough and objective economic, national security, and environmental cost-benefit analysis of mitigation, geoengineering, and adaption.
- Mitigation strategies should include an emphasis on promoting technological innovation and market-based policies instead of imposing mandates, taxes, and restrictions on consumers and businesses.

Cost-Benefit Analyses

Policy options stemming from mitigation, geoengineering, and adaption strategies should be evaluated and debated based on their costs and benefits for climate, the economy, human wellbeing, and national security. Accurate information and analyses from existing bodies like the Energy Information Administration (EIA) and the International Energy Agency (IEA) will be a cornerstone of this effort. Policymakers need to start with reliable data and information to make the best policy determinations.

Recommendations for Congress and the Executive Branch:

- Direct the IEA to return to conducting analyses that enable cost-benefit analysis. Specifically, the IEA must resume business-as-usual or Current Policies Scenarios that assume only existing policies in its long-term energy forecasts. The IEA should publish its underlying datasets and make all its analysis, except for mission-sensitive energy security information, available for free public consumption and outside inspection.
- Require the IEA and EIA to conduct energy market "stress tests." For example, analyze the impacts on energy prices if hydrocarbon supply remains constrained but demand trends upwards in the coming decade.
- Require UN and U.S. official analyses and their summaries to consider both the positive and negative economic impacts of various climate projections, including costs.

Rapidly abandoning hydrocarbons without replacements would lead to energy price spikes and shortages, enriching incumbent and authoritarian oil and gas producers like Russia, Venezuela, and Iran. Governments that increased energy prices triggered social upheaval including France, Chile, Brazil, Mexico, Iran, and most recently, Kazakhstan.

- Require the Department of Defense and Department of State to analyze the impacts on U.S. national security of sharp increases in oil and gas prices later this decade.
- President Biden's signature climate policy goal is to eliminate the use of natural gas and coal in the power sector by 2035, even though natural gas and coal currently comprise 40% and 20% respectively, of U.S. electricity generation. Require the EIA to analyze economic, energy security, and emissions impacts of President Biden's proposals to reduce carbon emissions.
- While hundreds of billions of dollars in capital has responded to climate change concerns by investing in a precipitous energy transition, it is by no means clear this capital is backstopped by durable political will to impose necessary transition costs. There is much discussion in energy circles of potentially stranded hydrocarbon assets if there is a successful rapid transition to renewables and minerals, but little examination of the potential stranded assets arising from a *failed* transition.
- Require the Treasury Department and the Federal Reserve to analyze the economic and financial impacts of a successful *and unsuccessful* transition this decade. Foster closer collaboration with the financial community to educate it on realistic scenarios and emissions reductions across industry to protect against dislocation of capital

Prevent Dependence on Chinese Controlled Critical Minerals

Officials, investors, and companies are moving aggressively to shift away from petroleum and the internal combustion engine to electric vehicles and batteries. While the commercial viability of this plan remains to be seen, its ultimate success depends on voluntary and unsubsidized mass adoption of electric vehicles and the buildout of the electric grid and electric vehicle charging infrastructure. Should vehicle electrification accelerate, U.S. and global dependence on OPEC+ oil reserve holders will increasingly shift to China, which already dominates the electric vehicle supply chain. U.S. policy must ensure neither China nor or any other power can dominate core global energy systems, including transportation.

FAL's Recommendations for Congress (also included in FAL's April 1, 2022 report "Creating an Arsenal of Energy"):

- President Biden invoked the Defense Production Act of 1950, a federal statute passed during the Korean War that gives the president broad authorities to influence domestic industry in the interest of national defense, to bolster critical mineral security. Congress should provide additional funds if needed to support the necessary surveys, site expansions, and new initiatives.
- Congressional leadership should make critical minerals legislation a higher priority. A bill introduced in January by Rep. Mike Kelly and Sen. Tom Cotton would establish a strategic rare earths reserve and restrict the use of rare earth elements from China in advanced defense technology in the US. This proposal and others could significantly raise the profile of critical mineral supply.
- Enact legislation that would force defense contractors to stop buying rare earth-enabled products from China by 2026 and use the Pentagon's Defense Logistics Agency, to create a permanent stockpile of rare earth minerals. The United States has only one producing rare earths mine at Mountain Pass, California, run by MP Materials, and currently has no capability to process rare earth minerals downstream of the mine.
- Smooth the path for companies to open new mining production and refining facilities. Companies must overcome numerous permitting hurdles and sparse sources of financing for upstream projects, leaving them less competitive against established producers in China.
- Congress should consider establishing tax credits, modeled after wind and solar credits, for the domestic production, processing, and recycling of critical minerals.
- Prioritize domestic development and direct any foreign assistance for clean energy mining towards friendly and stable sources of supply, particularly U.S. allies such as Australia and Canada and partners in Latin America.
- Ease and enable critical minerals recycling, recognizing that domestic mining will also be required to meet future supply.
- Mandate an update to the 2019 Critical Minerals Strategy and require subsequent updates every two years.
- Create streamlined permitting process for mineral extraction and processing in the United States. A major reason for our lack of mineral processing facilities is the difficult, costly, and time-consuming process involved.

Foster Innovation and Harness Free-Market Enterprise

Given its poor historical track record, governments should not be picking winners in the economy, especially in energy. Any policy responses should be fuel- and technology-neutral and account for market forces.

- Congress should repeal tax benefits and outlays benefitting mature but uneconomic technologies and redirect any futures taxes benefits and outlays toward activities the private sector may bypass, such as basic science and strategic endeavors, including

work on minerals dependence and nuclear energy, with the potential for disproportionate security benefits.
- Should fuel switching policies be required, all energy sources should be on the table including nuclear, natural gas, fusion, hydrogen, and advanced biofuels.

Legislate on Climate at Home and Lead Abroad

Congress Should Enact Durable and Effective Climate Policies

Congress should lead the debate and enact domestic and international climate policy. The current approach led by the executive and judicial branches yields only transient, legally vulnerable, and easily reversible domestic and foreign policies.

- Congress should enact legislation establishing a nonpartisan, expert national commission that integrates technology development, economics, and the strategies, principles, and policy options noted above to develop coherent options for Congress to debate and enact.
- The Senate should ratify any international energy or environmental agreements in which the President has joined.

Lead the World toward a Better Climate Strategy

Emissions policies are inherently global and directly impact trade, energy security, national security, development, human rights, and the environment. World governments and investors are moving ahead with climate policies and seismic investments, with or without the United States. The United States must engage and lead internationally on this issue.

The text of the 2015 non-binding Paris Agreement includes unwarranted alarmism and unrealistic targets. The United States should push for changes that would strengthen the agreement by depoliticizing science, embracing all strategies and fuels, leveraging sound principles, and clearly messaging that the purpose of the agreement is not to establish transfer payments from wealthy countries to poorer ones.

- The improved Paris Agreement should be submitted to the U.S. Senate for ratification, ensuring that U.S. climate policy enjoys a strong and durable political and legal foundation.
- The United States should withdraw from the International Renewable Energy Agency. It serves no purpose that is not already fulfilled by the IEA.
- The United States negotiators should insist that China and other major countries similarly enact legally binding, verifiable policies to backstop their international commitments. The United States must develop both cooperative and non-cooperative methods to verify other countries' emissions reductions.

Eight Necessary Steps to Defend U.S. Critical Energy Infrastructure from Cyber Attacks, October 4, 2021

A prolonged disruption in energy flows caused by foreign cyberattackers could quickly inflict catastrophic harm to American lives, health, and national security. The May 7, 2021, Colonial Pipeline cyberattack highlighted the importance of engaging in strategic deterrence against future, potentially catastrophic, attacks on our critical energy infrastructure and exposed significant national security gaps that require timely legislative and executive branch remedies. Congress must work with the Executive Branch to take robust steps to deter and punish cyberattacks on critical energy infrastructure while preparing the country to manage future attacks better than it did in May. Actions to date have fallen far short.

Recommendations for Congress to Restore Deterrence and Bolster Our Domestic Defense

1) Toughen penalties and sanctions for foreign cyber attackers who target critical energy and other vital U.S. infrastructure.
2) The United States should deter, preempt, and punish foreign cyber attackers targeting U.S. critical energy infrastructure as it would Al Qaeda, ISIS, or any other similar foreign-based terrorist planning or using WMD to inflict catastrophic harm to the homeland and Congress should provide additional clear authority, funding, and direction to the Executive Branch to do so.
3) Require the President to notify Congress of countries that support cyber attackers who have, are, or are likely to plan or execute cyberattacks against critical energy infrastructure.
4) To prevent a loss of deterrence, Congress should first require the President to immediately formulate and implement a more robust and cyber defense and offense strategy. Elements of such a strategy should include bolstering our active defense, persistent engagement between the executive and legislative branches and with our allies and defend-forward efforts. Likewise, it should provide clear authority, funding and direction for activities that go beyond such efforts in the case of further attacks on critical energy infrastructure. In addition, the President should be required to inform Congress of the Administration's implementation progress no later than six months after completion, and upon submission, Congress should allocate additional funds to ensure swift and safe implementation of the new strategy.
5) Declare it shall be the policy of the United States to regard any future attempts to disrupt or dismantle U.S. critical energy infrastructure by cyber attackers an act of aggression that shall warrant swift and commensurate retaliation against the attackers and any foreign governments deemed to sponsor them.
6) The federal government should spend more money on human capital and training for public-private cybersecurity programs, which will improve the government's capacity to help companies that are managing critical energy infrastructure assets. Specifically, federal agency representatives should embed with the nation's most essential infrastructure nodes to facilitate intelligence and real-time information sharing on

industrial security threats (including attacker methodologies) and defensive countermeasures, as well as to create the ability for the government and industry to collaborate in real-time prior to, during, and in the aftermath of a potential attack.

7) Require critical energy infrastructure owner-operators to immediately inform the federal government of major cyber or any other type of attacks that could impact domestic supply. TSA has mandated reporting for pipelines, but Congress should provide incentives—like liability and regulatory protection to encourage robust and rapid reporting. Reporting mandates should protect the identity of reporting organizations and provide liability and regulatory protection.
8) Require the owner-operator of a critical energy infrastructure asset to consult and obtain the permission of the appropriate federal authority before taking any discretionary action, including the prolonged shutdown of energy flows, that could threaten the economy or national security. Provide an exception in cases when operators do not have sufficient time to consult with federal officials, i.e., to prevent a chain reaction, leak, staving off an ongoing attack, or the like. Normally, the appropriate initial response to a detected cyberattack is to immediately shut down the asset to contain and assess damage. Normally, the federal government should not dictate how private companies operate. However, in the case of a foreign attacks on vital energy infrastructure that could quickly inflict catastrophic damage to the homeland, the federal government must have the final say about whether to implement any prolonged, discretionary shutdown of critical energy flows.

Analysis and Further Information

The Russian-based attack on the Colonial Pipeline differed critically from the thousands of prior cyberattacks on U.S. persons, businesses, and government agencies. For the first time, foreign attackers directly, if temporarily, disrupted physical energy flows vital for the societal functioning and national security of the United States. The pipeline supplies 45-50% of East Coast liquid fuel supplies, 90 military bases and installations, and seven major airports. The Colonial Pipeline attack resulted in by far the biggest loss of domestic energy supplies due to hostile foreign action against the U.S. homeland.

The most dangerous consequence of the Colonial Pipeline attack, if it is not met with an appropriate response, could be the loss of strategic deterrence. If adversaries believe they can attack critical energy infrastructure without significant cost, further and more extensive attacks are likely. Because energy storage is limited, a prolonged disruption in electricity and liquid fuels would quickly cascade into other critical sectors, inflicting catastrophic harm on American health, public safety, and national security. The Cybersecurity and Infrastructure Security Agency noted in November 2019:

> Energy stakeholders provide essential power and fuels to stakeholders in the communication, transportation, and water sectors, and, in return, the energy sector relies on them for fuel delivery (transportation), electricity generation (water for production and cooling), as well as control and operation of infrastructure (communication).

The Colonial Pipeline attack requires that national leaders review and revise how we protect the homeland from the mounting and dangerous threat posed by foreign cyber attackers. The Constitution assigns responsibility for protecting the homeland from foreign attack to the President as commander-in-chief. At the same time, Congress has a role to play through its constitutionally assigned roles as the body responsible for raising and supporting military forces and authorizing war. Cyberattacks are a potentially lethal and increasingly common means for foreign enemies to attack the United States. A weak or incomplete response to the Colonial Pipeline attack, and failure to rectify the weaknesses and vulnerabilities it exposed, would invite further attacks with potentially devastating consequences. The President has so far not demonstrated that Russia – the state that harbored the Colonial Pipeline attackers – will pay any substantive price nor taken any public action to deter future attacks other than public statements and a conversation with Russia's leadership that doesn't appear to have had any substantive impact.

While investigation is still ongoing, public reports and five congressional hearings have already revealed the attack exposed important challenges in how with address cyberattacks targeting critical energy infrastructure. For example, the federal government was neither informed nor consulted about Colonial Pipeline's decision to pay the ransom, despite FBI recommendations against victims doing so, nor was it consulted beforehand about Colonial Pipeline's decision to protect the pipeline systems by shutting it down temporarily.

There are legitimate concerns about what role the federal government should play in directing how a private company operates its assets, including when to shut down or pay ransom in response to a cyberattack. However, in the case of foreign cyberattacks *on critical energy infrastructure*, the federal government's responsibility to protect the homeland suggests that, at a minimum, the government ought to be consulted when such major actions are taken with respect to such systems, particularly given the enormous potential harm a prolonged energy outage poses to American lives, health, and national security.

Since May, Congress and the Biden administration have taken some reasonable, if only preliminary, steps. They include:

- On May 12, President Biden issued an Executive Order aimed at improving cybersecurity and federal government networks.
- On May 27, DHS issued a security directive that required critical pipeline owners and operators to report confirmed and potential cybersecurity incidents to the federal government.
- On July 20, a second DHS security directive required further steps by hazardous liquids and natural gas pipeline companies.
- On July 28, President Biden signed a National Security Memorandum directing various vanilla interagency actions.
- Several bills working their way through the House and the Senate aim to improve cybersecurity after the Colonial Pipeline attack. They would increase requirements for private companies to report on cybersecurity incidents and provide funding for states and local governments to increase cybersecurity measures.

These measures fall far short of addressing some of the challenges noted above. Bolstering our defenses against catastrophic attacks on critical energy infrastructure requires clear and robust legislative changes that establish clear thresholds for the U.S. response.

Congress is Key to Restoring Realism in U.S. Energy Policy, September 28, 2022

Among its many destabilizing and disastrous effects, Russia's war in Ukraine serves as a wake-up call for the need for realism in U.S. and allied energy policy. For too long, those driving the conversation have embraced wishful but misinformed thinking about the pace of energy transitions.

Hydrocarbon energy—oil, gas, and coal—supply 83% of the world's energy and are the lifeblood of modern civilization. While renewable energy's 6% share will continue to grow, energy transitions take many decades and are driven by technological innovation and the private sector, not by government central planning. The current energy crisis highlights the geopolitical infeasibility and economic costs of policies that distort markets and would abruptly starve the United States of hydrocarbon energy that is critical for national security and economic growth.

Those advocating for unrealistic energy transitions have so far not been held accountable by having official energy agencies analyze their unrealistic, extreme proposals. Specifically, the International Energy Agency (IEA) has recently skewed its forecasts to please climate extremists while depriving policy makers of the ability to evaluate costs and benefits of energy and climate proposals. And the U.S. Energy Information Agency (EIA) has failed to provide Congress with feasibility and cost-benefit analyses of President Biden's executive orders and legislative proposals that would mandate abrupt, massive bans on hydrocarbon energy.

To help right the ship and avoid costly and dangerous policy errors, Congress should reassert its role in setting U.S. energy policy. As a starting point, Congress should insist that taxpayer-funded agencies provide unbiased forecasts of energy markets, as well as objective evaluations of proposed energy and climate policies. Toward this end, the Forum for American Leadership Energy Working Group calls for more congressional oversight of the International Energy Agency (IEA) and U.S. Energy Information Administration (EIA) and recommends the establishment of a National Commission on Energy Transition Realism, an expert, non-partisan commission of renowned energy experts to advise government officials and evaluate policy options for energy transitions.

The IEA Must Return to its Security Mission and EIA Must Analyze Extreme Climate Proposals

The Paris-based International Energy Agency is composed of 31 member countries (the U.S. and primarily European nations) and nearly a dozen association countries, including China, India, Brazil, and Argentina. Established in the wake of the 1973 oil crisis, IEA is a forum under the Organization for Economic Co-operation and Development with a mandate to respond to disruptions in the global oil supply and provide policy recommendations, as well as data analysis, on global oil and energy supplies to bolster global energy security. In recent years, however, the IEA has strayed from its assigned role as a watchdog for energy security and instead has transformed into a lap dog for climate zealots advocating for unrealistic energy transition targets. Congress should steer U.S. policy to walk IEA back to its security mission.

Specifically, in recent years the IEA's analyses have veered from security and unbiased analysis to feeding the newly fashionable myth that the United States, and the world, can afford to immediately ban investment in fuels that compromise over 80% of global energy. Moreover, IEA began skewing its energy forecasts to hide the costs of extreme climate policies while depriving elected officials of the ability to make informed cost-benefit assessments of energy and climate proposals. For example, the IEA 2020 World Energy Outlook abolished its "business-as-usual" (BAU, formally known as Current Policies Scenario) reference case scenario, an unwarranted break with decades of forecasting convention that makes it impossible to evaluate the costs and benefits of climate proposals.

Meanwhile, the non-partisan EIA is being underutilized as an agency that was created to deliver accurate energy data and transparent, objective forecasts and analyses to policymakers and the American people. For example, the EIA has so far not provided Congress with analysis of the energy security and economic implications of recent presidential executive orders and legislative climate proposals. The list includes President Biden's executive order calling for the federal government to reduce its emissions by 65% by 2030 and reach net-zero emissions by 2050, as well as a bill (CLEAN Future Act) introduced by Democrats, mirroring the President's campaign proposal, to ban the use of fossil fuels—which currently account for 60% of U.S. electricity generation—in U.S. power plants by 2035.

We recommend Congress seize an opportunity to use hearings, letters, legislation, and other oversight tools to restore realism in U.S. energy policy by focusing on U.S. policy at the IEA and the EIA. Furthermore, the creation of an expert, non-partisan National Commission on Energy Transition Realism will further enable Congress to access fact-based data, analysis, and counsel from energy sector experts.

Congressional Oversight Recommendations

Pressure for a Return to Unbiased Forecasting at the IEA

There is a dire need for unbiased global energy forecasting. To please climate change advocates eager to reduce hydrocarbon investment, starting in 2020, the IEA cancelled including a "business as usual" (BAU) reference case scenario in its long-term forecasts and instead only included scenarios that assume quick and easy peak demand. For decades, forecasters used a BAU reference case because it has proven historically accurate, and it enables policymakers to evaluate, compare, and contrast energy and climate policy proposals. The IEA provided no analysis or justification for this radical shift in forecasting convention. Congress should pressure the Biden administration to use the voice and vote of the United States to return the IEA to resume the long-held forecasting convention of including a BAU reference case among its annual forecast scenarios.

Insist the IEA Retract Its Call to Ban Oil and Gas Investment to Reach an Impractical Net Zero Target by 2050

The IEA's calls for the immediate termination of all new investments in oil and gas run counter to U.S. economic, energy security, and geopolitical interests. U.S. diplomats should push IEA to retract these statements and clarify that the agency does not oppose new investment in oil and gas, which remains critical for the foreseeable future to meet demand and ensure secure, affordable energy for U.S. citizens and the world. The IEA must produce an unbiased, objective assessment of the costs and benefits of achieving net zero emissions by 2050 through government taxes, regulations, and subsidies, using BAU as a reference case scenario.

To date, the Biden administration has been complicit in the IEA policy shift from its security mission to climate policy advocacy that is both unrealistic and threatening to U.S. energy security. Following a March 2022 IEA Ministerial Meeting, chaired by U.S. Energy Secretary Jennifer Granholm, the organization issued a press release announcing that the meeting marked "the launch of a new phase of the Agency" in which "the IEA has a new guiding principle: supporting countries in the global effort to attain net zero greenhouse gas emissions in the energy sector by mid-century." The vow to change the Agency's focus to the energy transition and global fight against climate change was also expressed in a joint communiqué.

The IEA must return to focusing on its original objectives and obligations under the International Energy Program treaty. If the IEA does not do so, then the United States should evaluate whether continued membership in the IEA as currently structured is consistent with long-standing U.S. domestic energy policy priorities. If not, Congress should shrink U.S. participation in the IEA to purely energy security matters such as coordinating the use of strategic petroleum reserves, its original task.

Instruct the IEA to Bolster the Transparency of Natural Gas Markets

As Putin's war in Ukraine enters its sixth month and forecasts show the run-up to a grim winter for European energy supply and security, the time is ripe for the IEA to begin publishing transparent data on natural gas markets and trade flow, especially in Europe. This will exhibit both the effects of European divestment from fossil fuels and the extent of its reliance on imports from undependable authoritarian regimes.

Require the EIA to Analyze Biden Climate Proposals as a Condition for Congressional Funding

Congress should insist the EIA do its job by providing timely, astute, and unbiased analyses of the energy and economic consequences of policies proposed or implemented by the President and members of Congress. The EIA has yet to release a cost-benefit and feasibility study of President Biden's executive order requiring the federal government to achieve net-zero carbon emissions by 2050. President Biden has also called for Congress to ban the use of natural gas and coal in U.S. electricity generation by 2035, despite those fuels accounting for 60% of electricity generation. Congress must have the means to analyze the practicality, costs, and benefits of such ambitious policies. To this end, Congress should tie funding for the EIA to its timely issuance of transparent, unbiased modeling of major energy and climate proposals, including analyses of their impact on the U.S. economy and energy security. The American people deserve to know how new energy proposals will affect their daily lives and security, and right now, the EIA is being underutilized and not delivering on its intended and critical mission.

If the EIA refuses to perform its duty to provide timely and objective analyses of congressional and administration policy proposals, Congress should explore leveraging other statistical and analytical possibilities, such as the General Accounting Office, or recreating an entity like the Office of Technology Assessments.

Establish a National Commission on Energy Transition Realism

The Commission should be composed of independent energy experts to advise Congress and the executive branch on the physical, economic, and technological realities involved with energy transitions. Composed of reputable experts selected in a bipartisan fashion, the Commission should furnish policy makers with objective analysis and recommendations. It would serve to identify, challenge, and correct shortcomings in current official data, analysis, and forecasting agencies, helping to avoid the misinformation currently leading to costly policy errors, and to bolster U.S. energy security.

Setting U.S. Climate Policy Straight: Recommendations for the 118th Congress, December 5, 2022

The Forum for American Leadership's Energy Working Group has produced an agenda for Congress to set U.S. climate policy straight, particularly in international fora such as the IEA and at home in the EIA, drawing on its Blueprint for a Serious and Sound Climate Policy and plan for Congress to Restore Realism in U.S. Energy Policy. The new Republican House majority in the 118th Congress has promised to implement a bold energy policy agenda, and that agenda should include the following climate and critical minerals priorities:

- *Insist the IEA Retract its Call to Ban Oil and Gas Investment to Reach an Impractical Net Zero Target by 2050.* U.S. diplomats should push IEA to retract these statements and clarify that the agency does not oppose new investment in oil and gas, which remains critical for the foreseeable future to meet demand and ensure secure, affordable energy for U.S. citizens and the world.
- *Require the EIA to Analyze Biden Climate Proposals as a Condition for Congressional Funding.* The EIA has yet to release a cost-benefit and feasibility study of President Biden's executive order requiring the federal government to achieve net-zero carbon emissions by 2050. Congress should tie funding for the EIA to its timely issuance of transparent, unbiased modeling of major energy and climate proposals, including analyses of their impact on the U.S. economy and energy security.
- *Pressure for a Return to Unbiased Forecasting at the IEA.* To please climate change advocates eager to reduce hydrocarbon investment, starting in 2020, the IEA cancelled including a "business as usual" (BAU) reference case scenario in its long-term forecasts and instead only included scenarios that assume quick and easy peak demand. Congress should pressure the Biden administration to use the voice and vote of the U.S. to return the IEA to resume the long-held forecasting convention of a BAU reference case in its annual forecast scenarios.

- *Depoliticize and Accurately Represent the Science and Technologies, Including:*
  - Require that authors of Summaries for Policy Makers (SPMs) of UN IPCC reports be selected by independent, non-government experts and scientists without conflicts of interests.
  - Require settlement of differences arising in the peer-review of assessment reports by an independent referee, as is the case for research papers, instead of allowing the assessment report's authors to discard criticisms without explanation, as is the case now.
  - Require UN and U.S. assessment reports to undergo a formal review by a group of independent climate experts tasked with challenging the assumptions, conclusions, and presentation, probing for weak spots, distortions, and exaggerations. Require report authors to rebut any points raised.
- *Apply Sound Principles and Leverage All Strategies and Fuels, Including:*
  - Require the DoD and State to analyze the impacts on U.S. national security of sharp increases in oil and gas prices now and in the future.
  - Require the Treasury Department and the Federal Reserve to analyze the economic and financial impacts of a successful and unsuccessful transition this decade. Foster closer collaboration with the financial community to educate it on realistic scenarios to protect against dislocation of capital.
  - Congress should repeal tax benefits and outlays benefitting mature but uneconomic technologies and redirect any futures taxes benefits and outlays toward activities the private sector may bypass, such as basic science and strategic endeavors, including work on minerals dependence and nuclear energy, with the potential for disproportionate security benefits.
- *Legislate on Climate at Home and Lead Abroad*
  - Congress should lead the debate and enact domestic and international climate policy. The current approach led by the executive and judicial branches yields only transient, legally vulnerable, and easily reversible domestic and foreign policies.
  - Congress should enact legislation establishing a nonpartisan, expert national commission that integrates technology development, economics, and the strategies, principles, and policy options noted here to develop coherent options for Congress to debate and enact.
- *Critical Minerals Recommendations:*
  - Legislate a streamlined permitting process for mineral extraction and processing in the United States to bolster our competitiveness against established producers in China.
  - Provide funds for bolstering critical mineral security through the Defense Production Act of 1950, a federal statute passed during the Korean War that gives the president broad authority to influence the domestic industry in the interest of national defense.
  - Make critical minerals legislation a higher priority. A bill introduced in January 2022 by Sen. Mark Kelly and Sen. Tom Cotton would establish a strategic rare earths reserve and restrict the use of rare earth elements from China in advanced defense technology in the United States.

- Enact legislation that would force defense contractors to stop buying rare earth-enabled
- products from China by 2026 and use the Pentagon's Defense Logistics Agency, to create a permanent stockpile of rare earth minerals.
- Consider establishing tax credits, modeled after wind and solar credits, for the domestic production, processing, and recycling of critical minerals.
- Prioritize domestic development and direct any foreign assistance for clean energy mining towards friendly and stable sources of supply, particularly U.S. allies such as Australia and Canada and partners in Latin America.
- Mandate an update to the 2019 Critical Minerals Strategy, issued by the Department of Commerce, and require subsequent updates every two years.
- Establish a "one-stop-shop" consolidating and expediting the processes across the Department of the Interior, Department of Energy, and USDA for permitting, grants, and loans to produce, process, and recycle critical minerals.
- Create an Export-Import Bank program that supports exportation of U.S.-origin rare earth mining equipment and services and debt and risk insurance for U.S. rare earth investments.

Restoring U.S. Energy Security: Recommendations for the 118th Congress, December 5, 2022

The Forum for American Leadership's Energy Working Group has prepared concrete recommendations for Congress to help right the ship in U.S. energy policy, drawing on its Steps to Defend U.S. Critical Energy Infrastructure and plan to Create an Arsenal of Energy. The new Republican House majority in the 118th Congress has promised to pursue an ambitious energy policy agenda, and that agenda should include the following priorities:

- *General Energy Recommendations:*
  - Until the enactment of a serious and sound national and international climate policy, suspend executive branch regulatory initiatives, including by independent federal agencies, pertaining to greenhouse gas emissions emanating from the exploration, production, transmission, refining, and trade in all forms of energy.
  - Abolish the requirement for a national interest permit for any cross-border energy infrastructure projects unless the President finds that it would gravely imperil the national security of the United States.
  - Remove executive branch restrictions/bans on U.S. government support, technical assistance (such as at the Departments of Energy, Commerce, Treasury, and State), and financing (U.S. International Development Finance Corporation and Export-Import Bank) for fossil fuel projects and related infrastructure to enable the U.S. Government's ability to support energy security and development of partners and allies.
  - Modernize and improve the leasing and management of the federal estate to ensure that taxpayers realize the full and complete benefit of the resource base of the United States.

- Amend the National Environmental Policy Act to enforce timelines and provide expedited permitting for critical national security energy projects.
- *Oil Recommendations:*
  - Repeal non-emergency drawdowns from the Strategic Petroleum Reserve (SPR) and direct the Department of Energy to expand the SPR from its current 568 million barrels to one billion barrels and fill it.
  - Approve all necessary permits for the Keystone XL pipeline.
  - Reverse all Biden Administration policies restricting oil and gas development in Alaska, a critical region where Russia already excels, and establish a powerful U.S. Arctic presence founded on energy projects.
- *Natural Gas and LNG Recommendations:*
  - Deem every LNG liquefaction project currently before the Department of Energy to be approved.
  - Congress should amend the Natural Gas Act of 1938 to eliminate the public interest determination for LNG exports to non-Free Trade Agreement countries.
  - Consider loan guarantees, co-funded with European countries, to expedite the construction of LNG liquefaction and regasification facilities.
  - Require international lending institutions receiving federal government funding to prioritize natural gas infrastructure, including trade-related terminals, ships, and barges, decentralized production, processing facilities, and pipelines.
  - Facilitate government and the unlock of private sector financing to enable USG's ability to support the energy security and development of allies.
  - Remove/overturn executive branch restrictions/bans on U.S. government technical & financing assistance for fossil fuel projects (including at the Departments of Energy, Commerce, Treasury, and State).
  - Create/reinstate programs financing tools and mechanisms to support fossil and infrastructure projects abroad (e.g., LNG regasification) via loan guarantees, risk insurance, and other financing supports (at the Department of State, U.S. International Development Finance Corporation, and Export-Import Bank).
- *Restore Deterrence and Bolster Defense of Critical Energy Infrastructure, Including:*
  - Toughen penalties and sanctions for foreign cyber attackers who target critical energy and other vital U.S. infrastructure.
  - Deter, preempt, and punish foreign cyber attackers targeting U.S. critical energy infrastructure as it would Al Qaeda, ISIS, or any other similar foreign-based terrorist planning or using WMD to inflict catastrophic harm to the homeland; provide additional clear authority, funding, and direction to the Executive Branch to do so.
  - Require the President to notify Congress of countries that support cyber attackers who have, are, or are likely to plan or execute cyberattacks against critical energy infrastructure.
  - Require the President to immediately formulate and implement a more robust and cyber defense and offense strategy; the President should be required to inform Congress of the Administration's implementation progress no later than six months after completion, and upon submission, Congress should allocate additional funds to ensure swift and safe implementation of the new strategy.

- Declare it shall be the policy of the United States to regard any future attempts to disrupt or dismantle U.S. critical energy infrastructure by cyber attackers an act of aggression that shall warrant swift and commensurate retaliation against the attackers and any foreign governments deemed to sponsor them.
- Spend more money on human capital and training for public-private cybersecurity programs, which will improve the government's capacity to help companies that are managing critical energy infrastructure assets.
- Require critical energy infrastructure owner-operators to immediately inform the federal government of major cyber or any other type of attacks that could impact domestic supply.
- Require the owner-operator of a critical energy infrastructure asset to consult and obtain the permission of the appropriate federal authority before taking any discretionary action that could threaten the economy or national security.

Reducing Russian Involvement in Western Nuclear Power Markets, by Dr. Matt Bowen and the Honorable Paul Dabbar, May 2022

The Russian invasion of Ukraine has led to considerable sanctions being levied against Russia.[36] The nation is a major energy supplier to the world—including of oil, gas, coal, and nuclear fuel and reactors—and its energy sector is one area that has already been targeted. Since the war began, some members of the US Congress have called for banning imports of enriched uranium from Russia as soon as 2022 or as late as 2026.[37] This commentary discusses Russian involvement in the Western[38] nuclear power supply chain, particularly in the United States, as well as policy options to reduce—or end—that involvement.

International Nuclear Fuel Dependencies on Russia

Russia has exported more reactors in recent decades than any other major supplier. In 2021, there were 439 total nuclear power reactors in operation: 38 of them resided in Russia and 42 of them in operation in other countries were of the Russian VVER type (15 of which were in Ukraine). At the end of 2021, 15 Russian-designed reactors were under construction in other nations.

Even for countries that do not host VVERs—including the United States—Russia is a major supplier of several services involved with the manufacturing of nuclear fuel (illustrated in Figure 1). In brief, to make nuclear fuel, raw uranium must be mined out of the ground and

---

[36] Richard Nephew, "The Sanctions War Is Just Beginning: Targeting Russia Was the Easy Part," *Foreign Affairs*, March 31, 2022, https://www.foreignaffairs.com/articles/russian-federation/2022-03-31/sanctions-war-just-beginning.

[37] For example, Senator Barrasso introduced legislation with several cosponsors to ban Russian uranium imports 45 days after enactment: https://www.barrasso.senate.gov/ public/index.cfm/2022/3/barrasso-leads-bill-to-ban-russian-uranium-imports; Senator Manchin and Senator Risch introduced legislation that would not allow Russian LEU imports as of 2026: https://www.congress.gov/bill/117th-congress/senate-bill/4064.

[38] For the purposes of this paper, "Western" means North America, Europe, Japan, and South Korea.

milled into uranium-oxide (U_3O_8) before being shipped to facilities that convert it into uranium-hexafluoride (UF_6), which is suitable for enrichment.

Gas centrifuge plants are the type of enrichment facility in use today, where a single centrifuge will take an input stream of UF_6 and produce two output streams: one with a higher percentage of the isotope U-235 than the input stream, and a second with a lower percentage of U-235. Enrichment plants can have thousands of centrifuges, and those centrifuges are connected to one another in different ways depending on the company's goals. Ultimately, an enrichment plant will produce a stream of uranium that is enriched to the desired level (e.g., U-235 content enriched to between 3 percent and 5 percent for use in nuclear power plants) as well as other streams at lower enrichment levels.

In the final step of nuclear fuel making for US power plants, the enriched UF_6 is sent to fuel fabrication facilities where it is converted into UO_2 and fabricated into fuel pellets. Those pellets are stacked inside metal fuel rods that are connected to each other as part of a fuel assembly, and power reactor cores have many fuel assemblies inside of them.

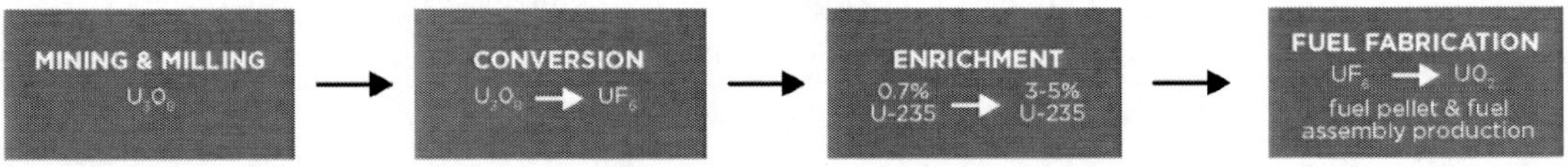

Figure 1. Creating nuclear fuel for reactors.

Table 1. Global production of uranium from mines and identified recoverable resources

Country	2020 production from mines (metric tons U)	2019 identified recoverable uranium resources (metric tons)
Kazakhstan	19,477	907,000
Australia	6,203	1,693,000
Namibia	5,413	448,000
Canada	3,885	565,000
Uzbekistan (est.)	3,500	132,000
Niger	2,991	276,000
Russia	2,846	486,000
China (est.)	1,885	249,000
US	6	48,000
All other	1,525	1,344,000

Source: World Nuclear Association, "World Uranium Mining Production," accessed March 30, 2022, https://world-nuclear.org/information-library/nuclear-fuel-cycle/mining-of-uranium/world-uranium-mining-production.aspx.
Note: US production has declined greatly from 1,919 metric tons in 2014, though could be increased again.

Russia is not one of the leading miners of raw uranium, as shown in Table 1, though it has substantial involvement in the uranium mining operations of some other countries that have greater production. There are multiple allied nations, such as Canada and Australia, who could increase their uranium mining production, if necessary, to make up for any shortfall that might result from cutting off Russian uranium.

The next two steps in the nuclear fuel making process, however, involve a large Russian presence in international markets. As Table 2 shows, Russia accounted for nearly 40 percent of global conversion services in 2020. It also shows that several of the existing Western conversion facilities are operating at low utilization rates, which could be ramped up if there was a need.

The ConverDyn conversion facility in the United States, for example, closed in 2017 due in part to challenging market conditions (e.g., decreases in demand from Japan and Germany following the Fukushima accident), but the company announced in 2021 plans to restart operations at the plant in 2023.[39] ConverDyn has also stated that it could potentially reinstate a capacity of 15,000 tons of uranium per year should there be a market signal to do so.[40]

Table 2. Global uranium conversion capacity and utilization in 2020

Country	Nameplate capacity (metric tons U)	Capacity utilization (metric tons U)
France	15,000	2,600
China	15,000	8,000
Canada	12,500	9,000
Russia	12,500	12,000
United States	7,000	0

Source: World Nuclear Association, "Conversion and Deconversion," citing the association's Nuclear Fuel Report (2021 edition), accessed March 30, 2022, https://world-nuclear.org/information-library/nuclear-fuel-cycle/conversion-enrichment-and-fabrication/conversion-and-deconversion.aspx.

Note: The US conversion facility plans to return to operations in 2023.

Similarly, Russia has a large presence in the international enrichment market. As Table 3 shows, Russia had around a 46 percent share of global enrichment capacity in 2018. The Russian enrichment entity Tenex provided 30 percent of enrichment services to European Union utilities in 2019.[41] South Korea's Korea Hydro & Nuclear Power (KHNP) recently signed a contract with Tenex to supply enrichment services out to 2030, reportedly bringing the total value of Tenex's contracts with KHNP to $2 billion.[42]

The price of enrichment services (described as "separative work units" or "SWUs") has been depressed for years.[43] The low price has led to existing enrichment facilities resorting to a practice known as "underfeeding": having less natural uranium shipped to them and devoting a greater amount of their capability toward enriching the material that has been produced with

---

39 "US conversion plant gears up for next 40 years," *World Nuclear News*, April 14, 2021, https://www.world-nuclear-news.org/Articles/US-conversion-plant-gears-up-for-next-40-years.

40 ConverDyn, HALEU Workshop, April 28–29, 2020, https://gain.inl.gov/HALEU_Webinar_ Presentations/10-Critchley,ConverDyn,Rev1-28Apr2020.pdf.

41 Euratom Supply Agency, "Annual Report 2020," table 6 on page 22, https://euratom-supply.ec.europa.eu/publications/esa-annual-reports_en.

42 "Russia's Tenex wins tender for fuel supply to South Korea," *Nuclear Engineering International*, January 25, 2020, https://www.neimagazine.com/news/newsrussias-tenex-wins-tender-for-fuel-supply-to-south-korea-7653727.

43 "Enrichment excess is here to stay," *Nuclear Engineering International*, October 13, 2015, https://www.neimagazine.com/features/featureenrichment-excess-is-here-to-stay-4691321/.

lower enrichment levels. In other words, the lower price for enrichment services in recent years has meant that the market is using less uranium mining and conversion services in order to produce the needed amount of low-enriched uranium (LEU) for Western power reactors.

Table 3. Global uranium enrichment capacity in 2018 by operator

Operator	(thousands of separative work units/year)
Rosatom (Russia)	28,215
Urenco (UK, Netherlands, Germany, USA)	18,600
Orano (France)	7,500
CNNC (China)	6,750
Other	46

Source: World Nuclear Association, "Uranium Enrichment," citing the association's Nuclear Fuel Report 2019, accessed March 30, 2022, https://world-nuclear.org/information-library/nuclear-fuel-cycle/ conversion-enrichment-and-fabrication/uranium-enrichment.aspx.

However, if there is a policy in place to reduce or eliminate Russian involvement in Western nuclear fuel markets, this would almost certainly lead to a realignment in the supply chain. In the near-term, the market would likely shift to mining more natural uranium (from non-Russian sources) and raising the low utilization rate of existing Western conversion facilities to closer to their nameplate capacities. This would lead to more expensive fuel for operating reactors and thus will not happen without clear policy direction. (There are no legal or regulatory rules preventing this, and thus if this were the cheaper approach, the market would have already done it).

No Russian fuel fabrication facilities are involved in the final step of making fuel for US nuclear power plants, but this is not the case for many of the countries hosting Russian VVERs. There is, however, some Western capability to fabricate fuel for those VVER reactors to supplant Russian involvement, if countries so chose. Westinghouse, for example, has the capability to make fuel for Russian VVER-1000s,[44] and could develop the capability to make fuel for other VVER models.

"Megatons to Megawatts" and the Russian Suspension Agreement

The high level of Russian enriched uranium in US nuclear fuel has a unique history and factors into how the United States got to where it is today in terms of diminished domestic capabilities. In 1993, the Russian Federation and the United States signed an agreement to eliminate excess highly enriched uranium (HEU) from dismantled Russian nuclear weapons.[45] From 1993 to 2013, downblended Russian HEU provided about half of the enriched uranium used in US power reactors, as part of the "Megatons to Megawatts" program.

Separate from the Megatons to Megawatts program, exports of Russian uranium products into the US fuel market have been limited for decades by what is colloquially referred to as "the Russian suspension agreement." The original agreement was signed on October 16, 1992, when the US Department of Commerce suspended an anti-dumping duty investigation on

---

[44] Westinghouse, "VVER," https://www.westinghousenuclear.com/operating-plants/nuclear-fuel/vver.

[45] Lawrence Livermore National Laboratory, "A Transparent Success: 'Megatons to Megawatts' Program," Science & Technology Review, April/May 2013, https://str.llnl.gov/ content/pages/april-2013/pdf/4.13.3.pdf.

uranium from the Russian Federation (hence, "suspension agreement").[46] The agreement restricted the amount of Russian uranium products that could enter the US market, and it has been amended several times since it was first signed. It had previously been scheduled to expire at the end of 2020, but the amendment signed that year extended the agreement to 2040.

In the most recent amendment, the US Department of Commerce and the Russian state nuclear corporation Rosatom agreed to lower the amount of Russian uranium products allowed for export into the US market, and to limit the percentage of US enrichment demand met by Russia to 15 percent starting in 2028. In other words, even before the war in Ukraine began, the US government had been taking steps to reduce Russian involvement in US nuclear fuel markets, both for strategic and commercial reasons.[47] Some of the limits in the suspension agreement are reproduced in Table 4, though only out to 2030.

Table 4. Partially reproduced suspension agreement with limits on Russian exports

Export limit year	Percentage of US enrichment demand	Total export limit in kilograms U as LEU (A)	Total export limit in kilograms U-235 content (B)	USEC export limit allocation in kilograms U-235 (E) (Subset of B)
2021	24%	596,682	26,254	7,780
2022	20%	489,617	21,543	7,430
2023	24%	578,877	25,471	10,700
2024	20%	476,536	20,968	10,300
2025	20%	470,376	20,697	10,300
2026	20%	464,183	20,424	10,700
2027	20%	459,083	20,200	10,600
2028	15%	344,312	15,150	4,100
2029	15%	340,114	14,965	0
2030	15%	332,141	14,614	0

Source: US Department of Commerce, "2020 Amendment to the Agreement Suspending the Antidumping Investigation on Uranium From the Russian Federation," https://www.federalregister. gov/documents/2020/10/09/2020-22431/2020-amendment-to-the-agreement-suspending-the-antidumping-investigation-on-uranium-from-the-russian.

Note: The table does not show two columns in the agreement, "C" and "D": "Cap for LEU exports pursuant to sales of EUP [enriched uranium product] (may include sales of SWU plus conversion) in kilograms U-235" and "Cap for additional LEU exports pursuant to sales of SWU plus conversion Only in kilograms U-235."

As the Table 4 shows, one US company is specifically mentioned in the agreement: USEC,[48] which is today named Centrus. Centrus (located in Ohio) is the sole company with installed enrichment capacity in the United States that is not foreign owned (Urenco, which operates the only commercial-scale enrichment plant in the United States, is owned by the British, Dutch, and German governments). Part of Centrus's current business model is purchasing enrichment services from Tenex in Russia and supplying Russian LEU to US

---

[46] International Trade Administration, "2020 Amendment to the Agreement Suspending the Antidumping Investigation on Uranium From the Russian Federation," Federal Register, October 9, 2020, https://www.federalregister.gov/documents/2020/10/09/2020-22431/2020-amendment-to-the-agreement-suspending-the-antidumping-investigation-on-uranium-from-the-russian.

[47] US Department of Commerce press release, "U.S. Department of Commerce Finalizes 20-Year Amendment to the Suspension Agreement on Uranium from the Russian Federation," October 6, 2020, https://2017-2021.commerce.gov/news/press-releases/2020/10/us-department-commerce-finalizes-20-year-amendment-suspension-agreement.html.

[48] USEC was the private entity created from the DOE's enrichment operation. See: https://www.centrusenergy.com/who-we-are/history/.

customers, nominally through 2028, and a congressional intervention to block out Russia, such as the bills mentioned earlier would do, would terminate those contracts.[49]

Fuel Supply for Non-Light Water Reactors

Previous to the Ukraine invasion, there was a different potential involvement that Russian enriched uranium might have had with some of the future advanced reactors under development in the United States. In recent decades, a variety of private companies have been founded to pursue commercialization of different advanced reactor designs. Some of these designs use uranium with significantly higher enrichments than light water reactors use: instead of 3–5 percent, the enrichments may be as high as 15–19.75 percent. Currently, the only commercial source of this high-assay low-enriched uranium (HALEU) is Russia.

In 2020, the US Department of Energy (DOE) announced a series of large cost-share awards with some of these private reactor developers.[50] For the biggest demonstrations, DOE would contribute a share of the demonstration costs, as long as private entities more than matched that investment. Given that Russia has been the only commercial source of HALEU, some advanced reactor developers were either planning to obtain—or at least considered obtaining—their first fuel load's worth of HALEU from Russia. In 2018, the Nuclear Energy Institute reported (based on company inputs) that estimated HALEU needs might potentially ramp up from tens of metric tons per year in the mid-2020s to over a 100 metric tons per year in the late 2020s.[51]

Existing enrichment companies, such as Urenco, Orano, GLE, and Centrus, could make HALEU, but these companies would likely be hesitant to invest too much in building HALEU infrastructure and completing NRC licensing without being confident there will in fact be a profitable market for the product. Industry estimates that establishing a commercial-scale production capability would cost more than $500 million.[52] On the reactor developer side, if a single company were to come to an enrichment company and ask to buy only the amount of HALEU they needed—perhaps at the level of tens of metric tons—the price per kilogram of HALEU would be much higher than if the associated development costs could be spread over a large order. Challenges—real and perceived—with HALEU fuel procurement could in turn deter investment in the deployment of some non-light water reactor designs. This is the "chicken and egg" dilemma that the US government is currently grappling with.[53] Buying the

---

[49] Centrus, 2020 Annual Report, https://investors.centrusenergy.com/static-files/77087849-8051-4919-a125-d0a8600bcc98.

[50] For example, DOE announced cost-share awards with TerraPower and X-energy on October 13, 2020, US Department of Energy press release, "U.S. Department of Energy Announces $160 Million in First Awards under Advanced Reactor Demonstration Program," October 13, 2020, https://www.energy.gov/ne/articles/us-department-energy-announces-160-million-first-awards-under-advanced-reactor.

[51] Nuclear Energy Institute letter to Secretary of Energy Rick Perry, July 5, 2018, https://www.nei.org/CorporateSite/media/filefolder/resources/letters-filings-comments/letter-perry-haleu-20180705.pdf.

[52] Nuclear Energy Institute, "Establishing a High Assay Low Enriched Uranium Infrastructure for Advanced Reactors," January 2022, https://www.nei.org/resources/reports-briefs/ establishing-a-haleu-infrastructure-for-advanced-r.

[53] See comments by the French enrichment company, Orano, in "Certainty of demand key to HALEU supply, says Orano," *World Nuclear News*, March 15, 2022, https://www.world-nuclear-news.org/Articles/Certainty-of-demand-key-to-HALEU-supply-Orano.

first core loads of HALEU from Russia would have enabled reactor developers to easily meet their stated timelines for when they needed fuel. It would also have allowed the federal government to gauge at some level how the construction and operation of the first non-light water reactor projects were executed before committing potentially large amounts of money toward domestic HALEU production.

If the Russian supply option were off the table, however, the United States would need to turn in earnest to remaining possibilities. Congress directed DOE to establish a HALEU availability program in the Energy Act of 2020. At the end of 2021, DOE had already put out a request for information regarding planning for the establishment of a program to support the availability of HALEU for civilian domestic research, development, demonstration, and commercial use,[54] and subsequently received a variety of responses.[55] Multiple bills in the 117th Congress have been introduced that would further authorize and direct DOE to pursue HALEU production programs.[56] Centrus has been working with DOE since 2019 to demonstrate a capability to produce HALEU and obtained NRC approval for HALEU production in 2021.[57]

Exposure of Global Nuclear Power to Russia for Nuclear Components, Services, and New Construction

In addition to a significant exposure to Russia in the uranium fuel chain, the global nuclear market relies on Russia for equipment and construction efforts. Many reactors in operation and under construction around the world are using Russian reactor technology. Given that most of the exported Russian reactors are of the VVER/pressurized water reactor design, we will focus on that technology for this analysis.

Existing/Operating Russian Designed and Built Nuclear Reactors

Operators of existing nuclear reactors can have significant supply chain exposure to original equipment manufacturers ("OEMs") of their reactor type. Reactors have unique components that their OEM designed and built for their specific reactor type, including (focusing on the pressurized water reactor design):

- Most of the internal components of the reactor vessel, such as the fuel assembly structure, coolant, and flow components; the reactor vessel and head; and the control rod structures

[54] US Department of Energy, "Request for Information (RFI) Regarding Planning for Establishment of a Program To Support the Availability of High-Assay Low-Enriched Uranium (HALEU) for Civilian Domestic Research, Development, Demonstration, and Commercial Use," Federal Register, December 14, 2021, https://www.federalregister.gov/documents/2021/12/14/2021-26984/request-for-information-rfi-regarding-planning-for-establishment-of-a-program-to-support-the.

[55] For example, the Nuclear Energy Institute's response is here: https://www.nei.org/ CorporateSite/media/ filefolder/resources/letters-filings-comments/NEI-02-14-22_DOE_HALEU-RFI-Response.pdf.

[56] S.4064 and S.4066 from the 117th Congress would support domestic HALEU production.

[57] "Centrus receives licence for HALEU production," *World Nuclear News*, June 15, 2021, https://www.world-nuclear-news.org/Articles/Centrus-receives-licence-for-HALEU-production.

- Components in the rest of the nuclear primary system (i.e., the system immediately connected to the reactor core) are also from the OEM, including the pressurizer, steam generators, and the primary water pumps and related systems

In addition to primary components, in various settings around the globe, many different parts of VVER power plant secondary reactor systems could be from Russian origin, including:

- Control room and reactor control systems
- Secondary pumps and their control systems
- Turbine generators and their control systems

Examples of services and components provided to global nuclear customers listed by Rosatom, the state nuclear power company of Russia and the OEM of the VVER reactors, include: assessing and developing key nuclear infrastructure components, large life extension projects, regular supply of spare parts and equipment, and power capacity expansion uprates.[58] In 2019 (the last year Rosatom published an annual report), Rosatom overseas revenues for nuclear fuel assemblies (excluding the uranium supply chain), reactor components, and services was $1.9 billion.[59]

While there is little data on annual maintenance and capital costs for Russian VVER reactors, public data from US nuclear reactors show that annual capital costs for nuclear power plants are $5.35/megawatt-hour (MWh), and annual maintenance costs are $18.27/MWh,[60] of which about a quarter are for physical supplies that have to be purchased.

The list of VVER reactors operating globally (excluding Russia, Belarus, Iran, and China, since those countries are likely to continue using Russian supplies) are listed in Table 5.

Exposure of existing nuclear operators to these OEM replacement components is more acute than for typical energy infrastructure for a couple reasons. First, the design requirements for nuclear systems are significantly more stringent than for typical power systems, given failure events are potentially much more acute. As a result, maintenance and procurement teams at existing operators typically return to the OEM for many critical system efforts, and alternative supply chains for replacement nuclear components don't tend to be well developed.

Additionally, nuclear operators culturally tend to be very conservative on procurement of maintenance and upgraded equipment, and tend to default to purchasing from the plant OEM.

There is little public data on maintenance and capital expenditure purchase costs for operating VVER reactors. However, VVER reactors in general are of similar designs, sizes, and capacity factors to US light water reactors, so using the robust US data available is a reasonable surrogate for VVER operating costs. Assuming an 80 percent capacity factor for operating reactors, US data would suggest purchases of capital supplies required to run the existing global fleet is $4.3 billion annually.

---

[58] Rosatom, "Performance of State Atomic Energy Corporation Rosatom in 2019," https://www.report.rosatom.ru/go_eng/go_rosatom_eng_2019/rosatom_eng_2019.pdf?msclkid=a5334dd3cf9f11ec964dc636df2b47ef.

[59] Ibid.

[60] Nuclear Energy Institute, "Nuclear Costs In Context," November 2021, https://www.nei.org/ resources/reports-briefs/nuclear-costs-in-context.

Table 5. Global operating VVER reactors in 2021, excluding those in Russia, Belarus, Iran, and China

Country	Reactor name	Model	Megawatt electrical net
Armenia	Armenian-2	VVER V-270	415
Bulgaria	Kozloduy-5	VVER V-320	1,003
Bulgaria	Kozloduy-6	VVER V-320	1,003
Czech Republic	Temelin-1	VVER V-320	1,027
Czech Republic	Temelin-2	VVER V-320	1,029
Czech Republic	Dukovany-1	VVER V-213	468
Czech Republic	Dukovany-2	VVER V-213	471
Czech Republic	Dukovany-3	VVER V-213	468
Czech Republic	Dukovany-4	VVER V-213	471
Finland	Loviisa-1	VVER V-213	507
Finland	Loviisa-2	VVER V-213	507
Hungary	Paks-1	VVER V-213	479
Hungary	Paks-2	VVER V-213	477
Hungary	Paks-3	VVER V-213	473
Hungary	Paks-4	VVER V-213	473
India	Kudankulam-1	VVER V-412	932
India	Kudankulam-2	VVER V-412	932
Slovakia	Bohunice-3	VVER V-213	466
Slovakia	Bohunice-4	VVER V-213	466
Slovakia	Mochovce-1	VVER V-213	436
Slovakia	Mochovce-2	VVER V-213	469
Ukraine	Zaporozhye-5	VVER V-320	950
Ukraine	Zaporozhye-6	VVER V-320	950
Ukraine	Rovno-1	VVER V-213	381
Ukraine	Rovno-2	VVER V-213	376
Ukraine	Rovno-3	VVER V-320	950
Ukraine	Khmelnitski-1	VVER V-320	950
Ukraine	Khmelnitski-2	VVER V-320	950
Ukraine	South Ukraine -1	VVER V-320	950
Ukraine	South Ukraine -2	VVER V-338	950
Ukraine	South Ukraine -3	VVER V-320	950
Ukraine	Zaporozhye-1	VVER V-320	950
Ukraine	Zaporozhye-2	VVER V-320	950
Ukraine	Rovno-4	VVER V-320	950
Ukraine	Zaporozhye-3	VVER V-320	950
Ukraine	Zaporozhye-4	VVER V-320	950
Ukraine	Zaporozhye-3	VVER V-320	950
Ukraine	Zaporozhye-4	VVER V-320	950

Source: International Atomic Energy Agency, "Power Reactor Information System," https://pris.iaea.org/ pris/."

While some of the countries listed in Table 5 are less likely to embargo Russian businesses as a result of the Ukraine war, the countries in Europe such as in Bulgaria, Czech Republic, Finland, Hungary, and Slovakia could face important impacts on their abilities to continue to operate their facilities due to parts and services needed from their VVER OEM. The VVERs in Ukraine are in a uniquely challenging position. Some non-Russian engineering and construction companies have built capabilities to replicate Rosatom VVER components, but replicating OEM designs and manufacturing can be a challenge for nuclear power plants.

New Nuclear Power Plant Construction

In addition, Russia, through Rosatom, has been a major potential provider of new nuclear power plants globally. Many plants around the world have been conducting multiyear design and site preparation efforts for the construction of VVER reactors, and many are already in construction. A work stoppage or abandonment of these projects would provide significant financial and energy supply impacts to those countries.

VVER reactors globally under construction or publicly identified as planned to be constructed (excluding those in Russia, Belarus, Iran, and China) are listed in Table 6.

Table 6. Global VVER reactors under construction or in planning in 2021, excluding those in Russia, Belarus, Iran, and China

Country	Reactor name	Model	Megawatt electrical net
Bangladesh	Rooppur-1	VVER V-523	1,080
Bangladesh	Rooppur-2	VVER V-523	1,080
Finland	Hanhikivi-1	VVER V-522	1,200
Hungary	Paks-5	VVER V-527	1,185
Hungary	Paks-6	VVER V-527	1,185
India	Kudankulam-3	VVER V-412	917
India	Kudankulam-4	VVER V-412	917
Slovakia	Mochovce-3	VVER V-213	440
Slovakia	Mochovce-4	VVER V-213	440
Turkey	Ankkuyu-1	VVER V-509	1,114
Turkey	Ankkuyu-2	VVER V-509	1,114
Turkey	Ankkuyu-3	VVER V-509	1,114
Turkey	Ankkuyu-4	VVER V-509	1,114

Source: International Atomic Energy Agency, "Power Reactor Information System," https://pris.iaea.org/ pris/.

Note: The IAEA lists two VVERs as having been under construction at the Khmelnitski site in Ukraine since 1986/1987, but they have not actually been under construction for many years.

While some of the above are less likely to halt new Russian designed VVER reactor projects as a result of the Ukraine war, ones in Europe such as Finland, Hungary, Slovakia, and possibly Turkey may reconsider development of those plants. Finland already seems to have decided against building a Russian VVER since the Ukraine invasion.[61] As a result, they might have to consider other reactor vendors or look at non-nuclear technologies to fulfill their power needs. Given that power plant development cycles take many years, this could have significant negative impacts on their energy supply and policies.

Policy Options to Move Away from Russian Nuclear Energy Supply Chains

Countries considering new reactors can simply choose other vendors if they don't want to be dependent on Russia for fuel, equipment, and services. The US, France, Republic of Korea, and China are all viable reactor suppliers.

Countries that currently host Russian VVERs have a more complicated set of choices to make. They are likely dependent on Russia not just for fuel but for reactor equipment and

[61] Richard Milne, "Finnish group abandons plans for Russia-backed nuclear power plant," *Financial Times*, May 2, 2022, https://www.ft.com/content/5e7730b5-be55-48a5-8f55-f67d6fb80d77.

services. There are some alternatives at least for replacing Russian fuel. Westinghouse, as mentioned, is able to make fuel assemblies for VVER-1000s, and it is possible the company will be able to manufacture fuel for other VVER designs in the near-term.[62] Westinghouse is also able to supply some services to VVERs.[63]

For all other countries with non-VVER power reactors, Russian-mined uranium is not the challenge in extricating themselves from involvement with Russia in procuring their fuel. As shown in Table 1, Russia only mines approximately 6 percent of the uranium produced each year, and other countries can expand production if necessary. Fuel fabrication is also not the challenge.

Instead, policy options to replace Russia's large presence in global conversion and enrichment services would be where the United States and its allies would need to focus their attention.

The conversion and enrichment capacities in Canada, France, Germany, the Netherlands, the United Kingdom, and the United States are enough to replace at least some of the Russian fuel services involved in fueling Western nuclear power plants with greater uranium mining. More investment in mining, conversion, and enrichment facilities may be necessary to fully extricate Western nuclear fuel chains from Russian involvement. However, adding sufficient new conversion capacity and enrichment capacity will take years to accomplish.

In any case, mining, conversion, and enrichment suppliers in the West will be looking to national governments to provide clear policies before they invest money in new facilities and capabilities. Their worry will be that in a year or two, perhaps less, Russian uranium products will be allowed back into national markets and will undercut them, causing them to lose out on their investments. National laws that impose a date for ceasing supplies from Russia—such as the bills cited earlier—would send a clear signal to private markets in the United States and elsewhere so they can adjust efficiently.

International meetings with Western governments could be held to discuss how best to coordinate a reduction or elimination of Russian involvement in their nuclear fuel supply chains. The natural uranium, conversion, and enrichment markets are internationally linked, and national decisions in one country will have impacts on other programs. It will be important for these governments to communicate to the public that all policy options to remove dependency on Russian supply chains are likely to raise costs for operating plants.

The recently created Civil Nuclear Credit Program, established by the 2021 Bipartisan Infrastructure Law,[64] may be able to assist in limiting these impacts, as could the material in DOE's American Assured Fuel Supply.[65]

The optimal timing of full extrication—for either all Western nations or the United States alone—is beyond the scope of this paper but would certainly be an issue DOE would need to study immediately if it intends to act. The total impacts of a shutoff of Russian uranium

---

[62] For example, fuel development for VVER-440 has been underway. See: "Westinghouse fuel assemblies delivered to Rovno NPP," *Nuclear Engineering International*, July 22, 2021, https://www.neimagazine.com/news/newswestinghouse-fuel-assemblies-delivered-to-rovno-npp-8919297.

[63] Westinghouse, "Westinghouse and NT-Engineering Sign MoC to Increase Nuclear Reactor Power," September 22, 2021, https://info.westinghousenuclear.com/news/westinghouse-and-nt-engineering-sign-moc.

[64] US Department of Energy press release, "DOE Establishes $6 Billion Program to Preserve America's Clean Nuclear Energy Infrastructure," February 11, 2022, https://www.energy. gov/articles/doe-establishes-6-billion-program-preserve-americas-clean-nuclear-energy-infrastructure.

[65] US Department of Energy, "Notice of Availability: American Assured Fuel Supply," Federal Register, vol. 76, no. 160, August 18, 2011, https://www.govinfo.gov/content/pkg/FR-2011-08-18/pdf/2011-21067.pdf.

products in 2022 are unknown to the authors; the relevant information may exist outside the public realm, though congressional hearings could uncover some of these details. Given the long lead times for ordering nuclear fuel assemblies, the authors do not believe it likely that such a decision would result in outages due to lack of fuel at US power reactors this year, but it is harder to assess impacts in 2023. Other Western governments may also want to use the same non-Russian mining, conversion, and enrichment facilities as part of a similar strategy to reduce Russian involvement in their power markets, raising the prospect of competition for services. The only US conversion facility is still on standby and will not return to producing 7,000 tons of uranium per year until 2023; increasing that capacity to 15,000 tons per year will take longer.

Given these factors and the unique historical context for large amounts of Russian material in US commercial reactors (i.e., the Megatons to Megawatts program), the authors reason that ramping down Russian supplies would happen more smoothly over a period of years not months. This is more consistent with the types of lead times that reactor owners employ for contracts involving different stages of the nuclear fuel cycle.[66] The US government could simultaneously facilitate meetings with Western governments that have mining, conversion, and enrichment facilities to discuss policy options to hasten a greater capability to supplant Russian conversion and enrichment services. Such efforts to strengthen capability may involve a move away from underfeeding existing Western enrichment capacity using increased uranium mining and greater utilization of conversion services. The US government could also consider supporting the construction of a national security compliant fuel chain, given the lack of one using only US-origin technology, equipment, and materials—a national security concern noted by external reports.[67]

It is possible that a nearer-term or even immediate suspension of Russian imports could produce disruptions to the US market that are acceptably small. If this were the case and would not cause existing plants to shut down or experience extended outages, an earlier cutoff is preferable to a later one to minimize dollars sent to Russia as well as US exposure to Russian supply.

Finally, with regard to HALEU production, a logical next step that DOE could pursue is to issue a request for proposals on how best to establish a HALEU production line and see what the private sector proposes. Reports have suggested a combination of government cost-sharing and off-take agreements could form a viable strategy.[68]

About the Authors

Dr. Matt Bowen is a Research Scholar at the Center on Global Energy Policy at Columbia University SIPA, focusing on nuclear energy, waste, and nonproliferation. He was formerly a

[66] "Public Comments on Behalf of the Ad Hoc Utilities Group," September 25, 2018, page 17, https://www.regulations.gov/document/BIS-2018-0011-0245: "In addition, on average, nuclear generators allow for lead times of 1.7 years for contracts involving various stages of the fuel cycle."

[67] Energy Futures Initiative, "The U.S. Nuclear Energy Enterprise: A Key National Security Enabler," August 2017. https://energyfuturesinitiative.org/wp-content/uploads/sites/2/2022/02/The-U.S.-Nuclear-Energy-Enterprise_A-Key-National-Security-Enabler_ Report_August-2017.pdf.

[68] Nuclear Innovation Alliance, "Catalyzing a Domestic Commercial Market for High-Assay, Low-Enriched Uranium," April 2022, https://www.nuclearinnovationalliance.org/catalyzing-domestic-commercial-market-haleu.

Nuclear Policy Fellow at Clean Air Task Force and a Senior Policy Fellow at the Nuclear Innovation Alliance.

Dr. Bowen has written reports on federal and state policies to encourage advanced reactor development, and has also published papers on reforming U.S. nuclear export controls. During the Obama Administration, he was an Associate Deputy Assistant Secretary in the Office of Nuclear Energy and a Senior Advisor in the Office of Nonproliferation and Arms Control at the U.S. Department of Energy (DOE). Previous to working at DOE, he was an AAAS/APS Science Fellow for Senate Majority Leader Harry Reid. Dr. Bowen received a Bachelor of Science degree in physics from Brown University and a Ph.D. in theoretical physics from the University of Washington, Seattle. He has held positions at the National Academies with the Board on Physics and Astronomy, the Board on Energy and Environmental Studies, and the Division on Engineering and Physical Sciences.

The Honorable Paul M. Dabbar is a Distinguished Visiting Fellow at the Center on Global Energy Policy at Columbia University SIPA. He is also CEO and Co-founder of Bohr Quantum Technology, developing and deploying technologies for the emerging quantum internet.

Prior to that in 2017, the U.S. Senate unanimously confirmed Mr. Dabbar to serve as the Department of Energy's fourth Under Secretary for Science, where he served from 2017-2021. He managed several areas of the Department, as well as serving as the Department's principal advisor on fundamental energy research, energy technologies, science, and commercialization of technologies. He managed over 60,000 people with a budget of $15 billion p.a. at over 100 sites, including managing the majority of the U.S. National Laboratories.

Areas of research he managed included basic energy sciences, nuclear and high energy physics, advanced computing, fusion, and biological & environmental research. He also led the largest environmental remediation program in the U.S., addressing the operations of nuclear weapons and commercial power production, completing several multi-billion dollar construction projects. He also led various new efforts to commercialize innovations arising from the National Labs. He co-led several new energy innovation efforts, including the Energy Storage Grand Challenge, as well as the passage and implementation of the National Quantum Initiative Act.

Mr. Dabbar was awarded in 2021 the Secretary of Energy's senior DOE award, the James R. Schlesinger Medal, for leadership on developing energy technologies, discovery science, environmental management, and the National Quantum Initiative.

During his time in government service, Mr. Dabbar traveled to both the geographic North and South Poles. He traveled to the North Pole by submarine to conduct environmental research while in the Navy, and to the South Pole in support of high energy physics astronomy missions of DOE at South Pole Station.

Prior to confirmation as Under Secretary, Mr. Dabbar worked in operations, finance, and strategy roles in the energy sector. As a Managing Director at J.P. Morgan, he had over $400 billion in transaction experience across all energy sectors. In addition, he had a senior leadership role for the company's commodity trading business, including energy. Before joining J.P. Morgan, Mr. Dabbar served as a nuclear submarine officer. He has been a lecturer at the U.S. Naval Academy, and conducted research at the Johns Hopkins University Applied Physics Laboratory. Mr. Dabbar is a member of the Council on Foreign Relations.

Mr. Dabbar has a bachelors from the U.S. Naval Academy, is a graduate of the U.S. Navy's nuclear power and nuclear engineer programs, and a masters from Columbia University.

About the Center on Global Energy Policy

The Center on Global Energy Policy at Columbia University SIPA advances smart, actionable and evidence-based energy and climate solutions through research, education and dialogue. Based at one of the world's top research universities, what sets CGEP apart is our ability to communicate academic research, scholarship and insights in formats and on timescales that are useful to decision makers. We bridge the gap between academic research and policy — complementing and strengthening the world-class research already underway at Columbia University, while providing support, expertise, and policy recommendations to foster stronger, evidence-based policy. Recently, Columbia University President Lee Bollinger announced the creation of a new Climate School — the first in the nation — to tackle the most urgent environmental and public health challenges facing humanity.

Visit us at www.energypolicy.columbia.edu

About the School of International and Public Affairs

SIPA's mission is to empower people to serve the global public interest. Our goal is to foster economic growth, sustainable development, social progress, and democratic governance by educating public policy professionals, producing policy-related research, and conveying the results to the world. Based in New York City, with a student body that is 50 percent international and educational partners in cities around the world, SIPA is the most global of public policy schools.

For more information, please visit www.sipa.columbia.edu

Russia's State Nuclear Company Aids War Effort, Leading to Calls for Sanctions, by Catherine Belton, January 20, 2023 at 2:00 a.m. EST

Russia's state nuclear power conglomerate has been working to supply the Russian arms industry with components, technology and raw materials for missile fuel, documents show, aiding Moscow's deadly onslaught on Ukraine and leading to calls for the company, Rosatom, to be put under sanctions.

A letter from a Rosatom department chief, dated October 2022 and obtained by Ukrainian intelligence, refers to a recent meeting with the Defense Ministry and representatives of Russia's military-industrial complex. It shows the state nuclear company offering to provide goods to Russian military units and to Russian weapons manufacturers that are under sanctions.

The weapons manufacturers include Almaz-Antey, a missile systems producer; NPK Tekhmash, which manufactures unguided bombs and multiple missile launch systems; Vysokotochniye Kompleksy, which makes Iskander missiles; NPO Splav, which produces the Uragan rocket launchers that fire cluster bombs; the state-owned aircraft conglomerate; and several Russian armored carrier and tank makers. Detailed descriptions of the products available for use by the Russian military and the arms industry are attached to the letter, which was provided to The Washington Post.

Rosatom has long presented itself as a civilian entity operating nuclear power plants in commercial partnerships across the globe, despite its role in also developing nuclear weapons for Russia. "Rosatom is known in the nuclear world as a civil nuclear company. That's basically what its brand name is," said Mark Hibbs, senior fellow in the Carnegie Endowment for International Peace's nuclear policy program.

But Russia's war against Ukraine is exposing how closely the company, which has a board stacked with current and former senior officials from Russia's security services, is intertwined with its military-industrial complex — and even Russian military operations in Ukraine.

After Russia invaded Ukraine last year, Rosatom's employees facilitated Moscow's illegal seizure of Ukraine's Zaporizhzhia nuclear power plant, according to Ukrainian officials and a Russian presidential decree. The plant, Europe's largest, was cut off several times from Ukraine's electric power grid, and the military standoff at the facility brought the world "one step away from a nuclear accident," according to International Atomic Energy Agency chief Rafael Mariano Grossi.

Witnesses said Rosatom employees stationed at the plant appeared to have directed some of the Russian artillery targeting the plant, while at the height of the shelling in August, Ukrainian President Volodymyr Zelensky decried the attack as "nuclear blackmail" by Moscow. The Ukrainian government said Russia was intentionally seeking to cut off the plant, which is now in cold shutdown because of the artillery fire, from supplying electricity to Ukraine.

Rosatom said in response to a request for comment that "all the claims in your request are completely untrue."

Previous calls by Ukraine — backed by Lithuania, Estonia, Latvia and Poland — to impose sanctions on Rosatom over its involvement in the seizure of the Zaporizhzhia nuclear power plant have not led to action, but disclosures about its role in aiding Russia's arms industry will now add to pressure to list the company.

"If Rosatom is providing support for the Russian arms industry, that's sanctionable," said Daniel Fried, the State Department's sanctions coordinator until 2017.

The company has so far escaped sanctions, in part amid concerns over the potential economic fallout because of its extensive involvement in the civilian nuclear power industry across the globe, including in Western Europe and the United States. There are 18 Russian-designed nuclear power plants operating in the European Union, "and that means the Russians ... have leverage over the continued operation of most of these facilities," Hibbs said, noting that in Slovakia, Russian-designed plants produce half of the country's electricity. The United States depends on Rosatom for about one-quarter of its enriched uranium supplies.

In addition, Rosatom controls about 30 percent of the global market for uranium enrichment and 17 percent of the market for reactor fuel, and out of the approximately 450 nuclear power plants around the world, about 20 percent of them are Russian-designed, Hibbs said. Rosatom is currently working on 23 nuclear power units across the globe, including in India, Turkey and Egypt, with its foreign order book standing at $200 billion.

An attempt to immediately unwind dependence on Rosatom for nuclear fuel supplies, waste management and other partnership agreements could cause major new energy price volatility at a time when the E.U. is only just emerging from a bruising year of soaring energy costs as it weaned itself off Russian oil and gas.

The Ukrainian government, however, has been ramping up the pressure for sanctions in recent weeks. Zelensky's chief of staff, Andriy Yermak, met the State Department's sanctions coordinator, James O'Brien, on Jan. 12 to discuss Rosatom and "strengthening sanctions against individuals and legal entities that continue to support Russia's aggression against Ukraine."

After talks this month with European Commission Vice President Frans Timmermans, Ukrainian Prime Minister Denys Shmyhal indicated that he expected sanctions on Rosatom to be included in the E.U.'s next package of sanctions against Russia, saying that Moscow "must be punished for attacks on Ukraine's energy industry."

Ukrainian Foreign Minister Dmytro Kuleba said revelations over Rosatom's involvement in supplying the Russian arms industry "ring the alarm bell even more and prove that the problem is larger than previously known."

In a statement to The Post, Kuleba said: "Nuclear cooperation with today's Russia is not only immoral because it sponsors Russia's war crimes in Ukraine, but also dangerous. … We call on all states to halt any nuclear cooperation with Russia, avoid getting into any new joint projects with Russia in this field, and impose personal sanctions on the key Rosatom officials and employees who had interfered with Ukraine's nuclear system." Kuleba said imposing personal sanctions is "an absolute bare minimum."

Among the materials Rosatom's subsidiaries offered to provide to the Russian defense industry, the documents show, are aluminum oxide, a vital component for missile fuel. Other goods include lithium-ion batteries to power tanks, missile defense systems and other armaments; 3D-printing technology; and chemical compounds used in aircraft and missile engineering, the documents show.

It's not clear which — if any — of these products were supplied to the Russian arms industry. But in an address to Rosatom on its 15th anniversary in December, Russian President Vladimir Putin praised the nuclear company for its "huge contribution … to developing advanced weapons systems and military hardware and putting them on combat duty."

Pavel Luzin, an independent Russian military analyst, told The Post that Rosatom had developed a long-term strategy to supply the Russian arms industry. The Russian Defense Ministry has also spoken, in official statements, of long-running cooperation on "nonnuclear" weapons with the state nuclear conglomerate.

Customs data provided to The Post by the Washington office of the James Martin Center for Nonproliferation Studies shows that at least one Rosatom subsidiary, Renera, has imported at least $1 million worth of lithium-ion battery components from South Korea since the war in Ukraine began.

Experts said they feared Russia could be using Rosatom's current non-sanctioned status as a way to obtain components that would otherwise be embargoed. "It's a big potential loophole," said Jacob Nell, former chief Russia economist for Morgan Stanley and now a member of an independent group of international experts working with the Ukrainian presidential administration on sanctions against Russia. "Specialized imports are being sourced through the nuclear sphere."

Ukrainian officials say Rosatom's involvement at the Zaporizhzhia plant is reason enough to sanction the company. Groups of about a dozen Rosatom employees, accompanied by a contingent of officers from Russia's Federal Security Service (FSB) and hundreds of armed Russian troops, have been present at the plant since mid-March, soon after the Russian army seized control of it.

Rosatom has sought to shield itself from possible sanctions by insisting that its employees were present only to provide technical advice for the safety of the plant. In August, Rosatom said its staff played no role in the plant's management or defense.

But Rosatom has since officially taken operational control of the plant after Putin unilaterally signed a decree on Oct. 5 transferring the Zaporizhzhia facility to Russian federal property in the wake of Moscow's illegal annexation of the region. The decree handed the use of its "financial, material and other resources" to a new operating company owned by a Rosatom subsidiary, "to ensure the safety of nuclear facilities, radiation sources, storage facilities for nuclear materials and radioactive substances, and radioactive waste storage facilities."

A key Rosatom employee present at the plant since late March, Oleg Romanenko, was appointed general director of the plant's new operating company two days before Putin issued the decree, company registration documents show. The plant's Ukrainian chief engineer was appointed general director of the power station itself. In response to the request for comment, Rosatom said the transfer to the new operating company was aimed at "ensuring the safe and reliable operations of the plant."

"Ensuring nuclear safety and security of nuclear facilities and installations is Rosatom's number one priority," the company said.

One eyewitness, Oleh Dudar, a former head of operations for the Zaporizhzhia plant's nuclear reactor and turbine unit, however, told The Post that Rosatom employees, although initially in the background, increasingly took control over the summer, at first issuing demands to inspect the plant's systems, including its electricity circuits. "The longer time went on, the more they began dictating to us," said Dudar, who has since fled to safety in Western Europe.

Dudar said he believed Rosatom employees, under the direction of Romanenko, had helped direct Russian artillery fire targeting parts of the facility and its territory, including its transmission lines, to cut it off from supplying the Ukrainian electrical grid while blaming Ukraine for the fire. "They were studying the electricity circuits, the transmission lines, our equipment and the communication systems. They were studying them closely over a week," Dudar said. "And then the next week, the objects they'd been studying were shot at and the equipment went offline. It seems to me they were looking. I had the impression that they directed and managed this."

There are also questions about whether Rosatom staff may have been involved in the summary detentions of hundreds of the plant's Ukrainian workers, who, according to Dudar and other witness accounts described by Ukrainian officials, were held for weeks in cellars and tortured by FSB officers, sometimes with electricity. The detentions first occurred over the plant workers' pro-Ukrainian sentiments, but increasingly they were used as a mechanism to force workers to sign contracts to work for Rosatom, according to Dudar.

Kuleba said that Ukraine has moved away from dependence on Russian fuel supplies for its nuclear industry and that other countries should follow suit. In recent years, he noted, Ukraine switched to alternative supplies from Westinghouse, an American company, which developed the capability to replicate Russian-made fuel supplies so it could service Soviet-era plants.

There are signs that the war is beginning to affect Rosatom's bottom line. Bulgaria in the past month signed two supply deals, first with Westinghouse and then with France's Framatome, to replace the Russian fuel its nuclear power plant currently relies on. Last year,

Finland walked away from a multibillion-dollar project with Rosatom to build the Hanhikivi-1 nuclear plant, which would have generated 10 percent of Finland's electricity needs.

An international working group on Russian sanctions working with Yermak, the Ukrainian presidential chief of staff, has proposed a gradual process that would first focus on personal sanctions and barring supplies of raw uranium from Russia, where Moscow has only about 5 percent of the global market, while preparing the ground for a longer-term shift from dependency on Rosatom.

"Serious moves to reduce the West's critical dependency on Rosatom across the nuclear fuel cycle are more important over time," Nell said.

Ukraine War to Accelerate Shift away from Fossil Fuels — BP

The oil major's new energy outlook says countries will turn more quickly to domestic renewables as they look to increase their energy security.

By: Carlos Anchondo | 01/31/2023 06: 58 AM EST

A liquefied natural gas tanker. | Steve Campbell/AP Photo, Houston Chronicle.

Energywire

Russia's invasion of Ukraine could speed up the global transition away from oil and natural gas, according to BP PLC's annual energy outlook.

The report, released Monday by the oil and gas major, says that countries could shift to domestic renewables, prompted by energy security concerns as the war continues with "no end in sight."

"The increased importance placed on energy security as a result of the Russia-Ukraine war leads over time to a shift away from imported fossil fuels towards locally produced non-fossil fuels, accelerating the energy transition," company analysts wrote in the outlook.

That could be particularly true for China, India and the European Union, the outlook said. All rely heavily on imports for their oil and gas consumption, with any decline offset by a "greater consumption of domestically produced renewables," according to the report.

BP considered three scenarios in the report. Two are broadly in line with keeping warming in 2100 to "well below" 2 degrees Celsius when compared to pre-industrial levels, as outlined in the Paris Agreement. A third, called "New Momentum," looks at the current trajectory of the global energy system.

The company found that the latter scenario has changed significantly since its outlook last year, which was released before Russia's invasion of Ukraine. Oil demand in 2035 would be 5 percent lower — and gas demand 6 percent lower — than the 2022 outlook predicted.

The events of the past year "show how relatively small disruptions to energy supplies can lead to severe economic and social costs, highlighting the importance that the transition away from hydrocarbons is orderly, such that the demand for hydrocarbons falls in line with available supplies," said Spencer Dale, BP's chief economist, in a statement released with the outlook.

BP says oil will continue to play a "major role" in the global energy system over the next 15 to 20 years. But its report also emphasizes that the "carbon budget is running out."

"Despite the marked increase in government ambitions, CO2 emissions have increased every year since the Paris COP in 2015 (bar 2020)," the company wrote in a section titled "Core Beliefs." "The longer the delay in taking decisive action to reduce emissions on a sustained basis, the greater are the likely resulting economic and social costs."

The outlook's new momentum scenario has carbon dioxide emissions peaking in the 2020s and dropping by around 30 percent from 2019 levels by 2050. All scenarios assume the use of carbon capture, utilization and storage.

The outlook also looks at the future of liquefied natural gas, finding that the war in Ukraine has "scarred" the prospects for Russian exports.

A growth in global LNG demand out to 2030 is thus met by "a substantial expansion of exports from the US and Qatar," according to the report. It finds that by 2030, the U.S. and the Middle East could together produce about half of global LNG supplies, up from around a third in 2019.

Energy Justice and Climate Change: Key Concepts for Public Health

Climate Change Is a Threat Multiplier

Climate change exacerbates health inequities, disproportionately harming the most vulnerable among us. That includes children and pregnant women, people with low income, the aged and people with disabilities and chronic illnesses, some communities of color, indigenous people and tribal communities, immigrants, marginalized people of all races and ethnicities and outdoor workers, to name a few.

Difficulty in accessing the political, economic, social and environmental resources that enable people to cope with climate threats, such as extreme heat and natural disasters, can contribute to a potentially unmanageable energy burden as climate change worsens.

U.S. communities that lack accessible, affordable household energy services often have poorer health, fewer educational and economic opportunities, limited political representation and inadequate access to health care. As climate change worsens, alternative energies, such as solar or wind are becoming competitive at similar, or even lower, prices than fossil fuels like coal and natural gas.

Alongside community leaders, public health professionals have a role to play in monitoring, evaluating and supporting this transition to a healthier energy supply. A thoughtful energy transition provides opportunities for mitigating climate change and promoting health equity at the same time.

How Do Structural Racism and Climate Change Contribute to Energy Burden?

In the same way that climate change disproportionately affects certain populations, energy is a health equity issue. For over 60 years, the environmental justice movement has raised awareness that people of color and low-income communities often bear the brunt of "brown" environmental issues. Meanwhile, white and more affluent communities enjoy the benefits of "green" environmental benefits.

In the context of energy, race – even more than class, is the number one indicator for the placement of toxic facilities in this country. Housing in general is one of the best known and most well-documented social determinants of health. The affordability, location and quality of housing all have been independently linked to health. It is important to acknowledge the effect that historical practices of redlining and discriminatory mortgage lending have had on our nation.

Spotlight: Heat Islands

Poor-quality housing and blighted neighborhoods diminish property values, increase crime and erode the cohesiveness and political power of communities. As a result, many people of color and low-income communities in urban areas lack the infrastructure investment for things like green spaces that are critical to cooling urban heat islands.

These neighborhoods get much hotter on average than other areas, negatively impacting the health of residents and increasing energy usage. Residents who rent their property lack the long-term incentive to invest in costly air conditioning units for a short-term residence. And with the exception of Dallas and Phoenix, there is no law requiring landlords to provide air conditioning the way they must provide heat. This places renters at higher risk.

Through ongoing racial discrimination and worsening climate change, communities of color face increasingly impossible trade-offs when managing the cost of their undue energy burden and other household needs.

Key Concepts

Energy Insecurity

An inability to adequately meet basic household energy needs, such as gas, appliances, plumbing, heating, cooling. In 2015, 31% of U.S. households reported struggling to pay energy bills or to maintain adequate heating and cooling in their homes. About one in five households reported reducing or forgoing basic necessities like food and medicine to pay an energy bill. Eleven percent of households reported keeping their home at an unhealthy or unsafe temperature due to costs.

Case Study
The U.S. Low Income Home Energy Assistance Program assists eligible low-income households with their heating and cooling energy costs and provides bill payment assistance, energy crisis assistance, weatherization and energy-related home repairs. For more information: www.benefits.gov/benefit/623

Energy Burden
The proportion of a household's monthly income that is spent on home energy costs. Low-income households often face an energy burden that is three times higher than other households.

Case Study
The Kentucky Utilities WeCare program stands for Weatherization, Conservation Advice and Recycling Energy. The program assists with home weatherization for low-income populations, leading to a decreased energy burden through increased energy efficiency. For more information: www.lge-ku.com/wecare.

Energy Equity
Fair distribution of benefits and burdens from energy production and consumption. Energy equity requires equitable distribution of risks and hazards, as well as available and affordable energy sources. While unfamiliar to many residents, equity-centered energy and utility policies significantly enhance household economic stability and improve the overall quality of air, water and other natural resources that affect our health and well-being. Working directly with communities to build their capacity around energy planning and decisionmaking is key to advancing equity.

Case Study
Partnership for Southern Equity's Just Energy Circle and Just Energy Academy works to eliminate energy insecurity by advancing policies and institutional actions that promote racial equity and shared prosperity for all in the growth of metropolitan Atlanta and the American South. For more information: www.sites.google.com/view/pse-just-energy-circle and www.justenergyacademy.org.

Energy Justice
The principle that all people should have a reliable, safe and affordable source of energy; protection from a disproportionate share of costs or negative impacts or externalities associated with building, operating and maintaining electric power generation, transmission and distribution systems; and equitable distribution of and access to benefits from such systems. A global energy system that 1) fairly disseminates the costs and benefits of energy services and 2) contributes to more representative and impartial energy decisionmaking.

Case Study
The Energy Democracy Project and the Initiative for Energy Justice develop resources to build collective power and advance the movement to democratize energy by providing law and policy resources to advocates and policymakers. For more information: www.energydemocracy.us and www.iejusa.org

Environmental Justice
All people and communities have the right to equal environmental protection under the law, and the right to live, work and play in communities that are safe, healthy and free of life-threatening conditions. Environmental justice is often associated with environmental racism, in which, whether by conscious design or institutional neglect, actions and decisions result in the disproportionate exposure of people of color to environmental hazards and environmental health burdens.

Case Study
WE ACT for Environmental Justice in New York City builds healthy communities by ensuring that people of color and low-income residents participate meaningfully in the creation of sound and fair environmental health and protection policies and practices through member participation in education, policy, advocacy, research and more. For more information: www.weact.org

Just Transition
A set of principles and practices focused on changing the energy industry from fossil fuels to alternative energy sources, which ensures workers' rights and offers economic opportunities for affected communities. This cultural, as well as economic, shift emphasizes sustainability in the face of climate change.

Case Study
Coalfield Development trains former coal industry workers in Appalachia for jobs in sectors such as solar energy, sustainable construction and mine-land reclamation. For more information: www.coalfield-development.org

Heat Equity
The development of policies and practices that mitigate heat islands and help people adapt to the impacts of extreme heat in a way that reduces the inequitable distribution of risks across different populations within the same urban area.

Case Study
In Dallas, the National Center for Healthy Housing worked together with the Health and Wellness Alliance for Children to update the "Minimum Urban Rehabilitation Standards" of the Dallas City Code Housing Ordinance, requiring landlords to provide air conditioning in all residences of Dallas the same way heat is currently required. For more information: www.nphw.org/nphw-2020/-/media/files/pdf/topics/environment/built_environment/housing_ordinance_dallas.ashx

Note: Key concept definitions were courtesy of a variety of sources. For references, please email climatechange@apha.org.

Improving Equity Outcomes for New Federal Investments in Clean Energy Infrastructure

The Infrastructure Investment and Jobs Act (IIJA), which passed Congress in 2021 with bipartisan support, makes a historic and necessary $1.2 trillion investment in modernizing America's infrastructure. Of that total investment, nearly $400 billion will be directed to clean energy infrastructure including initiatives for transportation, power and grid systems, technology demonstration programs, modernized permitting, and climate resilience. The American Association of Blacks in Energy (AABE) and the Bipartisan Policy Center (BPC) are working together to explore the opportunities and challenges inherent in these new federal investments.

Public investment has a poor record of keeping pace with infrastructure needs, and infrastructure disparities persist for many disadvantaged Americans, including low-income, rural, Black, Latino, Asian, and Indigenous communities. AABE and BPC believe that the IIJA, if implemented carefully and effectively, offers a once-in-a-generation opportunity to create jobs, accelerate the clean energy transition, and grow the U.S. economy in ways that are both sustainable and equitable.

Our organizations convened a series of conversations with environmental justice advocates, government officials, and business and industry leaders to discuss options for ensuring that IIJA investments create direct value and economic opportunity in underserved communities, support the build-out of an inclusive clean energy workforce, and address the negative effects of past infrastructure decisions. Building on these discussions, AABE and BPC identified three categories of clean energy investment—foundational investments, remedial investments, and resilience investments—that can support more equitable outcomes for communities in distinct and important ways. By addressing communities' investment needs in each of these categories, federal officials, private sector executives, and state and local leaders can help ensure that IIJA-funded projects improve equity outcomes.

1. Foundational Investments: Targeting areas of historic underinvestment with the aim of modernizing energy infrastructure and supporting communities' full economic participation and well-being.

In many historically underserved communities, a basic level of energy and technology infrastructure and service either doesn't exist, isn't affordable, or isn't accessible to a large portion of residents. For example:

- Black, Latino, Indigenous, and low-income households pay two to three times more for energy than the national average as a share of their monthly income.[69]

[69] World Resources Institute. 2021. *Addressing Energy Equity in the United States: Policy considerations for federal investments*. https://www.wri.org/research/energy-equity-united-states-federal-investment.

- Over 24 million Americans, 80% of which live in rural and Tribal communities, lack access to fixed broadband services at threshold speeds.[1]
- Rural, low-income, Island, and Indigenous communities pay some of the highest costs for electricity and heating fuel in the country. At the same time, many of these households face geographic, financial, and access barriers that limit their ability to use clean energy technologies, like roof-top solar or efficiency improvements to reduce their energy bills.[70]

Foundational infrastructure investments at both the system and household level are needed to alleviate energy burdens and equalize clean-energy-related economic opportunities. Examples of system improvements include expanding electricity transmission and distribution infrastructure and bringing high-speed internet to rural and other underserved areas to support advanced energy technologies. Foundational investments for low-income households and multi-family buildings may include the improvement to housing quality and the installation of new behind-the-meter water, gas, and electrical equipment to support energy-efficiency, expand clean-energy access, and create positive economic and health outcomes.

A Closer Look:

Expanding access to high-speed internet in rural communities can deliver economic and climate benefits.

The IIJA makes a historic investment of $65 billion in high-speed internet infrastructure. Funds will go toward making connectivity affordable and expanding broadband access in rural, Tribal, and underserved communities. Access to high-speed internet facilitates digital energy efficiency, more weather-resilient electric systems, increased reliability, and lower energy costs. It also supports precision land management strategies that use web-enabled measurement and data analytics to improve and optimize the performance of agriculture and forestry producers. These practices can reduce fertilizer use and lower production costs while abating emissions and improving water quality. It has been estimated that expanding broadband access in underserved communities could reduce residential electricity bills by as much as 20% while creating $47–$65 billion in annual economic benefits. Read more.

2. Remedial Investments: Aiming to correct for or eliminate existing infrastructure deficiencies resulting from neglect, harm, or obsolescence.

Low-income, rural communities, and communities of color disproportionately contend with the economic, health, and environmental impact from the inadequate or polluting legacy of energy infrastructure, leaving them underprepared for the clean energy transition. Remedial investments can address these burdens and remove barriers to access new clean infrastructure. For example:

[70] Krieger, E. M., Casey, J. A., & Shonkoff, S. B. (2016). A framework for siting and dispatch of emerging energy resources to realize environmental and health benefits: Case study on peaker power plant displacement. *Energy Policy*, *96*, 302-313.

- Black Americans are more likely than others to be exposed to harmful air pollutants from electricity generation, especially by "peaker" power plants.[2] Such plants run infrequently, typically during periods of high (or peak) electricity demand and are generally less efficient and more polluting than other power plants. Energy storage facilities (e.g., battery storage systems or solar-plus-storage projects) are being proposed and built in some regions to replace peaker plants and meet peak electricity demand while minimizing local air pollution impacts.[71]
- Most climate-change mitigation strategies call for the widespread adoption of clean distributed energy resources (DERs) such as household solar panels and battery storage. Yet local infrastructure that is in disrepair or has not been upgraded can be a barrier to siting DERs or interconnecting new residential solar projects. Local electric grid capacity limits have been found to limit low-income, Black, and Asian households' access to DERs in California.[72] Investments to expand DER hosting capacity in underserved communities can enable more equitable participation in and benefit from clean energy.

A Closer Look:

Targeted redevelopment with clean energy can address local legacy pollution and reduce carbon emissions.

The IIJA provides $1.5 billion in grants, technical assistance, cooperative agreements, and direct support to states and local and Tribal governments to repair brownfield sites and enable environmental and economic revitalization. The Environmental Protection Agency (EPA) is reviewing a series of broad policy actions to protect communities near industrial sites with water and air quality issues.[73] For example, one EPA initiative, called RE-Powering America's Land, encourages renewable energy development on current or formerly contaminated lands, landfills, and mine sites. When consistent with the host community's vision for future development, clean energy projects on underutilized sites like landfills and brownfields can provide significant economic recovery opportunities. In many cases, these sites already possess existing transmission, distribution, transportation, and other support infrastructure that can be leveraged for renewable energy development, reducing project installed costs by $45–$113 per kilowatt of generating capacity.[74] EPA has identified more than 130,000 sites across the country that are potentially suitable for solar, wind, biomass and geothermal energy development. For example, the developer of a 3.3-megawatt solar project on a former auto salvage yard in Vermont was able to remediate the land beneath the solar arrays with pollinator-friendly ground cover; meanwhile land lease payments made it possible for the site's owners to stay on land that had been in the family for generations. Read More.

---

[71] Pacific Northwest National Lab. (2021). Energy Storage for Social Equity: Capturing Benefits from Power Plant Decommissioning.

[72] Brockway et al. Nature Energy. *Inequitable access to distributed energy resources due to grid infrastructure limits in California*. (2021).

[73] Environmental Protection Agency. *EPA Administrator Regan Announces Bold Actions to Protect Communities Following the Journey to Justice Tour*. (January 2022).

[74] Environmental Protection Agency. *RE-Powering America's Land Initiative Discussion Paper The Value of Existing Infrastructure for Renewable Energy Development* (2020).

Remedial projects include both front-end and back-end investments. On the front end, assessments of the state of current infrastructure systems (e.g., electric, water, transportation) and impact on the well-being of vulnerable communities will be needed, together with determinations about the readiness of these systems for 21st century upgrades. On the back end, long-term mechanisms to fund the operation and maintenance of new infrastructure are needed to prevent future deficiencies.

3. Resilience Investments: Improving energy infrastructure so that communities can better withstand the impacts of extreme weather and climate-induced hazards.

Economic and infrastructure disparities make low-income communities, rural communities, and communities of color more vulnerable to extreme weather events and other adverse effects of a changing climate. For example:

Hispanic and Latino individuals are 43% more likely to live in areas with high labor losses due to high-temperature events, impacting both their economic and health outcomes.[75] Island and Tribal communities are more likely to live in areas at the highest risk of sea level rise, endangering the survival of their communities.[7]

Climate change exacerbates the deficiencies of aging critical infrastructure (e.g., water and sewage systems, roads, bridges, and power plants), and increases exposure to damage or loss from extreme weather. Low-income, Black, and Latino populations in particular experience disproportionate hardships from infrastructure service disruptions.[76]

Underinvested, low-quality, poorly maintained, or high-density housing infrastructure can make communities more vulnerable to the adverse impacts of climate-change-induced extreme weather events, including storms and extreme heat, and indoor air quality problems.[8]

Resilience investments represent an opportunity to deploy clean-energy and energy-efficiency technologies not only for their climate mitigation and economic benefits, but also to bolster the ability of historically underserved communities to withstand and recover from climate-change impacts. Examples of community-led resilience projects include making climate-smart upgrades to local critical infrastructure, installing microgrids with distributed generation or energy storage to provide electricity during disasters, and weatherizing homes to increase protection from extreme weather and use less energy.

A Closer Look:

Remote, Island, and Tribal communities can enhance their energy resilience using new building and grid technologies.

The Department of Energy's National Labs have launched several programs to build climate resiliency in local communities. These programs are being undertaken in partnership with residents and regional organizations.

---

[75] Environmental Protection Agency. *Climate Change and Social Vulnerability in the United States*. (Sep. 2021).

[76] USGCRP, 2016: The Impacts of Climate Change on Human Health in the United States: A Scientific Assessment. Crimmins, A., J. Balbus, J.L. Gamble, C.B. Beard, J.E. Bell, D. Dodgen, R.J. Eisen, N. Fann, M.D. Hawkins, S.C. Herring, L. Jantarasami, D.M. Mills, S. Saha, M.C. Sarofim, J. Trtanj, and L. Ziska, Eds. U.S. Global Change Research Program, Washington, DC, 312 pp. http://dx.doi.org/10.7930/J0R49NQX.

- The National Renewable Energy Laboratory's Energy Transitions Initiative Partnership Project supports remote and Island communities with tailored, community-driven approaches for transitioning to clean and resilient energy systems.
- Pacific Northwest National Laboratory and Sandia National Laboratory are working jointly on the Energy Storage for Social Equity Initiative, which helps urban, rural and Tribal communities assess and plan for energy storage projects.
- A team of national labs is supporting Puerto Rico's Grid Resilience and Transition to 100% Renewable Energy study (PR100). The study will assess prospects for deploying renewable energy and other technologies for energy storage, distributed generation, distribution control, electric vehicles, and energy efficient and responsive loads.
- The National Renewable Energy Laboratory's Cold Climate Housing Research Center is co-developing and deploying 20 culturally responsive, high-performance buildings in Alaska that incorporate resilience to severe climate impacts and local energy generation.

China Invests $546B in Clean Energy, Far Surpassing U.S.

The country accounted for nearly half of the world's low-carbon spending in 2022.
By: Sara Schonhardt
01/30/2023 06:47 AM EST

Beachgoers walk near wind turbines along the coast of Pingtan in Southern China's Fujian province last year. China is poised to be the top investor in clean energy for years to come. AP Photo/Ng Han Guan

Climatewire

China once again topped the world in dean energy investments last year, a trend that could challenge U.S. efforts to develop more homegrown manufacturing.

Nearly half of the world's low-carbon spending took place in China, according to a *recent analysis* from market research firm BloombergNEF. The country spent $546 billion in 2022 on investments that included solar and wind energy, electric vehicles and batteries.

That is nearly four times the amount of U.S. investments, which totaled $141 billion. The European Union was second to China with $180 billion in clean energy investments.

China also dominated in low-carbon manufacturing, accounting for more than 90 percent of the $79 billion invested in that sector last year, according to the report.

The findings come as the U.S. and Europe work to expand domestic manufacturing capacity. The United States has begun in recent months to roll out the benefits of the Inflation Reduction Act, which is packed with $369 billion of incentives aimed at building up the U.S. clean energy industry.

But competing with China's networks won't be easy.

"China has managed to nurture these really integrated, efficient value chains for making things like solar panels, for making things like battery cells," said Antoine Vagneur- Jones, head of trade and supply chains research at BloombergNEF.

Both now make up a substantial percentage of China's export revenues, he added. But China is also investing heavily in domestic supply chains, including wind turbine components for a massive expansion in offshore wind deployment, Vagneur-Jones said.

The Inflation Reduction Act aims to counter China's dominance by offering billions of dollars in subsidies to domestic manufacturing. One major provision, for example, will offer consumers thousands of dollars in tax credits when purchasing an electric vehicle - but only when the bulk of battery components are made or assembled in North America.

Those incentives have generated strong international criticism, particularly from the European Union. But President Joe Biden has been clear what he's aiming for, saying in a speech last week that "the supply chains are going to start here" (*Climatewfre,* Jan. 27).

BloombergNEF's figures don't account for the dozens of planned U.S. factories announced in recent months. But while Inflation Reduction Act incentives will start to make a dent in China's share of manufacturing, Vagneur-Jones said, it will happen gradually.

"Even in a sort of, say, optimistic scenario, I think China's still going to remain dominant for the next decade or more. But it could mean that we see sort of these parallel supply chains develop," he said.

Take battery cell manufacturing. It's expected to increase sixfold by 2030, but China's share of the market is projected to decline to just 70 percent from 79 percent today, BloombergNEF research shows.

That's partly because China plans to expand its battery manufacturing capacity; even as supply chains get built upelsewhere. There are also barriers to ramping up clean energy manufacturing in the United States, including higher costs and the challenges of developing a new sector from scratch.

The task could be even more difficult if tensions between China and the U.S. increase, curtailing the ability of U.S. companies to access Chinese expertise on clean energy development, Vagneur-Jones said.

"The offshoot is not just that production will be more expensive in places like the U.S., which it would be, but also that it might be very difficult to really scale things up fast, because of the fact that you can't tap into that Chinese expertise," he said.

There are risks to having supply chains so heavily concentrated in one geographical area. Geopolitical tensions can more easily disrupt trade, as can the impacts of climate change. For example, a major heat wave led to power shortages last August in a Chinese province that supplies a significant amount of the world's solar panels and components - leading to a price increase for those goods.

The next few years will be decisive, Vagneur-Jones said. The United States, Europe, India and Indonesia are all ramping up their domestic manufacturing efforts. The question is whether those plans bear fruit.

"We could see a turning point for this trend that we've seen, which has been China's dominance in not just the existing capacity that's been built, but also ongoing investments," he said.

The Honorable Paul Dabbar
CEO, Bohr Quantum Technology
Former Under Secretary
U.S. Department of Energy
52 Secor Road
Scarsdale, NY 10583

Kaitlyn Peterson
Legislative Clerk
Committee on Energy and Commerce
2125 Rayburn House Office Building
Washington, D.C. 20515

Dear Ms. Peterson:

Thank you for allowing me to appear before the Committee on Energy and Commerce on Tuesday, January 31, 2023, to testify at the hearing entitled "American Energy Expansion: Strengthening Economic, Environmental, and National Security."

Pursuant to the Rules of the Committee on Energy and Commerce, I have been asked to answer additional questions for the record. The questions you sent to me, and answers, are attached. As requested, the format of my responses to these questions are as follows: (1) the name of the Member whose question I am addressing, (2) the complete text of the question I am addressing in bold, and (3) my answer to that question in plain text.

Paul Dabbar

Attachment-Additional Questions for the Record

The Honorable Michael Burgess, M.D.

1. Mr. Dabbar, Texas' population has grown exponentially over the previous decade. With that large increase in population, our grid suffers under tremendous strain because of the lack of investment in baseload power generation. Regrettably, it seems that too much emphasis has been placed on the expansion of renewable capacity. Mr. Dabbar, in your opinion, what role can Congress play in creating a resilient and diverse baseload generation ecosystem?

The Independent System Operators ("ISO"), including ERCOT, the Texas ISO, have not implemented rules to maintain base load generation as intermittent renewables have been added to the systems. Congress should look to reform the Federal Power Act to direct all systems, including the intra-state ERCOT system, to better support base load generation, especially as intermittent renewables are added. There are various ideas on how to word that requirement. One idea is mandating the ISOs to better model the instability that renewables are adding to the grid, and properly estimate the increasing needs of dispatchable generation as a result. And to formalize contracting or compensation systems for that dispatchable power to ensure that it gets built.

2. In your testimony, you discuss reforming certain regulations regarding our nations electrical grid, specifically, reforming regulations concerning Independent System Operators (ISO) that would encourage more baseload generation capacity to be built. Can you speak to those reforms?

The ISO's and FERC are mismanaging the supply needs of power sector in many regions of the U.S. Leading in many places to more power plants being retired than are being built. And this increasing supply risk is be exasperated by growing power demand from electrification. This is leading to higher outages and increasing prices.

Given FERC is not currently setting market rules to properly incentivize base load power from being built or maintained, it is imperative that Congress look at FERC reform legislation, including reforming the Natural Gas Act and Federal Power Act ("FPA"). On topics such as strong requirements that natural gas infrastructure get approved, and not disapproved by supposed views on siting that fall outside the intent of Congress, and the West Virginia vs EPA standard. The FPA could be amended to more specifically mandate base load support in ISO rules approved by FERC.

Finally, strong NEPA reform legislation should be looked at by Congress. Including narrowing maximum time frames for reviews. And closing gaps opened by decades of court rulings expanding challenges on the review process.

3. What are the benefits of expanding America's LNG exports, both domestically and globally?

LNG exports domestically have great value to both the U.S. economy, and national security posture.

Energy exports have a significantly positive impact on the U.S. balance of trade.

In addition, jobs from natural gas production, processing, transportation, and liquefaction are very numerous, and they jobs are very high paying. And given these jobs are a form of manufacturing jobs in their regions, they tend to spawn many other local jobs in their communities. And jobs from energy production and sales of gas and oil are far more numerous and higher paying than jobs for renewables.

4. What are the current challenges to expanding the production and export capability of LNG?

The largest impediment is siting of transportation of national gas from production regions to the liquefaction sites on the coast. New pipeline siting is near impossible to accomplish today.

The Honorable Richard Hudson

1. What is the government's role in protecting critical grid infrastructures from all hazards, including physical attacks?

Government has several important roles:

a) National Intelligence monitoring potential attacks.
b) Collecting that through the various Intelligence Community entities, such as NSA, CIA, and through our allied intelligence partners
c) Communicating with grid owners (private, municipal, and federally owned utilities) about attacks. Including having classified channels that allow prompt notification from utilities when they are getting attacked. And for communicating to utilities when the IC identifies an impending attack.
d) Coordinating response to prevent or react to an attack.
e) Provide possible counter-attack efforts, including utilizing cyber-attack capabilities of DOD and IC elements (such as Cyber Command).
f) Helping provide security standards around power equipment manufactured by countries of risk.
g) Helping coordinate recovery from grid or power plant attack incidents.
h) Helping develop new software and hardware technologies that are more resilient.

2. What technologies or methods have been used to ensure the grid remains resilient to outages and disruptions? How should Congress support the industry as they look for more innovative methods?

Each utility has their own information security program, as does each of the regional Independent System Operators.

However, cyber security programs between the DOE and the utilities, led by the DOE Cybersecurity, Energy Security and Emergency Response Office, is the most important program for grid security for the country. Note, this is a Senate confirmed assistant secretary position, but the administration has decided to not nominate an assistant secretary.

I would note that the electric system cyber response coordinated by DOE is much better developed than pipeline security programs, which is split between the Transportation Department and DOE. And there is much less coordination and partnership between the federal government and the pipeline industry. As a result, the risks to natural gas, oil, and oil products transmission are less addressed.

The in addition to the cybersecurity efforts described in #1 above, the same office at DOE is tasked with helping coordinate emergency response support when storms or other natural disasters effect the grid, in conjunction with the Edison Electric Institute and other trade associations. In general, power industry emergency response performs well across the U.S.

3. In a piece for the Hill this past spring, you said the U.S. can concurrently increase energy availability, reduce costs and reduce emissions. As the Energy and Commerce Committee continues to look for legislative solutions that do all three, what are some the most cost-effective investments we should be looking to make?

Investment in innovation for new energy technologies is the most impactful and cost-effective investment that can be made by the federal system. The U.S. leads the world by far in innovation for new energy technologies. And those past investments have made significant impacts to concurrently increase energy availability, reduce costs and reduce emissions. Examples of American energy innovation that accomplished that include nuclear power, solar PV, lithium ion battery chemistry, material gas turbine improvements, 3-D seismic imaging, directional drilling and hydraulic fracturing.

While the last Congress focused on increasing support for deployment of already proven technologies, the private sector has significant resources and more experience for those efforts. But the private sector under invests in early stage innovation. The best place for the taxpayer to invest is where the private sector will never be focused, and that is in early stage innovation and discovery.

The Honorable Greg Pence

As NERC and MISO continue raising the alarm of a reliability crisis facing our electric grid, the Biden Administration is charging forward with actions to drive out baseload generating supplies. I am concerned that the forced transition to electrification will compound these issues and upend an energy system that has taken over a century to build. Meanwhile, energy demands across the country are expected to continue outpacing energy supplies during times of high demand. For Hoosiers in Southern Indiana, that means less reliable access to energy at a higher price.

1. Do you agree that the overreliance of intermittent energy sources at the expense of baseload generating units are deteriorating the reliability of our electric grid?

Yes I agree. It is an undisputable fact from a system engineering analysis that replacing dispatchable and far more available power supply (such as from gas, coal, and nuclear plants), and replacing it with far less available and dispatchable power supply (such as wind and solar) reduces reliability. One can debate by how much in any particular situation, but is a clear fact it reduces reliability, and any debate needs to start from that systems engineering fact. While wind and solar have definite advantages on marginal cost and emissions, they have several engineering, power density, and other downsides compared to other types of generation.

And while lithium-ion batteries have certain performance and cost effective deployment options for localized grid deployment, at $1.2 million per MW for only four hours of supply at rated capacity, it is far too expensive for large scale, longer term back up.

As renewables increase their penetration in markets, it is therefore important to increase dispatchable power from power plants such as nuclear and gas, or in certain geographies geothermal and hydroelectric could be added. Or we need to develop cheaper and better performing new technologies, such as new battery chemistries, or fusion.

2. What sort of actions can Congress take to ensure that grid reliability is prioritized when considering electricity market decisions at the ISO/RTO level?

The ISO's and FERC are mismanaging the supply needs of power sector in many regions of the U.S. Leading in many places to more power plants being retired than are being built. And this increasing supply risk is be exasperated by growing power demand from electrification. This is leading to higher outages and increasing prices.

Given FERC is not currently setting market rules to properly incentivize base load power from being built or maintained, it is imperative that Congress look at FERC reform legislation, including reforming the Natural Gas Act and Federal Power Act ("FPA"). On topics such as strong requirements that natural gas infrastructure get approved, and not disapproved by supposed views on siting that fall outside the intent of Congress, and the West Virginia vs EPA standard. The FPA could be amended to more specifically mandate base load support in ISO rules approved by FERC.

Finally, strong NEPA reform legislation should be looked at by Congress. Including narrowing maximum time frames for reviews. And closing gaps opened by decades of court rulings expanding challenges on the review process.

Indiana has a long, storied history in energy intensive industries like steel and manufacturing. Across Indiana's Sixth District, companies are taking advantage of the expertise of Hoosier innovators to bring back domestic industries and secure global supply chains. However, these entities will need access to reliable, affordable energy supplies like natural gas to stay competitive.

1. How are actions by the Biden Administration to inhibit adequate energy distribution impacting manufacturing and industrial sectors?

Manufacturing facilities many times must use coal or natural gas for their processes vs using electricity. They might be able to use hydrogen in the future, but for many facilities there are no alternatives for foreseeable future.

As a result, as coal is being utilized less because of emissions concerns, the demand for natural gas for manufacturing is going up. And as more manufacturing is being built in the U.S., this adds additional demand for natural gas. In addition, as coal fired power plants are being retired, with more dispatchable & base load replacement generation needed that is primarily using natural gas, that puts even more pressure on natural gas transmission that can affect natural gas availability for manufacturing.

Given FERC, EPA and other government agencies are restricting new pipelines that are needed for manufacturing increasingly, this is effecting manufacturing several ways:

- Costs for currently installed manufacturing using this type of energy is going up. As more users increasingly compete to use the same pipelines.
- Areas of the country that need new natural gas infrastructure for new manufacturing facilities are less likely to get new manufacturing sited in their region.
- Areas of the country with existing extensive natural gas infrastructure, or have accelerated intra-state-only needed regulatory approvals for that infrastructure, have an advantage for siting new manufacturing facilities.

2. Can industries such as steel and manufacturing adequately replace their energy needs with wind or solar?

Many industrial processes are hard to operate solely with electricity as their energy input. Many need far higher heat input than can be generated by electricity.

Currently many of those can only use natural gas or coal for their processes. There are prospects that hydrogen could be a replacement, currently that is not an energy source that has wide availability. In addition, the cost of hydrogen vs natural gas is materially higher.

Additional innovation in hydrogen production, transportation and storage will be needed to have it be a viable replacement resources.

The Honorable Russ Fulcher

While geothermal covers a small part of our population today, it offers baseload reliability at utility-level scale according to the Energy Information Administration (the EIA). For example, Idaho's Raft River facility can sell up to 10 MW of this clean, reliable energy source. Yet, permitting for new geothermal projects can take 7-10 years.

1. My bill, the Enhancing Geothermal Production on Federal Lands Act, would allow for geothermal development where there are already existing oil and gas leases. Can you speak to how changes like this that remove duplicative red tape can help America return to energy dominance?

The geology issues that need to be reviewed for oil & gas drill are very similar. Directing agencies reviewing geothermal applications to use already available information, reviews and conclusions from previous oil & gas approvals would be good policy.

2. I understand that transmission development is one of the key factors that could limit deployment of geothermal technologies across areas of the West, like Idaho. This is largely due to geothermal being located far from load / population centers and the sheer amount of federal land the transmission lines would have to cross in order to meet demand. What barriers exist to developing transmission to help clean, baseload technologies like geothermal to provide reliable power to the grid?

In general geothermal transmission access issues are similar to the issues all other types of powers plant have to endure today. The levels of review under NEPA and other siting legislation significantly impede approval processes, and result in many power facilities never being built.

I would recommend that as you look at potential reforms to the Federal Power Act, that you look at mandating accelerated reviews for baseload power such as geothermal.

Hydroelectric constitutes roughly half of my state's total in-state generation. And, it speaks to the importance of Idaho's dams.

1. Can you discuss any transmission challenges that may occur with getting this highly reliable and clean energy to population centers, given these dams are often nested between large sections of federal land?

For the most part, hydroelectric facilities in your state and around the nation were built many decades ago, and as a result the transmission to off-take that power was also adequately built quite a long time ago. However, in many areas of the west, additional wind and solar plants have connected to the grid that have historically been primarily used for those hydroelectric facilities, putting power transmission capacity pressures on those lines. As a result, in those areas, those transmissions lines are under more pressure than in the past. It is therefore incumbent on FERC and ISO's to concurrently authorize expansion of transmission as needed when renewables are added, so as the hydroelectric power does not become restricted. If there are restrictions to added transmission capacity on that federal, state, municipal or private land, that should be taken into account before wind or solar are permitted in that location.

The Honorable Mariannette Miller-Meeks, M.D.

Like any engineered system, wind facilities in Iowa and across the country experience some deterioration over the course of their lifetime. For U.S. wind fleets, this degradation involves a step-change in performance after a decade of operation, according to a 2020 study. Unique to the United States, the study shows a significant drop in performance by 3.6% after 10 years. This level of degradation reduces a wind facility's output, increasing the cost of the electricity it produces.

The federal government has supported efforts to reduce turbine degradation. From 2016 to 2022, the ARPA-E program at DOE supported a project led by the University of Virginia to design the world's largest offshore turbine that allows blades to deflect in the wind, much like a palm tree, to withstand hurricane-wind speeds, reducing rotor mass and fatigue.

1. Are you familiar with the issue of wind turbine degradation; how big of an issue is it; and are you aware of other opportunities or research to maximize wind turbine performance over time?
   I am familiar that some of the larger wind turbine operators have developed a significant turbine sensing network to real-time identify wearing of their turbines all over the U.S. And they have contributed all that real time data over time to a master artificial intelligence turbine operations and maintenance program that is quite powerful. Those systems have significant value in predicting degradation, and developing maintenance plans far more predictive than direction from the turbine manufactures. That wind turbine degradation data and AI planning has materially improved performance of wind turbines.

As you know, a controlled nuclear fusion reaction at Lawrence Livermore National Laboratory produced more energy than it consumed in December 2022. The milestone, known as "fusion ignition," is unprecedented, according to DOE. Iowa's last nuclear plant shut down in 2020, but from a national standpoint, nuclear fusion would be revolutionary for the United States.

2. Can you contextualize the recent nuclear fusion breakthrough and how long you believe it will take for the technology to be commercially viable?
   The fusion net-energy generation accomplishment at LLNL was a significant science advancement, and has significant value for the DOE Defense Program stockpile stewardship mission. But facility operational characteristics and net-energy efficiency is still quite distant for a power plant to made with that specific facility design. In addition, the accomplishment shows that the specific design, Inertial Confinement, has prospects.
   It does importantly provide visibility on the fusion plasma dynamics for power plants. Currently there are more commercial efforts for fusion power plants in magnetic confinement, and other designs. And this LLNL accomplishment shows that innovation in the fusion sector is making strides, and worth additional support.

As you may know, in 2021, Iowa was the second-largest wind power producer in the country and wind turbines generated 58% of Iowa's electricity, the highest wind power share for any state.

My colleagues and I are clear that we cannot flip the switch overnight to renewable energy, but I believe we can make investments and address policy gaps to support renewable energy technologies as part of the "all of the above" energy mix.

3. Can you please speak to some of areas Congress needs to engage to help support base load fuels and "complementary technologies" as we continue to integrate renewables into the national energy mix?
   In regards to what Congress needs to look at to support base load fuels: The Independent System Operators ("ISO") have not implemented rules to maintain base load generation as intermittent renewables have been added to the systems. Congress should look to reform the Federal Power Act to direct all systems, including the intra-state ERCOT system, to better support base load generation, especially as intermittent renewables are added. There are various ideas on how to word that requirement. One idea is mandating the ISOs to better model the instability that renewables are adding to the grid, and properly estimate the increasing needs of dispatchable generation as a result. And to formalize contracting or compensation systems for that dispatchable power to that it gets built.
   In regard to support of "complimentary technologies," in addition to the above, it is clear that in many places of the country, base load power plants name plate capacity are being retired faster than intermittent power plants are being added. And as a result, there has been a rapid increase in risks and prices in the wholesale power markets. Near term support for available dispatchable power plants at scale for most of the U.S. can only come from natural gas fueled power plants. And given the scale of retirements of coal and nuclear around the U.S., the amount of natural gas plants needed is material, and immediately needed.

Longer term, additional dispatchable might be provided from better battery chemistries than we have today, hydrogen for use in simple cycle or combined cycle turbines, or fusion.

Committee on Energy and Commerce, Tuesday, January 23, 2023, Hearing Entitled "American Energy Expansion: Strengthening Economic, Environmental, and National Security," Additional Questions for the Record for Robert McNally, Submitted March 7, 2023

The Honorable Richard Hudson

How Can the Grid's Reliability and Security Be Improved in the Short-Term and Long-Term?

I suggest three broad policy strategies to enhance grid security and reliability.

1) Enact permitting reform to enable timely and substantial investments in generation, transmission, and efficiency improvements. Permitting reform is required for all forms of energy resource development, including hydrocarbon, geothermal, and renewable sources. While energy investment decisions are based on various market and industrial factors outside the direct control of Congress, removing unnecessary and debilitating permitting delays is necessary. Permitting reform is especially important for transmission capacity projects due to their long lead times compared to generation projects.
   - NEPA review process alone takes an average of 4.5 years to complete and adds an estimated $4.2 million to project costs.[77] One astute analysis noted "[t]he NEPA process is tailor-made for NIMBYism."[78]
   - A recent Brookings study of renewable permitting delays aptly described the unacceptable situation the US currently faces for both renewable and thermal energy projects: "All told, the dizzying array of permits and other regulatory obstacles to renewable energy projects can create extremely long delays. Reports of 10-year or longer timelines for transmission lines are not uncommon."[79]
   - Reform should include imposing legally enforceable deadlines on federal infrastructure reviews, amending NEPA to disallow steps taken by President Biden's January 2023 CEQ Notice of Interim Guidance, such as including indirect emissions, imposing a social cost of carbon, and omission of cost-benefit analyses. Congress should also reduce litigation risk for project developers, create a unified, centralized permitting process, reduce the scope for state and local regulatory delay and obstruction, and establish programmatic and general permits for major categories of infrastructure projects.[80]

---

77 https://www.c3solutions.org/policy-paper/nepa-modernization-101/.
78 https://cei.org/wp-content/uploads/2022/09/Mario_Loyola_-_Unleashing_America_s_Energy_Abundance.pdf.
79 https://www.brookings.edu/research/how-does-permitting-for-clean-energy-infrastructure-work/.
80 https://cei.org/wp-content/uploads/2022/09/Mario_Loyola_-_Unleashing_America_s_Energy_Abundance.pdf.

2) Avoid unnecessarily burdensome, excessive, and duplicative regulations and policies that delay or prevent investment in new, efficient generation and transmission capacity.
   - Avoiding mistakes requires the Energy Information Administration to provide Congress with timely and unbiased analyses of the cost-benefit impacts of mandating extreme energy outcomes. For example, the EIA still needs to provide Congress with an analysis of the energy security and economic implications of recent presidential executive orders and legislative climate proposals, including
     - President Biden's executive order calling for the federal government to reduce its emissions by 65% by 2030 and reach net-zero emissions by 2050, and
     - A bill (CLEAN Future Act) introduced by Democrats, mirroring the President's campaign proposal, to ban the use of fossil fuels—which currently account for 60% of U.S. electricity generation—in U.S. power plants by 2035.
   - Transmission capacity expansion is essential for grid reliability. A recent draft DOE analysis concluded that the US likely needs 47,300 GW-miles of new transmission by 2035, a 57% increase compared to today. Significant new transmission capacity is required by 2030 in the Plains, Midwest, and Texas regions. By 2040, sizeable new capacity will be needed for the Mountain West, Mid-Atlantic, and Southeast.[81]
   - Meanwhile, as a February 2023 report, "*Energy Transition in PJM: Resource Retirements, Replacements & Risks*," noted, four trends create increasing reliability risks.
     - Strong electricity demand growth;
     - Thermal (gas, coal primarily) generation retirements due to government and private sector policies, as well as economic factors;
     - Insufficient new generation resources to replace retirements due to industry forces, including siting and supply chain challenges; and
     - The preponderance of yet still insufficient amount of intermittent (wind, solar) and limited-duration resources at the expense of always-on, long-duration thermal resources.
   - To address these risks, Congress should scrap the current climate strategy and embark on a new one that would depoliticize science, consider the cost-benefit analyses of all strategies, and result in legislative changes at home and leadership abroad. For more background, please see this April 2022 report from the Forum for American Leadership entitled "*Blueprint for a Sound and Serious Climate Policy*."
3) Deter cyber attacks on critical energy infrastructure. Amid reports that our electricity and oil systems are absorbing growing numbers of attacks, Congress must deter cyber attackers. The following May 2021 Forum for American Leadership report entitled "*Eight Necessary Steps to Defend U.S. Critical Energy Infrastructure from Cyber Attacks*" provides more detail.

---

[81] Howland, Ethan. (2023, March 6). DOE finds 'pressing' transmission needs, with inter-regional links offering the biggest benefits. *Industry Drive, reported in Bloomberg News*.

The Honorable Russ Fulcher

While geothermal covers a small part of our population today, it offers baseload reliability at utility-level scale according to the Energy Information Administration (the EIA). For example, Idaho's Raft River facility can sell up to 10 MW of this clean, reliable energy source. Yet, permitting for new geothermal projects can take 7-10 years.

1. My bill, the Enhancing Geothermal Production on Federal Lands Act, would allow for geothermal development where there are already existing oil and gas leases. Can you speak to how changes like this that remove duplicative red tape can help America return to energy dominance?

Energy investment decisions are based on many market and industrial factors out of Congress's control. But permitting reform is a necessary, but not sufficient, condition for the timely expansion of transmission capacity, which is essential for Idaho's geothermal development and other forms of energy production.

2. I understand that transmission development is one of the key factors that could limit the deployment of geothermal technologies across areas of the West, like Idaho. This is largely due to geothermal being located far from load/population centers and the sheer amount of federal land the transmission lines would have to cross in order to meet demand. What barriers exist to developing transmission to help clean, baseload technologies like geothermal to provide reliable power to the grid?

While I am not an expert in Idaho's geothermal and hydroelectric electricity sectors, unnecessary regulatory delays in transmission siting, permitting, and construction impede connecting electric power generation with demand centers. Therefore, permitting reform is essential, as noted above and in my testimony.

Hydroelectric constitutes roughly half of my state's total in-state generation. And, it speaks to the importance of Idaho's dams.

1. Can you discuss any transmission challenges that may occur with getting this highly reliable and clean energy to population centers, given these dams are often nested between large sections of federal land?

While siting electricity lines is inherently uncertain and involves complex economic, social, and land-use factors, the federal government can contribute to more successful outcomes by limiting its interventions to large, high-powered interstate transmission lines and repealing, reducing, and avoiding intrusions into the process.

Chapter 2

Burning the Midnight Oil: Why Depleting the Strategic Petroleum Reserve Is Not a Solution to America's Energy Problem, Part I*

Committee on Oversight and Accountability

Wednesday, March 8, 2023
House of Representatives Committee on Oversight and Accountability
Subcommittee on Economic Growth, Energy
Policy, and Regulatory Affairs
Washington, D.C.

The Subcommittee met, pursuant to notice, at 2:21 p.m., in room 2247, Rayburn Office Building, Hon. Pat Fallon, (Chairman of the Subcommittee) presiding.

Present: Representatives Fallon, Donalds, Perry, Boebert, Fry, Luna, Edwards, Bush, Norton, Porter, Brown, and Stansbury.

Mr. FALLON. The Committee on Economic Growth, Energy Policy, and Regulatory Affairs will come to order. I want to welcome everyone for coming.

Without objection, the Chair may declare a recess at any time, and I recognize myself for the purpose of making an opening statement.

It is an honor to chair the Subcommittee on Economic Growth, Energy Policy, and Regulatory Affairs. I would like to thank the witnesses for their testimony today, and I look forward to working with Ranking Member Bush to provide necessary oversight on these critical sectors.

Today's hearing will be on the Strategic Petroleum Reserve, or the SPR for short. The decisions that led to the largest drawdown of our vital oil in history requires oversight.

At its height, the SPR was at 695 million barrels under the previous administration. It now sits at 371 million barrels, which is a 47 percent reduction. It is nearly half.

On the first day in office, Joe Biden canceled the Keystone XL Pipeline, killing thousands of high-paying American jobs and openly attacking the oil and gas industry.

---

* This is an edited, reformatted and augmented version of Hearing before the Subcommittee on Economic Growth, Energy Policy, and Regulatory Affairs of the Committee on Oversight and Accountability, House of Representatives, One Hundred Eighteenth Congress, First Session, Serial No. 118–6, dated March 8, 2023.

In: Energy Policy and National Security
Editor: Kuzman Sukanya
ISBN: 979-8-89530-879-0

Less than a week later, Mr. Biden placed a moratorium on Federal on and offshore oil and natural gas leases. These actions increase the average price of gasoline to record highs by the summer of 2022, forcing Americans already struggling with out-of-control inflation to foot the bill for poor policy.

In an attempt to hide the failed results of an all of government war on our domestic energy industry the Administration recklessly turned to the SPR.

Joe Biden authorized the biggest drawdown in history, releasing more oil from the SPR than all other previous administrations combined. All other previous administrations combined.

In the last two years more oil has been siphoned off our reserves than in the previous 46 years. For a quick bit of history, the SPR was founded in 1975 for national security reasons to protect our country in times of war, national emergencies, or severe acts of God.

Now, seven months after President Biden authorized the first release of crude, gas prices reached an all-time high of just over $5 a gallon in June 2022. Then the Administration tried to claim it was Putin's fault.

Well, news flash, it wasn't. Despite the Administration's flimsy claim of the, quote/unquote, "Putin price hike" that didn't necessitate the draw downs. The truth is Mr. Biden set up failure long before Russia crossed the border into Ukraine.

Shockingly, instead of cutting red tape and allowing domestic producers to even the scales of supply and demand, which would have driven down prices for consumers not only here at home but also alleviated pressures on our European allies, Joe Biden instead attacked American companies. He asked mini Putins like Venezuela to increase production instead of unleashing the American energy sector.

President Biden accused them of price gouging and demanded they ramp up production while refusing to acknowledge his Administration's role in the energy crisis.

His cancellation of the Keystone XL Pipeline, the moratorium on Federal on and offshore oil and natural gas, and even attacks on the American energy sector have driven this crisis.

Rather than lowering energy prices and encouraging domestic energy industry and supporting good-paying American jobs, the Biden Administration used the SPR as a stopgap to lower energy prices before the November midterm elections.

Further, the Administration sold millions of barrels of our crude to companies controlled by the Chinese Communist Party. So, why is the Biden Administration depleting our own emergency stock-piles to benefit foreign adversaries?

In fact, Secretary of Defense Lloyd Austin just issued a memorandum to his forces last week, which said, and I quote, "An increasingly aggressive China is trying to shape the international rules-based system to suit its authoritarian preferences," and I agree.

So, why is President Biden draining our SPR to fill China's? China has nearly 1 billion barrels in their reserve, and we have just about a third of that. So, that is unacceptable.

So, this hearing is going to examine why the Biden Administration is draining our SPR, the national security concerns with the depleted reserve, the long-term implications of renewables, exclusive—renewables exclusive political approach, and the practical policy solutions to ensure energy affordability and security for years to come.

We know that the Strategic Petroleum Reserve was formed to protect our country in what—remember, times of war, national emergency, and effects of acts of God.

Did any of these happen in the last two years? No. Are we at war? No.

Did we have a severe act of God? No.

And did we suffer through a national emergency that would have necessitated tapping the SPR? No.

So, why under this Administration were 267 million barrels of oil depleted from our emergency reserve? So, the fundamental question that this hearing must help us answer is this. Did Joe Biden risk our national security and deplete our reserves by over 40 percent in a vain attempt to curry short-term political gain before mid-term elections?

I want to thank you again very much, and I recognize Ranking Member Bush for her opening statement.

Ms. BUSH. Thank you so much, Mr. Chairman.

St. Louis and I are here today to have a serious discussion about what energy security looks like for people, not corporations. This hearing is not that discussion. If House Republicans actually cared about preventing an energy and economic crisis from happening, they will commit to investing in renewable energy.

Unfortunately, there are no serious proposals being offered. Rising inflation caused by Russia's violent invasion of Ukraine and COVID–19 pandemic-related supply chain shortages, combined with dangerous corporate greed and our reliance on fossil fuels, has left the United States in a tenuous position of expensive, unreliable, and unsafe energy.

Let us be clear, the Strategic Petroleum Reserve was precisely created to be tapped in emergencies involving severe shortages of oil. Of course, I believe that the best way to avoid similar challenges in the future is to end our reliance on oil.

But there should be no doubt that President Biden employed it explicitly for its stated purpose, in an emergency when our Nation's families and people needed it most.

According to the U.S. Energy Information Administration, gas prices peaked at $4.88 in the Midwest region last year and are now down to $3.24 as of this week. People were struggling to get to doctor's appointments or to keep their heat on, and the Republican Party would have loved to do nothing.

While the drawdown of the SPR was necessary to bolster supplies in an emergency, future events can be best ameliorated or prevented by reducing our demand for and our reliance on fossil fuels.

But my Republican colleagues have unironically invited for-profit think tank and oil titans whose expertise is in maximizing profits, especially at the expense of our Black, our Brown, and our indigenous neighbors' health, safety, and well-being.

One of these so-called energy experts is a philosopher who has previously espoused white supremacist views. For instance, in his 2000 college newspaper he wrote, quote, "The African and American Studies Department has 23 classes. In many of these classes, African culture is presented not as a——"

Mrs. BOEBERT. Mr. Chairman, I demand the gentlelady's words be taken down. She just called the witness a white supremacist.

Ms. BUSH. No, I referred to the words, not to the person. Not to him, the words.

Mrs. BOEBERT. The gentlelady is referring to the witness' statement and referred to him as a white supremacist. I would like a parliamentary inquiry——

Ms. BUSH. No, I did not.

Mrs. BOEBERT [continuing]. Into having the gentlewoman's words taken down.

Mr. FALLON. The Chair—excuse me. The Chair recognizes Congresswoman Boebert.

Mrs. BOEBERT. Thank you, Mr. Chairman.

I would like a parliamentary inquiry into taking down the gentlelady's words—the Ranking Member's words—for insinuating that our witness made white supremacist statements using his words.

Mr. FALLON. OK. The Chair cannot take down words that are directed to our witness, only to other Members.

All right. The Chair recognizes Ranking Member Bush to continue her opening statement.

Mrs. BOEBERT. Mr. Chairman, parliamentarian inquiry.

Mr. FALLON. Say your inquiry.

Mrs. BOEBERT. Is it appropriate for the Ranking Members to show respect to the witnesses who are here present in the committee room today?

Mr. FALLON. I think it would obviously be warranted to show respect for any witness.

Mrs. BOEBERT. I would like that sentiment to be expressed and the Members to be reminded to show respect to our witnesses who are here today.

Thank you, Mr. Chairman. I yield.

Mr. FALLON. Thank you. The Chair recognizes Ranking Member Bush.

Ms. BUSH. As I was saying, "The African and African American Studies Department"—this is the quote, "has 23 classes. In many of these classes African culture is presented not as inferior to Western culture but as on equal footing with it. In other departments the same is done with Latin American, Indian, and American-Indian culture," end quote.

When confronted about these views over two decades later, rather than disavow them he doubled down on this narrative saying, quote, "It has nothing to do with skin color. I was arguing that those cultures overall are inferior to Western culture."

We are not inferior to any culture, speaking as a descendant of one of those cultures. This is the witness the Republicans invited to discuss issues of energy security. He came here to promote fossil fuels, which we know are disproportionately harming and killing Black and Brown people.

Last week, I joined Ranking Member Raskin and all of my Democratic Oversight colleagues in calling on Oversight Republicans to denounce white nationalism and denounce white supremacy in all its forms. Not one single Republican joined us.

So, I ask again today will my Republican colleagues condemn white supremacy and work with us to ensure this Committee engages in advancing good and reducing harm rather than perpetuating it. That is respect.

Our Committee needs to be focused on actions that will help people and prevent financial hurt and pain. Whether we deploy the SPR or we don't, the instability of our energy system will continue if we are reliant on oil.

Our committees need safe, reliable, renewable energy and more investments in public transit.

Thank you, and I yield back.

Mr. FALLON. Ladies and gentlemen, I am pleased to introduce our three witnesses today. I am looking forward to discussing the SPR and what policy initiatives should be implemented to ensuring Americans have access to clean, reliable, and affordable energy, and thank you all for your testimony.

Alex Epstein is an author, commentator, and president of the Industrial Progress, which he founded in 2011. He is a New York Times bestselling author of the books "Fossil Future" and "The Moral Case for Fossil Fuels."

His research focuses on the global benefits of fossil fuels as reliable cost-effective sources for both developed and developing the world.

And then we have Ilia Bouchouev is the former president of Koch Global Partners and is currently the managing partner at Pentathlon Investments and an adjunct university professor.

During his 20 years in the global derivatives trading, he pioneered several energy derivatives products and is recognized as a pioneer in this space. He holds a Ph.D. in applied mathematics.

And then we have Demond Drummer, who served as the managing director of Equitable Economy and was co-founder and executive director. He currently consults with the Institute on Race, Power, and Political Economy at the New School in New York City.

I look forward to hearing from each of you on this important topic.

Pursuant to Rule 9(g), the witnesses will please stand and raise their right hands.

Do you solemnly swear or affirm that the testimony that you are about to give is the truth, the whole truth, and nothing but the truth so help you God?

[Witnesses are sworn.]

Mr. FALLON. Thank you. Let the record show that the witnesses all answered in the affirmative. We appreciate all of you being here today and, once again, thank you for your testimony.

Let me remind the witnesses that we have read your written statements and they will appear in full in the hearing record. Please limit your oral statements to five minutes.

As a reminder, please press the button on your microphone in front of you so that it is on and the Members can hear you. When you begin to speak the light in front of you will turn green.

After four minutes the light will turn yellow, and when the red light comes on your five minutes has expired and if you could then just wrap up your thought at that moment. We will give you a little bit of discretion.

And then I recognize Alex Epstein to please begin his opening statement, and because you were directly mentioned, as the Chairman's discretion I want to give you an additional minute if you would like to address the allegations levied on you against the Ranking Member.

Statement of Alex Epstein, President, Center for Industrial Progress

Mr. EPSTEIN. Yes, it would be great if they put up the cards again.

So, thanks for giving me a minute to address the truly despicable and false attack on me, claiming that anything I have ever said is white supremacist and then trying to imply that somehow applies today.

So, I don't know if you want to put those cards back up. But what I argued and what I have argued for my entire adult life is that Western culture is superior in the sense that it promotes individualism and freedom and I have fought my entire life for freedom around the world, including in Africa, including in Asia, including in India, and I want everyone around the world to have the same opportunities that I do in the United States.

There are different places around the world that are in many ways inferior. They have female genital mutilation. They have slavery. These are not as good as not mutilating females' genitals and not enslaving people.

And so, I make no apology and the idea that this is associated with skin color is despicable and racist. Skin color has nothing to do with ideas.

Where you came from geographically is irrelevant. What is relevant is what is true. So, you can ask more about that. But that was an insane diversion from the truth.

So, let us talk about the truth. The Strategic Petroleum Reserve is a crucial tool to protect America's oil security, which is the foundation of our national security, because oil powers our military, and economic security—oil powers modern mobility.

The Biden Administration is, unfortunately, abusing the SPR and as a result they are threatening our oil security. So, the purpose of the SPR is to aid the U.S. in securing a reliable supply of oil and oil fuels by providing an emergency stockpile of up to one-tenth of U.S. oil consumption to be used during major interruptions of supplies such as sabotage or disaster, then refilled after.

The SPR is only an aid to oil security. The core of oil security, which many here oppose, is facilitating a reliable affordable oil supply by protecting industries' freedom to invest in, produce, refine, and transport oil. Without oil industry freedom the SPR's modest stockpile can't make us secure.

Now, instead of protecting our oil security by protecting oil industry freedom and backing it up with an ample reserve, President Biden has damaged it by, one, attacking the oil industry's freedom, then, two, dangerously depleting our reserve for political purposes. So, how President Biden has harmed oil security by attacking the industry's freedom? For over 15 years the global anti-fossil fuel movement with major leadership by Joe Biden has artificially reduced the supply of oil by opposing oil investment, production, refining, and transport. When global anti-fossil fuel policies make oil supply go down and oil demand goes up, oil prices go up. It is really that simple.

Now, Biden blames Putin's war for today's oil prices. But the root cause is global anti-fossil fuel policies supported by Biden, which made oil and other fossil fuel prices artificially high prewar and prevented the free world from quickly increasing production in response.

More broadly, had Biden and other Democrats spent the last four years liberating U.S. fossil fuel investment, production, refining, and transport instead of restricting and threatening them, America would produce significantly more oil.

So, to summarize, instead of protecting the core of oil security, oil industry freedom, Joe Biden and other anti-fossil fuel politicians around the world have gravely harmed it for over 15 years by restricting oil investment, production, refining, and transport.

How Biden—how President Biden has further harmed oil security by abusing the SPR. The proper response by President Biden to Europe's and our vulnerability to Russia and to rising oil prices should have been apologize—apologize for supporting anti-oil policies and then lead a reversal of them to minimize further damage to our security and economy.

But instead of responding to artificially high oil prices, which are a consequence of anti-oil policies, with the reversal of those policies Biden chose to keep his anti-oil policies, but then lower short-term prices by depleting our SPR, which is now down to its lowest level since the early 1980's.

Even when full at 714 million barrels the SPR contains only one-tenth of U.S. annual oil consumption and now at 370 million barrels, it contains one-twentieth. This means less oil for real emergencies such as wars, terrorist attacks, or cyber attacks on U.S. oil operations.

Now, the Administration claims Putin's war is exactly the kind of emergency the SPR is designed for. This is totally wrong. Putin's war is not an oil emergency that justifies depleting the SPR now. It is a potential oil emergency that justifies keeping the SPR as full as possible.

Now, Biden's main goal in dangerously depleting the SPR has clearly been to lower gasoline prices to help his party's election prospects. This is the same goal he had when he asked OPEC+ to delay production cuts for just one month so that subsequent price increases would happen post-election.

Now, a president's discretion over the SPR gives them a sacred responsibility to use their best judgment about when to deplete it, doing so only when that is in their honest judgment what is best for the country's oil security, and President Biden failed to uphold this responsibility.

So, the solutions to our crippled oil security are, one, focus on liberating the oil industry as much as possible, including withdrawing this Administration's many threats, and then, two, commit to using the SPR responsibly, not politically.

Now, unfortunately, this Administration and some in this body are, one, continuing the attacks on the oil industry's freedom calling for less freedom—even less freedom, which is a total disaster— and then, two, taking no responsibility for the abuse of the SPR. So, I hope that this hearing is a step toward reversing this dangerous state of affairs and I welcome any and all questions. Thank you.

Mr. FALLON. Thank you very much.

I recognize Mr. Bouchouev. Sorry about that. Help me out with your last——

Mr. BOUCHOUEV. Ilia Bouchouev.

Mr. FALLON. Ilia Bouchouev.

Mr. BOUCHOUEV. Yes. Thank you.

Mr. FALLON. All right. Awesome. Thank you very much. You are recognized.

Statement of Ilya Bouchouev, Managing Partner, Pentathlon Investments, LLC

Mr. BOUCHOUEV. Thank you, Chairman Fallon. Thank you, Ranking Member Bush, and the Members of the Subcommittee for inviting me. It is my great honor to be here today.

I am an academic and scientist even though I spent most of my career managing the trading business. I would like to share some thoughts on how the science of markets can help manage the SPR. The backbone of oil markets is storage. Storage, literally, buys time for the problem to go away. The problem of optimal storage management goes back to the very beginning of human civilization. It is a difficult problem to solve because a decision today depends on what might happen tomorrow, which, of course, we don't know for sure.

Let me illustrate with a simple example of Robinson Crusoe. Robinson Crusoe lives on a remote island and he has a finite supply of food. Every day he needs to decide how much food to eat today and how much food to keep for tomorrow.

If he decides to take his chances and eat the entire supply of food today, he will be risking his life. At the same time, not eating anything today is also suboptimal because he might get weaker and some food, like raw fish, may get rotten so you actually have to store it.

So, the optimal decision between today and tomorrow is somewhere in between. If you are managing the oil storage you are conceptually facing the same problem. A naive strategy would be to take the chances and sell oil out of the storage with the hope to replenish it when the price is lower.

This strategy is highly speculative, as the price of oil is unpredictable. A more prudent strategy, which is what commercial traders do, would be to lend oil to the market when the market needs it.

For example, you can—if the oil price today is 20 percent higher than the price in the future, you can lend 100 barrels today and receive back 120 barrels, which generates you 20 percent return for free.

The same two strategy of selling oil or lending oil are also available for management of SPR. Last year, unfortunately, the strategy of selling oil was chosen.

To put things into perspective, like the Chairman already mentioned, the magnitude was unprecedented. The sale not only exceeded all previous sales combined but it was three times larger than all of the emergency sales in the past 40 years.

While the strategy of lending barrels generate return with no risk, the strategy of selling oil generate a lot of risk with no return and I say no return because in the absence of actual supply disruption SPR cannot have any impact on the price of oil. The oil price is determined in the futures market, not in the physical market between buyers and sellers.

The world consumes 100 million barrels of oil per day. But on a daily basis, the daily trading volume of oil futures and options is 5 billion barrels a day, which is five times larger.

So, when it comes to the direction of oil prices, financial market is always going to dwarf and determine the direction of prices. For example, the recent weakening of oil prices is not driven by SPR but it is driven by macroeconomic conditions such as rising interest rate and increasing probability of U.S. recession.

One way to manage SPR more scientifically and efficiently would be setting up a rigorous rule-based nondiscretionary program of oil loans that takes advantage of market conditions and bring benefits to taxpayers. Essentially, there is no risk.

However, such a problem can only be set up once all the previous barrels are repurchased. At the moment, SPR strategy is a speculative trade. Fortunately, this trade is currently in the money, but the ultimate outcome of this trade will not be known until the trade is closed and the barrels are bought back.

This should be done sooner rather than later because the price of oil—before the price of oil rises again. All traders know that the risk can only grow in time. Waiting any longer would be similar to Robinson Crusoe playing with fire.

Thank you, and I will be happy to take your questions.

Mr. FALLON. Thank you, sir.

The Chair now recognizes Mr. Drummer for his opening statement.

Statement of Demond Drummer, Institute on Race, Power, and the Political Economy, the New School

Mr. DRUMMER. Thank you, Chairman Fallon, Ranking Member Bush, and Members of the Subcommittee for the invitation to give testimony during this hearing.

My name is Demond Drummer and I consult with the Institute on Race, Power, and Political Economy at the New School in New York City. The institute advances research to understand and ultimately undo the unjust identity group-based social stratification that harms individuals and communities.

I will use my time to show that, in light of the global energy price shocks resulting from the war in Ukraine, the Administration's decision to tap the Strategic Petroleum Reserve was the right decision in the short term to provide much-needed relief to the American people.

This is especially true for the nearly one out of every three people in America who live in or near poverty and pay a disproportionate share of their limited household income to meet their energy and transportation needs.

To be sure, the Strategic Petroleum Reserve is not designed to offer a long-term solution to the problem with energy security. If our shared goal is to achieve true energy security, the path forward is clear—strategic and substantial public investments to accelerate our transition to a clean energy economy.

During the height of the oil price spikes last year, energy costs alone accounted for approximately half of all inflation. One year ago this month, the price of oil—of fuel oil, rather, surged 70 percent and the price of gasoline was up 48 percent and natural gas 21.6 percent.

Today, overall energy costs are still up 8.7 percent compared to last year, and while gasoline prices are up only 1.5 percent, fuel oil is up 27.7 percent and natural gas is up 26.7 percent.

The evidence is clear. Whether for powering and heating their homes or refueling their cars, the economic costs of fossil fuel energy are unsustainable for American households, especially those who are paid the least.

Energy is a nondiscretionary household expense. Households below the Federal poverty line spend 18 percent of their income on energy, nearly 10 times the energy burden of higher income households.

Households at 200 percent of the Federal poverty line spend six percent of their income on energy. That is three times the energy burden of higher income households. This is despite the fact that lower income households consume less energy on average.

Further, households in the bottom two income quintiles pay anywhere from 15 percent to 25 percent of their income on transportation costs alone. A high energy burden is not only a symptom of poverty, but in siphoning resources from other critical needs such as health care and childcare, it prolongs and exacerbates poverty. The evidence is clear. American cannot drill our way to energy security. The geopolitical turmoil of the last year highlights the myriad economic risks of U.S. reliance on fossil fuels. The Russian invasion of Ukraine sent global fossil fuel commodities prices sky-rocketing.

In light of the globalized market for fossil fuel energy commodities, the U.S. has very limited ability to insulate American households from the economic fallout of geopolitical conflicts.

In this moment, it is critical to align monetary, fiscal, and regulatory policy to accelerate adoption of low-cost renewable energy and promote the development of high-wage sustainable industries. Wind and solar energy are by far the most cost-effective sources of power generation on the planet.

However, in 2021, renewables accounted for only 20 percent of electrical power generation and only 12 percent of total energy consumption in the U.S., and beyond power generation, enacting an equitable and sustainable industrial policy that prioritizes people and the places where they live is the most effective way to insulate American households from fossil fuel price shocks while addressing the existential consequences of climate change that are being brought on by the fossil fuel industry.

This requires leveraging the full suite of economic policy tools available to direct public and private investments toward developing America's productive capacity to deliver the goods and services that will power an equitable and sustainable future for us all.

Thank you again for the invitation to testify. It has been an honor and I do look forward to your questions.

Mr. FALLON. Thank you, all the witnesses, and just as a reminder to the Members, we are just starting out here. This is our first hearing, and the issues that we are debating are important not only to the Members but to the witnesses as well.

And that said, this hearing and in future hearings we have in the Subcommittee aren't about levying accusations and insults at our witnesses, at each other. I think that is disrespectful to everyone and part of the process and folks that are here as well listening in.

And quite frankly, it is a waste of time and I think it is unproductive. I ask the Members to please refrain from this kind of behavior and focus on the topic of the hearing, which today is the Strategic Petroleum Reserve and its depletions.

Thank you very much.

I recognize myself now for five minutes.

Mr. Epstein, does the decision to deplete the SPR to historically low levels mean the Biden Administration has no real plan for American energy policy?

Mr. EPSTEIN. I think saying no real plan is too charitable. I think they have a very bad plan. So, the basic plan is to rapidly eliminate fossil fuels as soon as possible and then pretend that they can be replaced by unreliable solar and wind.

Unfortunately, I am seeing this advocated here today. But what the energy crisis should have taught us is that this doesn't work. You can only try to replace fossil fuels once you have truly cost-effective alternatives, which means reliable alternatives and which means replacing all uses of fossil fuels, including heavy duty transportation, industrial heat, et cetera.

So, I mean, it is worse than no policy. I wish they had no policy.

Mr. FALLON. Presidents have only declared emergency releases of the SPR on three previous occasions: Operation Desert Storm, Hurricane Katrina, and following the military intervention in Libya. Did the Biden Administration properly use, in your opinion, the SPR?

Mr. EPSTEIN. No. As I tried to indicate in my testimony, I mean, it is wildly improper, and I think particularly you hear this rationalization about the conflict in Russia.

Whenever you have an escalating international conflict that is when you need to keep your reserve as high as possible. So, the idea of just depleting it to half of its level when we don't know what could happen, going forward, this is really, really precarious and I think it is pretty obviously politically driven.

And, you know, the obvious thing to do—you know, we are hearing talks about, well, this oil price increase has hurt low-income families. This is absolutely true. But if this Administration cared about that, why did every action they took lead to higher prices? Why did they cancel the Keystone XL? Why did they threaten the industry—you know, I guarantee you we are going to end fossil fuel. Why have they been focused on having the SEC be sicced on fossil fuel companies? Why was there, you know, a leasing ban on Federal lands?

Like, every action was directed at making oil more expensive and thus harming low-income families along with everyone else.

Mr. FALLON. Well, you just answered my next question. So, thank you for that.

Mr. Bouchouev, the witness—one of the other witnesses said that renewables are by far the most cost-effective source of energy. Do you agree with that?

Mr. BOUCHOUEV. Well, the sun doesn't always shine and wind doesn't always blow. As I said in my testimony, storage is the back-bone of financial markets. You can get renewables, but renewables can't really do a whole lot without the storage.

So, it is not a question about renewables. Renewables are great, but we need batteries. So, if you don't have storage, I mean, there is not a whole lot you can do.

Mr. FALLON. And, sir, how has China benefited from the decision to deplete the U.S. SPR?

Mr. BOUCHOUEV. Well, I don't know if China benefited. But let us kind of—the data in China is very difficult to get to. Nobody really knows. But one thing that I know that we used to have a billion barrels in the reserves. Half of it was SPR. Half of it was private.

Now we are down to about 800 million, and 10 years ago China had about one-fifth of what we had, about 200 million, and based on third party data, last year China surpassed a billion.

So, we are kind of going down. China inventory is going up. Unfortunately, there is no clean data to put two graphs—two lines on the same graph. But based on what I have seen, they are really a mirror image of each other. So, as our inventories are depleting the China inventories are growing.

Mr. FALLON. So, China used to have—we used to have five times the reserves that China had.

Mr. BOUCHOUEV. Correct.

Mr. FALLON. Now they have three times—most likely three times what we have?

Mr. BOUCHOUEV. Not necessarily. As I said, our inventory is about 800 million. Three hundred seventy million is SPR and the rest is private. But private to what extent you can rely on them because they are held by a refinery.

So, if there is a crisis refineries are going to use them. You are correct that China has three times more than the government stocked. That is accurate statement.

Mr. FALLON. OK. And, Mr. Epstein, just a little off topic, but Western culture that you mentioned could exist in countries, let us say, in Japan and——

Mr. EPSTEIN. Well, yes, and it has proliferated to the great benefit of all, and it is really about—it is not really about the location. It is about individualism and freedom and reason.

So, it is certain values that are universal human values. The idea that freedom and reason and individualism only apply to people of certain skin colors that is the actual original racist idea, versus no, no, no, whatever your skin color is these ideas are true and good for human flourishing.

Mr. FALLON. OK. Thank you.

The Chair recognizes Ranking Member Bush for five minutes.

Ms. BUSH. Thank you, Mr. Chairman.

St. Louis and I are here today to talk about the impact that fossil fuels have on our community. Let me start by telling you what energy security looks like in my community. In my community, energy security looks like having the lights on. It looks like families being able to stay warm.

It looks like children walking to school without contracting asthma from local air pollution. It looks like children being able to depend on stable internet to do their homework.

Energy security looks like this and so much more, and to my Republican colleagues here today I want you to know that this is what energy security looks like in your communities as well.

Energy security is about dramatically reducing our reliance on fossil fuels whose consumption disproportionately harms low-income families and communities of color. It is our communities that suffer the most from instability in the energy market.

Studies have repeatedly shown that Black and Brown communities pay a significantly higher share of our income in energy costs.

We are also more likely to live in communities where fossil fuels are extracted, where they are burned, and in places where the climate crisis is causing dangerous floods, hurricanes, and wildfires. Mr. Drummer, can you tell us more about how the $2 price jump in gas following Putin's invasion of Ukraine exacerbated the disparities and energy burdens for Black and Brown in low-income families?

Mr. DRUMMER. Absolutely. Thank you, Representative Bush.

We know that particularly in the larger metropolitan areas the lowest earners live the furthest away from the jobs. So, what does that mean? Those who make the least had to gas up and pay a very high percentage of their income just to get to work.

And so, when we talk about energy security, we know in 2020 when everybody was at home, fossil fuel prices, particularly oil prices, went through the—went down into the basement. They were in the gutter. What did that do? It put a lot of fossil fuel workers out of work.

So, Vladimir Putin put fossil fuel workers in the U.S. out of work and Vladimir Putin made it unaffordable to get to and from work in 2022.

And so, we cannot drill our way out of this situation. So, it is hitting folks' pocketbooks and that release was absolutely necessary.

Ms. BUSH. So, you describe that the price increases had a disproportionate effect on Black, Brown and low-income communities—families. How did the Administration's actions and the ensuring—the ensuing drop in gas prices—how did it help those same families?

Mr. DRUMMER. While there are a number of factors that affect the price of oil, in an interview with Oxford, I believe, Dr. B did concede that the strategic use of SPR did have an impact. It is hard to quantify and hard to tell, but it definitely had an impact in the price of oil.

Ms. BUSH. We need to focus on making more public investments to lower our energy demand by taking actions such as expanding public transit and installing energy efficient technology like electric heat pumps.

Mr. Drummer, how does making investments in people through safe energy systems and people benefit climate justice and energy security?

Mr. DRUMMER. One of the exciting things from the Inflation Reduction Act are the domestic content requirements, so private capital can receive additional tax credits by paying prevailing wages for the construction of renewable energy and battery storage projects, as well as domestic sourcing of the components for these projects.

That is what is exciting about our renewable energy future, which includes batteries.

Ms. BUSH. Thank you. We must make massive investments in climate jobs and environmental justice and, frankly, this is a need that is overdue. I believe that we are going to build high-speed rail. We are going to establish bus rapid transit systems in frontline communities. We are going to create green social housing running on renewable energy. We are going to win a green new deal that will benefit everyone and let me just say I cannot wait. Thank you, and I yield back.

Mr. FALLON. Thank you.

The Chair recognizes Mr. Donalds for his five minutes.

Mr. DONALDS. Thank you, Mr. Chairman. Spirited committee already. A couple of things.

Look, if you are going to look at the costs for any American, especially Americans on the lower end of the socioeconomic spectrum, rising energy costs cripple them more than anybody else. I would assume every witness can agree to that, correct? Can we all agree to that?

OK. Mr. Drummer, I was reading through your testimony during all of the opening statements that we have in our committees. You say that fossil fuel—fossil fuel costs are unsustainable. Why is that?

Mr. DRUMMER. It is the data that I just shared, right. And so, even today, yes, gasoline is—as of the latest February 14 or so CPR report up 1.5 percent.

But that is off of the historic high from last year, right. This is gouging folks' pocketbooks. Again, the bottom two quintile——

Mr. DONALDS. Mr. Drummer, I want to—I am going to reclaim time now because I am asking this question. Are fossil fuel costs more expensive in the United States just because of the costs of actually drilling for and refining fossil fuel products or is it also because of government regulation with respect to fossil fuel products?

Mr. DRUMMER. The bottom two quintiles of earners pay 15 to 25 percent of their income——

Mr. DONALDS. I am asking you what makes it—but I am asking you what makes it more expensive. Is it just purely the procurement costs of fossil fuel or is it also the regulatory burdens on fossil fuel?

Mr. DRUMMER. It is the global fossil fuel market that we do not control. When there is chaos happening in Europe and Ukraine because of Russia's invasion that hits Americans' pocketbooks.

Mr. DONALDS. OK. But Mr. Drummer, people have been drilling for oil and using gasoline long before Vladimir Putin went into Ukraine. We have seen rises and falls in gasoline prices.

Let me ask you this. You say in your statement that renewables are actually the most cost-efficient form of energy. That is what you say in your statement.

Is that because renewable energy is largely subsidized by most nations in the world or is that because the actual per unit cost is being borne by the consumer? Is it the subsidy or is it the full cost? Which makes it most cost efficient?

Mr. DRUMMER. This is a great point you bring up, Representative, because actually fossil fuels are heavily subsidized more so than renewables right now and still renewables are more price competitive.

Mr. DONALDS. I am not sure what you are talking about, Mr. Drummer. You need to go back and check your facts.

Mr. DRUMMER. I will be happy to share the data with you after this hearing.

Mr. DONALDS. Real quick. Let us—I want to move on to a couple of things. Do you think we need more refineries in the United States, yes or no?

Mr. DRUMMER. Absolutely not.

Mr. DONALDS. So, you do realize that without more refineries in the United States the cost of fuel for all citizens, especially those at the lower two quartiles of the socioeconomic spectrum, are going to suffer as a result? You do realize that?

Mr. DRUMMER. Fossil fuel markets are a global market. We do not control the price of oil in the West——

Mr. DONALDS. But you do realize——

Mr. DRUMMER. We cannot drill our way out of energy——

Mr. DONALDS. Hold on. But you do realize that the supply of oil that goes to refineries in the United States, if you limit the capacity of refineries or continue to not allow them to expand or to actually get the necessary maintenance they are going to have less output capacity in the United States, which means all the products we use with fossil fuels, not just gasoline—the very glasses you wear, the very contact lenses I am wearing those are derivatives of oil and petroleum as well.

So, you do realize that by not allowing for new refineries you are going to increase the costs on the very people we are all trying to help?

Mr. DRUMMER. I think, again, we are talking about energy security.

Mr. DONALDS. Well, I would agree. This is energy security——

Mr. DRUMMER. Right.

Mr. DONALDS [continuing]. Because if you want to make sure that the costs for the people who have to buy products—you know, poor people, they don't have escape velocity. You know what I mean?

They don't have that. So, is it prudent to not allow for the things necessary to keep costs low in the name of the green new deal? Is that prudence?

Mr. DRUMMER. It is imprudent to double down on a losing strategy of drilling our way to energy security. It is not possible——

Mr. DONALDS. Mr. Drummer, do you realize that America was energy secure? Literally, four years ago America was energy secure?

Mr. DRUMMER. America was not energy secure four years ago.

Mr. DONALDS. We were actually—Mr. Drummer, do you not—do you not acknowledge the fact that we were exporting oil to other parts of the world in the United States? We were exporting.

Mr. DRUMMER. Absolutely. We were exporting——

Mr. DONALDS. Four years ago.

Mr. DRUMMER [continuing]. Natural gas last year when prices were going through the roof. It did nothing for domestic consumers of that natural gas. This is a global market and to insulate American households and our country from all the vagaries and chaos of the world, the best thing we can do, the most prudent thing we can do, is invest in batteries, renewables——

Mr. DONALDS. Mr. Drummer, let me jump to batteries real quick because I am running out of time.

Mr. DRUMMER [continuing]. To secure our energy future and not be reliant on the global supply chain.

Mr. DONALDS. Let me come to batteries real quick. Mr. Drummer, reclaiming my time. Batteries are made up with elements. You need elements to actually create batteries.

Mr. DRUMMER. Rare earth minerals.

Mr. DONALDS. Do you acknowledge that most of the rare earth minerals and cobalt are being mined in other parts of the world, mostly with African child slave labor by the Chinese? Do you acknowledge that?

Mr. DRUMMER. I am crystal clear, and do you acknowledge that——

Mr. DONALDS. Do you think that is OK? Do you think that is OK?

Mr. DRUMMER. Absolutely—that is absolutely not OK.

Mr. DONALDS. You think it is OK for child slave labor——

Mr. DRUMMER. It is not OK.

Mr. DONALDS [continuing]. In Africa to provide batteries for us here in the United States?

Mr. DRUMMER. It is not OK.

Mr. DONALDS. So, which one is it? I am asking you. You just said it is OK.

Mr. DRUMMER. I didn't say it was OK. Let the record reflect I did not say it was OK. I said it is not OK.

Mr. DONALDS. Well, let me set the record straight because I am over my time. I don't think it is OK. I think it is wrong, especially considering——

Mr. DRUMMER. And I agree with you that it is not OK.

Mr. DONALDS [continuing]. That fossil fuels provide more than enough that what we need for our economy to survive.

I yield back.

Mr. FALLON. Thank you. Thank you, Mr. Donalds. And Ms. Brown is recognized for five minutes.

Ms. BROWN. Thank you, Mr. Chairman. I would like to use my five minutes to highlight what an excellent job the Biden Administration has done in lowering gas prices on behalf of the American people.

With the legitimate tools at his disposal, President Biden effectively addressed the severe energy crisis sparked by Vladimir Putin's illegal invasion of Ukraine.

Facing uncertainty related to oil production in Russia, the world's second largest oil exporter, in addition to the aftershocks of the COVID–19 pandemic, the Biden Administration worked with the international partner—energy partners to release oil from the Strategic Petroleum Reserve on an emergency basis.

That is precisely why the reserve was designed, to bolster American energy in times of need. In Cleveland alone gas prices have fallen by nearly $2 a gallon since their peak last year.

To my Republican colleagues, I would argue that is a good thing. When gas prices are lower working families can focus on the things that really matter and worry less about paying at the pump. To further elaborate on the reserve, I will turn to you, Mr. Drummer.

Mr. Drummer, why was the Strategic Petroleum Reserve created in the wake of the 1973 energy crisis?

Mr. DRUMMER. It was created for exactly this type of situation, to secure availability of oil for the American people in times of crisis.

Ms. BROWN. Thank you for that.

And Mr. Drummer, how did President Biden's efforts to lower gas prices benefit Americans, especially those most vulnerable among us?

Mr. DRUMMER. Right, and as even Dr. B concedes, the release from the Reserve did have an impact and, of course, the passage of the Inflation Reduction Act along with the others' packages are the long-term investment to really secure the energy security for this country and for households all across the country.

Ms. BROWN. Thank you so much. And I noticed one of our panelists didn't answer the question. So, I will ask you if you care to elaborate on your statement as highlighted by my colleague on the other side of the aisle.

The most cost-effective source of power generation on the planet is renewable energy. Would you care to explain that?

Mr. DRUMMER. Yes, we know—I mean, there is no argument that in terms of energy density nothing beats oil, right. But when it comes to the cost of drilling and the cost of pipelines and all that infrastructure it becomes less competitive with pointing what is literally a sheet of glass—it is not that, I am being simplistic here— to the sky and using the sun that God has given us and drawing down all that energy, converting it to electrons and sending it where it needs to go and then storing the energy when it gets dark. The same for the wind.

And so, just dollar for dollar, renewables are the most cost-effective form of energy, period. That is established fact. Notice Dr. B. said, but you need batteries, and yes, the Inflation Reduction Act does bring the batteries.

Ms. BROWN. Thank you for that.

So, I think we can agree that President Biden's decision to release oil from our Strategic Reserve was not only critical in a time of need but was perfectly suited to the Reserve's mission and if he had not done so, if President Biden failed to take decisive action when gas prices increase wouldn't we be here criticizing that decision?

On behalf of the folks back home in Ohio's 11th congressional District who can now more easily afford to fill up their tanks, I commend the President for his swift action.

And with that, I yield back.

Mr. FALLON. The Chair recognizes Mr. Perry.

Mr. PERRY. Thank you, Mr. Chairman.

Mr. Drummer, you just kind of mentioned your—first of all, were you alive during the 1970's era gas crisis? Just out of curiosity. I don't know. So, I am just asking.

Mr. DRUMMER. I wasn't born yet.

Mr. PERRY. You weren't. But you do acknowledge that that was the reason for the construction and the development—the advent of the Strategic Petroleum Reserve, right? You will acknowledge that, right?

Mr. DRUMMER. Yes.

Mr. PERRY. And do you know what caused—what precipitated that? Why did we do it? Why did America do it?

Mr. DRUMMER. I would like to follow your point here. Would you like to elaborate? I mean, it is—why did we do what?

Mr. PERRY. Why did we create a Strategic Petroleum Reserve? What was happening at the time?

Mr. DRUMMER. There was an energy crisis that was outside of our control.

Mr. PERRY. OK, and who was controlling it?

Mr. DRUMMER. The cartel that—of states that are the primary energy producers.

Mr. PERRY. In the Middle East, right?

Mr. DRUMMER. Right.

Mr. PERRY. The cartel states in the Middle East.

Now, they said there was an energy crisis. I was alive at the time, and they said there was an energy crisis and we were actually running out of oil. That is what they said. There was a limited supply and it was all going to run out and the prices were going up and you couldn't get it. You could only buy gas on odd and even days based on your license plate.

I lived through the whole thing. And so, we created the Strategic Petroleum Reserve to counteract the global market's control over it, right. Would you agree?

Mr. DRUMMER. I am following you.

Mr. PERRY. OK. You are following me. OK. So, and I would acknowledge you say it is a global market, which you said it is a global market, right?

But you just said with the last—with the last—my colleague from Ohio you just said that when the President released oil from the SPRO that it actually reduced cost, did you not?

Mr. DRUMMER. I am citing the Majority's witness on—who acknowledges that of all the factors that definitely did play a role. Not an outsized role, but it played a role.

Mr. PERRY. Did it reduce—did it reduce the cost or not? That is what I am asking. Did it reduce the cost or not?

Mr. DRUMMER. It did. But the Majority is arguing that that should not have been done.

Mr. PERRY. OK. So, it reduced the cost. But prior to this with Mr. Donalds you said it is a global market and there is nothing we can do to control it. If there is nothing we can do to control it how did releasing oil from the SPRO reduce the cost?

Mr. DRUMMER. It reduced the cost marginally. But the price was at an elevated level.

Mr. PERRY. I don't care if it is marginal or not. But it reduced the cost. The point is, is that we can do something about it because, sir, in your testimony you say that we can't drill our way to energy security.

But clearly, clearly, we can because we can influence the global market. That is exactly what just happened and you just admitted to it.

So, what we are here to talk about is the Strategic Petroleum Reserve and what it is used for. Do you know what it is used for? Do you know why it was designed? Do you know the definition why we have a strategic petroleum reserve?

Mr. DRUMMER. We have the reserve to insulate the U.S. from these types of energy shocks in times of crisis.

Mr. PERRY. In times of crisis. Emergency—national security emergency and natural disaster. What national emergency was occurring that required the release of so much, or literally any, of the SPRO? What national emergency was occurring during the Biden Administration that required us to do that?

Mr. DRUMMER. If I recall, there was a war in Ukraine.

Mr. PERRY. And that is a national emergency for the United States of America? Did it imperil the United States of America's ability to drill its own oil, produce its own fossil fuels, refine them, and get them to market?

Mr. DRUMMER. The President made a judgment that——

Mr. PERRY. Did it make—just answer that question.

Mr. DRUMMER. The war in Ukraine—what is driving up costs for oil to the extent to which it would mandate or even lead him to make the decision to make the release?

Mr. PERRY. Was it an emergency?

Mr. DRUMMER. It was a global—it was a global energy crisis.

This is well established. Are we debating the facts here?

Mr. PERRY. No, it is not. Sir, you cannot say it was well established when it is not well established.

Mr. DRUMMER. There was global—record costs in energy globally, record costs in food prices globally.

Mr. PERRY. You also say—sir, reclaiming my time. What established facts do you have? You say that wind and solar are by far the most cost-effective sources of energy on the planet and you say it is an established fact. What are those established facts and who established them?

Mr. DRUMMER. I invite you to read the footnote and source for that claim in my testimony.

Mr. PERRY. Just tell me. This is your testimony. You just tell me and tell us what the established facts are.

Mr. DRUMMER. Renewables, particularly wind and solar, are the most cost efficient and cost-effective means of producing energy on this planet?

Mr. PERRY. Says who? Says who? Says who?

Mr. DRUMMER. Says physics. Says scientists and engineers.

Mr. PERRY. No. No. Who is the person? Name and cite the study, sir.

Mr. DRUMMER. Give me a moment. I am going to look at the report that I cited and give you the name of the report that——

Mr. PERRY. Sir, have you ever worked in the energy securities market?

Mr. DRUMMER. I have worked at the grassroots levels. I have worked at the national level——

Mr. PERRY. The energy securities market. The energy securities market. Have you advised——

Mr. DRUMMER. I have not worked in the energy securities market, but it doesn't take an energy trader. It doesn't take——

Mr. PERRY. Sir, you have got as much experience as I do, and I pump gas for a living.

Mr. DRUMMER. It doesn't take an energy trader to see——

Mr. PERRY. That is what this is about. I yield the balance.

Mr. FALLON. Thank you.

The Chair recognizes Representative Holmes Norton.

Ms. NORTON. Thank you, Mr. Chairman.

Let me just say it is not surprising that during the Trump Administration my Republican colleagues had no problem with President Trump's authorizing releases from the Strategic Petroleum Reserve in response to an attack on oil refineries in Saudi Arabia. This is pure public opportunism. Let me get to my question. I have a question here for Mr. Drummer. Last year, we tapped the Strategic Petroleum Reserve to prevent a global economic crisis and temper the spike in the price of oil caused by Putin's invasion of Ukraine. I think that has been established.

But the Biden Administration's swift action revealed a key lesson about how to achieve energy independence. It is just not the lesson my Republican colleagues think.

After expanding—after years of expanding production, the U.S. is the world's largest oil and natural gas producer but we still we are still not energy secure. More oil and gas production has not led to energy security. American consumers fell victim to the global price shocks that came from the Ukraine war.

Mr. Drummer, I want to drill down, forgive the pun, on a point you made and ask can we drill ourselves out of global price shocks like my Republican colleagues are suggesting?

Mr. DRUMMER. Thank you for the question, Representative.

Again, I want to remind the Committee—let us recall 2020 when people were driving less, right. We were shut down for the pandemic, quarantining and whatnot, and the price of global oil went to the basement, and around then you needed about $60 a barrel for it to be profitable to drill in the U.S.

It was down to, like, $15, $16 a barrel, which means there were fossil fuel workers who were put out of work, not because of some law or regulation, but because at this point a king in Saudi Arabia and an autocrat in Russia could not agree on the production reductions. And so, there was a disagreement within OPEC and that drove fossil fuel workers in the U.S. out of jobs.

The reality is that this market is bigger than the U.S. We can be a player. We are a player. It doesn't matter. Folks are still paying 15 to 25 percent of their income to commute to and from work. So, we cannot drill our way out. The best way forward is to turn to renewables and make the massive public investments we need to achieve true genuine energy security, which is not vulnerable and prone to any foreign interference. That is what is being argued here.

Ms. NORTON. Exactly. The prices are determined by global demand. We can never drill enough to insulate ourselves from Russian invasions, of Saudi production cuts.

Mr. Drummer, why do renewable energy sources provide more price security for American consumers?

Mr. DRUMMER. It is the same argument for onshoring supply chains. The same can be said for onshoring energy. And when backed with batteries, we know with solar the highest costs with solar are—once it is built in and you are paying down, you know, your investors and creditors, it doesn't take much to maintain.

It is very inexpensive to provide solar energy, and so with battery storage with wind we are really positioned to drive down the costs. And not just—and I didn't put this in my report, but we add renewables—we add energy efficiency insulating homes and all the jobs that we created from that work as well.

So, there is a path forward with renewables because they are not vulnerable to the same geopolitical shocks that we are experiencing today.

Ms. NORTON. OK. As we can see from the chart behind me, if I can have that chart up, of the 10 congressional districts in the country with the most clean energy projects being developed right now, nine are Republican districts, including Speaker McCarthy's, Mrs. Boebert's, and Mr. Sessions' districts because investing in clean energy is a matter of good business and common sense. Just ask your own constituents.

I yield back.

Mr. FALLON. The Chair recognizes Mr. Fry for his five minutes.

Mr. FRY. Thank you, Mr. Chairman. Thank you for having this hearing today and thank you for the witnesses for taking your time.

I am struck a little bit by the purpose of the hearing, Mr. Chairman, that we are here to discuss Joe Biden's withdrawal of the SPR inappropriately, in my opinion. But we seem to hear a lot, unfortunately, I think, about distractions. Obviously, we have already attacked some witnesses.

Mr. Epstein, I am sorry about that. But, you know, the definition of the SPR is that it is intended as a cushion to, quote, "diminish the vulnerability of the United States to the effects of a severe energy supply interruption and provide limited protection from the short-term consequences of interruptions in supplies of petroleum products."

To me, if you look at the history of this, Mr. Chairman, we have only done this since its inception in the—before my time in the 1970's—we have only done this three times. We have only withdrawn from the SPR three times.

Mr. Epstein, in your opinion have you ever seen this level of withdrawal in any president?

Mr. EPSTEIN. Nobody has because it is unprecedented.

Mr. FRY. Right. So, we have three presidents that have tapped it for a limited purpose when there has been a disruption in the marketplace—for a limited purpose.

And now, under this President we have seen 226 million barrels. In your opinion, do you think that that was an appropriate use of the SPR?

Mr. EPSTEIN. I think it was wildly inappropriate because the underlying cause was the anti-fossil fuel policies that Joe Biden has supported for 15-plus years.

So, the solution was, obviously, to correct that root cause, which we could have encouraged production around the world and the shale revolution showed us that we can make a big difference in global production ourselves.

And then, in particular, because there is a lot of uncertainty about Russia, we need the SPR as full as we can get. It doesn't make any sense to drain it before a potential catastrophic conflict arises.

Mr. FRY. Dr. Bouchouev—help me here.

Mr. BOUCHOUEV. That is good.

Mr. FRY. That is good. Can you explain the magnitude as well to the answer the same question? In theory, I guess presidents can authorize emergency releases to buffer the market and also replenish an oil stockpile.

Why might have the Biden Administration acted improperly in 2022?

Mr. BOUCHOUEV. Well, there was really no emergency disruptions. There was plenty of oil. The price was high for sure. But as I stated in my testimony, the price could be high for a bunch of different reasons.

I think sometimes there is a misconception that as there is more demand than supply—price go up. Not always the case. The rule No. 1 of free markets price goes up because there are more buyers and sellers and there are more buyers and sellers for totally unrelated reasons such as inflation or interest rate hike, trying to control the price. That is very irrational decision.

As I said, you cannot do that. If I were making such a decision in private practice, I would have probably lost my job, and there were no actual disruptions in the physical market at the time.

Mr. FRY. Thank you.

Mr. Epstein, you tweeted last week on March 3, stating, "As I have documented extensively, had Biden and other Democrats spent the last three and a half years, including two of the Biden Administration, liberating U.S. oil and gas investment, production, and transport instead of strangling them energy would be far cheaper."

Mr. Epstein, has the Biden Administration offered any realistic solutions to bring the price of gas down?

Mr. EPSTEIN. No, they have been in total denial, and they are compounding the problem. I mean, this is a president who ran on, I guarantee you we are going to end fossil fuel.

The U.S. is the leader of the world. Again, we have huge capacity ourselves. We can also lead the world. Imagine instead of going to—you know, going to Egypt and saying, hey, let us all get rid of oil as quickly as possible, Biden had said, look, this energy crisis shows we are crazy—we need to reverse course, we should be leading the world on policy.

And just, by the way, in terms of security, I keep hearing this and it drives me crazy. People are talking about renewables as secure.

The entire supply chain is dependent on China. It is true that it is nothing like oil. It is incredibly more insecure than oil. China controls renewables far more than Russia controls oil or natural gas.

Mr. FRY. What do you see as possible solutions to creating more affordable energy, particularly more—or cheaper energy prices for consumers?

Mr. EPSTEIN. I mean, the main thing is you have to have actual free markets, not subsidize things that are unreliable and can't survive on their own, and I think we need—we have a whole of government attack on fossil fuels.

The first thing we need is a whole of government support for all—for freedom for all forms of energy. And until that stops the energy industry is going to be terrified. Many executives have spoken up and said, hey, it is the whole of government attack on our industry. That is the reason we are afraid to invest.

Mr. FRY. Thank you. I yield back, Mr. Chairman.

Mr. FALLON. Thank you. The Chair recognizes Ms. Stansbury for five minutes.

Ms. STANSBURY. All right. Thank you so much, Mr. Chairman. And I also want to take a moment to congratulate you as Chairman of this Committee. This is, of course, our first Subcommittee hearing this morning and I do want to welcome our witnesses.

I do feel like some of the behavior of our colleagues and how they have addressed our witnesses has been unbecoming of this institution. So, we try to welcome folks who come before us. But it is important, of course, to reveal, you know, what is going on here and I want to really take this opportunity to talk about what exactly this hearing is.

I think we have seen some interesting and bizarre theatrics this afternoon and some challenges in really understanding the basic facts of how the global energy market works, and leading the American people astray.

So, I want to just take a couple of minutes here and also ask our witnesses some questions and establish, you know, what is actually driving energy inflation and energy prices in the United States, namely, oil prices at a global level, and this last year the war in Ukraine and also, which we haven't spoken about as much, Putin and OPEC+'s actions in the middle of last year to constrain global production, which highly impacted our economy and which the President took emergency action to address—the issues of how Democrats and the President are trying to address these energy security issues here in domestic markets and, of course, the talk about the action that we have taken as a body here in this Congress and the President to address our energy security long term, namely, global climate change.

But I think it is important before we kind of dive into the details of these issues to talk a little bit about the spin of this hearing and some of the misinformation that we are hearing and I think it is important for us to acknowledge, as the Chairman himself said, that actually two of our witnesses here today are actually fossil fuel industry folks.

And in fact, Mr. Bouchouev, I know we have already established this, but you worked for the Koch Brothers organization for more than 20 years. Is that correct?

Mr. BOUCHOUEV. I worked for the company called Koch Supply & Trading.

Ms. STANSBURY. Yes, which is the global partnership and industry operated by the Koch brothers, correct?

Mr. BOUCHOUEV. It is a trading company that is owned by Charles and David Koch—majority owners, yes.

Ms. STANSBURY. Yes, of course. OK.

And Mr. Epstein, I do—you know, I know it is difficult and we had some, you know, bizarre commentary, I think, in this hearing. But it is important for us to really understand who the people are that are coming before us.

Mr. EPSTEIN. Sure.

Ms. STANSBURY. And is it not true that one of your clients is actually a coal trade association? Just a—just a——

Mr. EPSTEIN. So, I just want to correct—so you falsely said I was a fossil fuel industry member. That is what you are——

Ms. STANSBURY. Just a yes or no answer. Do you work for the coal——

Mr. EPSTEIN. No, I can't give a yes or no answers to a loaded question. So, you can—you can ask me——

Ms. STANSBURY. Do you—do you—do you—Mr. Epstein——

Mr. EPSTEIN [continuing]. And you get a real answer or you can just——

Ms. STANSBURY [continuing]. Do you work for a coal trade association, yes or no?

Mr. EPSTEIN. I don't work for anybody. I have dozens of clients whom I advise. I do not follow them.

Ms. STANSBURY. OK. Mr. Epstein—Mr. Epstein, do you advise a coal trade association, yes or no?

Mr. EPSTEIN. Proudly. I advise many people and I am proud of it.

Ms. STANSBURY. Yes. OK. OK, great. So, we have already established that two of our witnesses are here on behalf of the fossil fuel industry.

Mr. EPSTEIN. Not true. That is not true. I am not on behalf of them. I am here on behalf of myself. So, you are lying.

Ms. STANSBURY. I am reclaiming my time. Thank you.

So, you know, I think it is important to establish that two of our witnesses who are here—they have been called by the Majority— really lack just basic credibility around the facts. But to add insult to injury we have also got this issue about these past comments about racial superiority of Western cultures. And so, obviously, I think we are all very disturbed that the—

Mr. EPSTEIN. It wasn't racial superiority.

Ms. STANSBURY [continuing]. That the Majority would stoop to this level to call witnesses who are not fit to actually testify on these very important issues.

And as a representative for New Mexico, which is one of the major energy producers in the United States, I want to just be clear on the facts. So, let us talk about the facts.

Last year, Vladimir Putin invaded Ukraine. Months later, he went to Saudi Arabia and OPEC. They constrained oil production, it jacked up global oil prices, and President Biden and his Administration took an emergency action to lower gas prices.

The U.S. Treasury studied it. It lowered gas prices up to $.38 at the pump per gallon. It actually helped alleviate a massive economic crisis in the United States at the end of the summer as we were facing fuel and food shortages, and American production is at an all-time high.

So, those are the actual facts and I am deeply disturbed to see that the Majority is holding a false hearing to spin misinformation to the American people about what is actually happening in energy markets, the ways in which we have lowered prices, and our efforts to make America and the planet more secure by passing the largest single piece of legislation to address the global climate crisis not only in American history, but in the history of the planet.

Thank you, Mr. Chairman.

Mr. FALLON. Thank you very much.

We are going to have a second round of questions. Each Member will get five minutes if they choose to take them and, as such, I recognize myself for my five minutes.

So, let us talk—you know, we heard about facts. So, let us talk about some facts—indisputable facts. January 20, 2021, the price— the national price for gasoline on average was $2.39. That is when this Administration began. That is a fact.

How about another fact? On February—I believe it is February 24, 2022, Vladimir Putin invaded Ukraine. That was the Putin price hike, except to an inconvenient truth, to quote someone that wrote a book about that about 20 years ago, was the fact that the price of gasoline at that time was $3.52 national average.

It was 47 percent higher before Vladimir Putin crossed into Ukraine. So, who is going to own up on that? And at its peak it was $5 and a penny, which was 110 percent increase from what he inherited. And today, even today, after all this miraculous intervention by the Biden Administration, it is still $3.41. It is 38 percent higher.

So, another thing—another fact, an inconvenient one—this is why we are having a hearing about the Strategic Petroleum Reserve—is it was only drawn down three times and when it was drawn down—Desert Storm, 1991, 17 million barrels; 2005, Hurricane Katrina—that was a catastrophe in Louisiana—21 million barrels. And then the military intervention in Libya in 2011 was 31 million barrels.

And then Joe Biden, it is not 16 or 17 or 20. It is 227 million barrels. That is a multiply—10 times multiplier in the reduction. And then this interesting question about how there are—the energy security can be best achieved with renewables. I have no problem with renewables. I don't know any of my Republican colleagues that do.

I am all for wind and solar. I am for a comprehensive, realistic approach. Because I can tell you that I am going to get on my unicorn and ride it to a pool to visit my mermaid that was financed by a leprechaun because that is pie in the sky and it is not reality. The fact that the President of the United States in his State of the Union said, we are going to need fossil fuels for maybe 10 more years, well, heck, I would love to say that we are not going to need them next year.

But that is not the reality, and I can't wait if we are here 10 years from now because it is going to be—very little is going to change, and even when you make a transition that takes a decade or two. There is 200 countries on the planet.

Mr. Drummer, why isn't any of them doing what you just said? If it is more efficient and more cost affordable shouldn't—renewables should be dominating the industry right now—dominating the energy sector. Why aren't they? Yes.

Mr. DRUMMER. It is, honestly, because there is too much money to be made. It is why my co-witnesses are here. It is why you are holding the line that you are holding.

Mr. FALLON. Whoa. Whoa. No, no, no, no. That is—that is——

Mr. DRUMMER. There is too much money to be made in the short term and that is what it is. That is just reality.

Mr. FALLON. Oh, that is going to—hold on—hold on. I reclaim my time. I reclaim my time.

That is absolute leftist bunk and this is why, and respectfully, because not everybody is in the oil and gas industry. If you are in the energy sector and you can make dough on providing wind and solar, why wouldn't you do it? That is a rhetorical question. That is obvious.

Mr. Epstein? Mr. Epstein?

They do it when they have incredible subsidies. If they would just do it with their own money that would be one thing. Energy security—renewables are going to do it for us? We are going to be relying on China?

Mr. EPSTEIN. I mean, it is just crazy that we can talk about solar and wind being incredibly energy secure when the supply chain is just completely dominated by China, which at this—they can cut it off at a moment's notice. So, they can destroy the whole thing. So, that is one.

And then two is—this is very important when we hear these bizarre claims of cheapness—solar and wind cannot replace fossil fuels. So, solar and wind are unreliable intermittent sources of electricity. Not all energy, electricity. They can go to near zero at any given time, which is why they need near 100 percent backup. So, when you have solar and wind you can't just look at the price of the panels and the wind turbines. You have to look at the price of the backup and this is why in California where I live, in Germany and Denmark, you see skyrocketing electricity prices the more solar and wind you use.

So, they don't replace costs on the grid. They add costs to the grid. So, it is just a total—it is a proven fiction that solar and wind are cheaper and replacements for fossil fuels.

Mr. FALLON. Mr. Drummer, do you recognize that to move an 18-wheeler there is only two things that can do it right now, natural gas or diesel fuel? Do you agree with me on that? The only way it is going to move some big rig like that.

Mr. DRUMMER. I disagree with you on that. But I am following the line of questioning.

Mr. FALLON. What—oh, you could make a billion dollars. How do you move an 18-wheeler?

Mr. DRUMMER. So, there is a thing called research and development and there are companies——

Mr. FALLON. Does that exist right now?

Mr. DRUMMER [continuing]. Working—human ingenuity is something that we need to lean in on. There are companies developing——

Mr. FALLON. No. Can we—can we move an 18-wheeler right now? Sir, can we move an 18-wheeler right now?

Mr. DRUMMER [continuing]. Battery powered, electric powered rigs——

Mr. FALLON. Are they on the road right now en masse?

Mr. DRUMMER. There are companies developing a range of technology——

Mr. FALLON. Oh, they are developing. They are not on the road right now?

Mr. DRUMMER. It is a thing called research and development.

Mr. FALLON. OK. So, they are not. Thank you. I yield.

OK. Ms. Stansbury, I recognize you for five minutes. Go ahead.

Ms. STANSBURY. Thank you, Mr. Chairman.

Well, just on that last point we should clarify that yes, in fact, there are electric freight vehicles on the road and it is anticipated over the next 10 to 20 years our entire vehicle fleet will be electrical, and that that is one of the necessary steps that we have to take in order to address the climate crisis.

And so, I want to just use this second round to talk some more about the climate crisis specifically and the things that we are doing through the Inflation Reduction Act and talk to you, Mr. Drummer, about some of the work that you have done around climate justice and building more equity into the kind of clean energy economy that we are building.

So, I want to just reiterate that this August, Democrats here in the House of Representatives and in the U.S. Senate passed the largest single piece of legislation not just in the history of the

United States but in the history of the world to address our climate crisis, and I think some of the factual challenges that folks are having in this hearing today is that to address the climate crisis we have to address it across all sectors.

That is correct, Mr. Drummer, right?

And so, we have to do it across the electricity sector. We have to do it across vehicle sectors. We have to do it in terms of our emissions, our building materials, things like that, and the Inflation Reduction Act is the most comprehensive piece of legislation to ever be created to help invest in all of those sectors so that we can actually avert passing a tipping point in our global carbon emissions and we can make this transition that will be necessary so that our communities do not suffer more.

And I mentioned in my comments a few moments ago I am from New Mexico. This summer we just had the largest wildfires ever in the history of our state. Our state is experiencing a millennial drought right now.

So, climate change is already here. We are already experiencing it. I am a water resources manager. I have worked on these issues for many, many years and we already see the signature of climate change.

And what we know is that if we don't implement the IRA and the 40 percent emissions reductions that are in it, we will see even more catastrophic impacts for communities, which is why we are seeing such industry resistance and, of course, these strange attacks here.

So, Mr. Drummer, I wonder if you could talk to us about what exactly is meant by the idea of a just transition and how do we build equity into this movement?

Mr. DRUMMER. Right. The idea of a just transition is about, one, the places who are most affected and first impacted not only by the fossil fuel infrastructure. The emissions, the extraction—you know, it is a product that poisons people and our environment across its entire product lifecycle.

So, the folks who are most affected by that should be prioritized in how we transition, which is why the Inflation Reduction Act gives additional tax credits to private investors, right, who are siting renewable energy and battery storage projects in those types of communities.

A just transition also acknowledges that there are many, many people who work in the fossil fuel sector. I was talking to one the other day who put his family—all his children through college welding pipes on—it might not be the technical term but that is what he did—and the idea is that they should have a way into the clean energy economy.

And, by the way, there is a lot of work to be done to wind down this infrastructure, to seal this infrastructure, and to maintain this infrastructure safely so it doesn't leak and start poisoning our environment when it is no longer in use.

And so, that is the idea of a just transition that we can bring everybody along and everybody can be part of this clean energy economy.

Ms. STANSBURY. Thank you, Mr. Drummer.

And I personally really appreciate that, and some of some of the folks who are here on the Committee today know because I serve on the Natural Resources Committee and have talked about my family's story.

But my parents were both energy workers. My mother was a crane mechanic who helped build a coal-fired power plant in north-western New Mexico. My father was a boilermaker and a welder in the oil fields working for one of the major global oil companies.

And so, a just transition isn't just for communities who are living in natural, wild, beautiful places. It is for the communities that are really at the heart of this transition to make sure that they have dignity, that they have jobs, that they are able to build communities and economies that really reflect their values, their cultures, their languages, and their needs.

And so, I am really grateful that you are on the front lines of this effort as a thought partner to us and just appreciate the opportunity to have the conversation about it.

So, thank you, Mr. Chairman. I yield back.

Mr. FALLON. Thank you. The Chair recognizes Representative Boebert for five minutes.

Mrs. BOEBERT. Thank you, Mr. Chairman. So, obviously, the day that he took office Joe Biden unilaterally declared an all-out war on American energy production, from shutting down the Keystone XL Pipeline, which has been mentioned, on day one, imposing new rules to block pipeline projects, canceling oil and gas leases on millions of acres in Alaska and the Gulf of Mexico, imposing moratorium on new Federal oil and gas leases on Federal lands, and he single-handedly regulated our communities into poverty.

We want to talk about lifting people out of poverty, well, my communities in my district that I am proud to represent have been regulated into poverty because these good-paying jobs are no longer there. Now we are talking about affordable housing and childcare because the good-paying jobs are gone. They are not in my district anymore.

And so, what did Joe Biden do? He helped lead these record high gas prices right before an election by draining the Strategic Petroleum Reserves. He depleted them. He begged our enemies for oil. Now, a lot of this has been brought up today and, Mr. Epstein,

I would love to get to you. The Inflation Reduction Act—the IRA— was recently brought up, which is really the green new deal.

And, Mr. Epstein, I am curious, would the Inflation Reduction Act with its billions of dollars, hundreds of billions of dollars, going toward green new deal energy policies, is that going to protect us from a climate emergency?

Mr. EPSTEIN. Thank you. So, if you are thinking about emissions,—we will talk in a second about whether there is a climate emergency—but if you are thinking about emissions there is obviously only one way to reduce emissions if you are talking about CO2 emissions because this is a global issue, and that is to make low carbon energy globally cost competitive.

So, anything you are doing just to randomly subsidize the U.S., fleece taxpayers to make things more expensive, or to do these things that can't work on the market, that doesn't accomplish anything at all.

So, what we need to do is things like liberate nuclear, liberate natural gas, liberate development so we can have things like deep geothermal. But you need real market things.

So, first of all, the IRA doesn't accomplish anything in terms of globally cost competitive solutions and, in fact, it keeps over subsidizing solar and wind into the indefinite future. So, it is a total mess.

Also, this idea of a climate emergency is a falsehood. There is a difference between climate impact and climate emergency. This idea that we are all going to die very soon from climate is empirically false.

We are actually safer than ever from climate-related disasters. Fossil fuels have helped us do things like alleviate drought, heat our homes when it is cold, cool our homes when it is hot.

Like, fossil fuels have actually made the climate far safer for human beings. So, any concern with emissions should be long term and nonemergency.

It is actually this emergency mentality that led Biden to say, I guarantee you we are going to end fossil fuel, and led to all of these catastrophic policies that you are referring to.

So, what we need is energy freedom where all forms of energy can compete, and insofar as solar and wind can be cost competitive and actually provide reliable electricity, great. But forcing inferior energy on us is a disaster and it causes an energy emergency in the name of a phony climate emergency.

Mrs. BOEBERT. Thank you, Mr. Epstein.

And I have a simple question for you. Was America energy independent under President Trump's Administration?

Mr. EPSTEIN. So, it was much—I would put it as much, much more energy secure. That is how I would put it, much, much more energy secure, and in part because President Trump was saying to the industry, we are going to leave you free in the future, whereas I want to quote one of America's leading oil executives, Cody Campbell, who said, "yes, Biden's war on the oil and gas industry is driving shortages and high prices," and he cited specifically the Democrats' idea that the need for oil and natural gas is soon coming to an end.

Mrs. BOEBERT. And, Mr. Epstein—Epstein, excuse me—would you say then America is not energy secure now or is less energy secure?

Mr. EPSTEIN. Yes. We are less energy secure because we are more and more dependent on unreliable forms of energy, specifically electricity, and we have threatened the fossil fuel industry going forward, and where I live in California is the perfect embodiment where we said no more internal combustion engine vehicles and then five days later Newsom said, don't charge your electric car.

Mrs. BOEBERT. Correct.

Mr. Epstein, what should the Federal Government do to help increase our energy security as quickly as possible?

Mr. EPSTEIN. So, the first thing is just to apologize. Like, it is really, really important to say, we have been in the wrong, and I know that that is difficult, but it needs to happen.

And then this whole of government thing needs to change because the root cause is the whole of government attack on fossil fuels and the net zero commitment.

I think a first really good action would be to immediately withdraw from the Paris Climate Accords and encourage every other country to do the same. That would be a great action.

Mrs. BOEBERT. Thank you. I think an apology is necessary to the American people because I do see this as sabotage.

But, Mr. Epstein, you have illustrated the dangers of energy insecurity and abusing the SPR. Are we making other areas in addressing energy security?

Mr. EPSTEIN. Yes. So, we are making many. I know we are about to be out of time. We are making many, many other errors, and I would like to offer particularly to Representative Stansbury, who told several lies about me, if you want to know the truth about me in this issue, I will sign a copy of "Fossil Future" for you so you can learn the facts.

Mrs. BOEBERT. Great book. Thank you, Mr. Epstein.

Mr. EPSTEIN. Thank you.

Mr. FALLON. The gentlelady's time has expired. A very generous offer, by the way.

Mr. EPSTEIN. Thank you.

Mr. FALLON. The Chair recognizes Ms. Luna.

Ms. LUNA. I actually want to see if Mr. Epstein wanted to use some of my time to address some of the lies that Ms. Stansbury told about you.

Mr. EPSTEIN. Oh, sure. Yes, that is great. So, the—so there is an argument that I—and I don't have the exact transcript but the basic ideas that I am bought——

Ms. STANSBURY. Mr. Chairman, I would like to raise a point of order, please.

Mr. FALLON. OK.

Ms. STANSBURY. Mr. Chairman——

Mr. FALLON. The gentlelady will state a point of order.

Ms. STANSBURY. Yes. Mr. Chairman, I would like to inquire about the Chair's read on the rules for this Committee pursuant to what we adopted in terms of name calling and correcting——

Ms. LUNA. Can I—can I take back some of my time right now? Because I am interested in the facts.

Mr. FALLON. No, you are—we are going to——

Ms. LUNA. OK. I just want to get the facts from you.

Mr. FALLON. OK, hold on one second. There is a point of order.

Ms. Luna, your time is frozen.

Ms. LUNA. Thank you.

Ms. STANSBURY. Mr. Chairman, I would like to ask that the words be taken down by the gentlelady from Florida regarding her name calling of me and the assertion that somehow the factual exchange that I had between myself and the witness were somehow other than a normal asking of questions to a witness.

Mr. FALLON. All right. Just give me one second. Thank you.

Ms. STANSBURY. Thank you.

[Chairman reviewing the rules.]

Mr. FALLON. So, what I want to do is real quickly read our rules of witness decorum and I would remind them that this is a violation of the rules of the House of Representatives to engage in personalities regarding the President or make statements that are personally offensive to him, that the rules govern Members in debate and also applies to us as witnesses deliver their testimony.

And so, what I would ask is if Ms. Luna could perhaps rephrase to say maybe some untruths that were spoken. I think that would be fair and part of decorum, because we are going to disagree all day long on certain topics as the two parties and sometimes we do agree and then many times we don't.

And so, perhaps just rephrasing the question would satisfy everyone. And Ms. Luna, you are recognized for your full five minutes.

Ms. LUNA. Mr. Epstein, would you please correct some of the mistruths spoken about you? I would like to hear your perspective on what is factually accurate.

Mr. EPSTEIN. Sure. So, there are two things that Representative Stansbury brought up. So, one I think I addressed very clearly, this idea of any kind of racism, and she attributed—this as almost an exact quote—me talking about the racial superiority of Western culture and I clarified probably five times that Western culture has nothing to do with race and it is about ideas, particularly individuals and freedom and reason.

So, it didn't seem very honest to me to portray me as saying racial superiority. It seemed like another attempt to discredit me, and I think there is an inability to actually refute any of my arguments.

And I think the same thing is true with the attack that I am somehow representing and paid for by the fossil fuel industry. First of all, if I were, which I am not, we need representatives of the fossil fuel industry here because they have expertise on this issue, which is about oil. So, that is one thing.

But it is just factually false. So, I came to all of my ideas independently and I said, I—once I came to the conclusion that fossil fuels are good, of course, I proudly tried to advise fossil fuel companies, including coal companies, how to tell their story better.

But to use my expertise and my convictions and to claim that I am dishonest and should not be listened to, and I made a compelling case, I didn't think that was appropriate.

Ms. LUNA. Mr. Epstein, I just wanted to ask you real quick, is China somehow protected by an invisible shield that prevents their carbon emissions from escaping?

Mr. EPSTEIN. No.

Ms. LUNA. OK. So, if we have the ability to produce clean energy here or at lower carbon emissions versus China, don't you think that that would be better for the climate?

Mr. EPSTEIN. Yes, but I think the main thing is the world needs far more energy and if you care about emissions you need to find truly cost competitive ways to do that and the best way to do that is through freedom, through liberating nuclear, through liberating natural gas.

And, unfortunately, this Administration is focused on limiting the freedom of certain industries, namely, fossil fuels, including natural gas, and then imposing extremely expensive things that the market won't choose on its own, which is the whole IRA, which is just a total subsidy fest that does nothing to make alternatives truly cost competitive.

Ms. LUNA. No, I agree with you. I think from a perspective on also national security to say that somehow the United States is going to go to completely electrical grid would actually pose a massive threat to us——

Mr. EPSTEIN. Yes.

Ms. LUNA [continuing]. Because if you have a hacker they could shut down an entire first world country and, obviously, that is something that we don't want.

My final question for you is—first of all, thank you for coming to testify before Congress.

Mr. EPSTEIN. My pleasure.

Ms. LUNA. I appreciate you being here. But with everything that we just heard, I mean, what is the best way that we as a country can really prevent China from essentially owning us? Because as you had stated earlier, the Paris Climate Accord—I mean, us even engaging in that is hindering us while they are polluting and destroying our planet.

And I hear a lot about climate change but, you know, few people of action that are willing to actually address that.

Mr. EPSTEIN. Can I just comment on the grid, which you also mentioned?

Ms. LUNA. Yes, of course.

Mr. EPSTEIN. Because you mentioned the security threat of the grid and it is vulnerable to these kinds of attacks. I think the number-one thing that it is vulnerable to is we are catastrophically reducing the supply of reliable electricity and then catastrophically increasing the demand for reliable electricity.

And so, California, where I live, is a perfect example. We have attacked fossil fuels, we have attacked nuclear, so we have less reliable electricity, and then we are trying to mandate electric everything.

And this is why, as I mentioned, Newsom had this announcement, no more internal combustion engine vehicles, and then five days later he had to say, don't charge your EV, and we have signs all over the place saying don't use your electricity. We only have three percent penetration of EVs.

So, we are talking about this catastrophic, unthinking increase of EVs, and we are actually threatened—I have this documented— we are threatened to go to lose 20 percent of our reliable electricity in the next seven years under the Biden EPA.

So, that is a catastrophic threat. I can also answer about China, if you would like.

Ms. LUNA. Yes, please.

Mr. EPSTEIN. So, I mean, with China is China actually cares about energy and they are doing what is good for their people and also the power lust of the government in terms of energy.

So they, for example, use five times as much industrial electricity as we do, overwhelmingly coal, and the only way for us to be secure and out compete them is to have a very aggressive energy freedom policy and that includes—one, is liberating domestic development from things like NEPA, which, unfortunately, this Administration has done nothing to do and, in fact, reversed the good Trump Administration things on NEPA.

We need to end preferences for unreliable electricity and actually have all forms of electricity compete, so we can actually have a stable grid. We need to reform the EPA, so it stops shutting down reliable power plants.

We need to what I call 'decriminalize nuclear', not just subsidize it and this kind of thing, and in general we need an emissions policy of long-term reductions, not catastrophic emergency reductions through liberating American innovation, not punishing America.

What the IRA does is it punishes America and accomplishes nothing globally because, as you said, Chinese emissions influence the global emissions picture just as much as American emissions do.

Ms. LUNA. Thank you so much. I appreciate it. Obviously, that was very factual, and I think you are very well qualified to be here. Thank you.

Mr. EPSTEIN. Thanks.

Mr. FALLON. The Chair recognizes Mr. Edwards for his five minutes.

Mr. EDWARDS. Thank you, Mr. Chair.

Mr. Epstein, first of all, I want you to know that on the recommendation of a previous energy secretary I have read your book——

Mr. EPSTEIN. Oh, wow. Thanks.

Mr. EDWARDS [continuing]. And it is a pleasure to see the author of such a piece of work here with us today.

Mr. EPSTEIN. Thank you.

Mr. EDWARDS. So, thank you.

President Biden, in his recent State of the Union address, called out corporations for being profitable—energy corporations for being profitable and he called for additional taxes on oil and gas companies.

Are higher taxes on corporations a path forward, in your opinion, for American energy producers to increase production?

Mr. EPSTEIN. So, this is an example of the tragedy I mentioned toward the end of my testimony, which is that this Administration has, unfortunately, learned nothing and apologized for nothing, because the root cause is—of the problems is global anti-fossil fuel policies and that is what artificially suppressed the supply and drove prices up.

And so, what is the Administration's response besides dangerously depleting the SPR? It is to threaten even more punishment. So, what more taxes says is if you manage to succeed through the gauntlet of factors that we are using to restrict you we are going to make you pay even more.

Does anyone think that is attractive to oil investors? What we need is a free market that oil investors have confidence in. So, this is 180 degrees wrong policy.

Mr. EDWARDS. Thank you for that. And my follow-up question, which you might have answered but I ask you to elaborate if you can, is how are those Biden Administration policies and actions toward oil and gas companies discouraging future investment?

Mr. EPSTEIN. I mean, we could spend all day on this. But the high level is that—the high level is what they call the whole of government approach to climate change. But climate change just means—in practice it means attacking fossil fuels.

So, there is a whole of government approach to attacking fossil fuels—and I think there should be a hearing on this because this is just the root of all the destruction—is that every agency, you can go to every agency, subagency, they are trying to figure out a way to attack fossil fuels.

So, look at the SEC. The SEC has climate disclosure rules. That is just basically designed to intimidate companies into agreeing to get off fossil fuels as quickly as possible. You have pressure on investment institutions through different ESG policies, and you have the signal sent by opposing the Keystone pipeline by the leasing ban.

And so, what this says to investors and companies is if you pursue this space of oil and gas, in particular, you have a high likelihood of being punished.

Investors are investing because they want to make a profit, and if you are increasing the risk you are going to decrease investment. So, that is why we need an apology, and we need a reversal of the whole of government approach.

And this Administration or future administrations need to give long-term security to all forms of energy.

Mr. EDWARDS. Thank you.

And still Mr. Epstein—is there a physical reason that the oil industry can't meet rising demand?

Mr. EPSTEIN. Thank you for asking that. I wish I had thought to bring that up. No. So, why are prices going up? Is it—do we have a physical limitation?

No, we have endless oil in the world for the foreseeable future. Do we have a know how limitation? Did the industry forget how to produce things cost effectively? No.

It is obviously a political limitation and the obvious—it is bigger than an elephant in the room, it must be, like, a dinosaur in the room—is that we have a global anti-fossil fuel movement that has been for years suppressing investment, production, refining, and transportation.

This is obvious. This is what every country in the world is committing to under the Paris Climate Accords. This is what the investment institutions are committing to.

So, what happens is they have suppressed the supply of fossil fuels. That makes prices go up. That is what they wanted. But what they didn't want is for the electorate to get angry.

So, they want this consequence. They just don't want the political results, and the SPR is a way of hiding it. Basically, if you release oil from the SPR prices go down temporarily so people don't see the damage you have inflicted on the global economy and on Americans in particular.

Mr. EDWARDS. So, thank you for that.

Also, Mr. Epstein, I know this may seem like an elementary question, but I believe it is important for the American people to know the answer.

Can you explain why price increases exist when you restrict the supply of a valuable resource?

Mr. EPSTEIN. Sure. I mean, just if you are thinking about anything—I mean, if you think about the price of apples and you take half the apples away from the world, well, that is the supply, and then the demand is people still have the same proclivity toward wanting apples.

And so, what happens is there are fewer of them, so the people who are willing to pay more, those are the only people who are going to pay—be able to pay for it. And so, the price is going to go up.

So, it is just—you know, there is the details but it is the basic supply/demand dynamic and it is very important. The world, particularly the Biden Administration, have committed to restricting the supply.

They promised us the demand would go away because unreliable solar and wind would replace fossil fuels, the same promise we are hearing today from others. It was a false promise. Demand went up, so supply went down. Demand went up, prices go up.

Mr. EDWARDS. Thank you. Thank you for being with us. Thank all of our visitors for coming and sharing with us this afternoon.

Mr. Chairman, I yield.

Mr. FALLON. Thank you.

In closing, I want to thank our panelists once again for their important insights and testimony today. The most valuable resource any of us have are time. So, we thank you for your time.

And I would like to yield to Ranking Member Stansbury for her closing remarks.

Ms. STANSBURY. All right. Well, thank you, Mr. Chairman. It has been an interesting adventure today, I will say that.

So, welcome to the Subcommittee. I am sure we are going to have lots of vigorous debate this Congress on these issues.

I do feel bad that we didn't get to ask Dr. Bouchouev more questions. I could see him shaking his head in disagreement with many of the comments that Mr. Epstein said, and it is unfortunate because he is a—while a former Koch Brothers employee, he is an energy economy expert and there is a lot of misinformation that was shared in the name of—as testimony today.

So, I apologize that we didn't have the opportunity to actually hear the facts about global energy markets. And I do thank Mr. Drummer for joining us today to talk about and share your perspective on these issues.

Mr. Chairman, obviously, there is nothing more important than addressing our global energy security and the climate crisis. It is something that the Biden Administration and Democrats understand very well.

It is part of why the Biden Administration has aggressively pursued energy policies that are both lowering prices, protecting American access to energy of all kinds, and took emergency action to take down the SPRO.

There is discussion about what that impact has about overall energy market stability. We understand that. But it was an emergency measure that was taken at the time in order to address gas prices at the pump at a time when we were experiencing record inflation and food shortages here in the United States.

It was an energy crisis, just as the SPRO was designed to do, which was to address issues when OPEC, other cartels, and global energy prices impacted the American consumer.

So, we are grateful for the Biden Administration's proactive effort, the impact that it had for communities. But most importantly, I want to really emphasize the incredible action that we have taken here in Congress this last year to pass the Inflation Reduction Act to do as we heard today, to address climate change across all aspects of our economy, to invest in our communities, and to ensure that this energy transition that we are going through is not only good for business, good for America, good for our energy security that ultimately is just and equitable for our communities as we are trying to navigate through this global crisis.

So, with that, Mr. Chairman, I very much appreciate it and look forward to working with you.

Mr. FALLON. Thank you. I am going to yield back your two minutes.

No, thank you all again for coming.

Just a couple of things to point out. The United States is not a planet. We live on a planet. We are part of it. But we have issues with China and the coal plants that they are opening. They are increasing their emissions.

I am very proud of the fact that our country has reduced our emissions dramatically over the last 20 years and we have India as well has a growing economy that has also contributed, and I think it is fascinating to know that it is projected that coal—global coal use will never be higher than this year, which is really not, I don't think, good news for anyone. But it is interesting to know. We haven't talked a lot about and we should have future hearings about nuclear energy. I want to focus on solutions, not just rhetoric, but the carbon that is already in the atmosphere. Trillion Tree project, things like that, that will filter it. Carbon capture— there is ideas with putting it in concrete and limestone caves and things like that that are solutions that we should all get behind, particularly, again, with nuclear. I think a lot of folks that were on the political left have seen the light with nuclear energy of late.

Not all, but many.

Also, the electric grid. Can it handle the capacity of hundreds of millions of cars? It was never designed for that, and so that is an issue that we are going to have to discuss, moving forward.

And now—and I recall, like, these almost Thomas Malthus type—Thomas Malthus exaggerations that some folks have said in the public sphere and Members of this chamber as well.

One said that—not in this hearing but another one that I attended, that by 2026 there is going to be large swaths of the United States that will be uninhabitable, and I believe it was by 2034 we are going to major famines in the United States.

And former President—Vice President Gore said that global sea levels are going to rise by 20 feet and the—you know, there was going to be millions of people displaced. And even the U.N.'s own IPCC said that it could—global sea levels may rise as little as four inches in the next hundred years. So, interesting to know former presidents have purchased properties at zero sea level on the coast. So, the coastlines aren't being threatened right now at all.

So, it is going to be interesting over the next eight to 10 years to see the realities of what happens. Are all or some tractor trailers going to be electric? I don't know. But we are going to find out.

I don't know if I will be here in 10 years or not. But so, take note of what is being said. Take note of what was said in 2000. Take note what was said today, and we will see where we are, moving forward. I think that is very important.

But we, again, need to recognize the realities, and this Administration did not lower prices. We have empirical data. Everybody can agree the price of gasoline was $2.39 when Mr. Biden took office. Today it is $3.41. It is 38 percent higher, despite the fact that we have drained the SPR to historic levels.

Seventeen million in 1991, 21 million in 2005, 31 million in 2011, and then 227 million over the last year and change. That is remarkably different.

So let us just, again, talk about and focus more on empirical, undeniable facts. That is why we had this hearing and that is why we can conclude beyond a shadow of a doubt that there was no national emergency.

The price of gasoline was already up prior to Putin's invasion. It wasn't his fault. It was bad policy. If you want to reduce the costs of energy, then make more of it.

Yes, it is a global market. We are a major player in that global market that can affect the price. It is a supply and demand issue, at the end of the day.

So, I want to thank everybody for coming. I appreciate our first hearing and I look forward to many more.

And in closing, with that, and without objection, all Members will have five legislative days within which to submit materials and to submit additional written questions for the witnesses, which will be forwarded to the witnesses for their response.

If there is no further business, without objection, the Subcommittee stands adjourned. Thank you.

[Whereupon, at 4:09 p.m., the Subcommittee was adjourned.]

Chapter 3

Unleashing American Energy, Lowering Energy Costs, and Strengthening Supply Chains*

Committee on Energy and Commerce

Tuesday, February 7, 2023
House of Representatives,
Subcommittee on Energy, Climate, and Grid Security,
Joint with the Subcommittee on Environment, Manufacturing, and
Critical Materials, Committee on Energy and Commerce,
Washington, DC.

The subcommittees met, pursuant to call, at 10:01 a.m. in the John D. Dingell Room 2123 Rayburn House Office Building, Hon. Jeff Duncan (chairman of the Subcommittee on Energy, Climate, and Grid Security) presiding.

Members present: Representatives Johnson (Subcommittee on Environment, Manufacturing, and Critical Materials chairman), Burgess, Latta, Guthrie, Griffith, Bucshon, Walberg, Palmer, Curtis, Lesko, Pence, Armstrong, Weber, Balderson, Pfluger, Rodgers (ex officio), DeGette (Subcommittee on Energy, Climate, and Grid Security ranking member), Tonko (Subcommittee on Environment, Manufacturing, and Critical Materials ranking member), Matsui, Castor, Sarbanes, Cárdenas, Peters, Veasey, Kuster, Blunt Rochester, Schrier, Fletcher, and Pallone (ex officio).

Also present: Representative Dingell.

Staff present: Sarah Burke, Deputy Staff Director; Michael Cameron, Professional Staff Member, Innovation, Data, and Commerce; Jerry Couri, Deputy Chief Counsel for Environment; Lauren Eriksen, Clerk, Oversight and Investigations; Nate Hodson, Staff Director; Tara Hupman, Chief Counsel; Peter Kielty, General Counsel; Emily King, Member Services Director; Elise Krekorian, Professional Staff Member, Energy; Mary Martin, Chief Counsel, Energy and Environment; Jacob McCurdy, Professional Staff Member, Energy;

* This is an edited, reformatted and augmented version of Joint Hearing, before the Subcommittee on Energy, Climate, and Grid Security and the Subcommittee on Environment, Manufacturing, and Critical Materials, of the Committee on Energy and Commerce, House of Representatives, One Hundred Eighteenth Congress, First Session, Serial No. 118–6, dated February 7, 2023.

In: Energy Policy and National Security
Editor: Kuzman Sukanya
ISBN: 979-8-89530-879-0

Brandon Mooney, Deputy Chief Counsel for Energy; Kaitlyn Peterson, Clerk, Energy and Environment; Carla Rafael, Staff Assistant; Peter Spencer, Senior Professional Staff Member, Energy; Michael Taggart, Policy Director; Timia Crisp, Minority Professional Staff Member; Waverly Gordon, Minority Deputy Staff Director and General Counsel; Tiffany Guarascio, Minority Staff Director; Anthony Gutierrez, Minority Professional Staff Member; Caitlin Haberman, Minority Staff Director, Environment, Manufacturing, and Critical Materials; Mackenzie Kuhl, Minority Digital Manager; Kris Pittard, Minority Professional Staff Member; Kylea Rogers, Minority Policy Analyst; Andrew Souvall, Minority Director of Communications, Outreach, and Member Services; Medha Surampudy, Minority Professional Staff Member; Isaac Velez, Minority Intern; and Tuley Wright, Minority Staff Director, Energy, Climate, and Grid Security.

Mr. DUNCAN. The Subcommittee on Energy, Climate, and Grid Security and the Subcommittee on Environment, Manufacturing, and Critical Minerals will now come to order.

The Chair now recognizes himself for 5 minutes for an opening statement.

Opening Statement of Hon. Jeff Duncan, a Representative in Congress from the State of South Carolina

First of all, I want to thank you all for being here. And I thank all of our witnesses for being here, as well. I preemptively want to say that we all appreciate your patience, as this might be a long day.

I am excited that we are holding our first legislative hearing, a joint hearing with the Energy, Climate, and Grid Security Subcommittee and the Environment, Manufacturing, and Critical Minerals [sic] Subcommittee. Our goal is to enact policy that delivers affordable, reliable, and clean energy to all Americans, a goal I believe we all share on this committee, regardless of party.

In our hearing on restoring American energy dominance last week, we heard how the Biden administration's energy policies are making energy unaffordable and less reliable for American consumers. The aggressive Rush to Green agenda is compromising our security by creating vulnerabilities in our energy supply chain, making us more reliant on our adversaries for energy and critical minerals.

I believe in unleashing all sources of American energy, from nuclear, oil, and gas to hydropower, renewables, hydrogen—a truly all-of-the-above approach. We also believe in unleashing innovation by creating a regulatory structure that encourages investment and growth in the private sector.

We have said it before: American energy production and reducing emissions are not mutually exclusive. We produce energy cleaner than anywhere in the world.

Unfortunately, many of our energy policies coming out of the Biden administration prioritize climate goals over reliable and affordable energy. They compromise the ability for Americans to afford their power bills and keep on the lights. They also fail to address the significant permitting barriers to bringing more clean energy online.

The bills we are reviewing today offer solutions. They will bring down the cost of energy, reduce emissions, strengthen our energy supply chains, and pave the way for restoring American energy dominance. We did invite the FERC Commissioners, Secretary of Energy, and the EPA administrator, all who, unfortunately, were unable to attend. I am hopeful that we can have them in front of this committee soon to give the administration's perspective.

I am, however, pleased we are moving this legislation through regular order, with a full committee hearing last week to inform us of the state of American energy. The legislation in front of us today will address some of the issues and propel the United States into American energy dominance.

For example, my bill protects American energy production by prohibiting the President from declaring a moratorium on hydraulic fracturing. This is necessary because President Biden has repeatedly stated that he would end fossil fuel production in the United States.

Representative Pfluger's bill repeals the costly natural gas tax created in the Inflation Reduction Act. The Promoting Cross-border Energy Infrastructure Act encourages the construction of energy infrastructure across the borders of the U.S., Canada, and Mexico, helping us secure Western hemispheric energy security.

Several bills also address the importance of American energy exports in the global markets. The world is safer when America is energy dominant, and Representative Johnson's bill to unlock our domestic LNG would make it easier for FERC to approve export terminals to deliver clean energy to our friends and allies.

We also will be taking up a resolution that expresses support for the free trade and export of crude oil and petroleum products. This is necessary because President Biden and the Democrats on this committee have advocated for reinstating the crude oil export ban. Lifting the export ban in 2015 has lowered prices while also increasing our leverage globally. It would be shortsighted to reverse this.

We will also focus on securing our nuclear supply chain with a bill to wean off reliance on Russian uranium. Our grid and energy infrastructure increasingly have come under attack. The critical Electric Infrastructure Cybersecurity Incident Reporting Act will increase transparency between critical electric infrastructure owners and the Department of Energy to strengthen our systems.

Just over 2 years ago, America was energy dominant for the first time since 1952. We were the largest energy producer in the world, while also leading the world in emissions reductions. We can and should be a world leader, and these bills will help get us there. It is time to stop handing over leverage to the CCP, Iran, and the OPEC cartel. Not only leverage, but American dollars.

Every American should have access to reliable energy. The most recent blizzards underscore the need for resilient energy infrastructure and a diversified generation mix capable of responding to storms. It is time to flip the switch, unleash American energy production. These bills are the first step in achieving energy dominance.

I look forward to hearing from our witnesses on these bills.

[The prepared statement of Mr. Duncan follows:]

Chairman Jeff Duncan (SC-03) Opening Remarks, Joint Energy & Environment Committee Hearing, Joint Legislative Hearing: "Unleashing American Energy, Lowering Energy Costs, and Strengthening Supply Chains." February 7, 2023, as Prepared for Delivery

Thank you all for being here. And thank you to all of our witnesses for being here. I preemptively want to say we all appreciate your patience; this will likely be a long day.

I am excited we are holding our first legislative hearing, a joint hearing with the Energy, Climate, and Grid Security subcommittee and Environment, Manufacturing, and Critical Materials subcommittee.

Our goal is to enact policy that delivers affordable, reliable, and clean energy to all Americans. A goal I believe we ALL share on this committee (regardless of party).

In our hearing on Restoring American Energy Dominance last week, we heard how the Biden Administration's energy policies are making energy unaffordable and less reliable for American consumers.

The aggressive "rush to green" agenda is compromising our security by creating vulnerabilities in our energy supply chain, making us more reliant on our adversaries for energy and critical materials.

We believe in unleashing all sources of American energy—from nuclear, oil and gas, to hydropower, renewables, and hydrogen. We also believe in unleashing innovation by creating a regulatory structure that encourages investment and growth in the private sector.

We've said it before—American energy production and reducing emissions are not mutually exclusive. We produce energy cleaner than anywhere in the world.

Unfortunately, many of the energy policies coming out of the Biden Administration prioritize climate goals over reliable and affordable energy. They compromise the ability for Americans to afford their power bills and keep on the lights. They also fail to address the significant permitting barriers to bringing more clean energy online.

The bills we are reviewing today offer solutions – they will bring down the cost of energy, reduce emissions, strengthen our energy supply chains, and pave the way to restoring American Energy Dominance.

We did invite the FERC Commissioners, Secretary of Energy, and EPA Administrator—all who unfortunately were unable to attend. I am hopeful we can have them in front of this Committee soon to give the Administration's perspective. I am however pleased we are moving this legislation through regular order. With a full committee hearing last week to inform us of the state of American energy, the legislation in front of us today will address some of the issues and propel the United States into energy dominance.

For example:

- My bill, Protects American Energy Production by prohibiting the President from declaring a moratorium on fracking. This is necessary because President Biden has repeatedly stated he would end fossil fuel production in the U.S.
- Representative Pfluger's bill repeals the costly Natural Gas tax created in the Inflation Reduction Act.
- The Promoting Cross-border Energy Infrastructure Act encourages the construction of energy infrastructure across the borders of the U.S., Canada, and Mexico – helping us

secure Western Hemispheric energy security.

- Several bills also address the importance of American energy exports in the global market.
  - The world is safer when America is energy dominant and Representative Johnson's bill to "Unlock our Domestic LNG" would make it easier for FERC to approve export terminals to deliver clean energy to our allies.
  - We also will be taking up a resolution that expresses support for the free trade and export of crude oil and petroleum products. This is necessary because President Biden and Democrats on this Committee have advocated for reinstating the crude oil export ban.
  - Lifting the export ban in 2015 has lowered prices while also increasing our leverage globally—it would be shortsighted to reverse this.
- We will also focus on securing our nuclear supply chain with a bill to wean off reliance on Russian uranium.
- Our grid and energy infrastructure increasingly have come under attack. The Critical Electric Infrastructure Cybersecurity Incident Reporting Act will increase transparency between critical electric infrastructure owners and the Department of Energy to strengthen our systems.

Just over 2 years ago, America was energy dominant for the first time since 1952. We were the largest energy producer in the world while also leading the world in emission reductions.

We can and should be a world leader and these bills will help get us there. It's time to stop handing over leverage to the CCP, Iran, and the OPEC cartel.

Every American should have access to reliable energy. The most recent blizzards underscore the need for resilient energy infrastructure and a diversified generation mix capable of responding to storms.

Let's Flip the Switch, Unleash American Energy Production. These bills are the first step in achieving energy dominance.

I look forward to hearing from our witnesses on these bills and I now recognize Ranking Member DeGette.

* * *

Mr. DUNCAN. And I now recognize Ranking Member DeGette.

Opening Statement of Hon. Diana Degette, a Representative in Congress from the State of Colorado

Ms. DEGETTE. Thank you so much, Mr. Chairman. I am happy to be here today, and I would humbly suggest if we want to reduce our reliance on OPEC and other bad actors internationally, we should reduce our reliance on oil, since it is an international market.

And when I looked at these bills, they don't just unleash America's energy potential. They tether us more fully to a global commodity that strains our budgets and also causes real harm to our environment. They expand oil and gas drilling throughout the country, and they undo many of the bedrock environmental laws that we have put in place to ensure every American has access to the clean air, clean water, and a clean environment.

The bills that we are considering today bolster an industry that is already reporting record profits. Last year the United States produced an average of 11.9 million barrels of crude oil a day. Now, that is the second highest level in U.S. history. And the oil and gas industry is on track to produce even more in the years to come. According to the Energy Information Administration, the U.S. is on pace to produce 12.4 million barrels a day this year, which is an all-time record, and up to 12.8 million barrels a day in 2024.

So I am not really sure why the majority thinks we need to give even more incentives for big oil to produce more oil. So if—and the title of this hearing, "Unleashing American Energy," it really means—is giving big oil unfettered access to do as they please, well, I guess that is what these bills do.

But again, they do nothing to unleash our Nation's true energy potential. They don't do anything to protect the American people from the volatility of the global oil market and the skyrocketing prices we saw last summer. They don't do anything to combat the climate crisis or deliver environmental justice to some of the most vulnerable communities across this country.

So if we really, really want to unleash America's energy potential and drive down the cost of energy, then we need to break our addiction to oil and we need an orderly transition to cleaner, renewable energy sources. Not only will doing so help continue to reduce the harmful emissions that are driving the climate crisis, but it reduces our dependance on the global oil market, and it will reduce energy bills that many Americans are struggling to afford.

Clean energy is now one of the cheapest sources of energy, and it is why countries around the world are already making the transition to expand the use of new technologies that will serve as a foundation for them to continue to grow their economies into the future. And we cannot fall behind in that effort. We can either be the leaders of a global clean energy transition or our adversaries will, because they understand the risk as much as we do.

So, instead of furthering our dependance on oil, we should actually be fueling the innovation of new technologies that can provide clean, renewable energy that is both reliable and affordable, and also by increasing our grid security. We need to build the foundation and we need to invest in training for the energy workforce so they can take on the jobs of the future. This is how we unleash America's energy potential.

So, Mr. Chairman, I want to say this again, because I think there is some confusion. People seem to think that if we produce more oil and gas domestically, even though we are producing record amounts, this is going to make us independent from international energy levels or from OPEC price increases. That is simply not the case, because it is an international market. And we saw that last year when oil and gas prices went up so much, even though there was the ability to have increased domestic production. So I think that, by working together, we could find bipartisan solutions. We all have the same goal: a solid energy source, transition to renewable energy, and combating the climate crisis, and building the foundation to make sure that can happen.

I don't think these bills are the solution, so I think we should go back to the drawing board and get that right.

[The prepared statement of Ms. DeGette follows:]

Committee on Energy and Commerce, Opening State as Prepared for Delivery of Subcommittee on Energy, Climate, and Grid Security, Ranking Member Diana DeGette, Hearing on "Unleashing American Energy, Lowering Energy Costs, and Strengthening Supply Chains," February 7, 2023

Thank you so much, Mr. Chairman. I'm happy to be here today.

I would humbly suggest if we want to reduce our reliance on OPEC and other bad actors internationally, we should reduce our reliance on oil, since it's an international market.

When I looked at these bills, they don't just unleash America's energy potential, they tether us more fully to a global commodity that strains our budgets and causes real harm to our environment. They expand oil and gas drilling throughout the country, and they undo many of the bedrock environmental laws that we've put in place to ensure every American has access to the clean air, clean water, and a clean environment.

The bills that we're considering today bolster an industry that's already reporting record profits.

Last year, the United States produced an average of 11.9 million barrels of crude oil a day. Now that's the second highest level in US history, and the oil and gas industry is on track to produce even more in the years to come.

According to the Energy Information Administration, the US is on pace to produce 12.4 million barrels a day this year, which is an all-time record, and up to 12.8 million barrels a day in 2024.

I'm not really sure why the majority thinks we need to give even more incentives for Big Oil to produce more oil. If the title of this hearing, "Unleashing American Energy," really means "giving Big Oil unfettered access to do as they please," well, I guess that's what these bills do.

But again, they do nothing to unleash our nation's true energy potential.

They don't do anything to protect the American people from the volatility of the global oil market and the skyrocketing prices we saw last summer. They don't do anything to combat the climate crisis or deliver environmental justice to some of the most vulnerable communities across this country.

If we really want to unleash America's energy potential and drive down the cost of energy, then we need to break our addiction to oil, and we need an orderly transition to cleaner renewable energy sources.

Not only will doing so help to continue to reduce the harmful emissions that are driving the climate crisis, but it reduces our dependence on the global oil market, and it will reduce energy bills that many Americans are struggling to afford.

Clean energy is now one of the cheapest sources of energy, and it's why countries around the world are already making the transition to expand the use of new technologies that will serve as a foundation for them to continue to grow their economies into the future.

We cannot fall behind in that effort. We can either be the leaders of a global clean energy transition, or our adversaries will, because they understand the risk as much as we do.

Instead of furthering our dependence on oil, we should actually be fueling the innovation of new technologies that can provide clean, renewable energy that's both reliable and affordable, and also increasing our grid security.

We need to build the foundation and we need to invest in training for the energy workforce so they can take on the jobs of the future.

This is how we unleash America's energy potential.

Mr. Chairman, I want to say this again, because I think there's some confusion. People seem to think that if we produce more oil and gas domestically, even though we're producing record amounts, this is going to make us independent from international energy levels or from OPEC price increases.

That's simply not the case, because it's an international market.

We saw that last year when oil and gas prices went up so much, even though there was the ability to have increased domestic production.

I think that by working together, we could find bipartisan solutions. We all have the same goal: a solid energy source, transition to renewable energy, combating the climate crisis, and building the foundation to make sure that can happen.

I don't think these bills are the solution, so I think we should go back to the drawing board and get that get that right. With that, I yield back.

* * *

Ms. DEGETTE. And with that, I yield back.

Mr. DUNCAN. The Chair now recognizes the chair of the Subcommittee on Environment, Manufacturing, and Critical Minerals, Mr. Johnson, for 5 minutes for an opening statement.

Opening Statement of Hon. Bill Johnson, a Representative in Congress from the State of Ohio

Mr. JOHNSON. Well, thank you, Mr. Chairman, and good morning, everyone. I, too, want to thank our panelists for being with us this morning.

You know, my constituents back at home in Ohio know firsthand the importance of affordable American energy and that abundant, affordable energy is vital to economic growth, and it is a key factor in ensuring our national security. Last week we heard about some of the domestic problems we face in meeting these challenges.

Today we will discuss thoughtfully removing some of the red tape and delays that can prevent constructing new critical energy projects, keep capital on the sidelines, and—that are killing innovation dead in its tracks. I am eager to get going on proposals to streamline the process for building essential energy projects in America. Producing more American energy will help reduce global emissions, improve energy reliability, and lower costs for American consumers.

President Biden’s war on affordable and reliable energy—and the problems that war creates—is not limited to killing the use of oil, natural gas, and coal. His administration’s policies are blocking progress on the President’s own stated goal to develop domestic resources essential for the very energy alternatives he prefers, such as wind, solar, and batteries.

For example, the supply of minerals necessary to build these alternative energy sources is insufficient to meet some of this administration’s climate goals, including a carbon-free power sector by 2035 and ensuring widespread use of zero-emission vehicles.

In addition, many of these critical minerals and the refining and processing capacity for them is controlled by adversaries like China and Russia. We cannot stake our future on certain technologies that then rely upon our enemies for the minerals and mineral processing needs to develop them.

This administration must stop promising Utopia while prohibiting our own mineral production, like canceling leases for new nickel and copper mines in Minnesota, blocking new lithium mines in Nevada, and rescinding a land swap necessary for a copper mine in Arizona.

Fortunately, the legislation we are considering today would reorient the law to reestablish America's energy dominance without weakening America's global leadership in advancing our higher environmental and labor standards.

So today we will consider a bill amending the Toxic Substances Control Act to require EPA to review and make timely decisions on the manufacturing of a new chemical or a new use of an existing chemical that is a critical energy resource. This bill still emphasizes risk protection but will prevent the marketplace from waiting an excessive amount of time for critical materials needed to meet our emissions, climate, and energy expectations.

We will also review legislation directing the EPA Administrator to allow more regulatory flexibility in enforcing air quality permits for critical energy resource facilities like processing and refining facilities.

Another measure amends the Solid Waste Disposal Act to allow for critical energy resources engaged in mineral processing to receive interim permit status for the treatment, storage, or disposal of their waste, a permit that the EPA must still review.

There is also a bill which authorizes EPA only during national security and energy security emergencies to waive certain regulations necessary for processing or refining of critical energy resources.

Additionally, we will look at legislation preventing EPA from imposing expensive design analysis requirements on already constructed gasoline refineries, which would elevate the EPA's view of what makes sense above what industry experts and best practices prove makes sense.

Staying on refineries, we will examine a bill to have the Department of Energy and the National Petroleum Council assess and report on the importance of petrochemical refineries in the United States, including a review of opportunities to expand capacity of such facilities, risk of such facilities, and an assessment of Federal and State regulations or policies that have contributed to a decline in the capacity of such facilities.

Finally, we will review a bill repealing the wasteful Greenhouse Gas Reduction Fund, which was established in the Inflation Reduction Act.

So, as you can see, a lot of work needs to be done to establish an energy strategy that encourages innovation, that drives investment, and benefits our economic and national security while we remain good stewards of our environment.

I want to also note that we are moving these in regular order. It is really good to be back to legislating through regular order.

I believe the new Republican majority on the Energy and Commerce Committee is leading with solutions to our Nation's energy and critical resource challenges, and I look forward to hearing from each of you as we talk today.

[The prepared statement of Mr. Johnson follows:]

Bill Johnson, Chairman, Subcommittee on Environment, Manufacturing, and Critical Materials, Opening Statement for Joint Subcommittee Legislative Hearing "Unleashing American Energy, Lowering Energy Costs, and Strengthening Supply Chains," February 7, 2023, as Prepared for Delivery

Well… Good morning, everyone.

I want to thank our witnesses for being here today.

You know…my constituents back home in Ohio know firsthand the importance of affordable, American energy…and that abundant, affordable energy is vital to economic growth…and is a key factor in ensuring our national security.

Last week, we heard about some of the domestic problems we face in meeting this challenge.

Today, we'll discuss thoughtfully removing some of the red tape and delays that can prevent constructing new critical energy projects… keep capital on the sidelines, and kill innovation dead in its tracks.

I'm eager to get going on proposals to streamline the process for building essential energy projects in America. Producing more American energy will help reduce global emissions, improve energy reliability, and lower costs for American consumers.

President Biden's war on affordable and reliable energy – and the problems that war creates – is not limited to killing the use of oil, natural gas, and coal.

His Administration's policies are blocking progress on the President's own stated goal to develop domestic resources essential for the very energy alternatives he prefers – such as wind, solar, and batteries.

For example, the supply of minerals necessary to build these alternative energy sources is insufficient to meet some of this Administration's climate goals, including a quote "carbon-free power sector by 2035" and ensuring widespread use of quote "zero-emissions vehicles."

In addition, many of these critical minerals – and the refining and processing capacity for them – is controlled by adversaries like China and Russia.

We cannot stake our future on certain technologies that then rely upon our enemies for the minerals and mineral processing needs to develop them.

This Administration must stop promising utopia while prohibiting our own mineral production…like cancelling leases for new nickel and copper mines in Minnesota, blocking a new lithium mine in Nevada, and rescinding a land swap necessary for a copper mine in Arizona.

Fortunately, the legislation we're considering today would reorient the law to reestablish America's energy dominance without weakening America's global leadership in advancing our high environmental and labor standards.

So, today:

- We'll consider a bill amending the Toxic Substances Control Act (TSCA) to require EPA to review and make timely decisions on the manufacturing of a new chemical or a new use of an existing chemical that is a critical energy resource. This bill still emphasizes risk protection but will prevent the marketplace from waiting an excessive amount of time for critical materials needed to meet our emissions, climate, and energy expectations.

- We'll also review legislation directing the EPA Administrator to allow more regulatory flexibility in enforcing air quality permits for critical energy resource facilities...like processing and refining facilities.
- Another measure amends the Solid Waste Disposal Act to allow for critical energy resources engaged in mineral processing to receive interim permit status for the treatment, storage, or disposal of their waste – a permit EPA must still review.
- There's also a bill which authorizes EPA, only during national security and energy security emergencies, to waive certain regulations necessary for processing or refining of critical energy resources.
- Additionally, we'll look at legislation preventing EPA from imposing expensive design analysis requirements on already constructed gasoline refineries, which would elevate the EPA's view of what makes sense above what industry experts and best practices prove makes sense.
- Staying on refineries, we'll examine a bill to have the Department of Energy and the National Petroleum Council assess and report on the importance of petrochemical refineries in the United States...including a review of opportunities to expand capacity of such facilities, risks to such facilities, and an assessment of Federal and State regulations or policies that have contributed to a decline in the capacity of such facilities.
- Finally, we'll review a bill repealing the wasteful Greenhouse Gas Reduction Fund, which was established in the Inflation Reduction Act.

So, as you can see, a lot of work needs to be done to establish an energy strategy that encourages innovation, drives investment, and benefits our economic and national security, while we remain good stewards of our environment.

I also note...we're moving these bills through regular order so that we can gather feedback, including constructive criticism. Several of the bills are in discussion draft form for this very reason. Chairs Rodgers, Duncan, and I welcome input on these bills...it's good to be back to legislating with regular order.

I believe the new Republican majority on the Energy and Commerce Committee is leading with solutions to our nation's energy and critical resource challenges, and I look forward to hearing from each of our panelists and yield back the balance of my time.

* * *

Mr. JOHNSON. And I yield back the balance of my time.

Mr. DUNCAN. I thank the gentleman. I will now recognize the ranking member of the Subcommittee on Environment, Manufacturing, and Critical Materials, Mr. Tonko, for 5 minutes.

Opening Statement of Hon. Paul Tonko, a Representative in Congress from the State of New York

Mr. TONKO. Thank you, Mr. Chair. I strongly support this committee's efforts to examine and legislate ways to enhance our long-term energy security, affordability, and sustainability for the American people.

But, unfortunately, nearly all of the bills before us today continue to look backwards toward the energy needs of our past rather than embracing the energy opportunities of the future. And those opportunities are overwhelmingly about positioning the United States to become the global leader in the clean energy technologies and supply chains that will dominate the energy system over the next several decades.

We need our national energy policy to have vision, and that vision cannot solely be how to further enrich oil and gas companies, which are raking in record profits. That is why we should be celebrating the Inflation Reduction Act's nearly $370 billion in clean energy and climate investments, which are already beginning to support the deployment of new clean energy resources, commitments in domestic manufacturing, and a significant reduction in climate pollution.

But, sadly, two of the bills being considered today would repeal critical sections of the IRA, which were developed and enacted by the Democrats of this committee in the 117th Congress. The Greenhouse Gas Reduction Fund is going to facilitate historic investments to decarbonize our grid, our transportation system, and buildings by supporting well-paying jobs and guaranteeing benefits in disadvantaged communities. And the Methane Emissions Reduction Program provides industry with significant funding to adopt emission-reducing technologies before using a market-based approach to incentivize pollution reductions. This is a sensible program that provides certainty for industry while incentivizing the reduction of superpollutants from the oil and gas sector.

I am also concerned that several of the bills under consideration would create new loopholes and—in important environmental laws, allowing a broad and inadequately defined group of polluting industries to get fast-tracked for approval with little consideration for the potential harms they may pose to Americans' air, water, and safety. This is not the way to achieve our shared goals of a more secure, affordable, and cleaner energy system.

But there are steps that we could take together that would. We could have focused hearings to wrestle with complex energy issues. How should hydrogen pipelines be regulated? What reforms are needed to the hydropower licensing processes? How can we build more interstate and interregional transmission lines to improve the reliability and affordability of our electricity system while enabling greater deployment of cost-effective, clean energy resources?

I suspect these questions may interest Members on both sides of the aisle, and each of those topics could be the subject of a narrow, largely bipartisan hearing. This approach would certainly require work and negotiations, but that is surely true of any serious effort to enact bipartisan energy legislation.

Unfortunately, the approach being offered today will not achieve this goal. We are considering 17 Republican bills covering a wide range of topics and amending numerous statutes. Several of these draft bills were seen for the first time just a little over a week ago. During the Democrats' time in the majority, we often tried to give our minority counterparts an opportunity to contribute to legislative hearings' agendas. As far as I am aware, that—there were not discussions of potentially Democratic-sponsored bills that could have fit this hearing's theme.

I also expect we will hear criticisms of the administration for failing to attend today. I agree with my Republican colleagues that we should seek and expect to hear from the agencies at legislative hearings, but we should also make efforts to accommodate their participation, including by providing legislative texts well in advance and being flexible with the hearing calendar.

When the Republicans were last in the majority during the Trump administration, EPA did not testify at any legislative hearings in 2017 and only twice in 2018. We should be consistent both in our expectations that the administration provide witnesses and technical assistance on legislation, and that we need to be flexible to accommodate schedules to ensure that their participation is well informed and instructive to the development of legislation.

So, while I am disappointed in the process that has led us here today, I still believe there are bipartisan policies that we can and should work on together to achieve the goals of this hearing's title. And with that said, I look forward to the discussion on the 17 bills before us today.

[The prepared statement of Mr. Tonko follows:]

Committee on Energy and Commerce, Opening State as Prepared for Delivery of Subcommittee on Environment, Manufacturing, and Critical Minerals, Ranking Member Paul D. Tonko, Hearing on "Unleashing American Energy, Lowering Energy Costs, and Strengthening Supply Chains," February 7, 2023

Thank you, Mr. Chair.

I strongly support this Committee's efforts to examine and legislate ways to enhance our long-term energy security, affordability, and sustainability for the American people.

But unfortunately, nearly all of the bills before us today continue to look backwards, toward the energy needs of our past, rather than embracing the energy opportunities of the future.

And those opportunities are overwhelming about positioning the United States to become the global leader in the clean energy technologies and supply chains that will dominate the energy system over the next several decades.

We need our national energy policy to have vision. And that vision cannot solely be how to further enrich oil and gas companies, which are raking in record profits.

That is why we should be celebrating the Inflation Reduction Act's nearly $370 billion in clean energy and climate investments, which are already beginning to support the deployment of new clean energy resources, commitments in domestic manufacturing, and a significant reduction in climate pollution.

But sadly, two of the bills being considered today would repeal critical sections of the IRA, which were developed and enacted by the Democrats of this Committee in the 117th Congress.

The Greenhouse Gas Reduction Fund is going to facilitate historic investments to decarbonize our grid, transportation system, and buildings by supporting well-paying jobs and guaranteeing benefits in disadvantaged communities.

And the Methane Emissions Reduction Program provides industry with significant funding to adopt emission-reducing technologies before using a market-based approach to incentivize pollution reductions.

This is a sensible program that provides certainty for industry while incentivizing the reduction of super pollutants from the oil and gas sector.

I am also concerned that several of the bills under consideration would create new loopholes in important environmental laws, allowing a broad and inadequately defined group of polluting industries to get fast-tracked for approval with little consideration for the potential harms they may pose to Americans' air, water, and safety.

This is not the way to achieve our shared goals of a more secure, affordable, and cleaner energy system.

But there are steps that we could take together that would.

We could have focused hearings to wrestle with complex energy issues— How should hydrogen pipelines be regulated? What reforms are needed to the hydropower licensing process?

How can we build more interstate and interregional transmission lines to improve the reliability and affordability of our electricity system while enabling greater deployment of cost-effective clean energy resources?

I suspect these questions may interest Members on both sides of the aisle.

And each of those topics could be the subject of a narrow, largely bipartisan hearing.

This approach would certainly require work and negotiations, but that is surely true of any serious effort to enact bipartisan energy legislation.

Unfortunately, the approach being offered today will not achieve this goal.

We are considering 17 Republican bills covering a wide range of topics and amending numerous statutes.

Several of these draft bills were seen for the first time a little over a week ago.

During the Democrats' time in the majority, we often tried to give our minority counterparts an opportunity to contribute to legislative hearings' agendas.

As far as I am aware, there were not discussions of potentially Democratic-sponsored bills that could have fit this hearing's theme.

I also expect we will hear criticisms of the Administration for failing to attend today.

I agree with my Republican colleagues that we should seek— and expect— to hear from the agencies at legislative hearings.

But we should also make efforts to accommodate their participation, including by providing legislative text well in advance and being flexible with the hearing calendar.

When the Republicans were last in the majority— during the Trump Administration— EPA did not testify at any legislative hearing in 2017 and only twice in 2018.

We should be consistent— both in our expectations that the Administration provide witnesses and technical assistance on legislation, and that we need to be flexible to accommodate schedules to ensure that their participation is well-informed and instructive to the development of legislation.

So, while I am disappointed in the process that has led us here today, I still believe there are bipartisan policies that we could work on together to achieve the goals of this hearing's title.

With that said, I look forward to the discussion on the 17 bills before us today, and I yield back.

Mr. TONKO. And with that, Mr. Chair, I yield back.

Mr. DUNCAN. I thank the gentleman. It is now my honor to recognize the chair of the full committee, Mrs. McMorris Rodgers, for 5 minutes for an opening statement.

Opening Statement of Hon. Cathy Mcmorris Rodgers, a Representative in Congress from the State of Washington

Mrs. RODGERS. Thank you, Mr. Chairman, Chairman Duncan, Chairman Johnson. It is great to be kicking off our legislative agenda with you.

Our goal on Energy and Commerce is to ensure reliable, secure, and affordable energy, and that it is available to power homes and businesses across this country.

America has been blessed with abundant natural resources. We should be looking toward developing a predictable regulatory landscape across the board that inspires innovation, entrepreneurship, and technological leadership. Hydropower, nuclear, fossil energies, wind, solar, and batteries: we need all of them in order to secure a stronger, more prosperous America, reduce costs and emissions, address climate issues, and create more robust and resilient communities.

Rush-to-green energy policies, both at the State and Federal level, have curtailed reliable energy and infrastructure, resulting in everything from blackouts to spiking prices. We have seen the devastating impact these policies have had on people in Europe, where forced government transition away from reliable energy sources resulted in more dependance upon Russia. These policies are unsustainable and lead to greater reliance on countries like Russia or, in our case, China.

If we cede our energy leadership to countries like Russia and China, they will always leverage that influence to advance their own authoritarian agenda. This is not the future that any of us want. The best way to address future risks, whether they be climate change or global price shocks, is with a strong economy and a more secure, abundant energy supply here at home. We need to put energy security back at the center of energy policy.

The solutions we are discussing today reflect key steps to return from the path of shortages and high prices to a path of prosperity. We have several bills that will help unlock American natural gas and its delivery systems. To provide reliable, affordable, and clean natural gas is essential for heating our homes and businesses and strengthening America's global standing, all while continuing to lead the world in reducing carbon emissions.

These solutions build on the lessons of the shale revolution, which proved energy expansion can be unprecedented and bring energy security while also helping drive down American emissions.

We also need to restore America's leadership in clean nuclear energy. I am leading a bill that aims to eliminate our reliance, which is currently 24 percent, on Russian nuclear fuels for our nuclear reactors. Expanding our leadership and developing and expanding nuclear energy is going to be one of the top priorities of this Congress. And addressing our reliance on Russian fuel is just the beginning.

To unleash American energy, we also need a regulatory environment that doesn't hamper industry. Several measures improve regulatory flexibility to assist with the reshoring of industries that manufacture and process critical energy materials. These bills strengthen existing regulations, and provide new authorities to enable the EPA and States, working together, to permit new and expand existing manufacturing—manufacturing that is key for strengthening our energy security, national security, and ending our reliance on China. It is also a necessary step as we incorporate technologies like electric vehicles and renewables into our energy mix.

Additionally, we will consider bills that make sure the EPA focuses on its core mission, which does not include forcing a transition to more expensive, less reliable energy sources and systems. Many of these issues have been bipartisan in the past, and I do hope and expect us to come together. This hearing is an opportunity to provide feedback as we work through regular order.

And I should note that it is disappointing representatives from the Department of Energy, the Federal Energy Regulatory Commission, the Environmental Protection Agency declined to appear. Appearing before this committee is an important part of their obligation to Congress, and we expect them to fulfill it.

In closing, I look forward to our discussion today on how this committee can improve American energy leadership, security, and people's lives. It is time to get to work.

[The prepared statement of Mrs. Rodgers follows:]

Chair Cathy McMorris Rodgers, Joint E&E Legislative Hearing: "Unleashing American Energy, Lowering Energy Costs, and Strengthening Supply Chains," February 7, 2023, as Prepared for Delivery

Thank you, Chairman Duncan, Chair Johnson.

It's great to kick off our legislative agenda with you.

Our goal on Energy and Commerce is to ensure reliable, secure, and affordable energy is available to power homes and businesses across the country.

America has been blessed with an abundance of natural resources.

We should be working towards developing a predictable regulatory landscape across the board that inspires innovation, entrepreneurship, and technological leadership....

...hydropower, nuclear, fossil energies, wind, solar, and batteries.

We need all of them in order to secure a stronger, more prosperous America...

...reduce costs and emissions, address climate issues, and create more robust and resilient communities.

Rush-to-green energy Policies—both state and federal — have curtailed reliable energy and infrastructure, resulting in everything from blackouts to spiking prices.

We've seen the devastating impacts these policies have had on people in Europe...

...where forced government "transition" away from reliable energy sources resulted in more dependence on Russia.

These policies are unsustainable and lead to greater reliance on countries like Russia... or in our case, China.

If we cede our energy leadership to countries like Russia and China, they will ALWAYS leverage that influence to advance their own authoritarian agenda.

This is not a future any of us want.

The best way to address future risks, whether they be climate change or global price shocks, is with a strong economy and a more secure, abundant energy supply here at home.

We need to put energy security back at the center of energy policy....

The solutions we're discussing today reflect key steps to return from the path of shortages and high prices, to a path of prosperity.

We have several bills that will help unlock American natural gas and its delivery systems.

Providing reliable, affordable, and clean natural gas is essential for heating our homes and businesses, and for strengthening America's global standing...

... all while continuing to lead the world in reducing carbon emissions. . .

These solutions build on the lessons of the shale revolution, which proved energy expansion can bring unprecedented prosperity and energy security, while also helping to drive down American emissions.

We also need to restore America's leadership in clean, nuclear energy.

I am leading on a bill that aims to eliminate our reliance—which is currently about 24% — on Russian nuclear fuels for our nuclear reactors.

Expanding our leadership in developing and expanding nuclear energy is going to be one of our top priorities this Congress...

... and addressing our reliance on Russian fuel is just the beginning.

To unleash American energy, we also need a regulatory environment that doesn't hamper industry.

Several measures improve regulatory flexibility to assist with the reshoring of industries that manufacture and process critical energy materials.

These bills strengthen existing regulations and provide new authorities to enable the EPA and states – working together – to permit new and expand existing manufacturing...

Manufacturing that is key for strengthening our energy security, national security, and ending our reliance on China...

It's also a necessary step as we incorporate technologies like electric vehicles and renewables into our energy mix.

Additionally, we will consider bills that make sure the EPA focuses on its core mission, which does not include forcing a transition to more expensive, less reliable energy sources and systems.

Many of these issues have been bipartisan in the past—I hope we can continue to come together.

This hearing is an opportunity to provide feedback as we work through regular order.

I should note that it is disappointing representatives from the Department of Energy, the Federal Energy Regulatory Commission, and the Environmental Protection Agency declined to appear.

Appearing before the Committee is an important part of their obligation to Congress...and we expect them to fulfill it.

In closing, I look forward to our discussion today on how this committee can improve American energy leadership, security, and people's lives.

It's time to get to work.

With that, I yield back, Mr. Chairman.

Mrs. RODGERS. And with that, I yield back, Mr. Chairman.

Mr. DUNCAN. I thank the chairwoman.

I want to pause and just say that Congress's thoughts and prayers are with the folks in Syria and Turkey after the devastating earthquake.

And I now recognize the ranking member of the full committee, Mr. Pallone, for 5 minutes.

Opening Statement of Hon. Frank Pallone, Jr., a Representative in Congress from the State of New Jersey

Mr. PALLONE. Thank you, Chairman Duncan.

Committee Republicans are showing today that their top energy and environmental priorities are to do the bidding of Big Oil, and to undermine our Nation's bedrock environmental laws. And these are not the same priorities of committee Democrats.

Over the last 2 years, Democrats delivered historic wins for the American people. We enacted laws that are already creating good-paying jobs, cutting costs for working families, and advancing homegrown clean energy, all while tackling the worsening climate crisis. And while we want to build upon these successes for the American people, House Republicans are stuck in the past and failing to address the energy challenges and opportunities we face today.

I would like to start by highlighting my serious concerns with some of the fossil-focused bills that we are discussing today. The Cross-border Energy Infrastructure bill is nothing more than a shadow approval of the Keystone Pipeline masquerading as legislation. It establishes that every single cross-border energy project is in the public interest, a radically higher bar than exists now. And this is not serious legislation.

Representative Burgess's bill would put the Federal Energy Regulatory Commission in charge of permitting reviews that it doesn't have the expertise or the time to lead, and the Unlocking our Domestic LNG Potential Act would eliminate the requirement that the Department of Energy determine that exporting natural gas from a U.S. facility is in the national interest. This legislation would effectively greenlight sending unrestricted amounts of LNG to adversaries like China. It is just more proof that committee Republicans are more interested in doing the bidding of their fossil fuel friends than actually protecting our energy security.

And I am also deeply disappointed with the legislative proposals being considered in the Environment Subcommittee's jurisdiction. Two of the bills would revoke programs enacted as part of the Inflation Reduction Act that are projected to cut climate pollution, reduce the deficit, and leverage private-sector investment in clean energy projects across the Nation.

H.R. 484 targets the Methane Emission Reduction Program, which establishes a suite of incentives to drive down excess methane pollution and remediate the effects of the pollution that does occur. This program fundamentally ensures polluters pay for wasted methane, and not customers.

Republicans also target the Greenhouse Gas Reduction Program, which invests $27 billion in non-State and local climate finance institutions that support the rapid deployment of low- and zero-emission technologies. My Republican colleagues claim to support all-of-the-above policies, yet they oppose a program that invests in clean energy projects.

We are also considering bills that allow so-called critical energy sources to bypass commonsense environmental protections. One bill would circumvent consideration of safer technologies to avoid chemical disasters under the risk management program, while another bill would undercut protective health policies that were developed and passed on a bipartisan basis by this committee in the Frank Lautenberg Act. And we should not be putting polluters over people by waiving critical public health and environmental protections that keep American communities safe.

Now, if Republicans really want to unleash American energy, I invite them to stop trying to tear down critical climate and environmental programs, and work with us to build a better future for all by investing in clean energy and bolstering our environmental safeguards.

I can't find much value in the legislation before us today, which is unfortunate, since many Members are interested in working on clean energy permitting. And instead, my majority colleagues scheduled a hearing on 17 Republican bills or discussion drafts without even asking Democrats if we have any bills that would address the underlying topic. If they truly want to enact legislation that addresses energy security and affordability, this is not the path. The American people deserve better.

Finally, I must set the record straight about why the administration could not testify at today's hearing. The majority claims they gave the administration 2 weeks' notice and that the administration said this was not enough time to secure witnesses.

What they left out is that the 2 weeks' notice came in an email right before midnight on January 24th. It did not contain the list of bills for the hearing, the legislative text of the bills, or any information about other invited witnesses. How were the agencies supposed to prepare for a hearing when they have not been told what bills they are supposed to be commenting on?

And to be clear, only 2 of the 17 bills had been introduced when this hearing was noticed 1 week ago. If the bills were not ready to share with the administration or with Democratic Members, then this hearing should have been postponed until a later date.

[The prepared statement of Mr. Pallone follows:]

Committee on Energy and Commerce, Opening Statement as Prepared for Delivery of Ranking Member Frank Pallone, Hearing on "Unleashing American Energy, Lowering Energy Costs and Strengthening supply Chains," February 7, 2023

Committee Republicans are showing today that their top energy and environmental priorities are to do the bidding of Big Oil and to undermine our nation's bedrock environmental laws. These are not the same priorities of Committee Democrats.

Over the last two years, Democrats delivered historic wins for the American people. We enacted laws that are already creating good paying jobs, cutting costs for working families, and advancing homegrown clean energy – all while tackling the worsening climate crisis.

While we want to build upon these successes for the American people, House Republicans are stuck in the past and failing to address the energy challenges and opportunities we face today.

I'd like to start by highlighting my serious concerns with some of the fossil-focused bills we're discussing today. The Cross-border Energy Infrastructure bill is nothing more than a shadow approval of the Keystone XL pipeline masquerading as legislation. It establishes that every single cross-border energy project is in the public interest – a radically higher bar than exists now. This is not serious legislation.

Representative Burgess' bill would put the Federal Energy Regulatory Commission (FERC) in charge of permitting reviews that it doesn't have the expertise or the time to lead.

And the Unlocking our Domestic LNG Potential Act would eliminate the requirement that the Department of Energy determine that exporting natural gas from a U.S. facility is in the national interest. This legislation would effectively greenlight sending unrestricted amounts of LNG to adversaries like China. It's just more proof that Committee Republicans are more interested in doing the bidding of their fossil fuel friends than actually protecting our energy security.

I am also deeply disappointed with the legislative proposals being considered in the Environment Subcommittee's jurisdiction.

Two of the bills would revoke programs enacted as part of the Inflation Reduction Act that are projected to cut climate pollution, reduce the deficit, and leverage private sector investment in clean energy projects across the nation. H.R. 484 targets the Methane Emission Reduction Program, which established a suite of incentives to drive down excess methane pollution and remediate the effects of the pollution that does occur. This program fundamentally ensures polluters pay for wasted methane, not customers.

Republicans also target the Greenhouse Gas Reduction Program, which invests $27 billion in nonprofit, state, and local climate finance institutions that support the rapid deployment of low- and zero-emission technologies. My Republican colleagues claim to support all of the above policies, yet they oppose a program that invests in clean energy projects.

We are also considering bills that allow so-called critical energy sources to bypass commonsense environmental protections. One bill would circumvent consideration of safer technologies to avoid chemical disasters under the Risk Management Program. While another bill would undercut protective health policies that were developed and passed on a bipartisan basis by this Committee in the Frank Lautenberg Act. We should not be putting polluters over people by waiving critical public health and environmental protections that keep American communities safe.

If Republicans really want to unleash American energy, I invite them to stop trying to tear down critical climate and environmental programs, and work with us to build a better future for all by investing in clean energy and bolstering our environmental safeguards.

I can't find much value in the legislation before us today. Which is unfortunate, since many members are interested in working on clean energy permitting. Instead, my majority colleagues scheduled a hearing on 17 Republican bills or discussion drafts without even asking Democrats if we have any bills that would address the underlying topic. If they truly want to enact legislation that addresses energy security and affordability, this is not the path. The American people deserve better.

Finally, I must set the record straight about why the Administration could not testify at today's hearing. The majority claims they gave the Administration two weeks' notice, and that the Administration said this was not enough time to secure witnesses. What they left out is that the two weeks' notice came in an email right before midnight on January 24. It did not contain the list of bills for the hearing, the legislative text of the bills, or any information about other invited witnesses.

How are the agencies supposed to prepare for a hearing when they have not been told what bills they are supposed to be commenting on? And to be clear, only two of the 17 bills had been introduced when this hearing was noticed one week ago. If the bills were not ready to share with the Administration, or with Democratic members, then this hearing should have been postponed until a later date.

* * *

Mr. PALLONE. And with that, Mr. Chairman, I yield back.

Mr. DUNCAN. I thank the gentleman. This now concludes with Member opening statements.

The Chair would like to remind Members that, pursuant to committee rules, all Members' opening statements will be made part of the record.

I want to thank all the witnesses for being here today and taking time to testify before the subcommittees.

Each witness will have the opportunity to give a 5-minutes opening statement, followed by a round of questions from Members.

There are some lights in front of you. Green means go, yellow means you have got 1 minute, a 1-minute warning, and red means you need to wrap up.

Our witnesses today are the Honorable Mark Menezes, Mr. Jeffrey Eshelman, Mr. Raul Garcia, Ms. Katie Sweeney, Mr. Tyson Slocum, and the Honorable Bernard McNamee.

We appreciate you being here today. I will now recognize Mr. Menezes for 5 minutes to give an opening statement.

Statements of Mark W. Menezes, Former Deputy Secretary, Department of Energy, and Former Chief Counsel, Energy and Environment, House Committee on Energy and Commerce; Jeff Eshelman, President and Chief Executive Officer, Independent Petroleum Association of America; Raul Garcia, Legislative Director for Healthy Communities, Earthjustice; Katie Sweeney, Executive Vice President and General Counsel, National Mining Association; Tyson Slocum, Energy Program Director, Public Citizen, Inc.; and Bernard Mcnamee, Former Commissioner, Federal Energy Regulatory Commission

Statement of Mark W. Menezes

Mr. MENEZES. Good morning. Madam Chair McMorris Rodgers, Ranking Member Pallone, Subcommittee Chairmen Duncan and Johnson, Ranking Members DeGette and Tonko, and members of the subcommittee, thank you for the invitation to testify on legislative solutions designed to promote U.S. energy production, lower energy costs, and to strengthen our supply chains of critical minerals and energy resources. Today's hearing features bills designed to achieve these important goals.

Now, in order to unleash American energy, it is important that Congress ensures the legislative will of the people is carried out by the President and the executive branch. Frequently, the executive branch is at odds with the laws of Congress, and takes actions without clear congressional authority. Several of the bills under consideration today make congressional intent clear to the executive branch.

The Protecting American Energy Production Act clarifies that States have primacy regulating hydraulic fracturing. Congress made this clear in the overwhelmingly bipartisan Energy Policy Act of 2005. Our global friends and allies are grateful for our production and export and have come to rely on U.S.-produced oil and natural gas as a reliable source of energy and an alternative to Russian supply. This bill prohibits the President from issuing any moratoria on hydraulic fracturing, thus preserving States' rights.

Similarly, the sense of Congress resolution opposes the executive branch placing restrictions on the export of oil and petroleum products. This makes clear the congressional intent that the President and all Federal agencies follow the law as written when a bipartisan Congress lifted the oil export ban in 2015.

The Unlocking our Domestic LNG Potential Act removes redundant reviews and the need for multiple Federal permission slips to produce and liquefy LNG for export. The bill makes clear that it is FERC and not the Department of Energy which has the exclusive authority to approve or deny requests to export natural gas to a foreign country.

The Promoting Cross-border Energy Infrastructure Act makes clear that it is the DOE and FERC, the Federal agencies with expertise in energy, that have the authority to grant or deny the interconnections and modifications of cross-border electricity lines and pipelines without the need for a presidential permit.

Turning now to lowering energy costs, it is important that the Members appreciate that, with the shale revolution, the U.S. now has an abundance of cheap natural gas which is replacing baseload coal, complementing the increased deployment of wind and solar, and is the primary reason why the U.S. leads in actual emission reductions.

The Promoting Interagency Coordination for Review of Natural Gas Pipelines Act ensures increased access and delivery of supply to lower natural gas costs by authorizing FERC to be the lead agency to coordinate other agencies, establish reasonable timelines, and keep track of progress of the permitting and environmental reviews required under NEPA and other laws.

Reducing costs of government can save taxpayers dollars to offset the cost of energy. The repeal of the Greenhouse Gas Reduction Fund repeals the $27 billion appropriated to EPA to provide grants and financial assistance to States, municipalities, Tribal governments, and nonprofits for zero-emission technologies. While laudable and generous, there is little oversight of EPA required by Congress to administer this fund. Remember, EPA's requested budget in 2022 was only $11.4 billion.

Another bill to lower energy costs is the Natural Gas Tax Repeal Act, which repeals the methane waste fee and statutory methane regulations included in the nonbipartisan Inflation Reduction Act. EPA has proposed a supplemental rule to regulate methane emissions open for public comment now, which essentially does the same thing without the methane waste fee but with Clean Air Act penalties.

Turning now to strengthening our supply chains, we know the U.S. relies on imports for 31 of 35 of our critical minerals necessary for the U.S. defense and our clean economy, 14 of which we are totally dependent on imports. To ensure critical energy resources, Congress should designate DOE with the responsibility to do this. That is what Securing America's

Critical Mineral Supply Act does. It amends the DOE Organization Act to give DOE the responsibility of securing our supply of critical energy resources.

Congress should ensure our critical energy facilities produce our critical energy resources during emergencies and threats to our energy security. Several of the bills under consideration here today do just that.

The National and Energy Security Waiver bill authorizes the EPA Administrator, in consultation with DOE and the Governor, to waive certain requirements, sanctions, or fees during times of threats to our national or energy security to maintain an adequate supply of gasoline and diesel and other critical refined resources. Another bill addresses EPA's backlog of pending applications to complete risk assessments of chemical substances necessary today to produce our critical energy resources. EPA's backlog is so great that U.S. companies are hesitant to make capital investments to produce critical battery components necessary for the deployment of EVs. This inexplicable delay is a problem China and our global competitors simply do not have.

Likewise, the interim permit bill for safe storage and disposal of critical energy resources allows temporary onsite storage and disposal similar to past practices and, like other bills, a commonsense approach to accelerate U.S. development of critical mass——

Mr. DUNCAN. I am going to ask the gentleman to wrap up his opening.

Mr. MENEZES. [continuing]. Our transition to battery technologies, EV use, and grid scale battery storage.

I will. I am in the course of wrapping up.

Since 2009, EPA has had the use of its flexible air permitting rule. EPA limits its use, and so we have a bill to allow that.

Congress also needs to ensure that interstate electric transmission infrastructure——

Mr. DUNCAN. I am sure a lot of this will be covered in the question-and-answering, so I am going to ask you to wrap up.

Mr. MENEZES. It is. Well, with that, I will, in fact, thank you for the opportunity to testify today.

I ask that my written statement be included in the record, and I look forward to your questions.

[The prepared statement of Mr. Menezes follows:]

Testimony of the Honorable Mark W. Menezes, Former Deputy Secretary, U.S. Department of Energy, Former Chief Counsel, Energy and Environment, U.S. House of Representatives Committee on Energy and Commerce, before the U.S. House Committee on Energy and Commerce, Subcommittee on Energy, Climate, & Grid Security and Subcommittee on Environment, Manufacturing, & Critical Materials, Hearing on "Unleashing American Energy, Lowering Energy Costs, and Strengthening Supply Chains," Tuesday, February 7, 2023

Chair McMorris Rodgers, Ranking Member Pallone, Subcommittee Chairmen Duncan and Johnson, Ranking Members DeGette and Tonko, and members of the Subcommittees:

Thank you for the invitation and opportunity to testify before you today at this joint legislative hearing to discuss legislative solutions to promote US production, use and export of its energy resources, to lower energy costs to the American people, and to strengthen our supply chains of critical minerals and energy resources. Today's hearing features bills designed to help achieve these important goals.

Unleashing American Energy

To help achieve the Congressional goal of American production, use, and export of its energy sources, it is important that Congress ensures that the legislative will of the people is carried out by the President and the Executive Branch. While our Founders created our three independent branches of government and the Constitution protects separation of powers, the Executive Branch is frequently at odds with the laws that Congress has passed and takes actions without clear Congressional authority to either implement new policies or to impede Congressional intent. Several of the bills under consideration today make Congressional intent clear to the Executive Branch.

H.R. 150 by Mr. Duncan and others, the "Protecting American Energy Production Act," clarifies Congressional intent that States have primacy regarding the regulation of hydraulic fracturing. Each State decides whether to permit or ban hydraulic fracturing. Under current law, each State has primary regulatory authority over oil and natural gas production using hydraulic fracturing. Congress made this clear in the overwhelmingly bipartisan Energy Policy Act of 2005. It is important that all the members of this Committee understand that following that bipartisan act of Congress in 2005, along with improved drilling and production technologies, the use of hydraulic fracturing in the production of both natural gas and oil caused the "Shale Revolution"[1] propelling the US to become the world's leading producer in both by 2019. Our global friends and allies are grateful for our production and export and have come to rely on US produced oil and natural gas as a reliable energy provider and an alternative to Russian supply.

Notwithstanding the dramatic success of hydraulic fracturing, the Federal Government has sought to regulate it ignoring the law and Congressional intent.[2] Thus, to ensure an abundant source of supply of clean, US-produced natural gas and oil, the bill prohibits the President from issuing any moratoria on hydraulic fracturing, thus preserving States rights and curtailing autocratic actions of the Executive Branch in contravention of clear Congressional intent.

Similar opposition to the Executive Branch placing restrictions on the export of oil and petroleum products is expressed in the proposed "Sense of Congress" Resolution. Our global partners' growing reliance on exported US energy products has increased dramatically since Russia's invasion of Ukraine. Our global friends and allies are looking to the US to replace Russia oil on the global markets as they have announced commitments to no longer use Russian natural gas or oil. The Resolution makes clear that the "Federal Government" should not impose any moratoria or other restrictions on the export of oil or petroleum products. Congress made

---

[1] See Daniel Yergin, The New Map: Energy, Climate, and the Clash of Nations (New York: Penguin Press 2020) at 26-30.

[2] *Wyoming vs. U.S. Dept. of Interior*, U.S. District Court, 2015; Case No. 2:15-CV-043-SWS (Lead Case) Case No. 2:15-CV-041-SWS. Wyoming v. U.S. Dep't of the Interior, Case No. 2:15-CV-043-SWS (Lead Case) | Casetext Search + Citator.

clear the conditions under which oil and petroleum products exports might be limited when it lifted the ban of US oil exports in 2015. The Resolution expresses the clear intent that the President and all Federal Agencies follow the law as prescribed by Congress when it lifted the oil export ban.

The "Unlocking our Domestic LNG Potential Act" also helps our friends and allies around the world to provide a steady source of clean natural gas to replace Russia gas and to help them meet their net-zero emission goals. The bill removes redundant reviews and the need for multiple federal permission slips to produce and liquefy LNG for export. It removes the anachronistic distinction between countries with Free Trade Agreements and those without as those distinctions are no longer indicative of the true extent of our global friends, partners and allies.[3] The bill makes clear that it is the Federal Energy Regulatory Commission (FERC), not the Department of Energy (DOE), which has the exclusive authority to approve or deny requests to export (or import) natural gas to a foreign country. The bill ensures US natural gas will go to only our friends and allies as it prohibits exports to any country that is an enemy of the US, is designated as a state sponsor of terrorism, or that is subject to sanctions, like Russia.

The "Promoting Cross-Border Energy Infrastructure Act" sets forth clear authority for energy interconnections with Canada, Mexico and the US for both electricity lines and pipelines.

Congress is making clear that the Federal Agencies with expertise in the energy operations and energy needs of the three countries have the authority to grant or deny the interconnections and modifications—without the need for a Presidential permit. These clear lines of authority stop autocratic decision making impeding Congressional intent to maintain an interconnected North American energy system necessary for a reliable, affordable, and sustainable supply of energy for the US and our neighboring countries. Plus, the bill will allow the three countries to be able to better withstand the global energy shocks of autocratic actions, like Putin's invasion of Ukraine. Had the Keystone Pipeline not been cancelled by President Biden, the US and its allies would have had access to Canadian oil to lessen the import and use of Russian oil. The proposed House Concurrent Resolution recognizes that sentiment by expressing disproval over President Biden's revocation of the Keystone Pipeline on his first day in the Oval Office.

Lowering Energy Costs

It's important for the US to maintain the necessary infrastructure to deliver its critical energy resources to the American people in sufficient supply at affordable prices. With the "Shale Revolution" the US has enjoyed an abundance of cheap natural gas which has lowered the energy prices for the American people for years. Because of its relatively low production and delivery costs, natural gas has replaced coal as the primary source of electric power in the US. Additionally, because of the environmental benefits and ramping abilities of combined cycle natural gas turbines, natural gas has complemented the growth of wind and solar in the US being the primary reasons why the US leads in actual emission reductions since 2005. However, as more and more electricity customers are using natural gas and renewables, it is necessary to ensure the adequacy of natural gas pipelines for its delivery at affordable costs. In our bid-based electricity markets today, it is typically the cost of natural gas which sets the price of electricity

[3] Free Trade Agreements | United States Trade Representative (ustr.gov). The US has an FTA with only 20 countries.

delivery throughout the day. Greater pipeline capacity serving the FERC-approved electricity markets will help lower the cost of electricity and home heating for consumers.

The "Promoting Interagency Coordination for Review of Natural Gas Pipelines Act" by Mr. Burgess both ensures domestic supply of natural gas for both domestic consumption and export and increases access and delivery of supply to lower the costs of natural gas to homes and businesses and the cost of electricity to consumers. This bill authorizes the FERC, an independent agency within DOE, to be lead agency to coordinate other agencies, establish reasonable timelines, and keep track of progress of the permitting and environmental reviews required under the National Environmental Policy Act (NEPA) and other applicable laws for the siting and construction of natural gas pipelines. This bill improves the accountability, transparency, and efficiencies of the processes involved in the siting and construction of natural gas pipelines without changing any permitting or environmental standards. Importantly, the bill also requires the FERC to consult with the Transportation Security Administration (TSA) regarding pipeline security and cybersecurity measures.

Not only is it imperative that all Americans have clean and reliable energy, Congress should prioritize efforts to reduce energy costs. Reducing costs of government can save taxpayers' dollars for use to offset the costs of energy.

One of the bills under consideration today is the "Repeal of [the] Greenhouse Gas Reduction Fund," section 134 of the non-bipartisan Inflation Reduction Act,[4] which appropriates $27 billion to EPA. The intended goal of this fund is to provide grants and financial assistance to States, municipalities, Tribal governments, and non-profits for zero-emission technologies to be developed and deployed over a 10-year period. But there is little oversight required by Congress. While the Act appropriates $25 million to GAO to "help support the oversight" of the $369 billion expenditures contained in the Act, Congress did not expressly require GAO to audit EPA's use or expenditure of these funds. Nor did Congress provide much in the way of direct oversight, accountability, or reports to keep the public informed as to how EPA is using the $27 billion. Keep in mind that EPA's annual budget request in 2022 was $11.4 billion.[5]

Another bill to lower energy costs to the American people is the "Natural Gas Tax Repeal Act" by Mr. Pfluger (H.R. 484). This bill repeals the methane waste fee and the statutory methane regulations included in the non-bipartisan Inflation Reduction Act.[6] Methane (natural gas) is a valuable energy product and industry is taking action to better detect, address, monitor, and measure methane leaks. EPA has a proposed supplemental rule to regulate methane emissions that is in the public comment period. These proposed regulations address methane leaks in a similar manner as does the statutory scheme. EPA's supplemental proposed regulations do not include the methane leak fee in its proposed rules, although a violation of the supplemental regulations, once final, carries the full penalties and fines set forth in the Clean Air Act.

---

[4] untitled (congress.gov).

[5] $27 billion is actually quite a significant amount of money to the American people. To put $27 billion in perspective, EIA estimates the US per capita average energy expenditures per year is around $3500. The $27 billion amount given to EPA would be equal to Congress footing the bill for over 7.7 million people for an entire year— that's the combined population of almost 9 states!

[6] See FN 4.

Strengthening Our Supply Chains

Members of the Subcommittees know well that the innovative technologies necessary to provide abundant, affordable, clean energy and to achieve emission reduction goals will depend on critical minerals. Whether wind and solar power, safe battery technology, or EVs virtually all sources of energy and products necessary to achieve net zero goals require critical minerals, which are produced primarily in other countries. And several of these countries, like China, do not share our values. The severity of this overdependence on other countries for critical minerals was brought to the public's attention by the report released during the Trump Administration.[7] The US relies on imports for 31 of the 35 critical minerals necessary for the US defense and economy. Many Americans were surprised to learn that for 14 of the listed critical minerals, the US relies completely on imports from other countries; having no US production at all. Following the release of this report, this Committee recognized the need for action and considered several bills to address this vulnerability. Congress funded critical mineral research and development in the Energy Act of 2020[8] but the COVID pandemic disruptions in international supply chains further highlighted the urgency to do more. The Biden Administration performed updated assessments on critical minerals with its 100-day report.[9] In response, Congress created an interagency critical minerals task force and incented recycling in the Bipartisan Infrastructure Law (Infrastructure Investment and Jobs Act).[10] Congress then provided tax incentives to stimulate the development of critical minerals necessary for battery components in the non-partisan Inflation Reduction Act.[11] But more needs to be done.

We need to designate an agency with clear authority to address these issues and to serve as the experts in the interagency process for energy. That's what "Securing America's Critical Minerals Supply Act" does. It amends the DOE Organization Act to add to its list of functions the responsibility to securing the supply of critical energy resources necessary to develop energy technologies and the operation of energy systems. This bill is a needed clarification of DOE's responsibility and leadership throughout the interagency process in ensuring a sufficient US supply of critical minerals to maintain the robustness of our nation's energy systems.

While plenty of attention is on the supply chains of critical minerals, we mustn't forget that, as a nation, we should continue to ensure the operation of our critical energy facilities producing our critical energy resources during emergencies and threats to our energy security. Congress has recognized the importance and efficiency of executive agency action necessary to ensure the speedy recovery from natural disasters.[12] But more needs to be done to provide sufficient energy and fuel supplies to the American people. Several of the bills under consideration here today do just that.

In order to ensure an adequate supply of gasoline and diesel and other critical refined resources, the national security and energy security waiver bill authorizes the EPA Administrator, in consultation with DOE, and the Governor of the State in which the critical energy facilities are located, to waive certain requirements, sanctions or fees during times of threats to our national security or energy security.

---

[7] See Critical_Minerals_Strategy_Final.pdf (commerce.gov).
[8] PUBL260A.PS (congress.gov).
[9] 100-day-supply-chain-review-report.pdf (whitehouse.gov).
[10] PUBL058.PS (congress.gov).
[11] See FN 4.
[12] For example, the Disaster Relief Act of 1974. STATUTE-88-Pg143-2.pdf (congress.gov).

Similarly, in order to get EPA to act on its backlog of pending applications to complete risk assessments of chemical substances that are being used today to produce our critical energy resources, the bill mandating EPA action is necessary. The backlog on routine chemical risk assessments is so great that US companies are frustrated and hesitant to make significant capital investments in the production of critical battery components necessary to accelerate the deployment and adoption of EVs. This inexplicable delay is a problem China and our global competitors do not have. The private sector is ready to invest billions in the US but our governmental assessment processes are not keeping up with our race for global leadership in critical energy resource use and deployment. Any risk in chemical use should not be presumed to be an "unreasonable" risk. Government inaction or delay has societal costs and global leadership repercussions. It's a common sense effort to get our government bureaucracy to implement the law in a responsible and timely manner.[13]

Likewise, the bill allowing an interim permit for the safe storage and disposal of critical energy resources allows for their onsite storage and disposal consistent with existing environmental laws during EPA's review. Like the other bills, this is a common sense approach to allow the US development of critical mineral refining and processing to accelerate US development of critical supply chains during our transition to battery technologies, EV use, and grid-scale battery storage.

Since 2009, EPA has allowed the use of its "Flexible Air Permitting Rule." EPA promulgated this rule "to promote flexible air permitting (FAP) approaches that provide greater operational flexibility and, at the same time, ensure environmental protection and compliance with applicable laws."[14] EPA, however, limits FAP's use. Thus, in order for the US to accelerate its development of critical energy resources, a bill is necessary to require the EPA to allow FAP's use by the owner or operator of a critical energy resource facility. This bill will facilitate the development and deployment of energy resources necessary to accelerate the production and deployment of materials necessary for the energy transition to a more decarbonized economy.

In order to ensure we have abundant, affordable, and clean energy, Congress needs to ensure that our infrastructure is safe from malign cybersecurity attacks. In 2005, Congress mandated cybersecurity standards for our nation's bulk power system to be set by our nation's electric reliability organization, the North American Electric Reliability Corporation or NERC, and enforced by FERC. Ten years later, Congress designated DOE as the sector specific agency for cybersecurity in the electricity sector. In carrying out these responsibilities, DOE has set up a process for it to be notified of cyber incidents but there lacks clarity for which federal agency should be notified when cyber incidents impacts critical electric infrastructure. That's why the Critical Electric Infrastructure Cybersecurity Incident Reporting Act is necessary. It designates DOE as the Federal Agency to receive such notices. It authorizes DOE to promulgate rules to "facilitate the submission of the timely, secure, and confidential notifications" regarding cybersecurity incidents and potential incidents. The bill applies to the owners and operators of critical electric infrastructure and other Federal Agencies.

As members of this Committee know well, available fuel supply is necessary to maintain downward pressures on retail gasoline prices. All elected officials care greatly when the price of gasoline increases to historic highs. There are several reasons for high gasoline prices but a

[13] CCI: Commercializing Innovative Chemical Products (chemicalinnovations.org).
[14] E9-23794.pdf (justia.com).

key factor is our nation's ability and capacity to refine its own gasoline. Our nation's refining capacity has been on a decline since the 1970s due to several reasons not the least of which is our inability to build new refineries. For policymakers to more fully understand all the various impacts on the state of our refining industry today, a bill is needed to require the Secretary of Energy to direct the National Petroleum Council to analyze the matter and produce a report available to Congress and the public. The National Petroleum Council has been in existence since President Truman established it recognizing "the importance of petroleum in the life of the Nation and the consequent necessity for assuring the adequate and continuous availability of this vital resource."[15] The council is composed members of the refining industry as well as all stakeholder groups making it well-equipped to provide such an assessment and offer recommendations for policymakers to consider.

In addition to the NPC study, to help address our decreasing refining capacity, is the proposed bill to prohibit the phase out of gasoline. This bill ensures the continued environmentally-compliant operations of a widely used refining process. This bill is necessary to keep fuel costs affordable and fuel available to the American people. EPA currently is proposing changes to its Risk Management Plan (RMP) which is targeting a well-known chemical used in the refining processes to manufacture almost 40% of US fuel. Should these refineries choose to make changes due to the modified RMP, it is estimated it will cost billions of dollars to redesign or replace current facilities. Such a cost would likely cause some refineries to close or go off line for significant periods of time further restricting gasoline and diesel supply leading to increase costs to the American people.[16]

Lastly, the Russian uranium import ban bill phases out and prohibits the US import of enriched uranium from Russia. For years, the US has been dependent on significant amounts of imports from Russia and other countries for enriched uranium to fuel our civilian nuclear fleet which is our largest producer of emission-free energy in the US. Recognizing the decline of our abilities to provide nuclear fuel and maintain our technological leadership in nuclear energy, the DOE released a report in 2020 recommending action to restore our competitive edge in nuclear energy.[17] Bold action is needed especially since Russia is no longer viewed as a reliable partner by NATO countries for critical resources. Compounding our lack of production, processing, and conversion facilities in the US for nuclear fuel is energy pricing in the FERC-approved electricity markets. Most of our civilian fleet sells emission-free electricity into these markets which inadequately values clean nuclear energy. As a result, our nuclear fleet is very sensitive to fuel costs. Of course, the Russians know this which is why they continue to provide cheap uranium fuel and a reason why we continue to import enriched uranium from Russia today.

Thank you for the opportunity to testify before you today. I look forward to your questions.

---

[15] NPC Letter from President Truman.

[16] "Replacing 100% of the existing HF alkylation unit capacity with sulfuric acid capacity across 41 refineries would require a total capital investment cost of $15 to 45 billion, excluding unforeseen inflationary pressure. On a facility-basis, cost of investment would be between $200 to $850 million, depending on the size of the unit, and excluding recent cost escalation and inflationary pressures. These costs are only direct capital costs and do not factor in the costs associated with unit downtime to demolish the HF units and build sulfuric acid units and the ability to replace alkylate during the transition. EPA failed to consider any of these costs in its Proposal." See AFPM's Comments on EPA's Accidental Release Prevention Requirements: Risk Management Programs Under the Clean Air Act; Safer Communities by Chemical Accident Prevention (EPA–HQ–OLEM–2022–0174), at 43-44, citing "Critical Considerations and Factors for US Refiners to Address as Part of an HF Alkylation Technology Conversion, Becht at 4 (Oct. 28, 2022) ("Becht Paper")."

[17] Restoring America's Competitive Nuclear Advantage_1.pdf (energy.gov).

Mr. DUNCAN. Thank you for that, and I apologize, but we do have a broad panel. So the Chair will go to Mr. Eshelman.

Statement of Jeff Eshelman

Mr. ESHELMAN. Good morning, and thank you for having me here today. It is a pleasure to be here. I am Jeff Eshelman, president and CEO of the Independent Petroleum Association of America.

I would like to thank Chairman Duncan and Chairman Johnson for gathering this hearing today. We really appreciate being a part of it.

As you know, the American natural gas and oil industry is very diverse and consists of many sectors. My organization, IPAA, represents the exploration and production part of the industry. The independent companies who don't have refineries or gasoline stations, these companies specifically search for and produce the Nation's gas and oil wells.

There are about 6,000 independent producers exploring and producing for the Nation's energy supply each day. They are spread across 33 States. And here is the kicker that a lot of people don't recognize: Collectively, these small businesses, the 6,000 of them, are responsible for developing 91 percent of the Nation's natural gas and oil wells, accounting for 83 percent of America's oil supply and 90 percent of our domestic natural gas supply.

Through good times and bad, our companies invest billions in new projects, searching for America's energy. So the characterization that this industry is Big Oil is actually a big myth. It is the independent, smaller companies that are drilling most of the wells and providing for most of the energy in this country.

Through the effort of independent producers, today America is a world leader in natural gas and oil production. And we are doing it responsibly. In this time of continuing uncertainty, one thing is certain: A healthy oil and natural gas industry is good for America. It is good for our economic and national security, as well as for our allies across the globe.

Let me take a few minutes to address some of the strengths, weaknesses, opportunities, and threats that we see at IPAA.

Strengths. Natural gas is actually good for the environment. Today the Nation has its cleanest air in 20 years. In fact, total greenhouse gas emissions continue to decline, despite production and consumption of natural gas increasing.

Our industry is committed to reducing leaks and improving pipeline infrastructure. American producers are taking the right, responsible approach to these issues.

Other strengths of our industry include that we have America's vast natural gas supply right under this ground, about 100 years' worth. Oil and gas will remain America's largest fuel source through 2050. And coupled with wind, solar, hydro, nuclear, coal, batteries, America has a strong, reliable energy portfolio.

The weaknesses we see. Inflation, which affects all Americans. Our industry is not just producing energy, but it is also consuming it. A weak economy results in a weakened industry.

Labor and service costs—for example, purchasing tubular goods—have driven up the costs of drilling and completing wells by 30 percent year over year. It is often difficult for us to find a new workforce, to even find trucks.

And the need for a takeaway capacity through pipelines and gas storage are essential.

But there are some good opportunities to keep in mind. America's natural gas and oil production are vital here at home.

On the legislation being discussed today, IPAA strongly supports H.R. 150, the Protecting American Production Act, sponsored by Congressman Duncan. This legislation prohibits the President from declaring a moratorium on the use of hydraulic fracturing unless Congress authorizes such a prohibition.

IPAA also supports H.R. 484, the Natural Gas Tax Repeal Act, sponsored by Congressman Pfluger. This legislation would strike language designed to establish a tax on natural gas imposed on America's independent oil and natural gas producers.

IPAA recognize that the importance of managing our emissions of methane and other volatile organic compounds, and we are committed to working diligently to comply with State and Federal agencies.

Now the threats. And mostly it is about uncertainty in our industry. Uncertainty breeds inaction, and that is not an option for us. So what are the threats that create uncertainty for our businesses? Well, we have lawsuits, we have new permitting infrastructure regulation threats, we have proposed threats of new regulations with the Endangered Species Act and increased taxes on methane, we have the Securities and Exchange Commission looking at climate plans, we have bans and setbacks, we have electrification proposals, like for natural gas stoves that we have seen so much about in the news, and there is delayed-lease sales in onshore and offshore. And there is so much more.

But I want to work with this committee—or IPAA wants to work with this committee to make sure that we can address these threats, strengths, weaknesses, and opportunities. Thank you for the opportunity.

[The prepared statement of Mr. Eshelman follows:]

Testimony of Jeff Eshelman, President & Chief Executive Officer, the Independent Petroleum Association of America, Regarding House Energy & Commerce Subcommittee on Energy, Climate, and Grid Security and Subcommittee on Environment, Manufacturing, and Critical Materials, February 7, 2023 Joint Legislative, Hearing "Unleashing American Energy, Lowering Energy Costs, and Strengthening Supply Chains"

I am Jeff Eshelman, President and CEO of the Independent Petroleum Association of America (IPAA). I would like to thank Chairman Duncan and Chairman Johnson and all the members of the Energy and Commerce Committee for holding this important hear today.

IPAA represents thousands of America's independent oil and natural gas producers. Our members are the primary producers of the nation's oil and natural gas and account for 83 percent of America's oil production and 90 percent of its natural gas output. These independent producers are a driving force in our economy and support roughly 4.5 million jobs in the United States. IPAA member companies are innovative leaders that broke the code to usher in the shale oil and natural gas revolution in the United States.

As the United States and the world struggle to rebound from the economic hardship caused by the COVID-19 pandemic, it is essential for America to continue to be a leader in energy development. All forms of energy will be needed in the coming years and natural gas and oil produced in the United States will be a key component of that energy mix. Oil and natural gas will not be the only energy sources for the United States, but they will be essential to the American economy for years to come.

The choices the nation makes regarding its energy mix will have a huge impact on its economy and its international position. If America does not pursue a thoughtful energy policy, the nation will suffer economically. Unless demand for fossil energy changes dramatically, efforts to suppress U.S. oil and natural gas production will be counterproductive to the goals of addressing greenhouse gas emissions, increasing job growth and expanding America's impact around the globe.

Energy is a geopolitical issue. For the last half-century, American foreign policy has been predicated on the nation's vulnerability to oil and natural gas supply disruptions. The shale revolution turned the United States into an energy superpower, enhanced American national security and created significant geopolitical advantages for this nation around the globe.

Additionally, natural gas production and use has created the cleanest air quality the nation has seen in two decades. The United States has become the envy of nations for its dedication to reliable, affordable, responsible energy production.

Independent producers recognize the need to manage their emissions, including methane emissions. Over the past several years, as methane regulations have been developed, IPAA has been active in trying to assure that the regulations are designed appropriately for the diverse elements of the industry, including the small business operations that dominate ownership of low producing wells.

While oil and natural gas greenhouse gas emissions must be managed, their use provides key environmental benefits. America's success in reducing its greenhouse gas emissions comes from its expanded use of natural gas. Internationally, expanded use of natural gas promises to help the world improve its greenhouse gas emissions. The environment and public health challenges across the world are large and complex and failure to address fundamental health challenges limits nations' ability to address their greenhouse gas emissions.

IPAA supports the committee in its efforts to pass legislation that will increase American energy production, lower energy costs, strengthen domestic supply chains and protect America's energy grid. The various pieces of legislation being considered today will move the nation forward in our effort to enhance and strengthen American energy security. The Energy and Commerce Committee should be commended for these efforts.

Although IPAA supports all legislative efforts to enhance American energy and security, we would like to especially focus on two pieces of legislation being considered by the Committee today. First, IPAA strongly supports H.R. 150, the "Protecting American Energy Production Act" sponsored by Congressman Duncan. This legislation prohibits the President from declaring a moratorium on the use of hydraulic fracturing unless Congress authorizes such a prohibition. The bill also expresses the sense of Congress that states should maintain primacy for the regulation of hydraulic fracturing for oil and natural gas production on state and private lands.

Hydraulic fracturing is a decades old well completion technology that is often coupled with horizontal drilling to develop oil and natural gas resources from tight rock formations. Hydraulic fracturing occurs after drilling has been completed and involves pumping fluid – typically 99 percent water and sand – into the target formation at pressure in order to open small fractures in the rock, which allow oil and natural gas to flow out of these tight formations.

Advances in hydraulic fracturing technology have reversed the U.S. trajectory from that of energy scarcity to being the leader of oil and natural gas production around the globe. There is no doubt that hydraulic fracturing has allowed the United States to increase oil and natural gas production and enhanced American energy security. In addition, through the use of increased natural gas, made possible by hydraulic fracturing, air quality has also dramatically improved.

Despite all of the facts and data showing that hydraulic fracturing has enhanced America's quality of living and made the nation energy secure, relentless attacks on this technology continue from the environmental community. IPAA applauds Congressman Duncan for introducing H.R. 150 to ensure that any efforts by the Executive Branch to prohibit the use of hydraulic fracturing technology, must first be authorized by Congress. Hydraulic fracturing is too important to the energy security of the nation and the world to be subject to the whim of a President determined to handcuff America's energy producers.

IPAA also supports H.R. 484, the "Natural Gas Tax Repeal Act" sponsored by Congressman Pfluger. This legislation would strike language designed to establish a tax on natural gas imposed on America's independent oil and natural gas producers as part of the "Inflation Reduction Act" passed by Congress last year.

IPAA recognizes the importance of managing air emissions of methane and other volatile organic compounds. The American oil and natural gas production industry participates in voluntary programs to identify and implement cost effective management technologies. Our members work diligently to comply with state and federal regulations.

The Methane Emissions Reduction Program (MERP), which was passed as part of last year's Inflation Reduction Act, is an inappropriate and unworkable methane emissions tax. This tax was included despite not ever being considered in a hearing, receiving expert testimony in favor or opposition, no economic analysis, and no consideration of efficacy. Instead of looking at this issue holistically, the MERP was on the simple premise that if something is taxed, less of it will be produced.

The methane emissions tax proposes to collect emissions data, which has historically reported under Subpart W of the Clean Air Act, and use that information to assess a tax for emissions exceeding 25,000 tons of CO2 equivalent. Beginning in 2024, companies will be required to collect this data and begin paying a fee of $900 per metric ton beginning in 2025. That cost ramps up to $1200 per metric ton in 2026 and ultimately settles at $1500 per metric ton for 2026 and beyond.

Currently, the Environmental Protection Agency (EPA) is in the process of revising regulation of methane emissions under OOOO. Comments are due next week, and that process is ongoing.

The methane tax is a redundant effort to lower methane emissions that goes beyond EPA's historical jurisdiction and mandate. Regulating emissions under the CAA through the regulatory process is within the scope of the agency's jurisdiction, experience, and expertise. The methane tax would add the burden of moving EPA into tax collection, including audit processes which involve a degree of accuracy in measurement and tax assessment that goes well beyond the agency's capacity.

Further, the MERP formula requires an operator to calculate the number of tons of methane emitted in a year from applicable facilities. The formula then subtracts that tonnage from a number called the waste emissions threshold. The aforementioned fee schedule applies to any overage.

The IRA establishes a waste emissions charge for methane from applicable facilities that report more than 25,000 metric tons of CO2 equivalent per year to the Greenhouse Gas Reporting Program (GHGRP) petroleum and natural gas systems source category (GHGRP Subpart W) and that exceed the statutorily specified waste emissions thresholds. None of the tools that the law uses to generate the tax were ever designed to be used for this purpose. Moreover, this law creates a tax collection function within EPA that triggers complex audit challenges and the potential for abusive use of Clean Air Act (CAA) enforcement authorities by OECA.

The methane tax established as part of the IRA will impose financial and filing burdens on independent American oil and natural gas producers that will jeopardize many of their operations. It will add another complexity to these small businesses and divert their attention from what they do best, produce the cleanest and safest oil barrels of oil and natural gas in the world.

We thank Congressman Pfluger for introducing H.R. 484 and urge the Committee to take quick action to stop the implementation of this misguided and unreasonable tax on natural gas.

Conclusion

Oil and natural gas will remain a key component of energy supply in the world for the foreseeable future. Their emissions will need to be managed, but no modern economy will function without them. This is clearly true in the United States where oil and natural gas contributes approximately 70 of the energy consumed in the country now and in 2050. Growth in other energy sectors will occur, but more energy will be needed to maintain a robust American economy.

Artificial political efforts to suppress American supply will not reduce demand; it will only lead to a return to an import dependent energy structure with attendant energy security risks. False attacks targeting American oil and natural gas producers will reduce supply while hurting independent producers, particularly small businesses, and royalty owners. They will not reduce greenhouse gas emissions. The ultimate beneficiaries of these actions would be foreign national oil companies producing with less emissions management than those in the United States.

Congress should oppose all efforts to reduce American energy supply. IPAA commends the Energy and Commerce Committee for holding this hearing today and looks forward to the Committee acting on these important pieces of legislation that will protect and enhance American energy security.

Thank you for this opportunity and I look forward to answering your questions.

* * *

Mr. DUNCAN. I thank the gentleman. The Chair will now recognize Mr. Garcia for 5 minutes.

Statement of Raul Garcia

Mr. GARCIA. Thank you, Chairman. Thank you, all the ranking members, for the invitation to speak. My name is Raul Garcia. I am the legislative director for healthy communities at Earthjustice. In giving an opinion about the 17 bills that we are considering today, I can't—I don't have enough time to go one by one. But there are some narratives that the bills overall establish for us.

Overall, there are litany of exemptions, a litany of go-arounds, and a litany of ways that big industries get to go around laws that we have in the books already that were put in the books by Congress in bipartisan support in order to protect the communities that this very Congress represents.

And so, when we talk about already having industries that act responsibly, one has to beg the question: If they are acting responsibly, why do they want to waive the laws that hold them accountable to acting responsibly? And we have not gotten an answer on that front.

But we also have to remember that what is at stake here is broader than the simple choice that the proponents of these bills give us. So they give us a false choice between having healthy communities, a healthy environment, and energy security. And that is simply not true. We can have both. But it is a clever twist.

I mean, it is a clever ploy, even if it is a cruel one. Because on the one hand, some of these bills actually repeal parts of the laws they have made—that have made it into the books that would speed up a transition to clean energy, that would give us energy security in a clean and healthy way, like the—like parts of the IRA, particularly the Greenhouse Gas Reduction Fund. Now, that would speed up distributed solar energy so that people can have energy at their own homes so that grids going down are not a problem for an entire State. But we want to take that away in these bills.

And then, on the back end, we actually want to give industry loopholes that they can use in order to not comply with the Clean Air Act, with the Toxic Substances Control Act. These laws were established by this Congress in bipartisan fashion to protect the air that we breathe, and the toxics that are in our environment, the water that we drink, the makeup that we put on our faces, everything that goes into our stomachs. And we want to peel that away. Now, we hear a lot about energy security and this phrase of critical minerals, critical energy sources. So let's talk about critical energy sources. Few bills of the 17 that we have here today actually establish a definition for critical energy sources. What they actually say is, "Let's leave that definition up to the Secretary of Energy." That is a Trojan horse. That means everything can suddenly become a critical energy source.

And so what are we talking about when we are talking about these sources? We are talking about making sure that we have a responsible way to get to clean energy that establishes safe protections for our communities. And these bills, frankly, do exactly the opposite.

And so, when we talk about the Toxic Substances Control Act, for example, one of the bills would have us consider the economic impacts, the economic costs of—when determining whether a substance is toxic or not. So that would mean that, if we put poison in three cups of water, we are going to drink them all and figure out what the economic cost is going to do to us.

So the—another question that I have for the proponents of the bill is, what is the cost of a human life? What is the cost that that poison is going to inflict on a human being? Because it is good to talk about everything that is on paper and laws and exemptions and procedures. How do we explain the emissions coming out of the fossil fuel industry or the mining industry to our communities who are suffering from cancers, from asthma, from cardiac conditions? But we haven't talked about that here yet.

And so I believe that that is what we need to focus on, and that is what we need to do. And so these bills, by and large, fail, flatout fail to address what communities across the country are dealing with.

Thank you.

[The prepared statement of Mr. Garcia follows:]

Testimony of Raul, Garcia Legislative Director for Healthy Communities, Earthjustice, before the United States House of Representatives, Joint Energy, Climate, & Grid Security Subcommittee and Environment, Manufacturing, & Critical Materials Subcommittee Legislative Hearing: "Unleashing American Energy, Lowering Energy Costs, and Strengthening Supply Chains," February 7, 2023

Good Morning Chairman Duncan, Ranking Member DeGette, Chairman Johnson, and Ranking Member Tonko:

Thank you for inviting me to testify today. My name is Raul Garcia, and I am the Legislative Director for Healthy Communities at Earthjustice, the nation's oldest and largest nonprofit public interest environmental law organization. Please accept this testimony for the hearing's official record. My testimony addresses serious concerns and strong opposition to the draft bills discussed in this hearing. Overall, these bills create a false narrative aimed at circumventing critical environmental and public health protections in order to greenlight harmful corporate polluters that profit at the expense of frontline communities across the country.

The bills attempt to create exemption after exemption for harmful industries to avoid complying with laws like the Clean Air Act (CAA), the Toxic Substances Control Act (TSCA), and the Solid Waste Disposal Act (SWDA), among many others. Before addressing the problems with some of the specific proposed legislation, we must remember that the laws they seek to waive are important for the health and safety of all people in this country. They were passed for a reason. The CAA, for example, is the most important legal tool available to protect the air we breathe from harmful pollutants caused by dirty industries. In a similar way, TSCA sets up protections so that toxic substances like lead, asbestos, and PFAS are not manufactured and consumed in ways that harm our communities. In turn, the SWDA establishes requirements for polluting industries to clean up areas that they have contaminated to such toxic levels that they pose a serious threat to human life.

Concretely speaking, these laws are at times the only line of defense that our communities—your constituents—have to prevent even more devastating and ever-increasing numbers of respiratory diseases like asthma, cardiovascular conditions, and other serious illnesses like cancer.

The bills in question in today's hearing are a litany of waivers that would allow dirty corporate industries to violate the environmental and health requirements in our laws. Many of them would waive protections for what they label as "critical energy resources" or some similar phrasing.

However, none of these bills actually define what those "critical" resources would be. Instead, they rely on the discretion of the Department of Energy to make that determination. This means that, at any given point, anything can be deemed as "critical," subject to the whims of whoever is at the Department of Energy. There is nothing in these bills that would stop any random and harmful industry from being falsely labeled as "critical." The bills before the subcommittees today are a trojan horse that would bypass critical health protections to line the pockets of dirty polluters and toxic chemical manufacturers. To be clear, this is not about critical resources that our country needs; it is about giving a free pass to deadly chemicals like lead, asbestos, sulfuric acid, mercury, and PFAS that are not critical to our national energy interests. In effect, these waivers would allow even the most harmful polluters to manufacture and spread the deadliest poisons into our air, water, and even our homes with impunity.

Together, many of the bills presented before the subcommittees create a narrative that our environmental and health protections exist to stop or significantly hamper energy and other development. This is utterly false. Our environmental and health protections are designed to ensure that energy production and other industrial activities are safe for our communities, especially those living at the fenceline and in close proximity. We have the capability to advance our technologies, create responsible clean energy development, and deliver it across the country in a way that protects our health and our children's future. Congress should avoid advancing bills that eliminate these critical public health protections and facilitate harmful energy production and poisonous chemical manufacturing. Lawmakers should instead focus on strengthening protections and holding bad actors who violate our laws and betray your constituents accountable for their actions. We reject the false choice between energy creation and human health. We can and must have both.

I offer the following as clear examples of the problems that plague the bills being proposed today:

H.R.__, To amend the Toxic Substances Control Act with respect to critical energy resources, and for other purposes.

We express our strong opposition to the draft bill: "To amend the Toxic Substances Control Act for critical energy resources, and for other purposes." In just three short pages, the bill would reverse and eviscerate several of the core reforms to the nation's chemical safety law, even after TSCA passed the House and Senate overwhelmingly with bipartisan support just a few years ago. It was reported out of this Committee unanimously and passed the House 403-12. In fact, nine of the nay votes were from Democrats, many of whom believed the final bill wasn't strong enough. Every Republican on this Committee who was in Congress in 2016 and cast votes voted in favor of the legislation, including Chairs McMorris-Rogers, Duncan, and Johnson, and Representatives Burgess, Latta, Guthrie, Griffith, Bilirakis, Bucshon, Hudson, Walberg, Carter, Palmer, and Weber.

The draft legislation would make it virtually impossible for EPA to meaningfully review the safety of new chemicals that are classified as "critical energy resources," regardless of their health risks. The bill promotes cursory assessments, followed by default approvals, of any new chemical that is deemed necessary for a "critical energy resource," no matter how toxic, how persistent, or how mobile in the environment. These chemicals could be used in anything from fracking to petrochemicals to mining.

The revisions to TSCA would sacrifice the health and safety of the public – including children, workers, the elderly, and fenceline communities – to expedite production of any potentially toxic chemical that the industry can persuade the Department of Energy, which is not charged with reviewing the health and safety of chemicals, to deem a "critical energy resource." We already know the limitless scope of what the chemical industry is likely to claim as "critical" based on their recent insistence that some of the most toxic chemicals in existence are "critical" for renewable energy or energy security, including PFAS,[18] asbestos,[19] and lead.[20] TSCA's failure to protect the public from asbestos, as well as other toxic chemicals like TCE and methylene chloride, served as Congress's catalyst to strengthen the law. To now roll back TSCA's protections in order to fast-track approval of chemicals like asbestos would be a sad irony.

The draft bill would:

- Mandate that EPA's risk evaluation of chemicals, rather than continuing to focus on the potential health risks, must also include the consideration of all cost and other "non-risk factors" when evaluating whether the chemical substance poses an unreasonable risk (as opposed to basing safety determinations solely on risks to health or the environment). The prioritization of economic considerations over public health protection was the major flaw that had stymied progress under the old TSCA, and Congress's deliberate shift to risk-based evaluations and decision-making was the fundamental reform that brought the law back to life after being ineffective for decades and badly in need of reform.
- Allow new chemicals to begin production before EPA has completed its determination whether they pose an unreasonable risk to human health or the environment. Because Congress wanted EPA to make an affirmative determination of safety for all new chemicals, TSCA explicitly provides that no new chemical can enter production until that determination has been made. The bill would completely reverse this vital health-protective policy.
- In addition to the newly-added consideration of costs and any other non-risk factors to EPA's analysis, which will lengthen the time necessary for review, the bill simultaneously prevents EPA from extending the review period for chemicals designated "critical energy resources." The inevitable result will be rushed and superficial reviews that fail to identify risks to health and the environment or incomplete reviews that result in default approvals of unsafe chemicals. As we have seen over and over, where a toxic chemical begins manufacture without a thorough review by EPA, it is almost impossible to end its production, or retrospectively establish sufficient protections from the chemical to protect the public.

---

18 ACC, *PFAS: Critical to Renewable Energy*, https://www.americanchemistry.com/chemistry-in-america/chemistries/fluorotechnology-per-and-polyfluoroalkyl-substances-pfas/pfas-critical-to-renewable-energy.

19 ACC, *ACC Urges EPA to Reconsider its Flawed Chlor-alkali Proposal*, https://www.americanchemistry.com/chemistry-in-america/news-trends/press-release/2022/acc-urges-epa-to-reconsider-its-flawed-chlor-alkali-proposal.

20 International Lead Association, *Using Lead Responsibly is Critical to Achieving a Sustainable and Low Carbon Future*, https://ila-lead.org/sustainability/).

- Create a limitless loophole from TSCA's chemical assessment and health protection requirements. "Critical energy resource" is an open-ended and undefined concept that could apply to virtually any chemical that plays a role in the production, refining, distribution, and use of energy and is designated as "critical" by the Department of Energy. Once a substance is deemed to be a "critical energy resource," and therefore is fast tracked through the TSCA pre-manufacture notice process, there is no limit on how the substance can then be used, beyond its ostensible "critical energy resource" use, and no constraint on non-energy applications that could also be harmful to health and the environment.

The bill would establish a precedent for enacting further loopholes to gut the health protective provisions of the Act. If it is acceptable to gut health reviews of chemicals for "critical energy resources," what is the principle that will prevent other broad categories or uses of potentially toxic chemicals from also getting special treatment under Section 5 of TSCA?

Notably missing from the draft bill are any findings demonstrating the need for the bill. In fact, there is no evidence that the public must sacrifice health protections from toxic chemicals in exchange for clean energy. We can develop and deploy new energy technologies without waiving chemical review requirements or placing the communities burdened by PFAS and other toxic chemicals at risk. The draft bill's rejection of that clean and health-protective energy future sells American innovation short. Overwhelmingly, the public wants more, not less, protection from toxic chemicals.[21] Yet the draft bill would roll back critical public health protections and weaken the nation's bedrock chemical safety law.

H.R. __, the "Securing America's Critical Minerals Supply Act."

This draft legislation modifies the organization and jurisdiction of the Department of Energy, which seems rather mundane at first glance. However, the draft legislation is about far more than critical minerals, as it uses the amorphous definition of "critical energy resource," which could include essentially anything pertaining to energy. The Environmental Protection Agency (EPA) and Department of Interior (DOI) would have their authority to protect communities and the environment removed, even though they have the expertise and mandate to protect our health, water, and lands. The Department of Energy's new authority would effectively have the only metrics for consideration in production be economic or "security," which will mean that our environment and communities will be left behind.

H.R. _ , To amend the Clean Air Act to prohibit the phase out of gasoline and prevent higher prices for consumers and for other purposes.

This legislation is an attack on the Environmental Protection Agency's Risk Management Program (RMP), which are safeguards intended to protect workers, fenceline communities, and first-responders from chemical disasters. The bill would amend the Clean Air Act to exempt refineries that use dangerous hydrofluoric acid (HF) from conducting a safer alternatives assessment. Safer technology and alternatives assessments are commonsense best practices that help facilities innovate and identify opportunities to prevent disasters and save lives. Alternatives identified in such assessments can prevent chemical disasters, such as the one that occurred in 2019 at the HF alkylation unit at the Philadelphia Energy Solutions Refinery in Philadelphia, Pennsylvania. This devastating series of explosions injured five workers and a

---

[21] University of California San Francisco Program on Reproductive Health and the Environment, *Public Opinion on Chemicals,* https://prhe.ucsf.edu/public-opinion-chemicals.

first responder, led to the evacuation of 4,000 people, and culminated in over 1,000 workers being laid off with no severance and almost no notice when the plant shut down. During the incident, over 5,000 pounds of highly toxic HF were released, and the estimated property damage loss was $750 million.

As the climate rapidly changes, the risk and severity of chemical disasters and releases are being exacerbated by extreme weather, making EPA's Risk Management Program even more important in preventing catastrophic disasters that lead to injury and loss of life and property. When industrial facilities, including HF refineries, located in these areas fail to adequately prepare for extreme storms, wildfires, earthquakes, heat waves, floods, rising sea levels, and other natural disasters this can lead to a cascading series of harms, including toxic chemical exposures, on top of the effects of the weather event itself.

This bill would disregard recommendations and information from the U.S. Chemical Safety Board and the Government Accountability Office and instead caters to the American Petroleum Institute and other fossil fuel interests. The dangers of HF are well-established, and millions of Americans live in the worst-case scenario zone for a toxic release. Congress should reject this bill and support EPA as they update the Risk Management Program Rule and work to better protect industrial workers, environmental justice and fenceline communities, and first-responders.

H.R. _ , To authorize the Administrator of the Environmental Protection Agency to waive application of certain requirements, sanctions, or fees, with respect to processing or refining of critical energy resources at a critical energy resource facility, and for other purposes.

This legislation would allow the Environmental Protection Agency (EPA) to waive the Clean Air Act (CAA) and Solid Waste Disposal Act (SWDA) requirements for waste produced during production of energy related products. The CAA and SWDA are essential for protecting our health and the environment from the hazards of countless types of industry waste. The draft legislation would exempt certain energy facilities from provisions of the SWDA, potentially including everything from fracking wastewater, to mine processing facilities and tailing sites to nuclear facilities. The SWDA was passed to protect human health and the environment from hazardous waste; exempting these types of waste from these laws threatens public health. It would destroy drinking water, increase negative health impacts on Indigenous and frontline communities, and pollute our public lands as well as the air we breathe. The EPA consistently classifies energy production waste as the most toxic, and often the waste site is impossible to completely remove contamination from, or it exists for hundreds of years.

H.R. _, To amend the Solid Waste Disposal Act to treat the owner or operator of a critical energy resource facility as having been issued an interim permit for the treatment, storage, or disposal of hazardous waste, and for other purposes.

This legislation bypasses normal permitting processes, by advancing interim permitting for certain energy facilities. The draft legislation would thus allow the facilities to operate before securing a permit, whose entire purpose is to review the facility's impact on air, water, and the community. By preemptively providing these interim permits, the result would almost certainly be an increase of negative health consequences and the destruction of cultural resources due to toxic pollutants being released into our air and water. The facilities that could receive an interim permit without a full understanding of their impact include everything from mining waste in tailings ponds, to fertilizers and petrochemicals.

H.R. _, To require the Administrator of the Environmental Protection Agency to authorize the use of flexible air permitting with respect to certain critical energy resource facilities, and for other purposes.

This draft legislation would take the science out of decisions around air permitting, regarding flexible permitting for certain energy facilities. It would allow the Environmental Protection Agency (EPA) to circumvent the transparent process of approving or denying flexible air permitting. Doing so could insert politics into the decision and potentially allow the EPA Administrator to increase air pollution from vaguely defined "critical energy resource facilities," subsequently harming public health and contributing to climate change. Similar to the other bills I've discussed, the facilities this could apply to are vast. If up to the discretion of the Department of Energy, they could include some of the worst offenders of pollution, including oil and gas, mining and petrochemicals.

Conclusion

Earthjustice represents and partners with communities on the frontlines of the climate and environmental justice crises across the country. Each day we see the real human consequences of what happens when polluting industries fail to follow the law. From the petrochemical plants lining Cancer Alley in the Gulf to the legacy of mining on Indigenous lands, we know what happens when environmental protections are removed, and people are left at the mercy of industry to regulate themselves. The bills before the sub-committees today will exacerbate the problem. Under the guise of guarding our national energy interests and security, these bills seek to gut commonsense protections and safeguards that all Americans, but especially environmental justice communities, depend on to protect them in their neighborhoods, workplaces, and homes. We look forward to continue working with the Committee to ensure that our country remedies existing environmental injustices, strengthens environmental protections to prevent those injustices from occurring in the future, and ensures that our clean energy future proceeds in an equitable and just way. Thank you for giving me the opportunity to testify today.

Sincerely,
Raul Garcia

Legislative Director for Healthy Communities
Earthjustice

* * *

Mr. DUNCAN. Thank you. I now recognize Ms. Sweeney for 5 minutes.

Statement of Katie Sweeney

Ms. SWEENEY. Thank you. Good morning, members of the subcommittee. I appreciate being here on behalf of the National Mining Association's mineral and hardrock mining companies to talk about the need to strengthen our mineral supply chains to unleash American energy and lower energy cost.

Domestic mining, conducted under world-leading environmental safety and labor standards, is critical to securing virtually every key supply chain, especially energy. But the right policies are needed to unlock our full potential.

Minerals are an integral part of all current forms of energy and for those we hope to rely on more in the future. From copper, nickel, and silver, and renewables to cobalt, and lithium, and EVs to barite, and molybdenum that keep oil and gas moving, to uranium and coal, which produced over 40 percent of our electricity, minerals security makes energy security.

As we enter the most mineral-intensive era in human history, the International Energy Agency estimates demand for some minerals required for energy generation transitions could grow by more than 40 times by 2040. Urgent action is needed to secure these essential supply chains. But recent U.S. Geological Survey information shows our country is headed in the wrong direction.

Despite the rhetoric around securing our mineral supply chains, we are at a crisis point. In 2022, the U.S. reached its highest recorded mineral import reliance. Imports made up more than one-half of U.S. apparent consumption for 51 nonfuel mineral commodities, up from 2021, when only 47 commodities met that metric.

We are more dependent than ever before on China and others for minerals essential to modern life. And each new announcement of a blocked mine, such as Twin Mines—Twin Metals Project in Minnesota, or foreign sourcing agreements with countries with documented problematic labor practices locks in our position of competitive weakness.

The U.S. must focus on supplying these minerals at home as well as restoring domestic smelting, refining, and processing capabilities. In a 2019 hearing, Benchmark Minerals talked about growing mineral demand for EV batteries and lack of domestic production, cautioning that those who control these critical raw materials and those who possess the manufacturing and processing know-how will hold the balance of industrial power in the 21st century auto and energy storage industries.

Automakers understand that truth and worry that the coming battery minerals shortfall will decimate the EV revolution. Ford's president and CEO recently highlighted the need to focus on domestic supply chains all the way to the mines to reduce our reliance on minerals sourced from countries with documented child labor practices and corruption.

Without permitting reform, the U.S. will be watching the global competition for energy dominance from the sidelines. Providing additional funds or incentives for projects that will never be approved does nothing. As the IEA concluded in a recent report, governments must leverage private investment in sustainable mining but also ensure clear and rapid permitting procedures to avoid potential supply bottlenecks.

Opening or expanding a U.S. mine typically involves multiple agencies and tens or even hundreds of permitting processes at the local, State, and Federal levels. Delays arise from duplication among agencies, absences of firm timelines, and failures in agency coordination. Necessary authorizations take an average of 7 to 10 years, one of the world's longest permitting processes. Valid environmental concerns should be fully addressed, but permitting processes should not serve as an excuse to trap mining projects in a limbo of duplicative, unpredictable, endless, and costly review.

We can build on important work done by this committee to support new domestic production and processing. Chair McMorris Rodgers and Natural Resource Chair Westerman's Securing American Mineral Supply Chains Act offers commonsense solutions to reestablish a domestic mineral supply chain. The act prioritizes responsible development and would provide certainty to mining companies, investors, and manufacturers; establish lead agencies and improve permitting timeliness; maintain access to mineralized Federal lands unless withdrawn by Congress and unless the USGS can assure that the withdrawal does not threaten supply chains; support research, development, and demonstration funding; and workforce development and training.

It is time for the United States to walk the talk on mineral security. As our minerals needs skyrocket for everything from EVs to advanced energy technologies, the U.S. is stumbling when it comes to our supply chains.

NMA appreciates this committee's prioritization of these issues and is eager to help craft solutions. Thank you.

[The prepared statement of Ms. Sweeney follows:]

Testimony of Katie Sweeney, Executive Vice President & General Counsel, National Mining Association before the United States, House of Representatives, Committee on Energy & Commerce, "Unleashing American Energy, Lowering Energy Costs, and Strengthening Supply Chains," February 7, 2023

Good morning members of the subcommittees. I am Katie Sweeney, Executive Vice President and Chief Operating Officer of the National Mining Association (NMA). America's mining industry supplies the essential materials necessary for nearly every sector of our economy – from technology and healthcare to energy, transportation, infrastructure and national security. The NMA is the only national trade organization that serves as the voice of the U.S. mining industry and the hundreds of thousands of American workers it employs before Congress, the federal agencies, the judiciary and the media, advocating for public policies that will help America fully and responsibly utilize its vast natural resources. We work to ensure America has secure and reliable supply chains, abundant and affordable energy, and the American-sourced materials necessary for U.S. manufacturing, national security and economic security, all delivered under world-leading environmental, safety and labor standards. The NMA has a membership of more than 275 companies and organizations involved in every aspect of mining, from producers and equipment manufacturers to service providers. I am pleased to testify this morning on behalf of the mineral and hardrock mining companies in the NMA.

Ever-Increasing Demand for Minerals

There is widespread recognition that we are entering the most mineral and metal intensive era in human history.[22] Consequently, the right policies to secure new domestic mineral production and our supply chains are more important than ever.

The international competition for minerals will be fierce. The European Union (EU) recently unveiled its "REPowerEU Plan" to scale-up and speed up renewable energy in the next decade.[23] Late last year, the United Kingdom (UK) released its "Resilience for the future: The UK's critical minerals strategy."[24] In December, Canada released its "Canadian Critical Minerals Strategy," a generational "plan to position Canada as the global supplier of choice for critical minerals and the clean technologies they enable."[25] Of course, China, with its much longer planning horizon, moved earlier and more quickly to address the risks to its mineral supply chains. In 1999, the Chinese government announced its aggressive "go global" campaign to secure raw materials. The policy, which was fully implemented around 2002-2003, articulated three main objectives: (1) to support national exports and expand into international markets; (2) to push domestic firms to internationalize their activities as a means of acquiring advanced technologies; and (3) to invest in the acquisition of strategic resources.[26]

Many public analyses evaluate the demand for minerals for new technologies but especially energy generation. Last year the International Energy Agency (IEA) issued a cautionary report about risks related to the mineral supply chains required for energy generation transitions.[27] IEA estimates and others show that demand for some minerals could grow by more than 40 times by 2040. According to IEA:

- Lithium demand is anticipated to grow by more than 40 times by 2040, followed by graphite, cobalt and nickel at around 20-25 times;
- Copper demand for grid infrastructure and electrification more than doubles by 2040
- Demand for cobalt is expected to be anywhere from 6 to 30 times higher than today's levels; and
- Rare earth elements may see three to seven times higher demand in 2040 than today.[28]

---

[22] Google results for the term "critical minerals" return nearly 24,000 responses (7,000 news specific) for the last month alone.

[23] Communication from the Commission to the European Parliament, The European Council, The Council, The European Economic and Social Committee and the Committee of the Regions: REPowerEU Plan, {SWD(2022) 230 final}, May 18, 2022. https://eur-lex.europa.eu/resource.html?uri=cellar:fc930f14-d7ae-11ec-a95f-01aa75ed71a1.0001.02/DOC_1&format=PDF.

[24] Department for Business, Energy and Industrial Strategy, "Resilience for the future: The UK's critical minerals strategy, 22 July 2022. https://www.gov.uk/government/publications/uk-critical-mineral-strategy/resilience-for-the-future-the-uks-critical-minerals-strategy.

[25] Natural Resources Canada News Release, "Countries Commit to the Sustainable Development and Sourcing of Critical Minerals," Dec. 12, 2022. https://www.canada.ca/en/natural-resources-canada/news/2022/12/countries-commit-to-the-sustainable-development-and-sourcing-of-critical-minerals.html.

[26] CRS," China's Mineral Industry and U.S. Access to Strategic and Critical Minerals: Issues for Congress," R43864, March 20, 2015, p. 2. https://crsreports.congress.gov/product/pdf/R/R43864/6.

[27] International Energy Agency, "The Role of Critical World Energy Outlook Special Report Minerals in Clean Energy Transitions," May 2021.

[28] Id at pp.8-10.

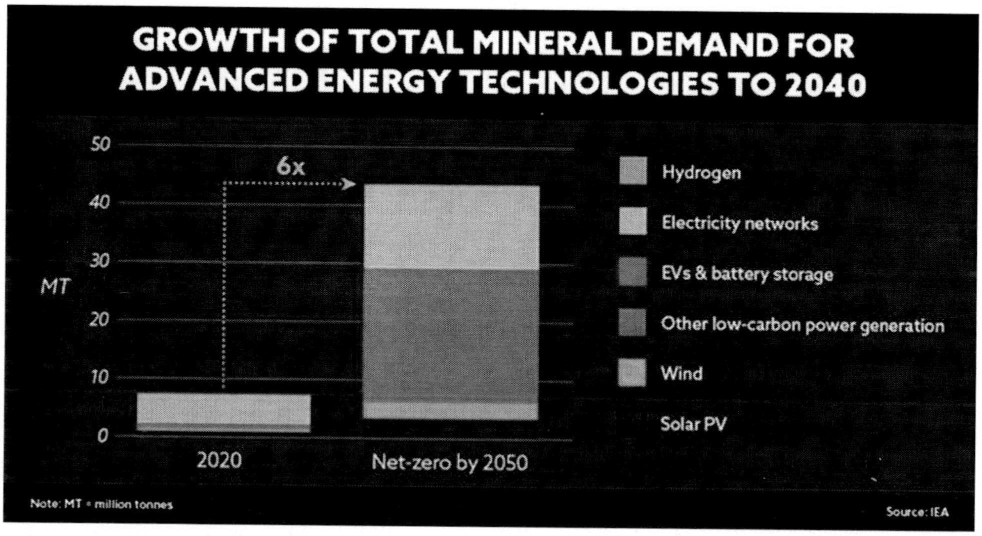

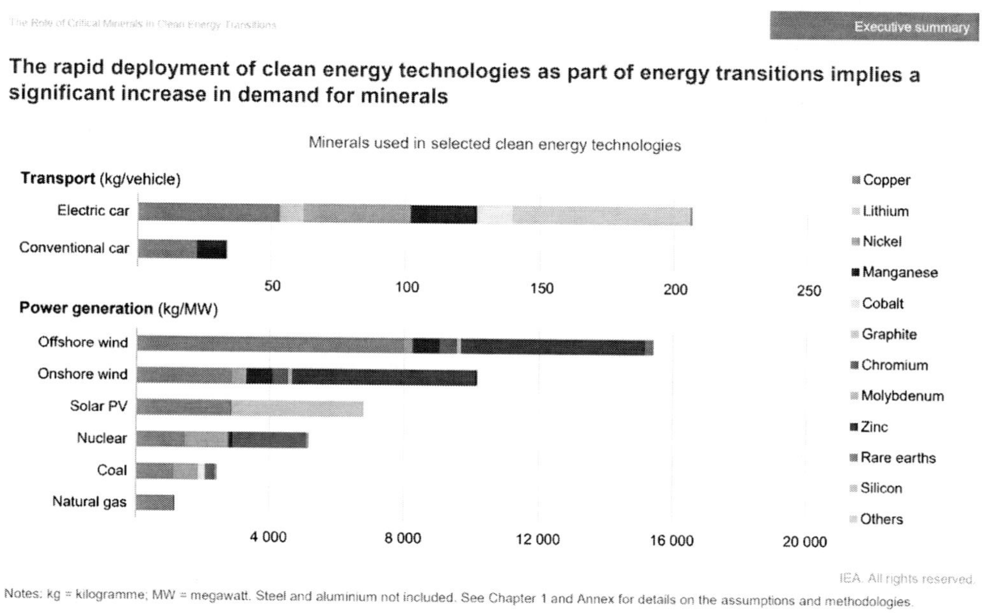

Other major reports echo the findings of the IEA. Wood Mackenzie, the World Bank,[29] the Wilson Center[30] and others outline staggering demand increases that are likely to outplace the available minerals supply. According to Wood Mackenzie:

- Demand for copper and aluminum is anticipated to increase by a third by 2040,
- Nickel demand grows by two-thirds and cobalt and lithium by 200 percent and 600

[29] World Bank Group, "Minerals for Climate Action: The Mineral Intensity of the Clean Energy Transition," 2020. https://pubdocs.worldbank.org/en/961711588875536384/Minerals-for-Climate-Action-The-Mineral-Intensity-of-the-Clean-Energy-Transition.pdf.

[30] D. Wood, A. Helfgott, M. D'Amico, and E. Romanin, Woodrow Wilson International Center for Scholars, "The Mosaic Approach: a Multidimensional Strategy for Strengthening America's Critical Minerals Supply Chain," Oct. 12, 2021, https://www.wilsoncenter.org/sites/default/files/media/uploads/documents/critical_minerals_supply_report.pdf.

percent, respectively.[31]

Matching the speed and scale of this rising demand requires a permitting regime that enables the mining sector to respond to market signals. Current U.S. permitting timelines do not.

As the IEA recently concluded in a July 2022 battery supply chain report:

> Electrifying road transport requires a wide range of raw materials. While all stages of the supply chain must scale up, extraction and processing are particularly critical due to long lead times. Governments must leverage private investment in sustainable mining and ensure clear and rapid permitting procedures to avoid potential supply bottlenecks.[32]

Impacts Down the Supply Chain

End users of minerals have awoken to the challenge of securing mineral supply chains, a development perhaps most pronounced by the automotive sector as it advertises a transition to electric vehicles (EVs). Over the last few years, many of the major U.S. car makers have made ambitious announcements about their EV plans. As examples, General Motors has announced it will invest $35 billion in electric and autonomous vehicle product development until 2025 and that it will phase out petrol and diesel cars by 2035; Volkswagen wants half of its vehicle sales to be electric by 2030 and nearly 100 percent electric sales by 2040; and Audi will launch fully electric models only from 2026 and aims for all car sales to be electric by 2030.[33]

At the same time, automakers are warning with ever greater frequency that the coming battery material shortfall could stop the EV revolution in its tracks. As recently noted by RJ Scaringe, CEO of EV start-up Rivian, the auto industry's current semiconductor problems "are a small appetizer to what we are about to feel on battery cells over the next two decades."[34] No wonder, as the battery supply chain is already facing the pinch of rising material prices as the gap between demand and supply widens. Battery pack costs – which had been on a long downward trend – are now rising. Metals accounted for 40 percent of battery costs in 2015. Today, they account for 80 percent. Where the price of these metals goes, so does the cost of batteries and EVs. According to EV automaker Stellantis CEO Carlos Tavares, there will be a shortage of EV batteries by 2024-2025, followed by a lack of raw materials for the vehicles that will slow availability and adoption of EVs by 2027-2028 as the global automotive industry pivots to EVs to meet an expected increase in consumer demand and government regulations.

---

31 Gavin Montgomery, Wood Mackenzie, "COP26: Why battery raw materials are a highly-charged topic:- Aggressive EV uptake is needed to meet a 2° C target, but metals supply will struggle to meet demand." 13 October 2021, https://www.woodmac.com/news/opinion/cop26-why-battery-raw-materials-are-a-highly-charged-topic/.

32 IEA, "Global Supply Chains of EV Batteries," July 2022. https://www.iea.org/reports/global-supply-chains-of-ev-batteries.

33 van Halm, I. and Mullan, C., Feb. 14, 2022, " Booming EV sales challenge critical mineral supply chains," *Energy Monitor* https://www.energymonitor.ai/sectors/transport/booming-ev-sales-challenge-mineral-supply-chains.

34 Wall Street Journal," Rivian CEO Warns of Looming Electric-Vehicle Battery Shortage," April 2022. https://www.wsj.com/articles/rivian-ceo-warns-of-looming-electric-vehicle-battery-shortage-11650276000.

He recently cautioned that the "speed at which we are trying to move all together for the right reason, which is fixing the global warming issue, is so high that the supply chain and the production capacities have no time to adjust."[35]

Automakers have been seeking solutions, including inking deals directly with mining companies. For example, last year Tesla addressed its concern about obtaining the nickel for its EVs by entering into an agreement with BHP to obtain nickel from Australia and more recently with Talon Metals to buy quantities of nickel directly from a mine the company is building in Minnesota. Ioneer has signed a binding offtake agreement with the Ford Motor Company to supply lithium from its Rhyolite Ridge lithium-boron project in Nevada.[36] Only last week, General Motors announced it was investing $650 million in Lithium Americas to secure access to production from its Nevada operations, which General Motors estimates will contribute to one million EVs annually.[37] For this deal, General Motors was one of more than 50 automakers and companies seeking a secure supply of minerals from Lithium Americas.[38]

At the same time, automakers are urging the ramp up of domestic mining. Last year, the Alliance for Automotive Innovation wrote President Biden expressing concerns that "neither the current trajectory of consumer adoption of EVs, nor existing levels of federal support for supply- and demand-side policies, is sufficient to meet our goal of a net-zero carbon transportation future."[39] One of the specific policy recommendations offered by the Alliance is to promote national security and economic security enhancements through the development of U.S.-based supplies of critical minerals (extraction, processing and recycling), battery and fuel cell manufacturing, and other critical components, including semiconductors.[40] And as succinctly stated recently by Jim Farley, President and CEO of Ford Motor Co.:

> We have to bring battery production here, but the supply chain has to go all the way to the mines. . . So are we going to import lithium and pull cobalt from nation-states that have child labor and all sorts of corruption or all we going to get serious about mining?" . . . We have to solve these things and we don't have much time."[41]

We have our work cut out for us to build our domestic mineral supply chains quickly. As recently reported by *The New York Times*, how automakers will obtain enough materials for an all-electric lineup remains unclear. Just last month, Farley told analysts that only 50 percent of

---

[35] Media interview, May 24, 2022, https://www.cnbc.com/2022/05/24/stellantis-ceo-warns-of-ev-battery-shortage-lack-of-raw-materials.html.

[36] PR Newswire," Ioneer Signs Binding Lithium Offtake Agreement with Ford," July 21, 2022.

[37] Lithium Americas General Motors Transaction Announcement, January 31, 2023, https://www.lithiumamericas.com/news/lithium-americas-provides-general-motors-transaction-details-and-update-on-construction-plan-for-thacker-pass. Cecilia Jamasmie, January 31, 2023, "GM invests $650m in Lithium Americas to develop Thacker Pass mine" www.mining.com, https://www.mining.com/gm-lithium-americas-to-jointly-develop-thacker-pass-mine-in-nevada/.

[38] The Electric, "The New 'Elephants'—GM Grabs the Biggest Lithium Deposit in the U.S., Feb. 2, 2023. https://subscriptions.theinformation.com/newsletters/the-electric/archive/the-electric-the-new-elephants-gm-grabs-the-biggest-lithium-deposit-in-the-u-s.

[39] Alliance for Automotive Innovation letter to President Biden, March 29, 2021. https://www.autosinnovate.org/posts/communications/Auto%20Industry%20EV%20Policy%20Letter%20to%20P resident%20Biden%20March%2029%202021.pdf.

[40] Id. at 4.

[41] Jim Farley remarks, Detroit Homecoming VIII, Live-streamed interview with Mary Kramer (director of the annual event). Sept. 25, 2021. https://detroithomecoming.com/livestream-events/.

the raw materials needed to meet the auto industry's announced EV targets were actually available.[42]

Demand Cannot Be Met without New Mining

The automakers are just one stakeholder group that acknowledges the role of domestic mining in securing our supply chains. Certainly, the federal government has repeatedly noted that boosting sustainable domestic mining must be part of the solution. For example, in May 2021, the White House rebutted reporting from *Reuters* claiming that President Biden will primarily rely on ally countries to supply the bulk of the metals needed to build EVs. In its clarification, the White House noted that the reporting incorrectly characterizes the Biden-Harris administration's approach:

> President Biden is focused on seizing the electric vehicle (EV) market, sourcing and manufacturing the supply chain here in America, and creating good-paying, union jobs. Building American-made EVs and shipping them around the world will include leveraging American-made parts and resources. This includes responsibly pursuing, developing, and mining critical minerals and materials used for EV batteries. As we strengthen our supply chains, we will pursue strong environmental standards and broad, rigorous consultations with local and indigenous communities to support a responsible, fair, and sustainable EV industry."[43]

Working with our allies, like Canada, to build these supply chains is smart. But that must complement the essential work of standing up production and these supply chains at home. It cannot come in place of it.

Recent withdrawal decisions this month locking up more than 225,000 acres in federal Forest Service lands from mining for two decades after also withdrawing federal leases nearly sixty years old from projects in the same areas known for some of the nation's largest reserves of nickel, cobalt, copper, platinum, and palladium could only be described at best as short sighted and at worst self-sabotage.[44]

Current Permitting Process Discourages Investment in U.S. Mining

With over $6 trillion worth of mineral resources here in the United States, a highly trained and highly compensated workforce, and world-class environmental and safety standards, the U.S.

---

[42] Boudette, Neal E. 2022. "California E.V. Mandate Finds a Receptive Auto Industry." The New York Times, August 25, 2022, sec. Business. https://www.nytimes.com/2022/08/25/business/energy-environment/electric-vehicles-automakers.html.

[43] Statement from Ali Zaidi, Deputy National Climate Advisor, Reuters, Epoch Times etc.: https://www.theepochtimes.com/white-house-denies-report-that-biden-looks-overseas-for-electric-vehicle-metals_3832373.html?welcomeuser=1.

[44] Wall Street Journal, "Biden's Green-Energy Mineral Lockup. The feds block mining that is essential for making EV batteries" January 29, 2023, https://www.wsj.com/articles/biden-administration-mining-duluth-complex-minnesota-superior-national-forest-deb-haaland-electric-vehicles-11674860178.

mining industry is essential to helping the nation meet ever-increasing demand for minerals for electrification, infrastructure and manufacturing needs.

However, there is real room for improvement. To improve supply chain security, we must also have a robust domestic mineral supply chain. That includes more smelting, processing and refining capabilities in the U.S. necessary to claw back these essential processes from geopolitical adversaries like China, which controls more than 80 percent of global rare earth element production, nearly 90 percent of global mineral processing capabilities as well as the market prices for rare earth elements at each step of the process.

Permitting delays have been, and continue to be, one of the most significant risks to meeting domestic mineral production goals. As the permitting process for important projects across the U.S. drags on, geopolitical rivals are taking advantage of our bureaucratic inertia. Opening or expanding a mine in the U.S. typically involves multiple agencies and the navigation of tens or even hundreds of permitting processes at the local, state and federal levels, with little transparency into status, delays arising from duplication among federal and state agencies, an absence of firm timelines for completing environmental assessments, and failures in coordination of responsibilities between various agencies.

Necessary government authorizations now take an average of seven to 10 years to secure – one of the longest permitting processes in the world for mining projects – a time period that is completely out of step with the dramatic increases in minerals production that will be needed in the coming decades to keep up new technologies, infrastructure, manufacturing and even with the administration's goals.

In the U.S., necessary government authorizations place the U.S. at a competitive disadvantage in attracting investment for mineral development. By comparison, permitting in Australia and Canada, which have similar environmental standards and practices as the U.S., take between two and three years. The NMA believes that valid concerns about environmental protection should be fully considered and addressed but permitting processes should not serve as an excuse to trap mining projects in a limbo of duplicative, unpredictable, endless and costly review without a decision point. Moreover, there is little evidence that such delays yield commensurate environmental benefits. The length of the permit process should not be confused with the rigor of review. Ironically, it takes about two years to build a new battery gigafactory, but it takes at least eight years (sometimes more than 10 years) to build a new lithium mine.[45]

Nearly two decades ago, the U.S. attracted almost 20 percent of the world's total mining investment. Unfortunately, in the time since, there has been a sharp decline in U.S. exploration investment. This is not due to lack of resources, but rather a lack of confidence in the U.S. as a viable mining jurisdiction in which to invest hundreds of millions of dollars in upfront costs due to duplicative, inefficient and costly permitting timeframes, making the U.S. more dependent on other countries for metals.

[45] Comments of Dr. Qichao Hu, founder and CEO of Massachusetts-based battery maker SES, in an interview with Charged.

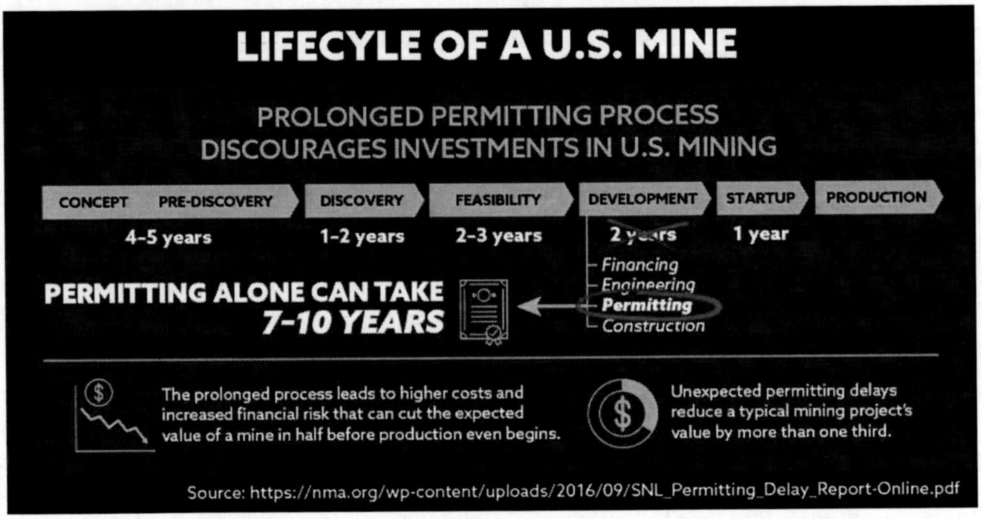

Current Permitting Process Encourages Foreign Dependence

The U.S. is increasingly vulnerable to supply chain disruptions and retaliation from geopolitical adversaries due to our ever-increasing reliance on imports for these essential resources. Less than half of the mineral needs of U.S. manufacturing are met by domestically produced minerals, which leaves our economy and national security at a strategic disadvantage. The U.S. Geological Survey's annual commodity summary released only last week makes some key findings:

- Last year, imports made up more than one-half of the U.S. apparent consumption for 51 nonfuel mineral commodities, and the United States was 100 percent net import reliant for 15 of those.
- Of the 50 mineral commodities identified in the "2022 Final List of Critical Minerals," the United States was 100 percent net import reliant for 12, and an additional 31 critical mineral commodities (including 14 lanthanides, which are listed under rare earths) had a net import reliance greater than 50 percent of apparent consumption.
- Underscoring the vulnerability of U.S. mineral supply chains, China was the leading source of mineral commodities with a greater than 50 percent import reliance providing 26, with significant imports of other essential commodities also coming from Russia.
- The estimated value of U.S. metal mine production in 2022 was $34.7 billion, six percent lower than the revised value in 2021. In 2022, the capacity utilization for the metals mining industry was 61 percent, less than the 63 percent capacity utilization in 2021.[46]

[46] U.S. Geological Survey, 2023 Commodity Summary, https://pubs.er.usgs.gov/publication/mcs2023.

U.S. Mineral Import Reliance

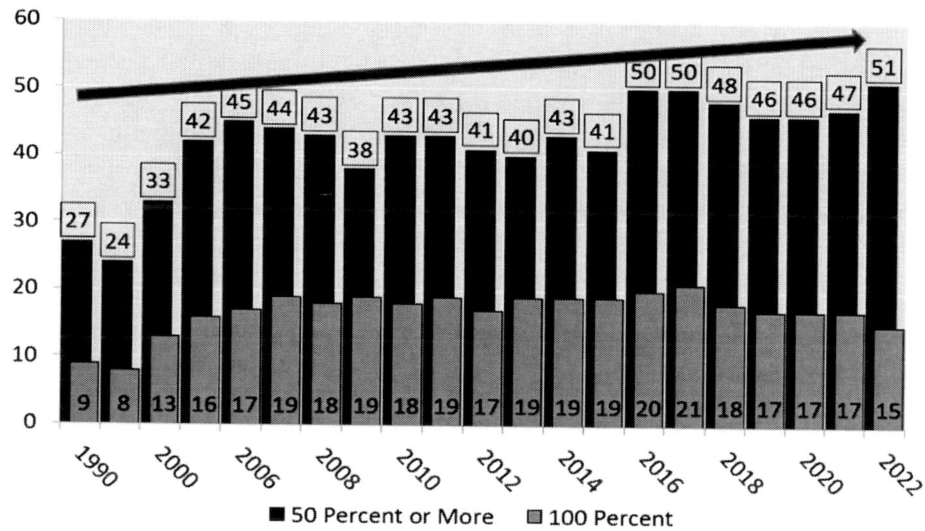

Source: USGS Mineral Commodity Summaries 1990-2023 editions.

While alarming, these findings are the latest in a 20-year trend of net imports that cost our country roughly $90 billion last year alone. Though the warning signs about our import reliance have been highlighted by a few key legislators for years, overall political concern about minerals supply chains has waxed and waned – with periods of frenzy following unexpected shortages, especially for military applications such as China's exercise of its dominance over the rare earths' minerals supply chain – followed by periods of complacency.[47]

Before the more recent exposure of supply chain vulnerabilities from the pandemic and geopolitical developments of the last few years, the most recent panic occurred in 2010, when China threatened global rare earth supplies. As the Congressional Research Service (CRS) explained:

> Chinese export quotas on a type of critical minerals referred to as rare earth elements (REEs) and China's curtailment of rare earth shipments to Japan over a maritime dispute in 2010 *represented a wakeup call* for the United States on China's near-monopoly control over global REE supply. The actions of the Chinese led to record high prices for REEs and, as a result, *began to shine a light on the potential supply risks and supply chain vulnerability for rare earths and other raw materials and metals needed for national defense, energy technologies, and the electronics industry,* among other end uses. U.S. legislators have introduced and deliberated on bills that would address the potential supply risk and vulnerability with respect to rare earth supply and bills that would promote domestic rare earth mine development.[48] (Emphasis added.)

---

[47] *See e.g.*, The Domestic Minerals Program Extension Act of 1953; the Mining and Mineral Policy of 1970; the Federal Land Policy and Management Act; the National Materials and Minerals Policy, Research and Development Act of 1980; and the 1984 National Critical Materials Act.

[48] Humphries, Marc. Congressional Research Service, "Critical Minerals and U.S. Public Policy." R45810, June 28, 2019, p. 5. https://www.everycrsreport.com/files/20190628_R45810_b3112ce909b130b5d525d2265a62ce8236464664.pdf.

Unfortunately, none of these past efforts or policies have reversed the U.S. overreliance on foreign sources of minerals despite widespread acknowledgement that this overreliance weakens our economy and endangers our national security. China's mineral dominance remains a major threat. Currently, China is the leading producer and/or supplier of 66 percent of mineral commodities listed as essential to U.S. economic and national security including lithium, rare earths and other battery metals.[49] According to USGS, production concentration has increased markedly over the past few decades for many mineral commodities with the most notable global shift has being the increasing production of mineral commodities in China.[50] As illustrated by the following USGS data, China's share of global mineral production and processing has grown markedly since 1990 for many mineral commodities, including aluminum, bismuth, refined cobalt, gallium, lead, magnesite, magnesium metal, mercury, REEs, silicon, steel (raw), titanium, vanadium and zinc.

China's Share of Global Primary Mineral Commodity Production over Time[51]

China's strong supply chain position stems, in large part, from state investment in processing and manufacturing, rather than an inherent advantage in reserves for most materials. China's "go global" strategy included $390 billion in outbound direct investments in the mining sector.[52] For example, as discussed in a recent White House report on supply chains:

- China is the primary global supplier of cobalt for batteries, despite having very limited reserves, through its aggressive investment in processing capacity coupled with foreign direct investment for ores and concentrates.
- China has a dominant position over the Democratic Republic of Congo cobalt reserves, which constitute half of the known global cobalt reserves.
- China has billions invested in nickel projects in Indonesia, home to one-quarter of overall global reserves.
- Mexican-based Sonora clay lithium deposit, operated by China-based Gangfeng Lithium, is currently under development, and would increase total lithium production

[49] Notably this reliance comes despite existing U.S. resources. In the 2022 Mineral Commodity Summaries, the USGS indicated the U.S. had an estimated 48 million metric tons (mt) of copper that can be mined and processed economically, 69 million mt of cobalt, 340 million mt of nickel and 750 million mt of lithium. Regardless, in 2021, the U.S. imported 48 percent of U.S. consumption of nickel, 76 percent of cobalt, 45percent of copper, and more than 25 percent of lithium.

[50] Nassar, N.T., Alonso, E., and Brainard, J.L., 2020, Investigation of U.S. Foreign Reliance on Critical Minerals—U.S. Geological Survey Technical Input Document in Response to Executive Order No. 13953 Signed September 30, 2020 (Ver. 1.1, December 7, 2020): U.S. Geological Survey Open-File Report 2020–1127, p. 4. https://pubs.usgs.gov/of/2020/1127/ofr20201127.pdf.

[51] For selected elements of the periodic table, the figure displays a time series of China's estimated share of global production for various associated mineral commodities for the years 1990–2018. In the periodic table, production refers to primary production or mine production. In the subfigure below the periodic table, multiple supply chain stages or forms are displayed for each mineral commodity. Elements not assessed are white. For a few mineral commodities (gallium, germanium, indium, selenium, silicon, strontium, and tellurium), data are not available for all years in the time series.

[52] Humphries, Marc. Congressional Research Service, "China's Mineral Industry and U.S. Access to Strategic and Critical Minerals: Issues for Congress," March 20, 2015. http://fas.org/sgp/crs/row/R43864.pdf).

by roughly half of today's production.[53]

- Chinese firms have also made multiple and large investments in mining operations around the world to ensure their supply of critical materials like cobalt, nickel and lithium.[54] Just last month, China based CATL, the world's largest EV battery manufacturer, beat out U.S. and Russian companies to develop the worlds largest lithium deposit in Bolivia.[55]

As a result of these tactics, China controls significant portions of the global mineral supply chain. The IEA reported in May 2021 that China was responsible for 60 percent of global rare earth elements production and nearly 90 percent of global processing for rare earth elements in 2019.[56] And this threat is not limited to rare earths. As noted in USGS criticality methodology, "of the 54 mineral commodities evaluated, China was the leading producer of at least one stage of the supply chain for 35 commodities."[57]

It did not used to be this way and it does not have to be our future. At every turn, our import dependence is both outsized and unnecessary. As explained in a recent opinion piece published in *The Hill*:

> In the 1980s, the U.S. was the mineral capital of the world. Since then, China has developed a juggernaut battery supply chain industry. The industry is centered around chemical processing of battery materials, backed by substantial government funding and coordination. These subsidies led to a wave of outsourcing by American companies across industries from semiconductors to steel. In addition, China has spent the last two decades investing in the mining industry abroad, including major investments and mineral rights in Australia, Africa, Asia and South America. This has led to an overreliance on China — and in turn vulnerable supply chains and a lost economic opportunity at home.[58]

---

[53] White House, "Building Resilient Supply Chains, Revitalizing American Manufacturing, and Fostering Broad-based Growth, 100-Day Reviews under Executive Order 14017," June 2021, p. 94.

[54] *See also,* USGS 2020 Investigation of U.S. Foreign Reliance on Critical Minerals (There are instances where the mineral deposit or mining and mineral processing operation of a commodity is partially or completely owned and (or) controlled by foreign companies with strong ties to their governments. For example, Chinese firms have purchased equity stake in lithium deposits and operations in Australia and Chile, niobium operations in Brazil, a rare earth deposit in Greenland, and cobalt operations in the D.R. Congo, Papua New Guinea, and Zambia (S&P Global Market Intelligence, 2020). Investigating China's investment in cobalt assets worldwide, Gulley and others (2019) found that when taking into account Chinese companies' ownership in foreign assets on an equity-share basis, China's share of global cobalt production increases from 2 to 14 percent for cobalt mine materials and from 11 to 33 percent for cobalt intermediate materials (figure 6). Furthermore, if the Chinese companies' equity shares of the production from these assets are assumed to be as secure as its domestic production, then these acquisitions have the effect of reducing China's NIR from 97 percent to an adjusted 68 percent, thereby reducing China's exposure to supply disruptions (Gulley and others, 2019).) p. 8.

[55] Reuters, "Bolivia taps Chinese battery giant CATL to help develop lithium riches," Jan. 20, 2023. https://www.reuters.com/technology/bolivia-taps-chinese-battery-giant-catl-help-develop-lithium-riches-2023-01-20/.

[56] International Energy Agency. "The Role of Critical Minerals in Clean Energy Transitions," 2021. https://iea.blob.core.windows.net/assets/24d5dfbb-a77a-4647-abcc-667867207f74/TheRoleofCritical MineralsinCleanEnergyTransitions.pdf.

[57] 2021 Methodology, p. 7.

[58] Ellen Hughes-Cromwick, Ph D. 2022. "How the US.. Can Secure a Resilient Electric Vehicle Battery Supply Chain." The Hill. June 8, 2022. https://thehill.com/opinion/energy-environment/3516265-how-the-us-can-secure-a-resilient-electric-vehicle-battery-supply-chain/.

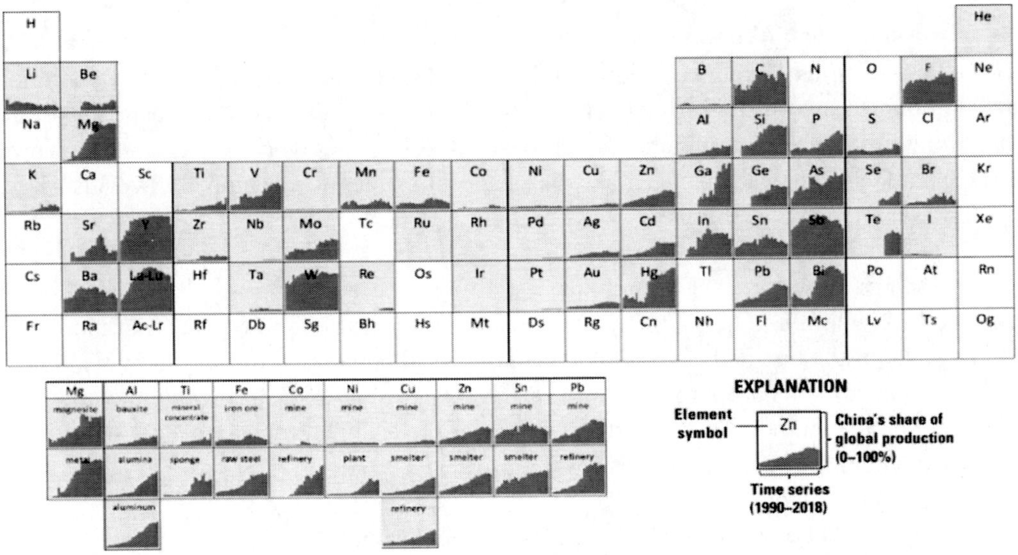

Our mineral import dependence will be our next Achille's heel. The U.S. must focus on supplying these metals at home as part of the solution "to diversify supply chains away from adversarial nations and sources with unacceptable environmental and labor standards."[59]

In order to support new domestic production, a robust domestic supply chain that includes minerals and metals sourced, refined, processed and smelted within our borders, we need to build on the important work done by this committee.

The following data from the mining program at the University of Missouri of Science and Technology is an important snapshot which allows us to better understand the domestic supply chain issues impacting production and refining and processing for simply one widely used metal.

United States– Copper Mining, Smelting and Refining

- 1995
  - 40 mines produce copper
    - 1.9 million tons production
  - 7 primary smelters
  - 4 secondary smelters
  - 7 electrolytic refineries
  - 15 electrowinning facilities
    - 2.3 million tons refined copper
- 2020
  - 25 mines produce copper
    - 1.2 million tons production
  - 3 primary smelters
  - 0 secondary smelters
  - 3 electrolytic refineries
  - 14 electrowinning facilities
    - 0.9 million tons refined copper

China – Copper Mining, Smelting and Refining

- 1995
  - 0.4 million tons mine production
  - 0.7 million tons refined copper
- 2020
  - 1.7 million tons mine production
  - 9.8 million tons refined copper

All information from USGS National Minerals Information Center's Mineral Commodity Summaries or Yearbook
All tons are metric tons

[59] "FACT SHEET: Biden-Harris Administration Announces Supply Chain Disruptions Task Force to Address Short-Term Supply Chain Discontinuities." 2021. The White House. June 8, 2021. https://www.whitehouse.gov/briefing-room/statements-releases/2021/06/08/fact-sheet-biden-harris-administration-announces-supply-chain-disruptions-task-force-to-address-short-term-supply-chain-discontinuities/.

This next graphic further illustrates the enormity of the supply chain issues that the U.S. and our allies continue to face if we do not take the necessary steps to support regulatory policies that encourage private investment rather than attempting to control it.

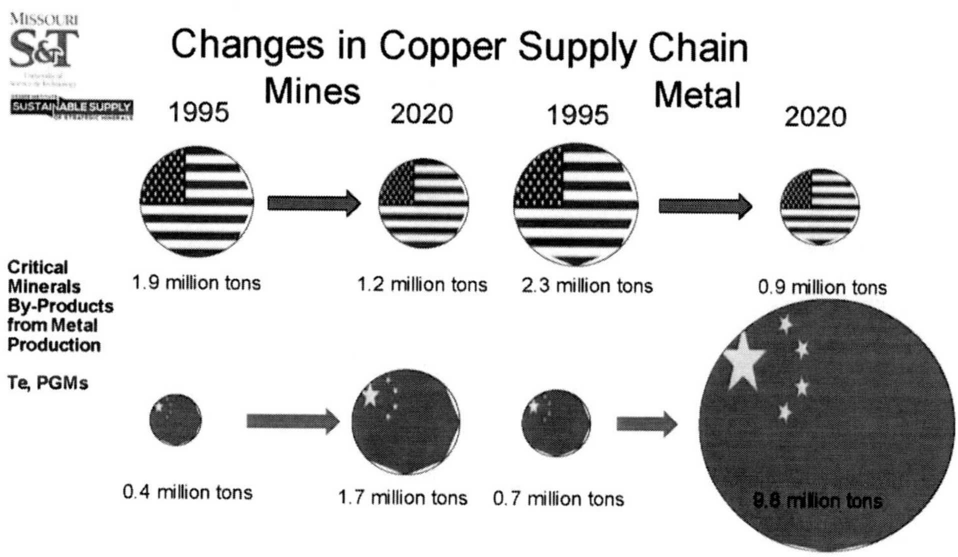

What Are the Solutions?

Chairwoman Cathy McMorris Rodgers and Natural Resources Committee Chairman Bruce Westerman's *Securing American Mineral Supply Chains Act* supports a robust domestic mineral supply chain that prioritizes responsible resource development through policies that provide certainty to all mining operations and manufacturers; sets lead agencies and improves the timeliness of the permitting process with timelines; maintains access to mineralized federal lands unless specifically withdrawn by Congress and unless the U.S. Geological Survey can assure that a withdrawal does not threaten supply chains; supports research, development and demonstration funding at the USGS and Department of Energy; new workforce development and training opportunities; and unlocks innovation by not supporting prescriptive mineral policies.

These policy recommendations are commonsense changes that would provide regulatory certainty to investors that the U.S. seeks to once again compete on a global scale in the mineral supply chain. Instead of only seeking to secure mineral supplies from foreign sources or exporting domestically extracted materials for further refinement, processing and smelting, these improvements in the permitting process would signal that the U.S. intends to secure the entirety of its supply chain, lessening vulnerabilities from outside sources, including geopolitical impacts.

Finally, other legislation on today's legislative agenda speaks to the need for domestic uranium production. NMA strongly supports the funding Congress has appropriated to develop a uranium reserve. However, more needs to be done to support and fund a uranium reserve at the Department of Energy. Although recently removed from the critical mineral list, uranium is no less critical to the U.S.

Other opportunities within the committee's jurisdiction to support a domestic mineral supply chain is to pass bipartisan Good Samaritan legislation. Last Congress, the NMA worked with groups like Trout Unlimited to develop legislation introduced in the U.S. Senate by nearly twenty bipartisan senators and supported by state departments of environmental quality, Tribes, and conservation organizations. This is the first bipartisan introduction of Good Samaritan legislation in twelve years. We are working toward an introduction of this legislation again this Congress in the House and Senate. This would not only allow the processing of tailings and other mine waste from historic mining operations long since abandoned and with no current liable owner, but also provide the added benefit of cleanup and remediation activities that if left alone, will continue to degrade the water quality conditions and pose ongoing safety risks to surrounding habitat and neighboring communities.

Conclusion

The U.S. is at a mining crossroads. Mineral demand is soaring, but our policies are lagging. We must encourage more domestic mining and processing to meet future demand and ensure that the materials required for everything from infrastructure to electrification are readily available. The NMA appreciates the prioritization of these issues by the House Energy and Commerce Committee and is eager to help craft important policy solutions for the future.

* * *

Mr. DUNCAN. Thank you for that. The Chair will now recognize Mr. Slocum for 5 minutes.

Statement of Tyson Slocum

Mr. SLOCUM. Thank you, Mr. Chairman, Ranking Member, members of the committee. It is my pleasure to be here today. I am Tyson Slocum, and for the last two decades I have been the director of the energy program with Public Citizen here in Washington, DC.

So liquefied natural gas exports are the disruptive event that is radically upending domestic energy markets. For the first time in history, American natural gas consumers—whether they be households, operators of power plants—are forced to compete with their counterparts in Berlin and Beijing on price. That is why, coast to coast, Americans are now paying significantly higher prices for electricity and to heat their homes.

There have been periods in the last 2 months where prices on the U.S. West Coast and in New England have been more expensive for natural gas than in Ukraine. That is because we are now—because, prior to LNG exports, our domestic markets were insulated from global calamities. War could break out in Europe, and there would be no bump in price. Now our domestic benchmarks are directly linked with global events like the same that has been the way with oil and gasoline markets for more than a generation.

So this is not a crisis of inadequate natural gas production. We are breaking records on natural gas production in the U.S. every month, according to the Energy Information

Administration. Twenty percent of U.S. natural gas production is exported out of the United States, and that is what is driving the imbalance.

But on the production side, we are far and away the largest producer on the planet. The number 2 and number 3 global producers of gas, Russia and Iran, combined don't produce as much as the United States every day. So we are not going to produce our way to lower prices as long as the export spigot remains open.

So, of course, a focus of policy on energy efficiency and promoting zero emission alternatives to gas has to be part of the solution. But in the meantime, we need stronger regulation over LNG exports, not less. And so I am just going to briefly cover 4 of the 17 bills. So H.R. 647 would eliminate the requirement that natural gas exports be consistent with the public interest. That is a standard that has been in place for 85 years, and we believe that it is crucial that all natural gas exports continue to be subject to that public interest standard.

The Cross-border Energy Infrastructure Act would require FERC to approve any natural gas export pipeline within 30 days of receiving the application. So that is a de facto approval. We are a party in a case at FERC right now involving a natural—a proposed natural gas cross-border pipeline, a 155-mile pipeline that would connect the Permian Basin at the Waha Hub to Mexico, and then directly send that U.S.-produced gas through Mexico to new LNG export terminals on Mexico's Pacific coast.

Thankfully, there is a public interest review, and we are an intervener in that FERC proceeding, where we are going to raise concerns about the threat to the public interest of exporting Permian gas directly to China. And removing that review would not be advisable.

H.R. 484 would eliminate the methane fee in the recently enacted Inflation Reduction Act. What is wild about this is the Inflation Reduction Act bends over backwards to accommodate the oil and gas sector. Congress appropriates a billion and a half dollars in grants to the oil and gas sector to help them invest in facilities to reduce their methane emissions. I don't see the oil and gas industry complaining about a billion and a half dollars in taxpayer grants to help their business. And in addition, the fee is waived if you successfully comply with EPA regulations on methane emissions. So that makes the industry a partner with effective regulation. And so I don't think that that legislation is advisable.

And last, I just want to touch on the bill that would prevent the EPA from requiring oil refineries using hydrofluoric acid alkylation to explore less hazardous alternatives. It is important to note that, in October of 2022, the U.S. Chemical Safety and Hazard Investigation Board, in response to a series of tragic accidents with a hydrofluoric acid alkylation, recommended and urges the EPA to subject these facilities to hazard reviews. So I think it would be imprudent to remove that opportunity.

Thank you so much. I look forward to your questions.

[The prepared statement of Mr. Slocum follows:]

Public Interest Review for LNG Exports Are Essential to Protect Americans, Testimony before the Joint Subcommittee on Environment, Manufacturing, and Critical Materials & Subcommittee on Energy, Climate, and Grid Security of the U.S. House Energy and Commerce Committee

February 7, 2023
Tyson Slocum, Energy Program Director
Public Citizen, Inc.
215 Pennsylvania Avenue SE Washington, D.C. 20003
(202) 588-1000
tslocum@citizen.org Twitter @TysonSlocum

Introduction

I am Tyson Slocum, and I direct the Energy Program at Public Citizen. We are a national consumer advocacy organization with over 500,000 members and supporters nationwide. I serve on two advisory committees to the U.S. Commodity Futures Trading Commission (Energy and Environmental Markets; and Market Risk); am a member of the faculty at the University of Maryland; for two decades have routinely intervened on behalf of households at the Federal Energy Regulatory Commission; and have provided testimony for Congress for years about the detrimental impacts natural gas exports have on raising prices for domestic consumers.[60] My organization's financial details are on our website.[61]

The purpose of today's legislative hearing is to consider several proposed bills, including two designed to eliminate or reduce regulatory oversight of fossil fuel exports:

- H.R. 647: To repeal restrictions on the export and import of natural gas.[62] The legislation eliminates the requirement that exports and imports be "consistent with the public interest"—a standard that has been in place to protect consumers for 85 years. This legislation would remove all routine regulatory review to ensure that exports are not increasing prices for American families, and would allow unregulated exports to China. I do not support this legislation.
- The Promoting Cross-border Energy Infrastructure Act.[63] The legislation is based upon H.R.575 from the 117th Congress, and S.23 in the 118th Congress. It would require FERC to approve any natural gas pipeline designed to import or export natural gas to or from Canada and Mexico within 30 days of receiving the complete application. This automatic approval eviscerates the Commission's current public interest determination, and will encourage the construction of cross-border pipelines to Mexico designed to re-export U.S.-produced natural gas from LNG terminals in Mexico. The

---

60 Including 2018 testimony before the U.S. Senate on problems of LNG exports, www.energy.senate.gov/services/files/E86FB560-F2B6-4D3D-B016-F92526D10CD7.

61 www.citizen.org/about/annual-report/.

62 www.congress.gov/bill/118th-congress/house-bill/647, text: https://d1dth6e84htgma.cloudfront.net/06_HR_46aa939aae.pdf.

63 https://d1dth6e84htgma.cloudfront.net/03_HR_66ea55d203.pdf.

legislation would also remove regulatory review of any modifications to existing cross-border oil and natural gas pipelines. I do not endorse this legislation.

And legislation repealing a fee on methane emissions that exceed standards established by the U.S. Environmental Protection Agency:

- H.R.484, The Natural Gas Tax Repeal Act.[64]

My testimony will address these proposed bills, as well as issues related to the impact natural gas exports have on increasing energy costs for American families; the threat of re-exports of U.S. produced gas from Mexican LNG terminals; the need to Congress to address FERC's recent loophole of oversight of smaller-scale LNG export terminals; the importance of FERC improving oversight of natural gas pricing markets; the benefits of the new fee on methane emissions; that no additional Congressional action is needed to encourage electric transmission siting; and legislative proposals to prevent effective EPA regulation of hydrofluoric acid alkylation is premature.

Record Natural Gas Exports Increase Energy Bills for American Families

In 2023, the United States is the world's largest natural gas and petroleum producer and exporter on the planet. Sixty percent of our domestically-produced petroleum is now exported,[65] and 20% of our natural gas is now allocated for export.[66] These numbers will only increase as domestic demand continues to flatten and export infrastructure capacity continues to expand. While oil markets—and domestic gasoline prices—have long been directly influenced by global calamities, natural gas had been insulated from Our record natural gas exports have radically upended domestic energy markets, forcing American families to compete with Berlin and Beijing for U.S. produced energy. Natural gas exports are directly responsible for Americans paying higher prices to heat and cool their homes. Current statutes and regulations clearly present little challenge to domestic gas production, and offer minimal impediments to their export to foreign countries. Congress should be strengthening, not weakening, public interest protections for gas exports.

The United States is far and away the largest natural gas producer in the world: we alone account for 25% of the entire world's production every day, outproducing the next two biggest (Russia and Iran) *combined*,[67] with U.S. natural gas production reaching an all-time high in 2022.[68] At the same time, natural gas exports have exploded. Exports via pipeline to Mexico and Canada, combined with Liquified Natural Gas (LNG) exports by ship today account for 20% of domestic gas production—up from 6% in 2015. And in 2023 the United States will claim the title as biggest LNG exporter in the world.[69]

---

[64] www.congress.gov/bill/118th-congress/house-bill/484/text.
[65] https://twitter.com/TysonSlocum/status/1617998886660112384.
[66] www.citizen.org/article/letter-to-dept-of-energy-to-protect-consumers-from-lng-exports/.
[67] www.eia.gov/international/data/world/natural-gas/dry-natural-gas-production.
[68] www.eia.gov/naturalgas/weekly/archivenew_ngwu/2023/01_12/.
[69] Stephen Stapczynski, "US Surges to Top of LNG Exporter Ranks on Breakneck Growth," January 2, 2023, Bloomberg, www.bloomberg.com/news/articles/2023-01-03/us-surges-to-top-of-lng-exporter-ranks-on-breakneck-growth.

These record exports have come with a tragic cost: American households, power producers and other consumers are now forced to directly compete with their counterparts in Berlin and Beijing, which has globalized domestic benchmark prices, exposing Americans to higher prices and increased volatility.

SEPTEMBER 6, 2022

U.S. LNG export capacity to grow as three additional projects begin construction

U.S. liquefied natural gas export projects: existing and under construction (2016–2025)
billion cubic feet per day

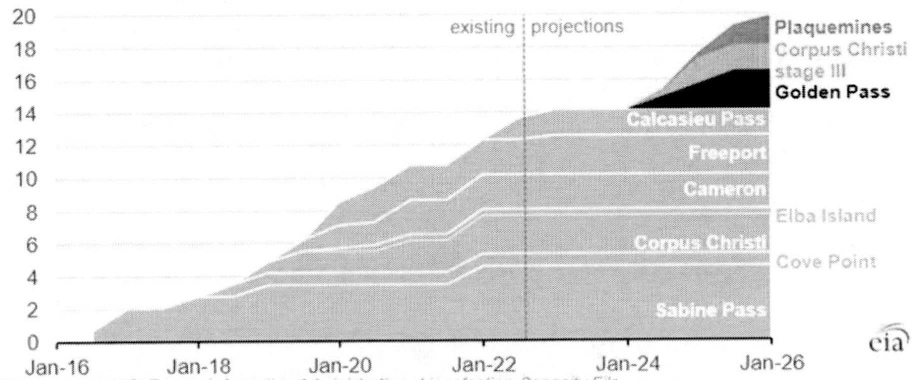

Data source: U.S. Energy Information Administration, *Liquefaction Capacity File*
Note: EIA estimates are based on information from Federal Energy Regulatory Commission and U.S. Department of Energy filings, company websites, trade press, and other industry sources.

The Federal Energy Regulatory Commission's 2022-23 *Winter Energy Market and Reliability Assessment* concludes that "continued growth in net exports, including from liquified natural gas (LNG) export facilities, will place additional pressure on natural gas prices this winter . . . Traditionally, domestic fundamentals drive U.S. natural gas prices; this winter, international markets will likely also affect U.S. natural gas markets and prices . . . the expansion of LNG export capability has integrated formerly disparate North American regional natural gas markets into the global market . . . In New England, high global LNG prices are contributing to higher winter natural gas futures prices."[70]

USA Today reports that record LNG exports are directly contributing to punishing high energy bills for American families.[71]

LNG is natural gas that has been cooled to –260° F, changing it from its natural vapor state into a liquid that is 1/600th of its original volume, making it more economical to transport via ship.

The U.S. Energy Information Administration notes that "2022 average wholesale U.S. natural gas spot price at the Henry Hub was the highest in real and nominal terms since

[70] https://ferc.gov/media/report-2022-2023-winter-assessment.
[71] Medora Lee, "Electricity bills may continue to shock you even as overall inflation eases," January 24, 2023, www.usatoday.com/story/money/personalfinance/2023/01/24/electricity-prices-inflation/11089430002/.

2008"—which was the era just prior to the fracking boom,[72] and reports that the "U.S. residential price of electricity will average 14.8 cents per kilowatthour in 2022, up 7.5% from 2021. Higher retail electricity prices largely reflect an increase in wholesale power prices driven by rising natural gas prices."[73]

The National Energy Assistance Directors' Association estimates that household heating costs will be 34.3% higher for families using natural gas and 6.9% higher for those relying on electricity this winter.[74]

> There are several legislative proposals for today's hearing that would suspend various environmental compliance for critical mineral mining and production. In general, such approaches should — at a minimum — be subject to notice and opportunity for public intervention and comment, so that communities and experts have a chance to challenge environmental exemptions.

The *Wall Street Journal* reports "that natural-gas exports are pushing domestic prices higher . . . The pinch shows a growing tension between exporters and buyers who have enjoyed cheap gas for more than a decade. Some manufacturing and chemical companies have built entire businesses around low U.S. gas prices. . . Utilities from the Pacific Northwest to New England have filed regulatory requests to raise rates for natural gas this winter, citing a supply squeeze as a result of higher global demand . . . the U.S. is exporting a larger share of its natural gas than it ever has and shale producers aren't quickly ramping up in response to high prices . . . some of the biggest natural-gas producers have vowed to keep investments in production growth low."[75] Therefore so-called capital discipline is keeping a check on domestic production not rising on pace with exports in order to ensure domestic producers will enjoy higher prices.

These high prices are creating significant economic hardship for tens of millions of American families. Twenty-six percent of respondents to a U.S. Census Bureau survey taken in the summer of 2022 said they had forgone necessities like food or medicine to pay their energy bills sometime during the preceding year.[76] Rising energy costs—anchored by higher natural gas prices stemming in part from record LNG exports—are the biggest factor driving inflation in the U.S.[77]

While sanctions in response to the 2022 Russian Federation invasion of Ukraine constrained Russian supply, raised European gas benchmarks above other global benchmarks, and led to a reorientation of U.S. LNG exports from Asia to Europe, 2023 will likely see diminished U.S. LNG exports to Europe and a return to increased flows to Asia. Europe has been reluctant to sign long-term LNG contracts, thereby disincentivizing commitments to the continent.[78] And at the end of the day, U.S. LNG exports will chase whatever country is willing to pay the highest price, which typically will be Asia:

---

72 www.eia.gov/naturalgas/weekly/archivenew_ngwu/2023/01_12/.
73 www.eia.gov/outlooks/steo/.
74 https://neada.org/wp-content/uploads/2022/09/winter2022-23PR.pdf.
75 Collin Eaton and Katherine Blunt, "Natural-Gas Exports Lift Prices for U.S. Utilities Ahead of Winter," November 7, 2021, www.wsj.com/articles/natural-gas-exports-lift-prices-for-u-s-utilities-ahead-of-winter-11636281000.
76 www.census.gov/data/tables/2022/demo/hhp/hhp48.html.
77 www.bls.gov/cpi/.
78 Justin Jacobs, "US companies say EU climate goals are deterring new gas deals," February 6, 2023, www.ft.com/content/5f13fee7-dc04-4f47-bd06-cfbb3c444011.

> Russia's invasion of Ukraine last February upended long-standing expectations about natural gas supplies to Europe and resulted in elevated global gas prices as countries bid for LNG to fill the void. But U.S. suppliers can only produce so much LNG, and how much of it ends up in Europe versus Asia or other gas-consuming regions in 2023 and beyond will depend largely on market forces — in other words, who needs the LNG more and is willing to pay up for it.[79]

How U.S. Department of Energy Currently Authorizes Natural Gas Exports

DOE is responsible for authorizing exports of U.S. produced natural gas, including LNG, to foreign nations pursuant to section 3 of the Natural Gas Act.[80] 1992 amendments to the Natural Gas Act deemed exports to countries with which the U.S. has a free trade agreement requiring national treatment for trade in natural gas are automatically deemed to be in the public interest. The U.S. has such free trade agreements with 18 countries, only two of which (South Korea and Singapore) are in Asia, with none in Europe.[81] From 2016 through November 2022, only 23.5% of all LNG exports are to nations with which we have a free trade agreement.[82]

The bulk of LNG exports (76.5%) are to nations with which we do not have free trade agreements, and therefore require DOE to only authorize them "it finds that the proposed exportation or importation will not be consistent with the public interest."

Seven currently operating LNG terminals (Sabine Pass, Cove Point, Elba Island, Corpus Christi I and II, Cameron, Freeport and Calcasieu Pass) have received authorizations from DOE to export to non-free trade agreement countries, and will have combined export capacity of 14 billion cubic feet per day before the end of 2022. Three additional terminals authorized to export and under construction (Plaquemines, Corpus Christi III and Golden Pass will boost export capacity to nearly 20 million Bcf/d by 2025.[83]

As a result, the U.S. catapulted from zero LNG exports prior to 2016 to the largest LNG exporter in the world today. Nearly 20% of natural gas produced in the U.S. was exported in 2022, up from 11.5% in 2017.

Courts have long interpreted the intent of the Natural Gas Act public interest determination "was to protect consumers against exploitation at the hands of natural gas companies."[84]

On June 8, 2022 the Freeport LNG export terminal in Texas experienced a massive explosion that has kept the facility offline for 8 months. PHMSA concluded that human error and equipment malfunctions caused the blast.

Congress left it to the executive branch to define what factors would determine exports to be consistent with the public interest. DOE assesses several variables, including net economic

[79] Richard Pratt, "Portfolio Players Take On Critical Roles In Rapidly Commoditizing Global LNG Market," February 1, 2023, https://rbnenergy.com/two-of-us-portfolio-players-take-on-critical-roles-in-rapidly-commoditizing-global-lng-market.

[80] 15 USC § 717b.

[81] The other 16 nations are Australia, Bahrain, Canada, Chile, Colombia, Dominican Republic, El Salvador, Guatemala, Honduras, Jordan, Mexico, Morocco, Nicaragua, Oman, Panama and Peru.

[82] www.energy.gov/fecm/articles/lng-monthly-2022.

[83] www.eia.gov/todayinenergy/detail.php?id=53719.

[84] *FPC v. Hope Nat. Gas Co.*, 320 U.S. 591 (1944).

impacts, international impacts, the security of domestic natural gas supply, and environmental impacts.[85]

DOE's 2018 *Policy Statement Regarding Long-Term Authorizations To Export Natural Gas to Non-Free Trade Agreement* Countries makes clear that Section 16 of the Natural Gas Act gives it authority to "amend, and rescind such [export] orders . . . as it may find necessary or appropriate . . ." to satisfy its statutory responsibilities.[86]

Over the years, DOE has commissioned macroeconomic studies to determine whether LNG exports provide net economic benefits, in order to be consistent with the public interest. These studies attempt to estimate the impact exports have on domestic energy prices, and the economic contributions that LNG exports have for employment and other contributions to gross domestic product.

The most recent of these reports was conducted in 2018 during the Trump Administration, when LNG exports were still in relative infancy. *Macroeconomic Outcomes of Market Determined Levels of U.S. LNG Exports* was prepared by NERA Economic Consulting for DOE.[87] This study has aged poorly, as it assumed that consumer welfare—which it defines as the present value measure of the standard of living of all U.S. households—was directly and beneficially linked with higher LNG exports.[88] The 2018 study gave only a 3% probability that significant LNG exports would result in domestic prices above $10/MMBtu, concluding that "increasing U.S. LNG exports under any given set of assumptions about U.S. natural gas resources and their production leads to only small increases in U.S. natural gas prices."[89] Furthermore, the study claims that "as U.S. LNG exports increase . . . households who hold shares in companies that own liquefaction plants receive additional income from take-or-pay tolling charges for LNG exports. These additional sources of income for U.S. consumers outweigh the income loss associated with higher energy prices."[90] DOE relies upon the conclusions of this discredited 2018 study to help determine whether exports will be consistent with the public interest.

DOE currently performs no distributional analysis to measure the impact that LNG exports may have on families at different incomes, and provides no assessment of the impact exports have on energy burdens of communities of color. Utility bill burdens are regressive, meaning lower-income families pay larger proportions of their income on such necessities compared to their more affluent neighbors. With natural gas representing the largest share of fuel (37%) for electric power generation in the U.S., combined with many families' reliance on natural gas for home heating, the export-driven energy spikes are resulting in profound energy insecurity for millions of Americans.

A distributional incidence analysis that measures the impact higher natural gas prices have on households at different income quintiles is necessary to demonstrate whether LNG exports are consistent with the public interest. DOE's failure to measure the price impacts for vulnerable populations renders its current methodological approaches inadequate to capture the adverse pricing dynamics impacting millions of households.

---

[85] www.govinfo.gov/content/pkg/FR-2018-06-21/pdf/2018-13427.pdf.
[86] www.govinfo.gov/content/pkg/FR-2018-06-21/pdf/2018-13427.pdf.
[87] www.energy.gov/sites/prod/files/2018/06/f52/Macroeconomic%20LNG%20Export%20Study%202018.pdf.
[88] At page 20.
[89] At page 55.
[90] At page 67.

Providing price impacts by population quintile is one necessary reform; the other must be a geographic assessment of these price impacts. Because different regions of the country have unique energy profiles—including the types of home heating fuels, and the proportion of gas used in regional power generation—geographic modeling of the price impacts of LNG exports must also be determined.

A central component of both of the approaches are quantifying the impact higher prices have on communities of color. The Biden Administration's energy justice initiatives must translate to assessing the impact LNG exports have on communities of color.

Congress Intended the 1992 Natural Gas Act Amendments to Promote a North American Gas Market for the Benefit of American Consumers—Not to Promote Unfettered Exports

The legislative history of the Energy Policy Act of 1992 demonstrates that the Natural Gas Act amendments do not endorse re-exports of U.S. produced gas from Mexico as qualifying for the automatic public interest designation.

The Natural Gas Act language designating exports to countries with free trade agreements deemed to be in the public interest were added as Section 201 of the Energy Policy Act of 1992.[91] At the time of passage, the United States only had a free trade agreement with Canada that included natural gas treatment, and there were active negotiations with Mexico on the North American Free Trade Agreement. The congressional record makes clear that the purpose of Section 201 was to promote a North American natural gas market that would benefit consumers—and not tolerate the use of a free trade agreement public interest determination to freely re-export to nations with whom no free trade agreement exists.

The Report of the Committee of Energy and Commerce (Rept. 102-474, Part 1) noted that Section 201 was intended to establish fewer restrictions on natural gas imports from Canada and Mexico, ensuring that such imports would be treated "more like domestic American natural gas production" by designating them as "first sale" status; barred FERC "or state regulators from treating these imports differently than domestic gas"; making "the current import approval process purely automatic, so that this procedure—which domestic gas does not undergo—cannot cause any delays"; and "ease regulation of Mexican gas imports if a free trade agreement with Mexico is reached."[92]

U.S. Rep. Phillip Sharp (D-Indiana) further elaborated congressional intent when he spoke on the floor of Congress in support of the conference report on the Energy Policy Act of 1992:

> the conferees did agree to expressly forbid discrimination against imported natural gas . . . [and ensures] a broad policy of free and competitive wellhead markets in North America by, in effect, deregulating Canadian natural gas imports in section 201 . . . As for section 201, we note it applies, for example, to imports of Canadian natural gas into the United States; exports of natural gas to Canada from the United States; and imports of liquified

---

[91] www.congress.gov/102/statute/STATUTE-106/STATUTE-106-Pg2776.pdf.

[92] Legislative history of the Energy Policy Act of 1992, prepared for the Committee on Energy and Natural Resources, United States Senate; by the Congressional Research Service, Library of Congress, November 1994, Volume 4 of 6, at pages 2731-2732. https://babel.hathitrust.org/cgi/pt?id=pst.000023406209.

> natural gas into the United States . . . Finally, as drafted, the new fast track process would not be available for LNG exports to, for example, Pacific rim nations other than Canada.[93]

And U.S. Representative Barbara B. Kennelly (D-Connecticut) spoke on the House floor on remarks May 20, 1992 that "section 201 of this bill eases existing rules for importing natural gas thereby protecting this region's [New England's] access to affordable, clean burning natural gas."[94]

The congressional record elaborated that Section 201 "is intended to increase the free flow of natural gas *throughout the North American market*" [emphasis added].[95]

U.S. Rep. Norman F. Lent (R-NY) noted the importance of Section 201 to protect his state's consumers:

> The Energy Policy Act of 1992 contains important provisions that remove regulatory barriers which hinder the importation of natural gas from countries with which the United States has entered into a free trade agreement requiring national treatment for trade in natural gas. Currently, this means Canadian gas must be treated the same as domestic gas. Once the North American Free Trade Agreement is ratified, this will also apply to Mexican gas. Section 201 of this act is vital to assuring that U.S. regulators do not interfere with the importation of natural gas to customers in the United States. Its provisions provide critical protection to the citizens of my home state, New York, who receive supplemental volumes of natural gas from Canada. The purpose of these provisions is not to give imported natural gas an advantage, but to ensure a level playing field for imported gas . . . Section 201(b) deems the importation to the United States, and exportation from the United States, of natural gas consistent with the public interest. By making this determination, applications for import of Canadian natural gas are granted automatic approval. The result is, imported natural gas is not subjected to burdensome import licensing proceedings that place it at a disadvantage relative to domestically produced gas . . . these provisions are good competitive policy. U.S. producers supply over 92 percent of the natural gas needs in this country. Fair treatment of imports helps maintain healthy competition in the United States without posing any threat to U.S. producers. Greater access to a variety of natural gas sources will help create a more stable natural gas market so that more U.S. consumers will benefit from this economic and environmentally sound source of energy.[96]

---

[93] Legislative history of the Energy Policy Act of 1992, prepared for the Committee on Energy and Natural Resources, United States Senate; by the Congressional Research Service, Library of Congress, November 1994, Volume 6 of 6, pages 4555, 4557 and 4560. https://babel.hathitrust.org/cgi/pt?id=pst.000023406032.

[94] Legislative history of the Energy Policy Act of 1992, prepared for the Committee on Energy and Natural Resources, United States Senate; by the Congressional Research Service, Library of Congress, November 1994, Volume 5 of 6, page 3868. https://babel.hathitrust.org/cgi/pt?id=pst.000023406063.

[95] Legislative history of the Energy Policy Act of 1992, prepared for the Committee on Energy and Natural Resources, United States Senate; by the Congressional Research Service, Library of Congress, November 1994, Volume 5 of 6, page 3729, https://babel.hathitrust.org/cgi/pt?id=pst.000023406063.

[96] at page 4578-4579.

Mexico Will Rely on U.S. Produced Gas to Emerging as a Major LNG Exporter to Asia—Demonstrating the Need for Public Interest Evaluations to Protect American Consumers

One of the proposed bills subject to today's legislative hearing—the *Promoting Cross-border Energy Infrastructure Act*—would require FERC to approve the siting and construction of a cross-border natural gas pipeline in 30 days, thereby repealing the current public interest standard of review. This would detrimentally impact American consumers, because the legislation would allow for unregulated U.S. gas exports to Asia. There are multiple LNG export terminal projects being built on Mexico's pacific coast to serve Asian markets. Exports from Mexico's pacific coast avoid the expense and time of scheduling travel through the Panama Canal faced by LNG export terminals located on the U.S. gulf coast. Because Mexico does not currently produce enough natural gas for its domestic needs—let alone to export—U.S. produced gas would supply these proposed LNG export terminals on Mexico's pacific coast.

Bloomberg notes that the LNG export terminals planned for Mexico's pacific coast will rely almost exclusively on U.S. produced natural gas for those exports:

> Mexico—which imports nearly all of the natural gas it burns—has laid out a somewhat surprising mission: to become one of the world's top exporters of the fuel, and fast. Although natural gas exports from Mexico are today non-existent, seeing as it produces too little of the power-plant fuel to supply even its own domestic needs, the country's physical proximity to booming US reserves positions it well to supply American gas to hungry buyers in Europe and Asia. With US shale in mind, a total of eight liquified natural gas export projects have been proposed south of the border boasting annual combined capacity of 50.2 million tons. Some of the operations aim to come online as soon as next year. If they're all completed, the Latin American newcomer would join a very small club of nations that ship abroad the superchilled fuel—commonly called LNG — clocking in at No. 4 behind only the US, Australia and Qatar. And unlike those other three export heavyweights, Mexico would mostly be shipping out gas that it imported in the first place.[97]

Companies developing LNG export terminals "have cited strong offtaker interest in Asia for their projects under development on Mexico's West Coast."[98] Sempra's Costa Azul and Vista Pacifico LNG export terminals; Saguaro LNG[99] and Amigo LNG[100] will all rely on Permian basin gas and are possible export conduits for Oneok's proposed border crossing.

This isn't hypothetical. On December 20, 2022, Oneok, Inc. submitted an application with FERC to construct a natural gas border crossing pipeline with Mexico. Oneok's application specifically mentions the free trade agreement with Mexico, and therefore requests that FERC automatically find it to be in the public interest. The border crossing would directly connect with the Waha hub in the Texas Permian Basin via Oneok's proposed 155 mile pipeline. Oneok

---

[97] Sergio Chapa, "Mexico Plans to Become an Export Hub With US-Drilled Natural Gas," August 12, 2022, www.bloomberg.com/news/articles/2022-08-12/mexico-plans-to-become-lng-export-hub-using-american-drilled-natural-gas.

[98] Andrew Baker, "U.S. Firms Advancing Mexico LNG Export Plans as Asian Buyers Line Up Supply," *Natural Gas Intelligence*, December 28, 2022, www.naturalgasintel.com/u-s-firms-advancing-mexico-lng-export-plans-as-asian-buyers-line-up-supply/.

[99] https://mexicopacific.com/.

[100] www.lngalliance.com.

states in the application that the gas would travel from its border crossing facility on a new pipeline to directly supply LNG export terminals on Mexico's pacific coast.[101] So Oneok's border crossing will bridge a seamless and interconnected export infrastructure that begins in Texas' Permian basin and extends to LNG export terminals on Mexico's pacific coast. Under Oneok's proposal, Mexico will only serve as a land mass conduit to export U.S. produced gas to foreign nations with whom we do not necessarily have free trade agreements, like China.

FERC's LNG Export Loophole

On March 25, 2022, FERC approved a petition by Nopetro LNG to exempt the planned export terminal in the Florida panhandle from the Commission's oversight.[102] The Commission justified the exemption because the facility's three liquification trains would be located 1,300 feet from the export dock, where trucks (rather than a pipeline) would traverse the quarter-mile with LNG-filled ISO shipping containers. FERC concluded that this 1300 foot gap effectively severs its oversight of the planned facility. We have sued FERC in the DC Circuit to challenge the Commission's order.[103] If the erroneous order is allowed to stand, we predict a wave of similar facilities designed explicitly to exploit the Commission's new design loophole. When large-scale LNG terminals cost billions of dollars and take years to build, FERC's Nopetro loophole will incentivize smaller scale facilities that cost $50-$80 million and can be built in 12 months absent FERC oversight, posing risks to communities.[104]

Congress Should Ensure FERC Protects Consumers from Natural Gas Index Price Gouging

A natural gas index price is derived from trades within specific geographical boundaries that market participants voluntarily report to a price index developer. Price index developers are private, for-profit companies that classify most of the voluntarily-reported data as proprietary, that the index developers then commodify and sell only to those that can afford the very expensive subscription fees.

These voluntarily-reported transactions determine the price of natural gas for millions of households and businesses across the country, as market participants reference index prices in their physical and financial transactions: natural gas pipelines and Regional Transmission Organizations feature natural gas indices in their FERC-jurisdictional tariffs for various terms and conditions of service; state utility commissions rely on natural gas indices as benchmarks when setting rates; and many natural gas financial derivative contracts used in hedging and speculation settle against the natural gas price indices.[105] In a way, hundreds of billions of dollars of energy transactions rely upon voluntarily-reported price indexes—a 21st century version of a smoke-filled, price-fixing establishment.

---

101 www.citizen.org/article/oneok-saguaro-natural-gas-export/.
102 www.citizen.org/article/rehearing-request-of-nopetro-lng-export-order/.
103 U.S. Court of Appeals DC Circuit docket # 22-1251, www.citizen.org/litigation/public-citizen-v-ferc-2/.
104 www.citizen.org/article/tyson-slocum-port-st-joe-nopetro-lng/.
105 NOPR, at 4.

Federal law requires the Commission to ensure that spot natural gas price indices feature adequate price discovery and market transparency. Spot natural gas price indices are structurally non-competitive and the voluntary nature of reporting trades renders them susceptible to market manipulation. The rest of the world has been replacing voluntary price indices for benchmarks with far larger economic impacts that U.S. natural gas spot prices (such as replacing the London Interbank Offered Rate (LIBOR) with the Secured Overnight Financing Rate (SOFR)). The Commission should therefore establish an electronic information system, as authorized by 15 USC § 717t–2(a)(4), which states that "the Commission shall consider the degree of price transparency provided by existing price publishers and providers of trade processing services . . . *The Commission may establish an electronic information system if it determines that existing price publications are not adequately providing price discovery or market transparency*" [emphasis added]. Such "an electronic information system" could be based on actual transactions, and not limited to those voluntarily reported, and would be freely available to all interested parties through a platform hosted by the Commission, rather than the proprietary, commodified data model of the index publishers.

Indeed, FERC conceived of the idea of having authority to create its own electronic natural gas price reporting system. In testimony before the House Committee on Energy and Commerce on February 10, 2005, FERC's general counsel Cynthia A. Marlette included in her prepared testimony a section entitled Price Transparency in Natural Gas and Electric Markets, where she declared:

> It would be helpful if the Congress clarified the Commission's authority to require the development of an electronic price reporting system, and if the Congress gave the Commission the ability to require all electric market participants to participate in such a reporting system . . . and make it publicly available.[106]

Methane Fee Repeal Unwarranted

Section 60113 of the Inflation Reduction Act (Public Law 117-169) establishes a fee on emissions of methane from the oil and gas industry, starting at $900 per ton in 2024 and increasing to $1,500 per ton by 2026. The section also provides $1.5 billion in grants to the oil and gas industry for compliance assistance for the new fee, including money to purchase equipment to capture methane emissions. Furthermore, companies are fully exempt from paying the fee if they are in successful compliance with EPA's methane emissions reduction rule.[107] While the fee covers some emissions that the EPA's rule does not—including offshore operations and LNG export terminals—the law provides ample incentives for industry to avoid paying the fee. In addition, companies like Range Resources have been able to underreport methane emissions by exploiting loopholes in the reporting of the use of certain oil-field devices.[108] Therefore, we view H.R.484 as unnecessary, and we oppose the legislation.

---

106 At pages 28-29, www.govinfo.gov/content/pkg/CHRG-109hhrg99906/pdf/CHRG-109hhrg99906.pdf.

107 www.congress.gov/117/plaws/publ169/PLAW-117publ169.pdf.

108 Zachary Mider, "Methane 'Loophole' Shows Risk of Gaming New US Climate Bill," August 10, 2022, www.bloomberg.com/news/articles/2022-08-10/methane-loophole-shows-risk-of-gaming-new-us-climate-bill.

Recent Congressional and FERC Action on Electricity Transmission Is More Than Sufficient to Ensure New Capacity

On April 21, 2022, FERC proposed a rulemaking *Building for the Future Through Electric Regional Transmission Planning and Cost Allocation* that would establish 20 year, dynamic long term planning for proposed transmission projects; would elevate states from stakeholders to decisionmakers on project selection and cost allocation; eliminate Construction Work In Progress incentive rates; and require incumbent utilities interested in new transmission to collaborate with non-affiliates.[109]

On November 15, 2021, President Biden signed H.R.3684 into law, establishing new authority for FERC to expedite transmission siting, and a $10 billion loan fund to help finance transmission projects like Dayton Power's.[110] And on August 16, 2022, President Biden signed into law the Inflation Reduction Act of 2022.[111] Section 50151 (*Transmission Facility Financing*) of the IRA appropriates $2 billion for a direct loan program for the development of transmission projects, and Section 50152 (*Grants to Facilitate the Siting of Interstate Electricity Transmission Lines)* spends $760 million for grants aimed at facilitating the siting of transmission lines.

In December 2020, FERC proposed a new rulemaking to adopt the statutory changes stemming from the infrastructure bill, including expansion of "national interest electric transmission corridors"; ensuring that states are involved in the pre-filing process; requiring an "Environmental Justice Public Engagement Plan" and an Applicant Code of Conduct to ensure the project's responsiveness to landowners.[112]

And in June 2022, FERC proposed a rulemaking to expedite interconnection procedures to bring new electric generation facilities to the grid, as current procedures often result in unnecessary years-long queues.[113]

The combination of these FERC actions will ensure that needed generation resources will be able to deliver affordable, reliable and clean energy for Americans.

Legislation to Prevent Effective EPA Regulation of Hydrofluoric Acid Alkylation Is Premature

One of the proposed bills considered in today's hearing would stop the EPA from requiring oil refineries utilizing hydrofluoric acid alkylation to explore less hazardous alternatives.[114] About 40% of America's refining capacity currently utilizes hydrofluoric acid as a catalyst to produce alkylate. Alkylate is valuable for its high octane, low volatility and low sulfur content for gasoline. But prior to the proposed EPA rule, some domestic refiners—including Chevron—have adopted ionic-liquid alkylation as a safer alternative to hydrofluoric acid.

---

[109] Docket No. RM21-17.
[110] www.congress.gov/bill/117th-congress/house-bill/3684/text.
[111] www.congress.gov/bill/117th-congress/house-bill/5376.
[112] Docket No. RM22-7, www.govinfo.gov/content/pkg/FR-2023-01-17/pdf/2022-27716.pdf.
[113] Docket RM22-14, www.ferc.gov/news-events/news/ferc-proposes-interconnection-reforms-address-queue-backlogs.
[114] https://d1dth6e84htgma.cloudfront.net/10_HR_59e00433e4.pdf.

Congress should permit the EPA to move forward with its proposal in an effort to work with domestic refiners to improve the public safety of their operations.[115]

This concludes my testimony.

* * *

Mr. DUNCAN. I thank the gentleman. The Chair will now recognize Mr. McNamee for 5 minutes.

Statement of Bernard Mcnamee

Mr. MCNAMEE. Thank you, Mr. Chairman. Chairman Duncan, Ranking Member DeGette, Chair Johnson, Ranking Member Tonko, Chair Rodgers, and Ranking Member Pallone, members of the committee, thank you for inviting me to participate in this hearing. I am Bernie McNamee, and I am here and want to make it clear that I am only expressing my own views, and they are not of my employer or any of its clients.

Today we are facing a new energy crisis. The people of the—the American people are facing a new energy crisis. Americans are now faced with energy scarcity, artificial shortages of natural gas and oil despite massive reserves in the United States, and an electric grid that is less reliable.

Nor can all this be blamed on Putin's war in Ukraine. Misguided government policies, as well as the politicization of capital, are causing much of the current energy crisis in this country. The energy challenges are wide ranging.

We have the means to reinvigorate our energy priorities for the benefit of the American people. Many of the bills here today will help do that. But recognizing the limited time for these opening comments, I will focus on three major issues: permitting reform, the importance of natural gas to energy security, and restoring reliability to our electric grid.

Permitting reform. Over the years we have seen a number of initiatives to speed up environmental reviews, including permitting reform. And the problem is not just agencies reviewing projects. The substantive aspects of various environmental laws contribute to the rejection, delay, and cost of energy projects. Therefore, attempts to make the bureaucracy work more efficiently may not result in more projects being approved or constructed.

To address permitting challenges, Congress should look at the substantive requirements in various laws and agency regulations that they are implementing, so as to ensure that they properly protect the environment but also do not create artificial barriers.

But there is another problem, and it is related to NEPA litigation. As enacted by Congress, NEPA does not provide for a private cause of action. But the courts have allowed agency actions on NEPA decisions to be challenged in court through the Administrative Procedures Act. The result is that agencies now spend an inordinate amount of time and effort trying to address every minor comment and issue raised in environmental reviews. And no matter how good an agency's review is, the agency's action can still be challenged in court, which can then

---

[115] Housley Carr, "For Refiners, Are There Viable Alternatives To HF Alkylation?" February 5, 2023, https://rbnenergy.com/whats-it-all-all-about-alky-part-3-for-refiners-are-there-viable-alternatives-to-hf-alkylation.

hold up a project for years. Such delays can end up killing a project or making it more expensive.

Congress should consider reforming NEPA and the EPA to limit how legal challenges can be made against agency actions. But of course, this is a two-edged sword. We want agencies to be accountable, so Congress will have to engage in a careful balancing of its authorities.

Next, natural gas energy security. American energy security and affordability is vitally dependent on access to domestically produced natural gas. Natural gas is important for home heating, manufacturing, but also provides about 38 percent of our electric generation. Unleashing natural gas production should be a priority, and natural gas pipelines are very important to get end product to users.

Furthermore, Federal and State policymakers need to recognize the interdependence of the electric grid and natural gas, especially natural gas pipelines. This means ensuring that pipelines are also safe from cyber and physical threats, which I also know is being considered by this committee.

Finally, restoring the reliability of the electric grid. Electric reliability is decreasing in many parts of the country. And we have seen this in California, Texas, and part of the East Coast this past December. And these failures have not been the usual causes for power outages, which are usually downed power lines. What we have been seeing is a lack of enough generation on the grid, dispatchable generation to keep the power going.

This is the result of policy choices, in particular the convergence of subsidized renewables and regional transmission organizations. Though described as electric markets, RTOs are actually complex regulatory constructs. And, unlike traditional utilities, generators and RTOs have no obligation to serve customers. Furthermore, RTOs and generators are not passing the full economic benefits of no-fuel and subsidized renewables to customers. The end result has been higher prices for customers, less reliability, and little accountability.

This can be contrasted with traditional utilities which engage in an integrated resource planning to provide reliability, have the rate set in a manner that is—provides the economic benefits of fuel diversity to customers, and is accountable to State public utility commissions and legislators.

As you look at the various issues about reliability, particularly the interaction between natural gas and electric, you may want to consider reforms to the Federal Power Act, FERC oversight, and enhancing the role of the States. Reliable, affordable, and abundant energy is essential for the American people and the Nation.

I am grateful for the committee's work. Thank you for having me here. And I ask that my written comments be put in the record.

[The prepared statement of Mr. McNamee follows:]

The U.S. House of Representatives, Energy and Commerce Committee, Joint Energy, Climate, & Grid Security Subcommittee and Environment, Manufacturing, & Critical Materials Subcommittee Legislative Hearing: "Unleashing American Energy, Lowering Energy Costs, and Strengthening Supply Chains," February 7, 2023, Opening Statement of Hon. Bernard L. McNamee, Former Commissioner of the Federal Energy Regulatory Commission

One Page Summary of Major Points; February 7, 2023; Hon. Bernard L. McNamee

Three ways to enhance access to reliable, affordable and abundant energy: _

1) Permitting reform.
   - The substantive aspects of various environmental laws contribute to the rejection, delay and costs of energy projects. Therefore, attempts to make the bureaucracy work more efficiently may not result in more projects being approved or constructed.
   - NEPA does not provide for a private right of legal action, but agencies' NEPA determinations are subject to legal action under the Administrative Procedures Act (APA). Permitting reform should consider limiting how APA challenges are applied in the NEPA and environmental contexts.
2) Enhancing natural gas production and natural gas pipelines.
   - America has abundant natural gas resources, but government policies need to support the production, transportation, and use of natural gas for the benefit of the American people.
3) Restoring reliability to our electric grid.
   - The reliability of the electric grid is declining. The convergence of regional transmission organizations (RTOs) and subsidized renewables result in price distortions, reduced reliability, and failure to pass along the full economic benefits of no-fuel, subsidized renewables to customers.
   - Reforms to the Federal Power Act, FERC oversight, and role of the states should be considered.

The U.S. House of Representatives, Energy and Commerce Committee, Joint Energy, Climate, & Grid Security Subcommittee and Environment, Manufacturing, & Critical Materials Subcommittee Legislative Hearing: "Unleashing American Energy, Lowering Energy Costs, and Strengthening Supply Chains," February 7, 2023, Opening Statement of Hon. Bernard L. McNamee

Chair Duncan, Ranking member DeGette, Chair Johnson, and Ranking Member Tonko—and Chair Rodgers and Ranking Member Pallone, Members of the Committee: Thank you for inviting me to participate in this hearing.

Before I begin, I want to make clear that the views expressed are my own and I am not representing anyone else in this hearing—neither my employer nor its clients.

New Energy Crisis

Today, the American people are facing a new energy crisis. Since the American Energy Renaissance of the early 2000s that transformed the United States from a net energy importer to energy independence and then energy dominance on the world stage, Americans are now faced with energy scarcity; artificial shortages of natural gas and oil despite massive reserves within the United States; and, an electric grid that is less reliable.

Nor can this all be blamed on Putin's war on Ukraine. Misguided domestic government policies, as well as a politicization of capital investment, are causing much of our current energy crisis. The results are reduced energy supplies and increased energy costs that hurt individuals and families; makes businesses that create jobs that drive our economy less competitive; and makes America less energy secure.

Though our energy challenges are wide-ranging, we have the means to reinvigorate our energy economy for the benefit of the American people.

Recognizing the limited time for these opening comments, I will focus on three major issues: 1) permitting reform; 2) the importance of natural gas to energy security; and 3) restoring reliability to our electric grid.

Permitting Reform

Over the years there have been a number of initiatives to speed up environmental reviews, including permitting reform. The problem is not just the agencies reviewing projects. Congressional actions such as the FAST Act[116], or One Federal Decision[117] have helped some, but the problems are much deeper than bureaucratic efficiencies.

Environmental reviews involve the interaction of NEPA (the National Environmental Policy Act) and other federal statutes, executive orders and guidance from CEQ. Agencies must apply substantive laws, such as the Clean Water Act and Endangered Species Act, as well as executive orders. Merely telling agencies to work faster, work together, or limiting page limits on environmental reviews, though helpful, will not resolve the underlying problems that make it so difficult to build anything.

The substantive aspects of various environmental laws contribute to the rejection, delay and costs of energy projects. Therefore, attempts to make the bureaucracy work more efficiently may not result in more projects being approved or being constructed.

To address permitting challenges, Congress should look to the substantive requirements in various laws and the agencies' regulations that implement them so as to ensure they properly protect the environment, but do not create artificial barriers.

---

[116] Fixing America's Surface Transportation Act (42 U.S.C. § 4370m et seq.) ("FAST-41").

[117] One Federal Decision started as an Executive Order under President Trump (Executive Order 13807, Aug. 15, 2017) and was also part of the Infrastructure Investment and Jobs Act (IIJA) (Nov. 15, 2021).

In addition, Congress should provide clear direction about the limits of CEQ's authority when issuing regulations implementing NEPA. Congress should consider repealing or narrowing regulations as to how an agency looks at cumulative impacts and indirect effects from projects. Furthermore, attempts to use the social cost of carbon in reviewing projects should be a decision made by Congress. Congress should also pay careful attention to how decisions related to environmental justice are made.

Another problem is NEPA litigation.

As enacted, NEPA does not provide a private cause of action. But courts have allowed agencies' NEPA decisions to be challenged in court using the Administrative Procedure Act (APA). The result is that agencies now spend an inordinate amount of time and effort trying to address every minor comment or issue raised in environmental reviews. And no matter how good the review is, the agency action can still be challenged in court and a project can be held up for years. Such delays can end up killing a project or making it more expensive.

Congress should consider reforming NEPA and the APA to limit how legal challenges can be made against agency actions. Of course, this is a two edged sword. We want agencies to be held accountable. So Congress needs to engage in a careful balancing of agency authority.

The Importance of Natural Gas to Energy Security

American energy security—and affordability—is vitally dependent on access to domestically produced natural gas. Natural gas is used for home heating and manufacturing; but also provides 38 percent of our electric generation[118] — though it varies by region and state. Unleashing natural gas production should be a priority and natural gas pipelines are needed to get the product to end-users.

The United States has abundant supplies of natural gas, particularly in Texas and in the Marcellus shale formations. Pipelines are needed to get the gas to local distribution companies, manufactures, electric generators, and LNG export facilities. Through a combination of permitting reform and FERC's adherence to its obligations under the Natural Gas Act when considering pipeline applications, all Americans will benefit from access to natural gas.

Furthermore, federal and state policy makers also need to recognize the interdependence of the electric grid and natural gas, especially natural gas pipelines.[119] Because natural gas fired generation is dispatchable, on-demand, with fast ramping, it can provide electricity quickly when intermittent renewable resources are not available. In addition, access to abundant natural gas helps keep electric prices down. This means ensuring that those pipelines are also safe from cyber and physical threats.

---

[118] U.S. EIA, "What is U.S. electricity generation by energy source?" 2021 data, https://www.eia.gov/tools/faqs/faq.php?id=427&t=3.

[119] NERC, "Electric–Gas Interdependencies, Potential Summer Energy Shortfalls are Focus of Board Discussions" (May 13, 2021), https://www.nerc.com/news/Headlines%20DL/Board%2013MAY21.pdf; NERC, "Special Reliability Assessment: Potential Bulk Power System Impacts Due to Severe Disruptions on the Natural Gas System" (November 2017) https://www.nerc.com/pa/RAPA/ra/Reliability%20Assessments%20DL/NERC_SPOD_11142017_Final.pdf.

Restoring Reliability to Our Electric Grid

Over 330 million Americans rely on electricity for their livelihood, quality of life, and survival. Yet, electric reliability is decreasing in many parts of the country—we have seen this is California, Texas and parts of the east coast. For example, the blackouts and shortages in California in August 2020 and summer 2022; Texas in February 2021 and summer 2022; and the December 2022 near misses in the eastern part of the United States. In fact, NERC warned of the potential for shortages in the summer of 2022[120] and this winter of 2022-23.[121] And these failures have not been the usual weather related causes of blackout—mainly downed power lines. These blackouts have been caused by a lack of enough generation to power the grid. This is primarily a result of policy choices. In particular, the convergence of subsidized renewables and regional transmission organizations (RTOs).

Though described as "electricity markets," RTOs are actually complex regulatory constructs regulated by FERC. RTOs generally use marginal price clearing auctions to set prices for wholesale electricity paid to generators throughout the day, but subsidized renewables are distorting price formation in the RTOs and undermining market fundamentals. These price distortions have made it harder for traditional baseload resources like coal and nuclear to stay in business. The price distortions are even undermining the ability of dispatchable natural gas to make the investments necessary to back up wind and solar. This does not mean wind and solar should not be part of the generation mix; rather, we need to have an all of the above approach that allows us to take advantage of the best attributes of each resource. Unlike traditional utilities, generators in RTOs have no obligation to serve customers. Furthermore, RTOs and generators are not passing on the full economic benefits of no fuel, subsidized renewables to customers. The end result has been higher prices for customers, less reliability, and little accountability.[122]

This can be contrasted with traditional utilities which engage in integrated resource planning to provide reliability; have there rates set in a manner that passes the economic benefits of renewables and fuel diversity to customers; and are accountable to state public utility commissions and state legislatures.

Congress may want to consider reforms to the Federal Power Act, FERC oversight, and enhancing the role of the states.

Conclusion

Reliable, affordable and abundant energy is essential to the American people and our nation. I am grateful that these committees are working on solutions to our energy challenges. I look forward to answering your questions.

---

[120] NERC, "2022 Summer Reliability Assessment" (May 2022), https://www.nerc.com/pa/RAPA/ra/Reliability%20Assessments%20DL/NERC_SRA_2022.pdf.

[121] NERC, "2022–2023 Winter Reliability Assessment" (November 2022), https://www.nerc.com/pa/RAPA/ra/Reliability%20Assessments%20DL/NERC_WRA_2022.pdf.

[122] For a more in depth discussion of the challenges with RTOs see Bernard L. McNamee, "Time to Update Wholesale Electric Markets – But Don't Forget the Benefits of Traditional Utility Regulation," RealClearEnergy, April 8, 2021. (https://www.realclearenergy.org/articles/2021/04/08/time_to_update_wholesale_electric_markets but_do nt_forget_the_benefits_of_traditional_utility_regulation_771956.html).

1Mr. DUNCAN. Well, let me thank you all for your testimony, and we will now move into the question and answering portion of the hearing, and I will begin the questioning, and I will recognize myself for 5 minutes.

Seeing Ranking Member DeGette's coffee this morning, I am reminded that America runs on Dunkin.

[Laughter.]

Mr. DUNCAN. I spell it a little differently.

The truth of the matter is, America runs on energy. For example, my bill, Protecting American Energy Production Act, clarifies congressional intent that States have primacy regarding hydraulic fracturing. It also prohibits the President from issuing any moratorium on hydraulic fracturing.

Two years ago, America was energy dominant for the first time since 1952. We went from the largest net importer to the—to a net exporter. We became number 1 in—oil and gas producer of the world. We became a global price setter, undercutting the leverage of OPEC. Finally, we achieved the—President Jimmy Carter's mission for the DOE: We were no longer reliant on foreign adversaries for our energy security.

We have now seen a reverse of this. The Democrats' energy policies have made us weaker and more reliant on other nations for energy and our supply chains. The Biden administration and congressional Democrats have taken over 100 actions to make it more difficult to produce oil and gas here in the United States. This may be why—that the administration didn't want to testify today. They know they have taken all executive action possible to undercut American energy production.

President Biden has then blamed energy companies for not producing enough, while pushing a rush-to-green energy agenda that would make us overwhelmingly dependent on China, Russia, and our adversaries for energy.

The bottom line is we have the resources here in America to meet all of our energy needs and help those around the globe, abundant natural resources.

The goal of the legislation we are reviewing today is the reverse of this: to make American energy and its supply chains more secure while driving down consumer costs and emissions. So, Mr. Eshelman, I appreciate your comments on the shale revolution. We are going to talk about the hydraulic fracturing and the moratorium today.

The innovation and entrepreneurial spirit helped make America energy dominant and a global leader in emissions reductions. This didn't happen in countries with price and supply controls. What would happen if this administration sought to curtail the use of hydraulic fracturing?

Mr. ESHELMAN. Well, it would——

Mr. DUNCAN. Mr. Menezes.

Mr. ESHELMAN. It would be——

Mr. DUNCAN. Or Mr. Eshelman, I am sorry.

Mr. ESHELMAN. Yes, so it would be devastating. I remember back in 1994, when I worked at IPAA, we issued a request to the Commerce Department called a Section 232 because we were so dependent on foreign oil that we wanted a national security, basically, ranking for America, that it is a threat to the United States, all this foreign oil coming into America.

But now, if we look back 20 years later, 30 years later, America has become energy independent, basically. We produce 12 million barrels of oil a day. We have—imports about 7 million, so it is not even half, because we also export about 3 to 4 million. So it has made us a global leader. It has created jobs. It has brought down prices. It has taken down the trade deficit, it has helped local communities.

I know, Mr. Joyce—I am from Pennsylvania, Dr. Joyce—and we are the number 2 natural gas producers. It has made our counties rich.

And so we operate in every community across the country, pretty much. We live in the communities. We are not headquartered all the time in Houston, or wherever you might think Big Oil is. So jobs, deficit reduction, helping local communities, it has done so much.

Mr. DUNCAN. Thank you for that. Yes, I meet with gas producers and pipeline managers all the time, owners.

And Mr. Slocum, you mentioned our natural gas production. I would argue that we have so much natural gas in this country we can't truly measure how much. You talked about, I think, 20 percent that we are exporting. Producers and pipelines tell me that they could provide more natural gas to the Nation if they had somewhere to put it. They just don't have anywhere for the gas to go. They are at capacity on the pipelines, what is produced in the Marcellus. They could produce more, they just don't have anywhere to send it.

My communities need it. My State needs it. The rest of the Nation needs it. Natural gas is what got us here in our emissions goals.

The goal of these bills today is to counter misguided energy policies that will offshore investment, make us reliant on countries that don't share our values. If they had their way, we would be relying on critical minerals primarily produced in countries with no regard for human rights or emission reduction goals, countries like China. These bills would reflect that the U.S. is serious about building out all forms of energy here, a true all-of-the-above approach.

What happens, Mr. Menezes, if we recede from the world on the energy production, would it be cleaner?

Mr. MENEZES. Well, we need to maintain our ability to produce because our global allies and partners have really become—to depend on U.S.-produced LNG. They prefer to do business with LNG. After Putin's invasion, Europe came to us asking us to increase production. So they have—really have come to rely on us.

Plus, with the agreement with the Biden administration to—that our U.S. LNG providers were going to produce more and provide Europe over this winter and next winter, also came the realization that natural gas helped Europe meet its net zero goals.

Mr. DUNCAN. Yes.

Mr. MENEZES. And indeed, so countries are looking to U.S. for LNG to help meet net zero goals, which, you know, the U.S. is proud to be a leader in that.

Mr. DUNCAN. Yes. My time is expired, but lessen their dependance on Vladimir Putin and others for their energy sources. And I will go to the ranking member, Ms. DeGette, for 5 minutes.

Ms. DEGETTE. Thank you so much, Mr. Chairman. The first thing I want to start with, Mr. Chairman, so we get this subcommittee off on the right foot, is I know the administration would love to come and testify on these bills.

And so, first of all, I think we should have another hearing with the administration on these bills. Secondly, if you expect the administration to come, then we would expect that you give them the same comity that we did when Mr. Trump was in the White House, and give them advance notice. But in addition, you have got 17 bills up today, and they only had notice of what 2 of those bills were. So if you really want fulsome testimony, you need to give them the bills.

The second thing I want to point out is that I have good news for you, Mr. Eshelman and everybody else and you, Mr. Chairman, which is, you know, we have a lot of fracking in my State of Colorado, too. And I have been dealing with it for some time. And as you know, fracking occurs a lot in places where traditional drilling did not. And so it is really important—so I don't think anybody is seriously saying we should eliminate fracking.

But what I am saying and what many are saying is we should have adequate environmental and security legislation to make sure that it is safe for the communities around it. And I don't think you could disagree with that. Is that right, Mr. Eshelman?

Mr. ESHELMAN. I believe you are correct, that the States——

Ms. DEGETTE. Thank you.

Mr. ESHELMAN. [continuing]. And local communities are doing their part.

Ms. DEGETTE. Right.

Mr. ESHELMAN. But I don't think the Federal Government can put——

Ms. DEGETTE. OK, thanks.

Mr. ESHELMAN [continuing]. A one-size-fits-all on——

Ms. DEGETTE. I have only got 5 minutes, unfortunately. So I would like to talk to you, Mr. Slocum, for a minute about this notion of increasing oil and gas production in the United States, whether it is traditional drilling or fracking, and this concept that that will somehow make us independent from foreign oil. And the chairman was talking to you about that.

Is that correct, that if we have increased oil and gas production here, that that will make us independent from international oil? And if not, why not?

Mr. SLOCUM. Yes. No, that is not correct, because, as you noted, oil and gasoline markets are globally priced, and they have been since the futures exchange opened in 1982 in New York.

And so that is why whenever—during the first Gulf War in 1990, we saw significant price spikes here in the United States, even though there were no domestic impediments to that. It was because the commodity is a globally priced commodity. And prior to the export boom of natural gas, natural gas was insulated. And now that we have got LNG exports, Americans are now exposed to global natural gas price shifts.

Ms. DEGETTE. So what would the best way to become independent from these—this foreign market?

Mr. SLOCUM. To get our economy off of volatilely priced, globally priced fossil fuels like oil and natural gas——

Ms. DEGETTE. And move more into renewable fuels that were domestically developed?

Mr. SLOCUM. Correct.

Ms. DEGETTE. OK. Now I want to ask you one more thing about H.R. 484, because one of the focal points I have been working on for some years is reducing methane emissions. And so H.R. 484 rolls back the Obama-era methane rule that had bipartisan support and industry support, including BP and Shell.

So I want to ask you why section 60113 of the Inflation Reduction Act is so important.

Mr. SLOCUM. It is so important because methane is an incredibly powerful greenhouse gas, far more destructive than CO_2. It is shorter-lived in the atmosphere, but capturing those methane emissions from the oil and gas sector is critical if we are going to successfully address the climate crisis.

And so that is why I thought that the methane provisions in the IRA were extremely fair. They provide——

Ms. DEGETTE. And it gives the——

Mr. SLOCUM. They accommodate industry——

Ms. DEGETTE. Yes.

Mr. SLOCUM [continuing]. To help industry achieve those emission reduction——

Ms. DEGETTE. They actually give funding to the industry to help them comply with the bill, right?

Mr. SLOCUM. Correct.

Ms. DEGETTE. And finally, Mr. Garcia, I want to thank you for your testimony. But I would particularly ask the committee to look at your written testimony, where you analyze every one of these 17 bills and talk about the fundamental problems. I don't have time in 5 minutes, like you didn't have time in your testimony, but it is really helpful analysis. And I think we should really try to work together to move this discussion forward rather than just rehashing tired debates that we have had over all the many years I have been on this committee.

With that, I yield back. And welcome to the new chairman.

Mr. JOHNSON [presiding]. The gentlelady yields back, and the Chair now recognizes himself for 5 minutes for the purpose of asking questions.

And let me—I want to touch on my legislation, the Unlocking our Domestic LPG Potential Act. And I mentioned last week how unleashing U.S. LNG exports strengthens America's geopolitical posture on the world stage, but it can also bring considerable benefits here at home.

Mr. Menezes, thank you for you—thank you for mentioning my legislation and for saying how America can lead the way in providing the world's energy. Can you explain how increasing U.S. LNG exports can have a positive effect on increasing production here at home, bringing with it the jobs and economic growth in places like my district in Ohio that sits atop the Utica and Marcellus Shale?

Mr. MENEZES. Thank you very much for the question. Indeed, you know, throughout the world, many of our friends and allies are coming to the United States and asking us to try to produce as much LNG for the purpose of export. They are aware that, when we produce natural gas, it is used, obviously, to drive our very powerful economy. But they have come to look to us, really, as part of the solution. They do not want to go to any of the monarchies if they have to. They do not want to go to Iran. And they certainly now have announced that they are no longer going to rely on Russia. So they have looked to the U.S. to replace this supply.

We need to look at a future where Russia will no longer be a reliable provider. Now, Russia will likely do deals with the monarchies and perhaps China, but it is important for our friends and allies to know they can depend on the United States.

And so with that demand for export, increases domestic production and lowers prices.

Mr. JOHNSON. Thank you, Mr. Menezes. Now let's shift gears a bit.

We have an impressive slate of bills from our Members on the Environment, Manufacturing, and Critical Materials Subcommittee that would address some of the major challenges that we are facing. Ms. Sweeney, please allow me to start with you, because you mentioned in your testimony that some critical materials needed for future industries and technologies like lithium and cobalt will skyrocket some 30 or even 40 times beyond current demand right now. Frankly, this increase is hard to even comprehend.

The bottom line is we need a lot more mining and a lot more mines, and it is going to have to be done somewhere. So, Ms. Sweeney, wouldn't we want to do more of this right here in America?

Can you tell us how, if we had the right permitting reforms and regulations in place, our domestic mining industry would be able to extract the minerals and metals we need in a safer and more environmentally responsible way than in other countries?

Ms. SWEENEY. Thank you so much for that question. And yes, we have about 6.2 trillion of value of minerals known in the U.S. today. And there is further mapping all the time. So there are more minerals that will be found, and we can mine here under the best standards—environmental, labor, and safety standards—in the world.

But we do need the right policies in place to really unlock that potential. And permitting is one of the impediments. We need more efficiencies. We are not asking for environmental shortcuts. We are asking for a more efficient process, similar to the processes that are similarly stringent in Canada and Australia, but they do more to focus on coordinating amongst the various agencies. They have firm timelines in place.

Mr. JOHNSON. OK. All right. Well, thank you.

And back to you, Mr. Menezes, very quickly. We are considering multiple bills today to allow America to protect its, quote, "critical energy resources" when America is under threat. Would you agree that, in addition to, say, oil and gas, this definition could also include critical minerals and rare earths?

Mr. MENEZES. Absolutely. They are going to be necessary for our transition to a cleaner——

Mr. JOHNSON. OK. Would you agree this is a broad definition? Could it also include things like battery and solar components, perhaps electrical, steel?

Mr. MENEZES. Yes, it can include all of those necessary for our energy production and use.

Mr. JOHNSON. OK. Well, thank you, because it is essential that we protect all of these components and their manufacturing supply chain.

I do want, in my remaining time, I want to go back. I heard, Mr. Garcia, you mentioned that these bills are replete with exemptions and go-arounds. These bills are actually the removal of burdensome regulatory barriers that are—to advancing a true, all-of-the-above, market-driven energy strategy, the kind that the President actually wants.

But yet these barriers for permitting, for mining, for exporting clean natural gas, you name it, those are things that would promote the President's agenda. But this administration doesn't want that. And that is what we are trying to break down today. Because good, solid energy policy is also good climate policy. And we believe Republicans are offering those up today.

And with that, I would yield back. And now I recognize our colleague, the ranking member on the Environment and Manufacturing and Critical Materials Subcommittee, Mr. Tonko, for 5 minutes.

Mr. TONKO. Thank you, Mr. Chair.

Mr. Garcia, thank you for your testimony. I am grateful to you and the Earthjustice organization for the partnership with frontline communities seeking protection from environmental and health threats.

Yesterday, Members received a letter from Earthjustice and seven other environmental organizations expressing opposition to five bills under consideration today. While I understand you have several concerns with each of these bills, the broad definition of "critical energy resource" seems to be a common issue. What is your understanding of what could be included in these bills' definition of "critical energy resources," and what risks might this pose to the communities with which you partner?

Mr. GARCIA. Yes, absolutely. As I mentioned, the bills themselves don't actually give us a definition of what critical, whether it be critical minerals or critical energy sources, actually are. And so it really defers to the Secretary of Energy in any given administration to determine what these are. And so that means that virtually anything can become a quote/unquote "critical energy resource."

And so, you know, the idea here is that we have to make sure that the energy that we produce is clean, but the manufacturing that goes into the creation of that clean energy also has to be clean, and it has to protect communities themselves.

And so, when we are talking about broad outlines, right, and broad schemes that do go around the laws, it is going to be particularly problematic because, again, we are talking about laws and regulations. And we certainly heard about red tape, but few people actually mentioned that it is not red tape, it is actually protections that people rely on. NEPA, the National Environmental Policy Act, is at times the only law that communities have on the ground to actually know what is going to happen in their back yards and then to be able to comment on what is going to happen, say, on a mining project or on a drilling project or an energy project.

And so we have to really think about how expansive this definition can really get, because, you know, when we are talking about the—for example, the TSCA bill, it says that for critical energy sources you have to study the economic costs to see if the substance is safe, which is backwards. A substance is either safe or not. It is either poisoning or it is not. And you can't figure that out afterwards.

It also puts these chemicals on the market before the determination of whether they are safe or not is actually made. And so you can imagine that, right? Like, you—hey, you got three cups in front of you. "Go ahead and drink them all. We will find out later if one of them is poisoned." It is backwards logic. And that is what we are doing to our—that is what we would be doing to our communities day in and day out, if we allowed this to move forward.

Mr. TONKO. Thank you.

Ms. Sweeney, your testimony highlighted the increasing international demand for critical minerals and the strategies developed by several foreign governments, including the UK and Canada, to address these growing demands. A common thread highlighted in these strategies is the importance of bringing greater transparency to supply chains.

I know that the National Mining Association has said that the Inflation Reduction Act's clean vehicle tax credit, coupled with other provisions in the IRA, can help U.S. energy security and supply chain resilience. As you know, a key component of this tax credit is verifying material sourcing, and I want to hone in on that verification effort.

What do you see as the need for increased supply chain transparency and tracking in helping verify and promote compliance with the domestic sourcing requirements in the recently enacted battery incentives?

Ms. SWEENEY. Thank you very much for that question. That is a great question, and transparency is needed.

I mean, there is information that we do get from the U.S. Geological Survey about where minerals come from, but it can be quite complicated to trace them back to the source, and additional resources are probably needed to be able to do that.

Mr. TONKO. So what——

Ms. SWEENEY. Of course, if we get them here, we wouldn't have to trace too far.

Mr. TONKO. Right. But what specifically should we do to improve that tracing, tracking opportunity?

Ms. SWEENEY. You know, I have not spent a lot of time thinking about that. I would love to get back to you with an answer.

Mr. TONKO. Sure. Thank you. And I also believe Congress should consider other methods for meeting our critical mineral demands, including investing in R&D to develop alternative battery chemistries that are less reliant on critical minerals, and putting a greater emphasis on the recycling of critical minerals already in commerce.

Mr. Garcia, do you believe more can be done to achieve more sustainable sourcing through recycling and reuse of critical minerals, provided this is done using safe and environmentally sound processes?

Mr. GARCIA. Yes, absolutely. And the laws that are in the books, far from being obstacles, are actually guides. They guide us in order to make sure that, while we are doing more recycling of these minerals and while we are figuring out what we want to do in terms of the next innovation, that we are doing it in a way that it is safe. And so, far from obstacles, these are guideposts that we need to make sure that we are keeping in mind as we move forward in innovation.

Mr. TONKO. Thank you very much. With that I yield back, Mr. Chair.

Mr. JOHNSON. The gentleman yields back. The Chair now recognizes the chairlady of the total committee, Energy and Commerce, Mrs. McMorris Rodgers, for 5 minutes.

Mrs. RODGERS. Thank you, Mr. Chairman. Our goal is to put security back at the center of our energy policy. That is what this, the package before us today—each of these bills is central to our energy work.

And flipping the switch for more American energy and energyrelated industrial activity, it is key, while maintaining America's highest labor environmental safety standards. There is nothing in any of the bills before us that would undermine those high standards.

And I—and, you know, Russia's invasion of Ukraine, it didn't cause the energy crisis. It just exposed what was going on in weakening American energy security. The risks extend way beyond oil and gas supplies to vulnerabilities in the civilian nuclear sector.

The sad fact is our domestic fuel infrastructure, from uranium mining and conversion to enrichment services, has eroded. Upwards to 24 percent of America's fuel this year—this year—will be Russian sourced, which places our fuel supplies at greater risk of disruption.

Mr. Menezes, during your time as Deputy Secretary of Energy, DOE highlighted the importance of restoring American nuclear leadership in the world. Would you speak briefly as to why a strong nuclear industrial base, from fuels to technological development, is critical to domestic nuclear development, quality energy supplies, and to our national security?

Mr. MENEZES. Well, thank you very much. I mean, we have always been aware of the need for the U.S. to maintain its global leadership in nuclear technology, but it became more apparent as I began to travel overseas and to see the role that Russia and China is playing, doing deals with our friends and allies across the world, trying to bring their nuclear technology to them.

Now, one might say, well, you know, that is not a big deal, but it is a big deal, because what we are doing is we are ceding the leadership, the technological leadership, the institutional knowledge, and the 50- to 100-year relationships with China and Russia—with our friends and allies. They are turning away from the U.S. So that is an important thing: We are losing the global leadership.

We have 123 requirements in our law. China and Russia do not. And so we should all be concerned about maintaining that. Regardless of how you feel about nuclear energy, we need to maintain that leadership. It is very real, and it is very threatening.

And then our own history is that we really have—we need to improve our mining, our milling, our conversion, our fuel fabrication, and, of course, enrichment. And so I really—I ask this committee to please look seriously at it.

We released a report in 2020 where we tried to look at it in a very broad view, and I would like to submit that in the record, as well, with my testimony. Thank you.

Mrs. RODGERS. Thank you very much.

Ms. Sweeney, just to build on this a little bit, my legislation is designed to send a clear signal to industry that a date certain America's nuclear industry can no longer rely on Russian-sourced, low-enriched uranium. What do you believe this signal would mean for building out our domestic fuel industry?

Ms. SWEENEY. It is a lifeline. Frankly, you know, talking about insecure supply chains, our reliance on foreign sources of uranium, over 97 percent is foreign. It is just crazy when you think about that. And that industry is on the precipice of extinction. They need that lifeline.

Mrs. RODGERS. So this week we are all talking about a Chinese spy balloon that made its way across America, underscored the importance of Americans and—having safe and secure systems, supply chains. It is real. The threat from China is real. It is active. So Ms. Sweeney, what do you believe this event says about the urgency for us building out our mineral, metal, and energy materials supply chains?

Ms. SWEENEY. It is such an important topic. You know, we need to be able to rely on ourselves for as much as we can, and we have so much here that we can be using. We have the great workforce, we have great transportation infrastructure. We can and should be doing it here, and that will protect us, you know, in energy crisis, in military crisis, in economic crisis.

Mrs. RODGERS. So would you speak to where do we get the minerals and the metals that we need right now? Where do they come from?

Ms. SWEENEY. Could you repeat that?

Mrs. RODGERS. The minerals, the metals that we need in energy, in——

Ms. SWEENEY. Oh, oh——

Mrs. RODGERS. Yes, yes.

Ms. SWEENEY. China, mostly China, right? And the processing, as well.

I mean we are 80 percent reliant on rare earths that come from China. Processing, they do about 90 percent of the rare earths processing there.

Mrs. RODGERS. And what has been happening recently in America, as far as making more of this available in America?

Ms. SWEENEY. Well, we have gotten a lot of funding for new projects in the U.S., but what we are lacking is the permitting reforms to accompany that. So what—we are seeing money going to projects, but we are not seeing approvals happening. So it is really delaying our ability to unleash our full potential.

Mrs. RODGERS. Yes, you can't build without permitting reform. I yield back, Mr. Chairman.

Mr. JOHNSON. The gentlelady yields back. The Chair now recognizes the ranking member of the Energy and Commerce Committee, Mr. Pallone, for 5 minutes.

Mr. PALLONE. Thank you, Chairman. I am going to try to get in some questions for Mr. Slocum and Mr. Garcia.

So, Mr. Slocum, in your testimony you talked about how increased LNG exports have increasingly tied American natural gas markets with global markets, meaning that American consumers are now sharing in the volatility that global markets are experiencing due to Russia's invasion of Ukraine. And American natural gas markets are currently reflecting that volatility.

Last year, American natural gas futures fluctuated wildly from a low of just under $4 per unit to nearly 10. I don't know how families in New Jersey are supposed to budget for gas to heat their homes when it could double in price on them with no warning whatsoever.

So first question, Mr. Slocum: Can you talk about how increased LNG exports have increased home energy costs for Americans?

Mr. SLOCUM. Yes. So the Department of Energy has authority to review export applications, and it has approved every one. And so, as a result, we went from zero LNG exports in 2016 to now we are the largest LNG exporter in the world. And what that has done is it has forced consumers in New Jersey and Texas and elsewhere to compete with our foreign counterparts for that gas.

And obviously, we want to help our allies in need after the Russian invasion, but I think that there needs to be a balanced assessment because, when there are physical shortages in New England and extremely high prices on the West Coast, it is clear that the level of export is creating detrimental impacts for energy affordability at a time when families are already stretched thin with high energy prices.

Mr. PALLONE. Well, thank you.

Now, I wanted to enter into the record—if I can ask unanimous consent to enter into the record—a CNBC article from last June detailing the decrease in U.S. natural gas prices after the explosion at a Freeport LNG terminal. If I could ask unanimous consent, Mr. Chairman.

Mr. JOHNSON. Without objection, so ordered.

[The information appears at the conclusion of the hearing.]

Mr. PALLONE. Thank you.

So if—my understanding is that, you know, because that Freeport terminal export was closed, that we had more natural gas here, and prices went down. Accurate?

Mr. SLOCUM. That is 100 percent correct. So on June 8th, 2022, there was a massive explosion at the Freeport LNG facility, which alone accounts for 20 percent of all of U.S. LNG exports, so it is a very large facility. And the futures market immediately reacted to the fact that that 20 percent export capacity was going to be offline for some time by sending domestic prices significantly lower, a recognition of the direct impact that exports have on domestic pricing.

Mr. PALLONE. All right. Let me just focus on Mr. Johnson's bill, Mr. Slocum. Then I have to get to Mr. Garcia.

Can you talk about how removing the requirement for DOE to find that LNG exports are in the public interest would worsen the problem of expensive natural gas?

Mr. SLOCUM. Correct. Right now, under the standard established in 1938, Congress dictated that no exports are allowed to occur unless they are found to be consistent with the public interest. And right now the Department of Energy performs that test.

We banded together with a number of other organizations in October, noting some methodology flaws in the way that the Department of Energy currently does that. But the statutory requirement is very important to ensure that exports are consistent with the public interest. And I think removing that public interest standard would not be advisable.

Mr. PALLONE. All right. Let me go to Mr. Garcia about methane. Consumers are also paying for market failures that make it cheaper for the industry to waste methane than to install or upgrade equipment to prevent leaks. And this leaked or intentionally wasted gas never makes it to consumers, but they are nevertheless stuck with the bill. So that is why we enacted the Methane Emission Reduction Program to ensure consumers stop paying for wasted energy or the harm its emissions cause. Cleaning up legacy damage and preventing future pollution from the oil and gas industry were also reasons for the program.

So Mr. Garcia, you have got less than a minute. Could you speak to how frontline communities would benefit from holding the oil and gas industry accountable for its methane pollution?

Mr. GARCIA. I mean, throughout we see, whether it be oil and gas production or petrochemical facilities across the country do some of the worst damage to communities across the country. And we can almost pinpoint the communities, the Ironbound community out in New Jersey, the Cancer Alley between Baton Rouge and Louisiana—sorry, and New Orleans in Louisiana.

All you have to do is really go—and I really encourage everybody on this committee—go and take a what they call the toxic tour of these communities to truly understand what they are dealing with left and right, day in and day out.

And people say, well, it is their choice to live there. They were there before. They were there before. And industry has come in, and they have utterly ravaged the health of these communities throughout.

And they will tell you about the cancer clusters. They will tell you about all of the concerns that they have about the respiratory health, cardiac health. And I honestly can't do it justice here.

But it is imperative that the protections that we have stay in the books. And what really needs to be the focus of this Congress is how do we strengthen protections so that not only are we addressing the energy needs, which we have heard a lot about here today—and what we haven't heard, many things, in fact, anything about—are the health protections that need to be in place for communities to actually thrive in the face of oil and gas, mining, and other dirty industries.

Mr. PALLONE. Thank you. Thank you, Mr. Chairman.

Mr. JOHNSON. The gentleman yields back. The Chair now recognizes the gentleman from Texas, Dr. Burgess, for 5 minutes.

Mr. BURGESS. Thank you, Mr. Chairman.

Deputy Secretary Menezes, let me just ask you a question, if I could. The statement was made that natural gas prices fell in June after the unfortunate accident at the Freeport facility.

Do you know what has happened since then, as far as the futures on natural gas?

Mr. MENEZES. Well, I think today, you know, they are back at their normal historic lows since the shale revolution. So that has been the constant thing. There was indeed a blip of the prices, as mentioned. But you have to look at the long-term view.

And so DOE has done, like, five studies using EIA and other experts in the field to look at if, in fact, all of the facilities that are pending there were actually permitted and built and exported. The studies have all been—the projections have been clear: modest increases in prices. And this is going out to 2040.

Mr. BURGESS. Yes. According to Bloomberg, the price did drop in late June and early July. It then immediately came back up for the balance of the summer, dropped again a little bit in September, when perhaps energy demands—air conditioning and electricity demands—fell off, and currently stands at 21⁄2 dollars per million BTU, I mean, which is in line with its historical precedent.

So—but there is probably another reason why they are paying higher prices in Boston. Can you help us with that?

Mr. MENEZES. So on the price of natural gas, there are several reasons for that. It should not surprise anyone that, due to limited capacity on the West Coast and the East Coast, that those prices are going to be high. Those areas of the country have taken the position to restrict natural gas going in there. So their constituents should be prepared to pay higher prices.

On pricing with respect to electricity, that is something actually different. So the pricing there is done, you know, at different points within the bid-based markets. And so those are—it is locational marginal pricing, and it really depends on the price of natural gas, really, to set the electricity.

But to be sure, when you restrict access to natural gas and our economy is growing on the use of natural gas, you will have higher electricity prices at the points where you can't get natural gas. And more and more of our generation is replacing coal and ramping renewables. So it is a good thing that it is growing, but you have got to be careful about how you characterize pricing.

And with respect to the jump in LNG going down, remember that was in—the prices went up for the unanticipated Putin invasion of Ukraine. That is what set the prices up. So when it came down because of the Freeport accident, it could only come down, to be frank about it, because the Putin invasion caused the global prices to go so high.

Mr. BURGESS. So what is the principal source for natural gas in Boston?

Mr. MENEZES. I don't know if there is a principal source. I do know that New England has, you know, the ability to import natural gas from Yamal and other places. Our Jones Act restricts, you know, our ability to get natural gas from the Gulf of Mexico up to New England. So that is——

Mr. BURGESS. So let me just ask Mr. Eshelman.

Are—those are your independent producers that are selling in Boston?

Mr. ESHELMAN. No, we are not up in Boston. I was going to say probably the biggest source of methane in New England is dairy farms.

Mr. BURGESS. Is what?

Mr. ESHELMAN. Dairy farms.

Mr. BURGESS. Well, look. And Mr.—or Secretary Menezes, I appreciate your thoughts on this and the restriction of natural gas. One of the bills we have under discussion is to help get product that is stranded in the Permian Basin and get it into the stream of commerce. So I appreciate your comments on that.

Mr. MENEZES. Just on a point on that, so, I mean, the President was taking the credit for increasing exports so they could get to Europe. You know where the exports were coming from? The exports were coming from the Permian Basin in——

Mr. BURGESS. Correct.

Mr. MENEZES [continuing]. Texas to the export facilities there with—beyond Federal overreach. That is the source. The President didn't make that point when he was taking credit for increasing production to help our European——

Mr. BURGESS. Well, maybe you could help him with that line in the State of the Union tonight. He likes to take credit for stuff.

But, look, I really appreciate your efforts on helping us with the bill to get the siting and the permitting for natural gas pipelines. Yes, I get the concern that you don't—in the production of natural gas, you don't want flaring and venting. The problem with stranded gas in the Permian Basin is you can't get it—sometimes you can't get it to market. So this is an effort to do that. And I appreciate you pointing out how we are helping the President in the process.

So thank you all for being here today, and I appreciate the lively discussion.

And I will yield back, Mr. Chairman.

Mr. JOHNSON. The gentleman yields back. The Chair now recognizes the gentlelady, Ms. Schakowsky, for 5 minutes.

Ms. SCHAKOWSKY. Thank you to our witnesses.

Drill, baby, drill is what Big Oil and Big Gas wants right now. And it seems as if, from the list of bills that are being considered today, that that is what my Republican colleagues want, as well. It is really a wish list, I think, and a laundry list of policies on the Big Oil agenda.

So, Mr. Slocum, I want to ask you, consumers have been on a natural gas roller coaster, with prices reaching their highest level since 2008. Do the bills that we have—or that we are considering today do anything to combat the high gas prices that consumers have suffered for this last year?

Mr. SLOCUM. I am unable to find any consumer protections in these proposed bills. In fact, I think that the expansion of exports would likely hurt consumers.

What is striking to me is the lack of any energy efficiency or demand reduction initiatives in this legislation. It is not always about supply, it is also about demand.

And how do we get more tools in the hands of consumers to help them avoid these high costs? More incentives. More funding for building efficiency, for building electrification, if municipalities and States want to go that direction. Weatherization. All of these types of tools can empower consumers to avoid their exposure to increasingly volatilely priced fossil fuels for energy.

Ms. SCHAKOWSKY. So I want to ask you this. So in that case, what can—tell us what we can do to encourage the utility companies to transition away from fossil fuels, while at the same time, of course, lower home energy costs and promote energy efficiency.

Mr. SLOCUM. Right. I think——

Ms. SCHAKOWSKY. I mean, we want to hear your ideas of what we can do going forward.

Mr. SLOCUM. Well, I think, you know, Congress already has directed a lot of financial incentives through the Inflation Reduction Act to try to spur the deployment of a number of different clean energy and energy efficiency technologies. And I think anything else that Congress can do to assist States with ensuring that utilities are making those investments in energy efficiency.

There are a number of States that have recognized that prodding their utilities to invest more in their consumers to increase energy efficiency is the best path forward. And so providing more regulatory incentives and financial incentives for utilities, for building owners, for landlords, and for homeowners to deploy and adopt energy efficiency technologies and clean energy technologies, that is only going to help reduce customers' bills and their exposure to volatilely priced fossil fuel energy.

Ms. SCHAKOWSKY. Can you give us some examples of things that are happening around the country that are delivering that kind of good outcome?

Mr. SLOCUM. Sure. So I live in Maryland, which has very proactive—it is called Empower Maryland, where the Public Service Commission, backed by the legislature, requires utilities to invest not necessarily in building new power plants to meet demand, but investing in consumers to help obtain energy efficiency initiatives. So, you know, I live in a house built in 1900 that needed a whole lot of weatherization renovation. And that was helped in part because of incentives through the utility.

And so I think looking at energy providers not as just providing energy, but assisting customers in avoiding energy use through improvements in energy efficiency, has to be a central component. And all investments in energy efficiency are typically far more cost effective, meaning the bang for your buck for investing in energy efficiency is always better than building new energy resources.

Ms. SCHAKOWSKY. I thank you so much for your comments——

Mr. SLOCUM. Thank you.

Ms. SCHAKOWSKY [continuing]. And I yield back.

Mr. JOHNSON. The gentlelady yields back. The Chair now recognizes the gentleman from Ohio, Mr. Latta, for 5 minutes.

Mr. LATTA. Well, thanks, Mr. Chairman, and thanks for our witnesses for being here today. I really appreciate it.

And one key area that is holding the United States back from reaching its full potential for energy production is refining capacity. Last year the Energy Information Administration estimated that North America lost over 1 million barrels of fuel per day in refining capacity in a 3-year period.

Where refining capacity has decreased, demand for energy has gone in the opposite direction. Our remaining refineries struggling to keep up with this demand are running at close to 95 percent of total capacity. Any economist will tell you that the situation—resulting in higher prices for the consumer. And my legislation, the Researching Efficient Federal Improvements for Necessary Energy Refining and Refinery Act, would address this.

If I could start with you, Secretary Menezes, how important is it for the Federal Government to drive the conversation towards increasing refining capacity in this country?

Mr. MENEZES. We do need a robust refining capacity here, simply so we don't have to import from any other country, and we can provide our consumers with what we need. But the fact is that there hasn't been another major—a new, major refinery built—greenfield—since 1977. We have built some small refineries, but we have been closing more of——

Mr. LATTA. Sorry, would you repeat that again? What was the date, again, for our major refinery?

Mr. MENEZES. In 1977, the last major greenfield oil production, I believe it was the Marathon refinery.

Now, there have been some small refineries that have been built, typically on brownfield sites, but we have been—our industry has been ratcheting out the inefficiencies back from the 1970s. And so we have been closing, we have been doing with what we can.

But the permitting process, essentially, is too difficult to overcome. I believe we have a North Dakota plant and a Utah plant that, I think, have basically stalled out because of lack of permits. So we are doing—we are making improvements with what we have had for many years, and that—it operates at the 95 percent efficiency that you say—does not leave much wiggle room for when the President wants us to increase refining. For example, we simply, A, don't have the facilities or it would be too costly for us to suddenly begin to take them out of mothballs and then try to get them going again.

So that is why we should look at this comprehensively to see what we can do to help, you know, get a robust, clean, environmentally compliant refinery.

Mr. LATTA. And just to follow up, what is the benefit going to be to the consumer? What is the benefit to the consumer?

Mr. MENEZES. Well, you would have a ready supply of refined products.

Mr. LATTA. Thank you very much.

You know, we have talked a little bit earlier today in regards to having a strong nuclear fuel security program, and I totally support it. And I also fully support our chair's legislation that would ban the importation of Russian uranium. And, Mr. Chairman, I ask unanimous consent to place into the record two letters of support in favor of this bill: one from the Uranium Producers of America and the other from the organization ClearPath.

Mr. JOHNSON. Without objection, so ordered.

[The information appears at the conclusion of the hearing.]

Mr. LATTA. Thank you very much.

Ms. Sweeney, if I could switch gears real quickly and talk about your testimony, because, interestingly enough, you know, just by chance—everybody saw the second page of The Wall Street Journal today. It is all about what? It is all about our—what we are going to do about EVs and our batteries in this country. And it is a very interesting article, and I thought right—apropos for where we are today.

But, you know, in your—this testimony that you talked—that you presented today, you know, you have that the lithium band is going to—demand is going to increase by more than 40 times by 2040, followed by graphite, cobalt, nickel at 20 to 25 percent in that timeframe. Our automakers are warning that the coming battery shortage could stop the EV revolution in its tracks.

And also in your testimony you state that, with over $6 trillion worth of mineral resources that we have right here in this country, you know, it is right in our own back yard.

But it goes right to—a question, then, is on the permitting and the delays out there, because also in your testimony you have your chart that shows that, you know, we could be looking at anywhere from 7 to 10 years to get a production site up. And I have been to the only lithium facility that we have in this country, out in Nevada.

But you also state that unexpected permitting delays could reduce that—mining projects by more than a third.

But do you also—when you think about all of these things, you know, what are we going to do in this country, especially when we look to our friends to the north in Canada, that their permitting is taking 2 to 3 years, and here in this country it is taking 7 to 10? What can we do?

Ms. SWEENEY. You know, they have some efficiencies in Canada and Australia, you know, which have very similar NEPA regulations and statute in place that do allow the permitting process to move a little bit faster.

One of those is actually allowing the project proponent to prepare the environmental impact statement. The Government then does a thorough review, and makes sure that, you know, that meets all the standards, and they allow the opportunity for the public to comment on the NEPA project—or the analysis, just like we do here.

But the project proponent is—has the best information. They are on the ground right there. They know what is happening on the ground. They have got the data in front of them, and they have the incentive to move more quickly. And they are only focusing on one project at a time. When the Government has to do it, they are looking at a lot of different projects, and it is just a matter of getting the resources to the project. That is part of the delays.

Another process that works better——

Mr. LATTA. Oh, excuse me. I am afraid my time has expired, but I would ask, if you could put that in writing, and I will address that to you then.

Thank you very much, Mr. Chairman.

Ms. SWEENEY. Sure.

Mr. JOHNSON. I thank the gentleman for yielding. The Chair now recognizes the gentlelady from California, Ms. Matsui, for 5 minutes.

Ms. MATSUI. Thank you, Mr. Chairman, and thank you for the witnesses for being here today.

Recent research from the Energy Innovation shows it is now cheaper to replace almost every coal plant in this country with new, renewable generation rather than pay to keep those old coal plants running.

In my community, our utility, the Sacramento Municipal Utility District, affectionately called SMUD, is on track to be zero carbon by 2030. At the same time, our electricity rates are among the cheapest of any competitors.

Mr. Slocum, fast forward to 2030. Based on current modeling, will the clean energy transition save consumers money?

Mr. SLOCUM. Yes.

Ms. MATSUI. OK. Again, Mr. Slocum, given the expected costs of climate change, will the clean energy transition save our Government money?

Mr. SLOCUM. Yes.

Ms. MATSUI. OK. I want to follow up on what you said about what utilities can do. Are utilities doing all those kinds of things, and making sure that—they are making sure that their customers have good investments so that we can transition?

As I mentioned, our municipal utility in Sacramento is on track to be zero carbon by 2030, 5 years ahead of the President's goal for decarbonizing the U.S. power sector.

The clean energy transition is achievable, it is cost effective, and the potential benefits are enormous.

Mr. Garcia, if every utility in this country were zero carbon by 2030, what kind of benefits would we see among the low-income and minority communities living adjacent to fossil fuel facilities?

Mr. GARCIA. Well, it is going to be a huge impact, and that is why the investments coming out of the IRA are so important, because we are basically reducing the process that is poisoning the communities across the country, right? And so, the less they have to rely on those dirty fuels, the less poisoned air they have to breathe.

Ms. MATSUI. And for those focused on economic impacts above all else, what are the potential economic benefits of improved health in those frontline communities?

Mr. GARCIA. Well, throughout the—you know, throughout all of the economic impacts, they would be quite substantial.

One, because, you know, I mean, you can think about, in very concrete terms, how much does an inhaler cost for—you know, for a child per se? And that is a saving. The trips to the emergency room, the hospital bills.

On top of that, it also makes electricity way easier for them to get, and not as expensive as it used——

Ms. MATSUI. OK. Nature-based solutions are among the most cost effective and underappreciated tools in our toolbox when it comes to mitigating and adapting to climate change. And I am very concerned to see a number of bills here today that would roll back important environmental protections and threaten fragile ecosystems.

Mr. Garcia, what would be the cumulative impact of those bills on our ecosystems, natural lands, and biodiversity across America?

Mr. GARCIA. It would be extremely destructive, not only because of the—I mean, one is the climate emissions that they would—that they bring about, and all of the way that they are going to make— the storms, the droughts, all of that is going to get worse.

But in addition to that, a lot of these projects are happening without any regard to nature, and I think that oftentimes we talk about nature as some distant place where we don't actually go. But we have to recognize that air knows no borders. It travels all across our States, and so does water. And so when we are polluting the water, when we are polluting the air, when the wildlife can't adapt quick enough, when we are throwing away grasslands that would protect our communities, say, from hurricanes or bigger storms that way, we are all going to end up suffering.

And unfortunately, communities of color and low income are at the front lines of that suffering.

Ms. MATSUI. Certainly, they are the most exposed, especially to nature's wrath and those dependent on the resources provided by the natural ecosystems.

How could we work better with low-income, minority, and frontline communities to protect these ecosystems and improve the potential of these ecosystems to help mitigate and adapt to climate change?

Many of these are in low-income neighborhoods, and it is very difficult to figure out how to work with them.

Mr. GARCIA. Absolutely. And, you know, the way that we work with them is by following the guideposts that laws like NEPA set in place, because NEPA is really about engaging the public, engaging frontline communities.

And so making sure that it is not industry's alternative that gets put front and center all the time, making sure that communities on the ground are actually being able to say, "That is actually a conflict of interest," making sure that communities on the ground are able to say, "That is actually going to affect my water, so you shouldn't do that," and, frankly, all of those protections that, again, many people here keep calling obstacles and red tape are those guideposts in order to engage those communities in a way that is going to be helpful to them and helpful to the project's sponsor themselves——

Ms. MATSUI. OK.

Mr. GARCIA [continuing]. To create better projects.

Ms. MATSUI. Well, thank you very much, Mr. Garcia, and thank you for the witnesses for being here today. Thank you.

Mr. JOHNSON. The gentlelady yields back. The Chair now recognizes the gentleman from Kentucky, Mr. Guthrie, for five——

Mr. GUTHRIE. Thank you——

Mr. JOHNSON [continuing]. Minutes.

Mr. GUTHRIE [continuing]. Mr. Chair. I appreciate the recognition.

And there is a cost. First of all, I will bring up Paradise, Kentucky, in Muhlenberg County. It is—TVA shut down its coal-fired power plant. It devastated that community, it devastated the people that live in that community, but it also devastated everybody in the TVA power area.

December 23rd, 24th, it was a cold day, it was unusually cold. But TVA wasn't prepared for it. They will give you about 10 different reasons because their board wants them to get out of fossil fuels. But I will tell you, people are suffering because of it. We had rolling blackouts, which is hard to believe in America during this time and this day.

The rising cost of gas, the rising cost of energy affects people at the low-income level at the most. I hate paying when we were paying $5—almost $5 a gallon worth of gas. I hate paying it. But I know people that had to change their lifestyle because they couldn't afford to pay it. And so this is serious stuff.

I mean, we don't need technology deniers. We need to understand that we have to have a system where people have access to affordable, reliable, and sustainable energy.

And for instance, you know, Germany tried it. We export coal, the price of coal has increased because the German economy decided they cannot continue down the path they were on. If we didn't export coal, then we wouldn't have the increased price of coal. Therefore, we wouldn't—they wouldn't have opened mines. And so now it has brought more coal into production because of the decisions that's happening in Europe.

You know, thank goodness Germany has had a mild winter. You know, I was praying for a mild winter for Germany because of some of the decisions they made. You are talking about people affected? People could have died from the cold weather if it had moved forward.

So it is important that we are part of this global economy. And for instance, I was in—I think Ms. Schakowsky has left—I was in her neighborhood, dropping my daughter off for college, when I heard that Iran had bombed a Saudi Arabian oil production facility. And being a child of the 1970s, my first thought is I better top my car off—because I can get home on one tank of gas if it is all the way to the full—expecting disruptions in the gas supply. But because of energy independence, we didn't have any. And I think it went up maybe a dime for a half a day, or a day or so.

So the point is we have got to do all the—and I grew up, I went to college on the Hudson River, about 45 miles from New York City. You couldn't swim in it when I was there. We don't want that, absolutely don't want that. Now, it is—fortunately, it has rejuvenated itself. We put in protections in place.

And so we need communities that are safe, we need communities that people can live in and enjoy the beauty of the Hudson Valley, which they can now because of laws that Congress put into place and efforts that people moved forward.

So we are not selling that, but we want people to have access to affordable, reliable, and sustainable energy. And one way to do it— I want to talk, Mr. Menezes and Mr. Eshelman, on—I have a bill in this, or a resolution with—that says we don't want export bans on petroleum and natural gas. And the reason is that it actually produces lower prices for everybody when producers can engage in the world marketplace. It allows expansion of supply. So that is why, even though the expansion of demand allows expansion of supply at the price that is sustainable, you may get some short-term lower prices, but not in the long term.

Also, do we want our friends and our neighbors and our allies to be dependent on dictators? If you are dependent on dictators, you are vulnerable to them. So when we choose to say we are going to keep it all here, we are going to say to our European friends, "You are going to have to buy from Putin, you are going to have to buy from Iran, you are going to have to buy from Venezuela."

And so, if the two of you would, kind of talk to why it doesn't make sense to ban exports of petroleum, natural gas, Mr. Eshelman and Mr. Menezes.

Mr. ESHELMAN. Well, Congressman, there are a few reasons.

One is if we continue to produce oil here at home, those are jobs that remain. If we would stop exporting the oil, those jobs would disappear. So it actually helps when we are producing more here at home and exporting to keep those wells pumping.

The other thing I mentioned before is the trade deficit. It has come down tremendously because of the export of oil and natural gas to other countries. About 68 percent of our LNG exports go to Europe. So it helps our allies, and it helps our own national security. It makes sure that we are on the world stage and being a player.

And so those are the big themes that I would hit.

Mr. MENEZES. And I look at it, really, from a separation of powers, you know, viewpoint, because having worked for Congress for many years—and I got, you know, accustomed to the fact that, unless Congress says it, you know, others can't do it, and we set the law of the land, and then you go over to the executive branch, and the first question you ask is, well, if Congress doesn't prohibit me from doing something, then I have all the authority I need to do something.

And it becomes relevant when emergencies occur, and the President wants to take action to solve a problem. What emergency authorities, you know, do I have? What can I do? What do I have in the Constitution? What is—what has Congress said I can and can not do?

And with respect to bans of oil, Congress has been—has set a process in place. The President can actually take decisions to limit exports, but he has to do it following a process that Congress clearly put in the bill when they decided to lift the export ban. So he can't simply unilaterally declare an emergency and take such action.

And this is what you do: You make him follow the law that Congress enacted. So that is my view of this and why I think it is important for Congress to express the clear intention to the executive branch to read the law and follow the law.

Mr. JOHNSON. The gentleman yields back. The Chair now recognizes the gentleman from Maryland, Mr. Sarbanes, for 5 minutes.

Mr. SARBANES. Thanks very much, Mr. Chairman.

Thank you all for your testimony today. As we know, we are considering several bills that would, according to our colleagues on the other side of the aisle, unleash American energy.

But, no surprise, a lot of these bills that are on the docket here are more of the same. They are really unleashing more profit-making by the big oil companies, and doing that at the expense of the American people, in my view. And it is beyond me why our colleagues, with these pieces of legislation, would seek to erode what are bedrock environmental laws for the sake of unleashing American energy, as if we need to choose between health and safety of communities on the one hand and promoting energy on the other hand.

One of the bills we have talked about—but I want to come back to it—that we are being—that we are considering today would create this new regulatory pathway for, quote, "critical energy sources" under TSCA, and it would circumvent what were bipartisan reforms that Congress passed in 2016. It would require EPA to consider nonrisk factors when determining the risks associated with a substance, which is very backwards thinking, you would think, in this day and age.

Congress deliberately and explicitly prohibited EPA from considering such factors when determining whether a chemical presents unreasonable risk. These factors are, however, considered in the risk management stage.

The bill would also provide a pathway for a, again, critical energy source to enter the market without regulation if EPA does not act in the review period, thereby potentially exposing communities and workers to toxic chemicals.

Mr. Chairman, without objection, I would like to enter into the record a letter that we received from the Natural Resources Defense Council and nine other environmental groups opposing this particular legislation.

Mr. JOHNSON. Without objection, so ordered.

[The information appears at the conclusion of the hearing.]

Mr. SARBANES. Thank you, Mr. Chairman.

Mr. Garcia, getting back to the broader frame here—and you have spoken to it, but I would like you to do it again, if you would. Is it necessary to roll these environmental protections back in order to secure our Nation's energy independence?

I mean, do we need to choose, make this choice between our environmental laws and energy, or can we do both?

Mr. GARCIA. Absolutely not. We don't have to choose. We can absolutely do both. We have the technology. And in fact, some of the legislation presented today would actually curtail that technology, which is sad to see.

Mr. SARBANES. Arguably, if you look historically, when we have leaned in with more—a more of an enlightened perspective on what we need to do with the environment, not only has that not compromised our economy, it ends up driving new economies that benefit. The pie grows from that. And I think the same can happen here. And if we go in the other direction, as is being suggested by these bills, we could undermine that kind of opportunity.

So I definitely agree with you, and it is one of the reasons that we passed in the last Congress the Inflation Reduction Act, which would put our Nation on track to reduce greenhouse gas emissions, accelerate the development of reliable and clean energy.

Unfortunately, again, our colleagues here seem unwilling to commit to a more sustainable future, as evidenced by their eagerness to get in there and start repealing all these things—again, many of them that came to pass and are on the books because of a bipartisan consensus and understanding that this is the right thing to do for our environment and for our economy and for our energy future.

So, Mr. Garcia, let me ask you one last thing while I have the time. What would erosion of our environmental laws mean for particularly disadvantaged and underserved communities, which, as you know, are already overburdened? If you could, speak to that. Mr. GARCIA. Yes, absolutely. You know, I don't know if I caught that right, but where—when Mr. Menezes mentioned, you know, we are not creating refineries and the old refineries that we had are going out, I am not sure if I caught that right, but I heard they were brownfield sites now.

Now, think about that. Brownfield sites, which are hugely contaminated sites that have not been cleaned up, that were contaminated by industry, are in that place now, and we want more of these things throughout the country? Right? That is what is compromising communities everywhere.

And so, unfortunately—and I have to say this very explicitly— these facilities are not in upper-class White portions of cities. These facilities exist primarily in neighborhoods of color, people—where people of low income live. And so that means that they are the ones absorbing this pollution, first and foremost.

And even though our laws are not perfect, they offer protection. We are now to weaken them further with these loopholes left and right, whether it is the Clean Air Act, cleanup laws, and planning laws, permitting laws that would avoid us having to clean up because we are planning correctly—that seems like it would multiply the damage.

Mr. SARBANES. I yield back, Mr. Chairman.

Mr. JOHNSON. The gentleman yields back. The Chair now recognizes the gentleman from Virginia, Mr. Griffith, for 5 minutes.

Mr. GRIFFITH. Thank you very much, Mr. Chairman.

I am so glad to have another Virginian here today, my old friend—I hope that doesn't hurt your reputation any; I know we have known each other at least 20 years, probably more than that—Bernie McNamee. Mr. McNamee, it is good to have you with us as an energy expert and somebody who teaches law at the Appalachian School of Law in Grundy, Virginia, in Buckhannon County. Now, I mention that because I recently had an interesting tour, which is right on your way as you drive there, as you get to that turn where 460 and 19 separate, and you turn west heading towards Grundy in Russell County—or excuse me, in Tazewell County, headed towards Richlands. Just off on your left behind the Food City is a CONSOL office, and they are doing some fascinating new work.

They have got a new technique to more efficiently and cleanly capture coal bed methane, and they are using it right now at Buckhannon No. 1. As you know, Buckhannon No. 1 is a huge underground mine for metallurgical coal. That means we make steel out of it, for those who don't know. The footprint underground right now is about the size of Washington, DC, and they are getting ready to open up a new section in the next couple of years. And so they are capturing, in a very clean and efficient manner, the coal bed methane out of that mine.

But what people may not realize is they also can use this technique to capture this from existing mines or mines that—or areas that have coal that may not have ever been mined, but because they may be close to the surface or whatever, they have escaping methane gas. We can use it on that, too, but they don't get any credit for having a clean, efficient way, because it is the dreaded fossil fuel. It is natural gas. Oh, my gosh, egads, it must be bad. But here is a way that American technology is helping us.

Do you think that is a good way that we should go, and that if we are going to do something with credits, that we ought to be looking at things that make it so that smaller steps forward can be made with existing fuel sources and baseload like natural gas, et cetera?

Turn your mic on.

Mr. MCNAMEE. Congressman Griffith, it is great to be back in front of you once again. And I think you are correct—is that the innovation—the American people have been the ones that have solved most of our energy crisis and energy challenges.

You think about the fracking revolution with George Mitchell in Texas, the innovation that CONSOL is doing, these resources—and what is neat about natural gas, what is great about the methane that is being produced is that it is something that can be dispatched and used to keep the grid growing. And these are very important things that we need to be focused on as a country.

It is great to talk about—you know, we think we can go 100 percent renewables, but the reality is, with technology we have today, we have to have dispatchable energy, and that is going to come from natural gas, from the methane that is captured at the coal seam, and these are things that we should not look negatively about. They have made our economy, people's lives, and the quality of life for all American people much better. And it is something we should embrace.

Mr. GRIFFITH. Yes, I appreciate that.

Mr. Menezes, you agree that we probably ought to be using this technology and rewarding it, instead of excluding it from being able to receive money from, you know, the Greenhouse Gas Reduction Program? Because it is a fossil fuel, I don't think they are eligible.

Mr. MENEZES. To be sure. I mean, there are a lot of technologies that we can be deploying right now to help reduce emissions. Remember, our quest here is not to choose one type of energy over another. Our quest here to solve the climate problem is to reduce emissions.

Mr. GRIFFITH. Yes, and I appreciate that.

Mr. MENEZES. So anything we can do to reduce emissions, it doesn't make any difference whether you are using coal or fossil, you have technologies to deploy, and the IRA actually encourages it in some in some ways to do that. So it is reducing emissions, not saying something should be antifossil or, you know, or renewables. That is the——

Mr. GRIFFITH. Yes.

Mr. MENEZES [continuing]. False narrative. It is reducing emissions.

Mr. GRIFFITH. And I appreciate that.

Back to you, Mr. McNamee. You know, it is interesting that that Buckhannon No. 1 mine, it is expanding. A lot of times people want to talk about, oh, we are going to have new sectors, and we are going to hire all these people in the renewables area. And while the CEOs may make good money, the frontline people don't make nearly what the miners in Buck 1 make. The new section, they estimate, is going to be about an average of $103,000 a year for people with sometimes not even a high school education. They are able to get in there, and they are able to learn a trade and go forward, and that is very exciting.

Also, as a former FERC representative, I would like to see us move forward on our pipeline reforms, including possibly even having collocation. But my time is out, so we will have to talk about that privately another time.

Mr. MENEZES. I would be happy to.

Mr. GRIFFITH. I appreciate. Always good to see you. Thank you for being here today, and for spending your time with us.

I yield back, Mr. Chairman.

Mr. DUNCAN [presiding]. The gentleman's time is expired. We will now go to Mr. Cárdenas for 5 minutes.

Mr. CÁRDENAS. Thank you, Mr. Chairman, and also Ranking Member, for holding this committee today.

As Members of Congress, we have been entrusted with the duty to protect and improve constituents' quality of life, their health, and overall well-being. And, as members of this committee, we have a unique opportunity to examine and put forth real solutions that advance our Nation's energy independence while ensuring a healthier future for our children and grandchildren.

Today we have convened to discuss 17 bills, all of which are partisan, none of which Energy and Commerce Committee Democrats were consulted about. Welcome to the

Republican rodeo, ladies and gentlemen. And this ain't my first rodeo. I have been in the minority in the House of Representatives once before, and here we go again.

The bills being discussed today are an attack on the environmental and public-interest laws that are most essential to ensuring that our constituents can breathe clean air and drink clean water. These bills will not serve the American people. They are intended to serve fossil fuel companies who continue to see record-shattering profits while everyday Americans pay higher prices at home.

Today's hearing is an—is indicative of my Republican colleagues' misplaced priorities, and their willingness to sacrifice the health, the safety of the American people, starting with frontline communities like the one that I represent.

My district and too many communities across our Nation know all too well the challenges of environmental injustices. In 2020 my district was impacted by a methane gas leak, a leak we later found out had been occurring for over 3 years before the actual community found out that it was going on right in their midst. In my district, residents already breathe some of California's most polluted air, and a chemical disaster can be a death sentence to vulnerable communities like the one that I represent.

There are far too many communities across the country that we were sent here to help protect and to make sure that their quality of life is secured. But yet, with some of these bills, they are really focused mostly on what industry prefers, rather than what the American people truly do deserve.

My first question is for Mr. Garcia. Can you elaborate on the implications of proposals that hinder efforts from the EPA's risk management program, and what would happen if they became law?

Mr. GARCIA. Absolutely. So EPA's risk management program essentially calls on refineries to be able to present safer alternatives or to try to study safer alternatives to their methods of production. And so it is a huge problem, because the bill that we are seeing today would essentially exempt—again, a loophole—from the Clean Air Act to refineries using hydrofluoric acid, which is incredibly dangerous for communities that live near the refineries and that really depend on the air around the refineries themselves.

And so this is something that it is very much common sense. In fact, much of the industry is already doing it. The fact that we are talking about a bill that would eliminate this is sort of puzzling, because large refineries are doing it. And again, it is common sense. It is the idea that you should study to see if there are safer and effective ways to go about the business that you have already been doing for quite some time.

Mr. CÁRDENAS. Thank you. And what would accountability look like for communities facing chemical disasters, particularly those that are low-income communities across the country?

Mr. GARCIA. Well, you know, on the foremost, we have to make sure that—we have seen disasters before. I mean, a couple of years ago, the plant in Philadelphia exploded and was a national disaster because of this same kind of lax enforcement of laws.

So what we really need to focus on is making sure that it doesn't repeat itself. And particularly for communities of color and low income, and those that live near those refineries, We need to make sure that the planning is done right and that it is reoccurring. We can't just pretend that a facility that is there and that is exempted from the Clean Air Act is suddenly going to be responsible enough to do its job. It has to be held in check, and that is why those laws exist.

Mr. CÁRDENAS. One of the reasons why I ran for office many years ago was to make sure that I am a voice for the community that I grew up in, and I grew up in one of the most impacted communities in all of Los Angeles or Southern California, with more dumpsites and plants around my home than most people would ever want to have to deal with.

In your testimony you explained that, under the bill that amends the Solid Waste Disposal Act, facilities could operate before securing a permit?

Mr. GARCIA. That is right. That is right. It essentially allows industry to roll out the red carpet, march in, do everything that it wants to do before we even know whether the practice is going to be safe, whether the appropriate precautions are being taken, whether the community's—the community alternatives, those projects that the community is actually bringing forward to accomplish the same goal, are being considered. All of that gets done through those laws, and somehow it is not going to matter.

And then it is like this, right? And then, once it is operational, we get the excuse that, well, it is already there, so we can't draw it back. It is like——

Mr. CÁRDENAS. So——

Mr. GARCIA. It is like pouring, like, food coloring in a cup of water.

Mr. CÁRDENAS. Thank you.

Mr. GARCIA. Good luck getting that out.

Mr. CÁRDENAS. Thank you.

Mr. Chairman, I just want to note that I think the clock was backwards on my——

Mr. DUNCAN. It was.

Mr. CÁRDENAS. So did it actually start at——

Mr. DUNCAN. Yes. You got 5 minutes and 44 seconds as of right now.

Mr. CÁRDENAS. OK. But I noticed that it was going up, in—rather than going down then.

Mr. DUNCAN. That is correct.

Mr. CÁRDENAS. OK, thank you.

Mr. DUNCAN. They didn't reset it, but you weren't shorted on time.

Mr. CÁRDENAS. OK, thank you so much.

Mr. DUNCAN. We apologize. They are going to make sure to reset it.

Mr. CÁRDENAS. No, no——

Mr. DUNCAN. And I will go to the crossroads of America, the gentleman from Indiana, Dr. Bucshon.

Mr. BUCSHON. Thank you, Mr. Chairman. Today I am speaking in support of the Securing America's Critical Minerals Act, a bill that our former colleague, Mr. Upton, introduced last Congress and I am introducing this Congress, and I am looking for Democrat cosponsors, if anyone is interested.

I do find it fascinating that some of the same groups that support total conversion to electric vehicles are also the same groups that are working to block the mining of minerals such as cobalt and lithium that are required for the batteries—block it here in the United States.

Just to mention, you know, that China is a major supplier of the lithium. And as far as the cobalt goes, I don't know if anybody has seen the video of the mines in Africa, but, essentially, slave labor in Africa to get the cobalt. And I would encourage everybody to look at the YouTube videos of those mines with little children digging through the dirt, trying to find cobalt.

So over the last few years we have discussed in these subcommittees the importance of critical minerals and other energy resources necessary to providing for our energy needs, as well as the potential vulnerabilities that exist in the supply chain and domestic production and capacity limitations.

This bill would ensure that the Secretary of Energy is engaged productively in addressing the issue. It would require the Secretary to conduct an assessment of our Nation's energy supply, identify resources that are critical to our economy and vulnerabilities in the supply chains of critical energy resources, and determine the extent to which critical energy resources play a role in developing new energy technologies.

The bill defines critical energy resources as those that are "essential to the energy sector and energy systems of the United States," and the supply chain of which is vulnerable to disruption.

The bill would also direct the Secretary of Energy to diversify energy sourcing and increase domestic production, refining, and processing of these resources.

As a supporter of an all-of-the-above energy approach, I appreciate our need for a diverse energy portfolio. We must take steps to ensure we safeguard our supply chains as well as prevent our adversaries from weaponizing potential vulnerabilities in these supply chains, and critical minerals is a large part of our vulnerability—I can't get that out today for some reason. As a country, we need more production here.

So, Ms. Sweeney, how potentially could securing America's Critical Mineral Supply Act help our country and the energy sector reduce this reliance on China and other foreign sources of these critical mineral needs?

Ms. SWEENEY. Thank you so much for the question, and that is a really important question.

I think that one area that the act really is important is having the Energy Information Agency actually look at that connection between minerals and energy. As I said in my testimony here earlier, there isn't any form of energy that doesn't rely on minerals as the—you know, as the base of that energy. So it is very important to focus on where these come from, where we are getting them.

And in particular, you know, you mentioned the processing and smelting and refining. You know, that is an area where we need to focus attention, as well. It is not just the mines themselves, but the processing also needs to take place here in the U.S.

Mr. BUCSHON. Mr. Menezes, do you have any comments on that?

Mr. MENEZES. Well, I think the bill is especially important because there are other agencies that also would like to get resources to look at critical minerals, and it is important to have the Department of Energy, in its organizational act, have the statutory authority to be the experts throughout the interagency process. Those agencies compete at—for Congress dollars.

And so when you can point to legislation that says we need this to do our jobs, it is important, really, and we hope that this would be a strong bipartisan bill to declare the Department of Energy— the Department of Energy has those national laboratories. I mean, they are better equipped than any other agency. But at the end of the day, it is resources.

And so this act is really important for that. Put that expertise at the Department. It will protect this committee's interest throughout that interagency process that can get pretty tough.

Mr. BUCSHON. Thank you. I do also want to emphasize what my colleague, Mr. Guthrie, mentioned.

We now have an example of what happens when you take an energy approach that this administration appears to be taking, and that is Europe, and that is countries like Germany. They are now building coal-fired power plants, importing coal from in the United States also, and it is probably going to set their clean energy agenda back decades by getting too far ahead of themselves and trying to restrict certain forms of energy rather than taking an all-of-the-above approach and advancing innovation and technology in every area of energy production.

I yield.

Mr. DUNCAN. I thank the gentleman. Now we will go to California, Mr. Peters, for 5 minutes.

Mr. PETERS. Thank you, Mr. Chairman. Last week I encouraged Republicans and Democrats on this committee to engage in a constructive bipartisan process to enact sensible permitting reforms to deliver energy security and environmental protection for the American people. I spoke about the need to reduce excessive process requirements necessary to build clean energy projects and reform environmental laws from the 1970s to meet the challenges of today. I talked about the dismal state of our electric grid and how we must build 200,000 miles of new transmission lines by the 2030s to keep the lights on, lower costs for Americans, and build clean energy projects like solar, wind, hydropower, and nuclear.

And my Republican colleagues have said publicly they are committed to advancing permitting reform, and that they are focused on an all-of-the-above energy agenda to secure our energy future, and that is why I am a little bit disappointed about the hearing today. We are discussing 17 bills, and almost all of them are focused on solely natural gas and oil. And doubling down on oil and gas will lead to more price uncertainty and financial pain for Americans.

Speaking of all of the above, the Energy Information Administration says that a very small amount of the planned projects on the ground today are going to be natural gas, and that 86 percent are zero-emission projects. That is what we are trying to build. But we don't really address that today. There is no focus today on key energy technologies like solar, wind, transmission, energy storage, advanced nuclear, hydropower, or hydrogen. And for a party that claims it doesn't want to pick winners and losers, Republicans are seeming to pick oil and gas every single time.

So I am not going to give up hope. But if we are going to pass permitting reform in this Congress, it has to be bipartisan. Today we are using our time to discuss partisan bills that I really don't think will be going anywhere, relitigating a pipeline that was terminated more than 2 years ago by the company developing it. I think we could do better.

I am a proud Democrat. I am a former environmental attorney. I am a climate hawk ready to have hard conversations about permitting reform. And we can compromise on NEPA, the Nuclear Regulatory Commission, hydropower relicensing, critical minerals, interstate electric transmission, and more, but I really want to get about that business.

Let me ask a couple of questions of the witnesses today, and thank you for being here.

Mr. Menezes, you helped negotiate the Energy Policy Act of 2005, which included language intended to streamline the construction of electric transmission lines. Can you elaborate on why we need to build these lines faster and the importance of advancing bipartisan legislation to permit these projects faster?

Mr. MENEZES. Well, even today, as it was back then, I mean, there are probably more difficult things to build and site—interstate transmission line, but it is hard for me to come up with what they are.

Mr. PETERS. Yes.

Mr. MENEZES. Really, it has opposition almost everywhere you turn. And although the goal of modernizing our grid, to make it green, et cetera, and to embrace the energy transition, it is one of the most difficult things to overcome, an interstate transmission line crossing States of low populations, et cetera.

So I know Congress has been looking at this. There are ways to go about trying to do this. We tried it with backstop authority. Two courts told us we didn't quite get it right.

Mr. PETERS. Right.

Mr. MENEZES. There's been some proposals over in the Senate to look at that. And so I think that this is something that certainly is within this committee's jurisdiction to take another look at.

Mr. PETERS. I drafted the POWER ON Act, which was put in by the Senate into the infrastructure bill to provide backstop authority. That is certainly helpful. But when we have a project that takes 10 years, and 7 of those years are permitting and processing, we will not be able to build the grid that we need to electrify this economy. And I think we are going to lose a lot of the benefit of the IRA if we do—if we don't.

Mr. Slocum, methane is a superpollutant responsible for about 25 percent of humanmade warming. Today we hear claims that we produce the cleanest energy in the world. I am not sure, if you consider methane, that that is true. But isn't it essential that oil and gas producers significantly reduce methane leaks to be the cleanest in the world?

Mr. SLOCUM. Yes.

Mr. PETERS. The IRA included billions of dollars of new funding to help large and small oil and gas companies reduce their methane emissions. The EPA is finalizing a new rule to reduce those emissions from oil and gas operations.

Will the oil and gas industry significantly reduce methane emissions in the absence of that regulation and strong funding?

Mr. SLOCUM. I don't think so. I think you need to have that regulatory structure and—in order for the industry to make those investments.

Mr. PETERS. I agree. And on this too I would reiterate my willingness to work in a bipartisan way.

One of the things I think we can offer to the small producers who are concerned about these costs is that the IRA provides funds to help those companies comply.

I also don't pretend that oil and gas is going away tomorrow. It is going to be around for a while. While it is around, we need to make it cleaner, and I am willing to work on that, as well.

Mr. Chairman, my time is expired. I yield back.

Mr. DUNCAN. I thank the gentleman. Now we will go to the vice chair of the Energy, Climate, and Grid Security Subcommittee, Mr. Curtis, for 5 minutes.

Mr. CURTIS. Thank you, Mr. Chairman. I, like many of you, view PFAS as a four-letter word. And I think on this committee it has been demonized frequently, and in some cases rightly so. But it might surprise all of us and my colleagues to know that a number of products needed for transformation and energy production require PFAS.

As a matter of fact, semiconductors, green hydrogen membranes needed for electrolyzers, hydrogen used in fuel cells, and lithium batteries all require fluoropolymers, especially plastics. Critical PFAS is used in EV charging infrastructures, batteries, powertrains. I have a list here of 21 uses in the semiconductor industry of PFAS. So frequently, when we quick rush to judgment and lump all of these together, it is probably a mistake.

I have a bill, one of the 17 that has been discussed today, that would make sure that these chemicals are approved in a timely manner. There is nothing about this bill that asks the agency to approve anything that is not safe, that is not healthy, simply to do it in a timely manner.

In short, chemicals are all around us and necessary for every industry, but especially to achieve decarbonization of our economy in the world. If we want a clean future, we need to approve chemicals more effectively and responsibly. My bill would help deploy clean energy technology more quickly and puts the following three conditions on there.

Hard deadlines on EPA's ability to make a decision on the risk presented by the critical energy resource. I believe, if we give them 180 days, it will take 280. If we give them 280, it will take 380. They need to stick to the guidelines that we have given them.

It prevents the EPA from telling an applicant to suspend their application unless EPA has reviewed the notice—it sounds reasonable—and make a determination.

And it requires EPA to consider cost and other nonrisk factors in determining if an unreasonable risk is present. We heard earlier from Mr. Garcia that that was akin to killing people, and I adamantly disagree with that. It does not say that they should make an unwise decision, simply that they should take that into consideration, and that seemed pretty melodramatic to me.

Mr. Menezes, you have experience as a former Deputy Secretary of Energy. Can you speak to the importance of chemicals in the energy sector?

Mr. MENEZES. Without critical—one of the two—I learned two lessons when I visited the labs. To achieve breakthrough technologies, we need two things: one, high-performing computing to do modeling and continue to do modeling—and it is modeling that is important—and another thing is we need to create chemicals and products that do not exist today. They do not exist, and we need to do that.

And to make that point, they actually gave me a new product that they had made.

[Holds up a coaster.]

Mr. MENEZES. This was at Argonne, and this is cesium aluminate. It didn't exist before 2017. It is going to be a key product that is going to be used in our green energy future. And to even make the point, they used the same product to make me a 3D printed replica of our Capitol building.

So this is what the future looks like. The problem is that these chemicals are bollocksed up at EPA. EPA reads risk as almost any risk is unreasonable, and so they can't seem to make decisions.

And so all the companies are asking, particularly those driving toward the EV technologies, et cetera, is, "Please tell us what the rules are, tell us what we need to do," and we need to do that.

But China is going to maintain the dominance in the electric vehicle space, OK? They are beating us. We need to get our act together, please. Just——

Mr. CURTIS. So, first of all——

Mr. MENEZES [continuing]. What we need to do.

Mr. CURTIS [continuing]. Everybody is not going to believe that this wasn't a setup, you came prepared with your props for my question. So you and I didn't rehearse this.

Mr. MENEZES. We did not rehearse it. In fact, as I was reading the bill, it dawned on me that this made a big impression on me. It was on my desk when I was preparing for this hearing. This was not set up.

Mr. CURTIS. So before we run out of time, my friends on the Democratic side often emphasize the importance of decarbonizing quickly, and I agree with them. But then they call bills like mine undermining environmental laws. Do you think it is possible to move in a reasonable speed and protect ourselves, as well?

Mr. MENEZES. Of course. We are trying to figure out ways to get the laws that are on the books to actually produce results. Please do your job. Stop with the delays.

The applicants go in and they are told—and they are—the bureaucrats are aware of the 90-day rule. So they will get you to withdraw and resubmit, because they can't meet the statutory deadlines. So they are sort of trying to do their job, but we have reached the point where the backlog now is so significant that those that want to take advantage of the IRA provisions can't do it because——

Mr. CURTIS. I am going to run out of time, but——

Mr. MENEZES [continuing]. They can't get the——

Mr. CURTIS [continuing]. So I just want to make two——

Mr. MENEZES [continuing]. Can't get their permits.

Mr. CURTIS [continuing]. Two quick points. There is a lot of parallel here with permitting reform.

Just tell us what the rules are, and then let us do it and make it timely and predictable. That is what people are asking for.

And Mr. Chairman, regrettably, I am out of time. I yield the balance of my time.

Mr. DUNCAN. I thank the gentleman. The Chair will now go to the gentlelady from an energy-producing area of the State of Texas, Mrs. Fletcher, for 5 minutes.

Mrs. FLETCHER. Thank you so much, Mr. Chairman. Thanks for holding this hearing. Thanks to you and, of course, Ranking Member DeGette, as well as our chairman—chairwoman, and ranking member, and others.

And I have listened to the testimony today and to the questions. I appreciate all of your time in being here. I think there has been a lot of really important information conveyed, as well, in your written testimony. And I am a little bit concerned about some of the things that I have heard this morning, specifically some of the bills that have been introduced for this hearing.

And I would, of course, like to note that we are considering 17 bills, many of them aimed at repealing some of, I think, the very good work that we did in the last Congress to try to address the complexities of the policy that we are trying to do here.

And so, you know, I would request, hopefully, that we will get a little more notice in the future in time to review these bills, because I think that, you know, we all know—and everyone on this committee should know and understand—that energy policy is complex. And we have had no better example than what we have seen happen over the last year, and the importance of all of us really having a depth of understanding on this committee as we work to make policy.

Certainly, what we have seen happening in Europe with Russia's unjustified and unconscionable invasion of Ukraine, what we have seen happen to our friends and allies in Europe, what we have seen happen here in the United States as a result of the market demands and some of the things that we have talked about today— I disagree with some of our witnesses about the importance of, for example, exporting natural gas and being able to help our allies reduce their dependance on Russian oil and gas at this critical moment.

What we know is this is domestic policy, it is foreign policy. It has real consequences in our communities. People who are living, especially in my hometown of Houston, people who are living near the largest petrochemical complex, arguably, in the world, there are real-world health impacts. There are real-world economic impacts. These are also our jobs.

So coming together and building consensus around what we can do, I think, is incredibly important for this committee. And I heard a few things today that I just—I want to take up.

Most important, I think, Mr. Menezes, you said—and I agree with you—that the issue here is that we want to reduce emissions. That is what we are trying to do, and that is what the good legislation that we passed in the last Congress really does. And so, you know, I think that what we did in the Infrastructure Investment and Jobs Act and, importantly, the Inflation Reduction Act is really important to accomplishing those goals.

And I am going to disagree with and I am disappointed to see the bill 484 that has been introduced by my friend from Texas, Mr. Pfluger, because I think it undoes the important work that we did on this committee just last year in trying to address the impacts of methane and deal with that in a way that is reasonable and workable for industry.

And we spent a lot of time on this committee and got a lot of criticism for what we have, a billion and a half dollars to help small operators employ the technology, this grant program. And so I want to kind of direct my question there, because I think, Mr. Eshelman, I saw you shaking your head during Mr. Slocum's testimony about the IRA and the methane fee. But there is funding there to help smaller operators in particular, because that was a real concern that we heard about, the ability to implement the technology to reduce methane emissions, which should be the goal of everybody here. And I think on both sides of the aisle we keep saying that is what we want to do.

So, you know, I would like to see if there are real concerns still about the implementation and the ability to implement that, something that we can do that doesn't involve repealing what many people in the industry have said is a very smart solution to try to address the complications but also address methane emissions. And so that is a concern that I have. And, you know, the question I have for you is, does this mean—I mean, it sounds like you don't support this, you want to see it repealed.

Are your members not going to take the grant money that we provided to try to assist them? What do you think should be happening with that?

Mr. ESHELMAN. As we speak with our members, it has been a very contentious relationship with EPA, especially the enforcement office. So we are mostly concerned that this grant money will come out of EPA. We think it should better come out of DOE and maybe the Petroleum Technology Transfer Council. So we are just mostly concerned that it is EPA that is making these regulations.

Mrs. FLETCHER. And so what I am hearing you say is the idea of the program, the idea of reducing emissions and having some coming——

Mr. ESHELMAN. Right.

Mrs. FLETCHER [continuing]. Through this legislation, it is designed to work in tandem with EPA, so that there can——

Mr. ESHELMAN. Our——

Mrs. FLETCHER. The concern you are expressing is the funding source, versus the idea of what we tried to do with this legislation. So it sounds like we——

Mr. ESHELMAN. That would be one——

Mrs. FLETCHER [continuing]. Don't necessarily need to repeal it.

Mr. ESHELMAN. That would be one of our concerns, yes.

Mrs. FLETCHER. OK. Well, I am going to run out of time. I have a ton of questions. We are going to cover them on this hearing. I hope I can work with Mr. Pfluger and my friends on both sides of the aisle to really work on understanding the depths of this incredibly complicated work we have in front of us and working together to achieve all of our shared objectives.

So thank you so much, Mr. Chairman, for the time, and I yield back.

Mr. DUNCAN. Agreed, and I look forward to working with you. The Chair will now go to Michigan, Mr. Walberg, for 5 minutes.

Mr. WALBERG. Thank you, Mr. Chairman, and thanks to the panel today for being with us. It is an important hearing.

Threats to our critical energy infrastructure have increased year after year. In 2022, attacks on United States power grids rose to an all-time high. More apparently needs to be done to protect our critical energy infrastructure, which is why I plan to introduce the Critical Electric Infrastructure Cybersecurity Incident Reporting Act.

Mr. Menezes, electric utilities and other energy infrastructure owners and operators are required to report critical infrastructure cyber incidents to DOE and FERC. Last Congress a law was passed that also required some of these entities to submit incidents to CISA. The FAST Act clearly establishes DOE as the sector-specific agency for energy cybersecurity and granted them authority to address grid security emergencies. As such, DOE has the expertise to best address some of these threats.

Mr. Menezes, do you agree that it makes the most sense for energy sector stakeholders to submit threat incidents to DOE and then have DOE share that information as necessary with CISA?

And secondly, how can Congress clarify DOE's role in the process?

Mr. MENEZES. Thank you very much. It is very important that you clarify DOE's role that it is the agency that private sector and other Government agencies need to report cyber incidents to. Because when Congress passed CISA, it created confusion as to where reports needed to go. When we were there, we knew the importance of cybersecurity. Of course, Congress had mandated cybersecurity standards. We created the Office of CESER, which remains today: the Cybersecurity, Energy Security, and Emergency Response.

But within the interagency—you are hearing this a lot from me today, I am bringing the experience that I gained in the executive branch—is that we need to designate DOE to have—to be the agency that all cyber incident reports on the bulk power system—this is electricity, it is not oil and natural gas, but it is electricity—to go to DOE. And that is important because we have all the expertise there, we have the information sharing there, we have some of the modeling that the labs develop to look for anomalies on data pools, et cetera.

So this is an important piece of legislation. It seems as though, you know, it is a simple thing to do. But we are going to need all support to get this through. And it is an important piece of legislation to give DOE clarity.

Mr. WALBERG. And I would hope that it would make more efficiency in the process for Members of Congress to understand what is going on and get information and more transparency, as well.

Mr. MENEZES. Absolutely.

Mr. WALBERG. Mr. McNamee, as a former FERC Commissioner, how will increased sharing and coordination of cyber incidents improve the safety and reliability of our electric infrastructure?

Mr. MCNAMEE. I think it is very important that anything that can be done to make the sharing information on the bulk power system and threats to it be done. The threats are real. As you mentioned in your comments, they are happening all the time, both cyber and physical securities.

And one of the things that frightens me the most is what happens with the limited natural gas pipeline capacity up to the northeast. If there is a physical attack on the pipelines up there, you are going to lose a lot of power. Then the problems that you have on the bulk power system in relation to transformers or the SCADA systems, these things are real threats. The utilities are being pinged every single day by foreign actors trying to get into the systems to be able to flip the switch off when we need it the most, on the coldest days of the year.

So I think legislation like yours makes sure that there is focus, and that is, as Secretary Menezes said, that there is one source in the Government that is absolutely responsible. Of course, FERC has responsibility for establishing, along with NERC, reliability standards, SIP standards, but I think it is important that reporting go through DOE.

Mr. WALBERG. Thank you. Being from Michigan, I am extremely concerned about the current backlog at EPA of hundreds of TSCA section 5 applications and the impact it has had on our auto supply chain. Last year my Michigan colleagues and I sent a letter to Administrator Regan imploring him for the bare minimum of timely review of two premanufacturing notice applications that were essential to the launch of an EV battery plant in the State.

Mr. Chairman, I would like to, if you would allow, to include that letter for the record.

Mr. DUNCAN. Without objection.

[The information appears at the conclusion of the hearing.]

Mr. WALBERG. Mr. Menezes, the Biden administration is forcing a transition to electric vehicles, yet it took almost a year for the EPA to approve this project that would supply the needed batteries. How will Representative Curtis's draft legislation improve the efficiency and timeliness of the TSCA review process so that auto supply chains remain in the United States?

Mr. MENEZES. Well, thank you, and we talked a little bit about that before.

It basically—it says that, look, not any risk is an unreasonable risk. Two, it stops them from forcing applicants to withdraw and resubmit so that they can reset the statutory deadline there. And basically, it allows them to be able to go forward after a certain time period that has elapsed while they were pending review. This will allow us to accelerate our move to the—through the energy transition, you know, to more EV use.

The letter that you mentioned, you know, expressed the frustrations clearly, and I think that is why Mr. Curtis's bill will go a long way to helping streamline the process, not remove any environmental protection.

Mr. DUNCAN. The gentleman's time is expired.

Mr. WALBERG. Thank you, Mr. Chairman. I yield back.

Mr. DUNCAN. Thank you. Now we will go to New York, Ms. Clarke, for 5 minutes.

Ms. CLARKE. Thank you, Chairman Duncan, Chairs Duncan and Johnson, and Ranking Members DeGette and Tonko for holding this hearing today. I would also like to thank our witnesses, as well, for being here to testify on these bills.

There are real challenges in America's power sector as our Nation begins an economywide transition to clean energy in the midst of the climate crisis. But I reject the premise presented today that the only way we can unleash American energy is through creating loopholes in our bedrock environmental laws and/or sacrificing the health of our communities. Many of the bills considered today would not only weaken our economic and national security but also turn the clock backwards on the progress that we have made combating climate change and protecting public health.

My first question is to Mr. Garcia.

In your testimony you state that many of these bills would circumvent bedrock environmental laws like the Clean Air Act, TSCA so polluters can profit at the expense of frontline communities. Could you elaborate on the connection between frontline communities and what waiving the Clean Air Act and Solid Waste Disposal Act would mean for communities sited near these facilities?

Mr. GARCIA. Yes, absolutely. I mean, when we are talking about the Clean Air Act, one of the—again, what they are seeking to waive with the bill that waives—that addresses the Clean Air Act is a commonsense practice that many industries already take on, which is simply to study how can we achieve the same goals that we already have in a safer way.

And we are dealing with hydrofluoric acid. And it exempts any refinery that uses hydrofluoric acid from that requirement to study if there is a safer way to do this. It is problematic, because hydrofluoric acid is extremely dangerous to the human body. It is— it can explode. And not only that, but there are already recorded alternatives that many across the industry already use in order to do the same thing in a safer way. And so there is really no other way to characterize this. It is something that industry absolutely doesn't need. And yet we are still seeing this bill being pushed through.

So, you know, unfortunately, the brunt of this comes down on those communities that live near these facilities and the workers that work in these facilities. Unfortunately, we have seen facilities like these blow up. And when that happens, it hurts the people who are there, first and foremost.

And so I think that, you know, Vice Chair Curtis said I was being melodramatic when I was talking about life and death. But it really is. It is really hard to not get dramatic when you are talking to the families of those that get devastated by the deaths of their loved ones. And so that is what is really at stake here, and that is why we have—what we have to keep front and center.

Ms. CLARKE. Thank you. My next question is for you again, Mr. Garcia, and Mr. Slocum.

So much of the discussion today has been focused on how burdensome regulations are. Can you tell us why it is so important to center community voices, especially those who have been historically marginalized, like communities of color and indigenous populations, when it comes to permitting decisions in the energy industry?

Mr. SLOCUM. It is absolutely essential that frontline communities that are being asked to host all of the hazards play a prominent role in the siting process.

We are working with an African-American community in the Florida Panhandle that is opposed to a liquefied natural gas export terminal that—we learned about it through a FERC regulatory process. Nobody in the local governments there had told these folks that this was the plan.

And this is a common, unfortunate occurrence that we see throughout the country, where the local community does not have involvement or consent. And it is crucially important that, as part of any sort of regulatory review, that those frontline communities play a prominent role in being able to have a say in the development within their own communities.

Ms. CLARKE. Thank you. Well, my time is winding down pretty quickly.

Let me just say that studies have continued to find that race, more than any other demographic, is the primary indicator for living near an energy facility emitting toxic pollutants. In fact, more than half of the individuals live—living life close to any hazardous waste site are people of color.

With that, Mr. Chairman, I yield back.

Mr. DUNCAN. I thank the gentlelady. The Chair will now go to Mr. Carter from Georgia.

Mr. CARTER. Thank you, Mr. Chairman, and thank all of you for being here.

Mr. Chairman, thank you for continuing on this theme that we have started with in the Energy and Commerce Committee this year, and that is about unleashing American energy. This is extremely important. We have all seen and witnessed what happens when we neglect American energy dominance and our own independence. It is to our own detriment. We—it results in high energy prices and diminished supply chains. And that is why I am really happy that we continue to focus on this.

And we know about supply chains, but there is perhaps nothing more important in supply chains when we talk about them than critical minerals. That has to be perhaps one of, if not the most, glaring weaknesses that we have. All of you have mentioned our dependance on China for critical minerals and how that needs to end. And we all recognize that. And we have got legislation in this package to fix that.

And I want to talk about some of that, because I am eager to talk about a bill that I am introducing, and it has to do with the Solid Waste Disposal Act, and it has to do with mining. As I understand it, when you get a mining permit, you get the first permit, then you have to—if you are going to keep the hazardous waste for longer than 90 days, you have to get a second permit. But while you are waiting on that second permit, what my legislation—what this legislation will do will be to give you an interim permit, if you will, until you can go through the process to get the second permit.

So I think it makes a lot of sense. It helps us with our supply chain for critical minerals, and it needs to be done now, because there is no time to waste. We have got to address this issue right now. No, it is not a silver bullet, but I think it is a fix that will help us tremendously.

Ms. Sweeney, I want to go to you, and I want to ask you, because it is interesting. You say in your written testimony—and I quote— "an average of 7 to 10 years to secure"—it takes "an average of 7 to 10 years to secure—one of the longest permitting processes in the world for mining projects"—to receive necessary permits to even begin to build the mine project. And then you compare this to Canada and Australia, who have kind of similar environmental regulations as we have, and there it only takes them a few years, 2 to 3 years, to complete.

How can—how have they been able to maintain comparable environmental standards to the U.S. and complete the permitting process for new mines in a fraction of the time that we do?

Ms. SWEENEY. They have a lot more coordination up front of the various agencies involved, whether they be provincial, territorial, or the overarching Canadian Government. They are seeking to do, like, one-stop permitting shopping.

They also allow the project proponent to prepare the environmental impact statement, which really involves a lot of efficiencies because you are not waiting for the agencies to have to do that. But the Federal—but the Government does oversee that to make sure that the rigorous rules are——

Mr. CARTER. So there are a lot of lessons we could learn from them, and a lot of good takeaways from them.

Ms. SWEENEY. Absolutely.

Mr. CARTER. What would be one of the most immediate that could help us?

Ms. SWEENEY. Up-fronting litigation. I think that is something in Canada that they are focused on, and getting that done, so you are not at the end of your 10-year process just entering into the litigation that could add another 10 years before you can actually start operations.

Mr. CARTER. Good, good. OK, I want to go to Mr. McNamee.

And when we talk about unleashing American energy, part of that is the structure of the market that the energy goes into. Can you help me out? Restructured electricity markets like regional transmission organizations, RTOs, do they lead to lower rates? Do they lead to greater reliability?

I mean, tell me what the advantage, if there is any, of these are.

Mr. MCNAMEE. They don't. And the RTOs were originally structured on the idea to use market forces to get efficiencies.

But the problem is that there are seven RTOs in the country, six of them regulated by FERC. The problem is that they use marginal pricing to set power prices, so you get bids in by each of the generators into—to bid to meet every 5 minutes of what the energy needed.

And the problem is usually natural gas does set the clearing price on that, but every other resource is getting paid that natural gas price. So if you are a subsidized renewable, you have no fuel costs, you have tax credits, yet you are getting paid the natural gas price. So the economic benefits of renewables are not passing through to customers. Hence the reason energy prices keep going up, despite adding all these renewables.

And then secondly, you have got a reliability problem, because none of these generators are like your utility. None of them have an obligation to serve. And so they bid in, and if they are picked, they run, if they don't—but then you have problems like you had in Texas, which you have seen also in this past winter, where there is no incentive to winterize your unit because you are, like, "Well, why should I do it if it is going to make it more expensive?"

What you want is a system that allows—that is designed to serve the people, and you need reliability as the number 1 thing.

Mr. CARTER. Good. Well, thank you all.

And again, this permitting process is crushing us. We have got to do something about it.

Thank you, Mr. Chairman. I yield back.

Mr. DUNCAN. The gentleman—the Chair will now recognize Ms.

Barraga´n for 5 minutes.

Ms. BARRAGÁN. Thank you, Mr. Chair. I am especially concerned about legislation under consideration today which says that no hazard assessment is required on the use of hydrofluoric acid at a refinery to understand the risk of an accidental release. This chemical is deadly. It is toxic. It is deadly toxic to people, and it is extremely corrosive. A hazard assessment is a common, safe safety measure the Environmental Protection Agency has proposed.

We also know there are safer alternatives available that many refineries already use.

The refineries in and near my district that still use hydrofluoric acid—and it is a safety—a serious safety concern. In 2015 there was a near miss at a refinery near my district in Torrance, California. An explosion at the refinery launched debris that landed close to two tanks containing hydrofluoric acid. People could have been killed.

Mr. Garcia, what safer technologies are available for oil refineries to replace hydrofluoric acid, and should the oil industry have to consider them?

Mr. GARCIA. Yes. I mean, there are quantities of alternatives. And that is the saddest part about this bill, is that there are quantities. One of them is sulfuric acid.

But the idea is that that is what the analysis should tell you, right? That is what the analysis that industry should have to do. That is the one that tells you, yes, there are better alternatives to do this and still accomplish the goal that we need to accomplish. And so when we hear about it, just this piece of the Clean Air Act being completely dismantled, it is really puzzling because you have everything—industry has everything that it could want in order to keep this in the books, and that—actually, compliance is relatively easy. So it is puzzling to see this.

Ms. BARRAGÁN. Thank you.

Mr. Chair, I would like to enter into the record a letter I led with Senator Booker to EPA on the proposed risk management plan rule signed by 47 Members and Senators. The letter asks EPA to finalize a rule that requires refineries to transition to safer chemicals and processes. This letter was supported by environmental groups and labor, including the United Steelworkers and United Auto Workers.

I will hand you that letter at the end of the questioning.

Mr. DUNCAN. Without objection.

[The information appears at the conclusion of the hearing.]

Ms. BARRAGÁN. Mr. Slocum, I am concerned about the Unlocking our Domestic LNG Potential Act, which eliminates the requirement that our Department of Energy find imports and exports to be consistent with the public interest.

What could the consequences of this bill be on domestic energy prices and our climate?

Mr. SLOCUM. Well, the public interest should be comprehensive. It should look at environmental justice considerations. It should look at climate and clean energy considerations. So it should be requiring the Department of Energy to perform those assessments, to require applicants to document how LNG exports—whether or not they are displacing dirtier forms of energy abroad, or whether or not they are displacing renewables by bolstering existing gas infrastructure.

Right now, the Department of Energy is not performing that assessment. To eliminate that assessment altogether, which has been in place for 85 years, would be a huge disservice to communities, especially in the Gulf, that are being asked to host these massive facilities.

Ms. BARRAGÁN. Well, thank you for that, and thank you for raising the issue of environmental justice issues.

I share the climate concerns, and I want to underscore the impact of energy prices because my constituents have been hit hard by this—by rising natural gas prices. And this shows how volatile fossil fuel prices are and why we need the Department of Energy to be a check on the fossil fuel industry's proposals to have countries like China competing with American consumers to buy gas. This is why the Industrial Energy Consumers of America, a coalition of manufacturing companies, has opposed this bill. I think it is bad for the U.S. manufacturing.

Mr. Garcia, legislation under consideration proposes to repeal the Greenhouse Gas Reduction Fund that was passed by Democrats in the Inflation Reduction Act. This fund would provide low-income communities with grants and loans to reduce greenhouse gas emissions with zero-emissions technologies. Can this fund help low-income residents to reduce their energy bills?

Mr. GARCIA. Absolutely. I mean—and we saw that—that fund is really meant to make sure that we don't make the mistakes that have happened in previous energy revolutions, where often people of color and low income get left out. And so this fund is specifically meant to make sure that low-income and people of color have access to those funds.

Ms. BARRAGA´N. And can the fund help create clean jobs, energy jobs, in low economic communities——

Ms. GARCIA. Absolutely.

Ms. BARRAGÁN [continuing]. Low-income communities?

Mr. GARCIA. Absolutely.

Ms. BARRAGÁN. Well, thank you for that. It is unfortunate, the Republican talk about lowering energy costs and creating jobs. They talk about it, but they want to repeal programs that empower communities to do just that.

With that, Mr. Chair, I yield back.

Mr. DUNCAN. The gentlelady yields back, and the Chair will now go to the gentlelady from Arizona, Mrs. Lesko, for 5 minutes.

Before I do that, there are going to be votes called around 1:30. Apparently, there's two votes. We are going to plow through and get as far as we can, but we will recess and come back and finish the hearing. Members are encouraged to come back right after the second vote is called. And vote, come back, and we will get back on it, because there is another hearing following this.

Mrs. Lesko, you are recognized for 5 minutes.

Mrs. LESKO. Well, thank you, Mr. Chairman. I love this committee, because we are talking about energy, critical minerals on the environment. What could be more important than that?

Mr. Chairman, I am proud to sponsor the House concurrent resolution expressing disapproval of the revocation by President Biden of the presidential permit for the Keystone XL pipeline. This is a simple resolution, and I hope my colleagues will support it. Many of my colleagues on the other side of the aisle have said they disapprove of the cancellation of the Keystone XL pipeline, and this resolution gives them and the entire House of Representatives the opportunity to show our disapproval.

On day one of his administration, President Biden canceled the Keystone pipeline. Soon after he canceled the Keystone, he removed sanctions on the Nord Stream II pipeline. This paved the way for Russia to hold the EU hostage to Russia's energy and not allow cleaner, U.S.-provided LNG.

Canceling the Keystone pipeline did not stop the drilling or exporting of the sand oils, as the extreme environmentalists thought. The 830,000 barrels per day of crude oil from Alberta, Canada, is still being produced and being shipped via rail or other pipelines. Canada's oil sands producers were able to export a record amount of crude in 2022 to overseas markets, including China and India. So instead of providing well-paying jobs for Americans, the Biden administration prefers those jobs to go elsewhere.

The cancellation of the pipeline also did little to stop emissions. CO_2 emissions will increase, since much of the oil is now transported by the railroads, a much dirtier method of transit, instead of the pipeline. Rail transport also increases the risk of derailment and ensuring environmental damage of spilled oil.

Mr. Menezes, thank you for pointing out in your written testimony another shortcoming of President Biden canceling the Keystone pipeline: the fact that the U.S. and its allies would have had access to Canadian oil to lessen the import and use of Russian oil. Canceling the pipeline was a horrible financial decision, in my opinion. Developers of Keystone XL are seeking to recoup more than $15 billion in damages connected to President Biden's decision.

Mr. Menezes, what other financial losses can you comment on, due to the cancellation of the XL Pipeline?

Mr. MENEZES. Well, I was going to add that, you know, we import about 700,000 barrels per day from Russia. So the Keystone Pipeline could have also been used to offset that amount as we, along with all other NATO countries and our allies, have said no to Russia resources. So it is very significant, as we have been talking about all morning.

The fact is that, if you can increase supply, you are going to have downward pressure on prices, you know, for our—you know, for your constituents and for the American people. And I know, you know, people want to have it both ways, right? They would like to stop—to have refineries operate, but they want their constituents to have cheap gas. You can't have it both ways.

Mrs. LESKO. Thank you.

Mr. MENEZES. We can have environmentally compliant refineries in operation and making rational decisions to remove the bottle-necks. But that's just commonsense approaches, really, is everything we are considering today.

Mrs. LESKO. Thank you. My next question is for Ms. Sweeney.

Ms. Sweeney, copper is not listed as a critical mineral, although it is—I think it uses four times as much copper in an electrical vehicle as in a standard vehicle and other things. Why do you think that copper isn't listed as a critical mineral, and do you think it should be?

Ms. SWEENEY. I definitely think that it should be. I think that it actually does meet the U.S. Geological Survey criteria. But when they were doing the latest list, they didn't have the most recent data in front of them. I think, using the data today, that copper would definitely make that list.

However, you know, in National Mining Association's viewpoint, anything that you need and can't get really should be critical. So maybe those kinds of criticality lists don't just make the most sense because, as technologies change, something that is critical today may not be critical tomorrow.

You know, people were talking about substitutions. Well, then you are creating a new critical mineral, because something that maybe we use now to substitute becomes the next big thing that everybody is going to need. So if we just had efficient permitting, criticality doesn't matter so much.

Mrs. LESKO. Well, good. And I hope that copper is added to the critical mineral list, because I am from Arizona, we produce lots of copper, and it is unfortunate that the Resolution Copper Mine has been put on hold. The final impact—environmental impact statement was approved under the former administration and now has been put on hold, plus a lawsuit.

So I yield back. Thank you.

Mr. DUNCAN. I thank the gentlelady. The Chair will now go to Dr. Ruiz for 5 minutes.

Mr. RUIZ. I would like to inform the gentlewoman from Arizona that I am very familiar with the copper mines in Arizona. My family actually worked in those copper mines in years past. I believe it is the Bisbee Queen's Copper Mines, is one of those big, big areas.

But thank you, Chairman. I would like to address a recurring theme that I have seen in multiple pieces of legislation before this committee today. In these pieces of legislation I see bills that are sacrificing key provisions of landmark legislation that help protect people's health, like the Clean Air Act, to—in order to increase critical mineral extraction. And that is the wrong approach.

Frontline communities already bear too much of the burden of the environmental injustice. To name an example, people living near fossil-fuel-drilling sites are at greater risk for preterm birth, cancer, asthma, and other respiratory diseases. I mean, it is a direct link to real people's health, childhood asthma, COPD, all these other things.

As a doctor, the health of my constituents is my top concern, and I have seen firsthand that we must do more to protect vulnerable communities from pollution and other environmental dangers.

I do want to be clear that I am not against critical mineral production. I have been unequivocal that we need to build our domestic supply chains. And as a prime example of this is right in my own district, in the Imperial Valley, or the Imperial County.

In the Imperial Valley, the Salton Sea has a massive supply of lithium. In fact, it is the fifth largest lithium deposit in the whole world. It has the potential to supply the lithium needed for electrical vehicle batteries and our clean energy future. We import the vast majority of our batteries and our lithium from other countries that are not aligned with our world views. We call the area Lithium Valley back home to emphasize the potential that this lithium has to transform the region.

And beyond powering our country towards its clean energy future, I believe that Lithium Valley can also provide a model for how we can both protect our community's health and get the critical minerals we need. And how do we do that?

So instead of hardrock mining or salt flat evaporation, breaking up the earth, putting more dust in the air, et cetera, in Imperial County the lithium is extracted from geothermal production. So it is in a closed loop cycle, and the brine that goes through this closed loop as they naturally produce geothermal energy—which is a good thing—they filter that brine out to extract the lithium. And so it is better for the environment and better for our communities. This shows that we don't have to sacrifice health and the environment. We can have a win-win for the environment, for our public health, and for places like the Imperial Valley.

Mr. Raul Garcia—I like your first name, by the way. We share that. In Spanish we would call each other tocayos. In your testimony you note that our country doesn't have to make this false choice between energy creation and protecting the health and safety of our vulnerable communities.

Tell me more about that, and as it applies to the critical mineral production, as well.

Mr. GARCIA. Well, I think that when we are talking about critical mineral production, we want to make sure that we are using the resources that are available to us that are the safest ways to use the critical minerals. And that, when that requires mining, it has to follow the law. It shouldn't—we shouldn't be seeking exceptions to the law in order to make sure that this happens safely. We actually need to make sure that it applies.

We also need to make sure that we, as you mentioned, have a circular economy on critical minerals. So making sure that they are being recycled, and that when they are being recycled they are being—the recycling is being done responsibly and in a clean way, the—whether it is extraction or recycling of it. And again, those laws, those bedrock laws are in place.

Now, we have talked a lot about permitting in some places and how long it takes. But I like to point out that a lot of the delays that happen on permitting are actually not done because of the requirements of the permitting structures themselves. They are done because the agencies are being starved from actual funding in order to carry out the permitting.

And so, if we want to speed up permitting, when we are talking about—whether it is a mine or whether it is something else, you actually need to fund the agency in order to make sure that the experts are going to do so, are going to get out the permits in a way that is going to protect communities on the ground.

Mr. RUIZ. Catch-22, right? Chipping away government in order to make it as ineffective as possible, and then complaining that they can't do their job and it is too slow. So the only result is going to be to eliminate government at all and eliminate these rules that protect the health. And that is not necessarily feasible or the right thing to do for the American people.

But with that my time is over, and I yield back.

Mr. DUNCAN. I appreciate the gentleman. The Chair will now go to another gentleman from the crossroads of America, Mr. Pence, for 5 minutes.

Mr. PENCE. I thank you, Chairmen Duncan and Johnson, and Ranking Members DeGette and Tonko for holding this hearing. And thank you to the witnesses for appearing today.

The bills before us begin to shore up our national energy strategy, lower energy prices for Hoosiers, and put our Nation back on track towards energy dominance.

The Biden administration's electrification-or-nothing approach is only deepening our reliance on China and leaving our country vulnerable in the event of a national emergency. Despite what the Biden administration is telling us, demand for oil and natural gas will only increase around the world.

Petroleum products are the lifeblood of the American economy, fueling businesses to bring good-paying jobs and lowering energy costs for families.

Over the past 2 years, the Biden administration has failed to put forth a coherent energy strategy. Hoosiers in southern Indiana deserve answers on why this administration has made it more expensive for families to heat their homes and for business and families to keep their lights on at the old prices. And I hope we hear a little bit of that tonight.

My legislation being considered today would provide waivers for critical energy resources in the event of a national security emergency. And I quote, "If the Administrator of the Environmental Protection Agency, in consultation with the Secretary of Energy, determines that the processing and refining of a critical energy resource at a critical energy resource facility is important to the national security or energy security of the United States, then the Administrator may waive application of any requirement, sanction, or fee under the Clean Air Act."

And by the way, I am one of the few—I am old enough, I actually read the Clean Air Act when it came back—came out many, many years ago.

This bill would ensure that the Federal Government can act swiftly to preserve access to energy supplies. And I look forward to hearing constructive feedback on how to improve this legislation and provide the Federal Government with tools to respond in the event of a crisis.

Mr. Menezes, I am going to direct my question and my comments to you at this point. Back in 2005 I went in with then-Governor Mitch Daniels when he became Governor in the State of Indiana, and I was the chief deputy commissioner of the Indiana Department of Environmental Management. And the biggest problem in the State of Indiana was overregulation by the EPA and IDEM. It was inhibiting business growth. It was holding farmers down. It was costing too much money. The regulations were just crushing the State of Indiana.

And that got changed. At that point the State of Indiana was $2 billion in the hole, and today they are $6 billion to the good.

My question to you: In the event of a national security emergency that threatens access to critical energy resources, what sort of emergency authority tools exist to address severe supply disruptions for different types of energy?

Mr. MENEZES. Right. Well, there—that is where the confusion lies here, because we face emergencies regularly. You go into the executive branch, you know, as a public official. So the President and his team typically says, "What do—what can we do at this point?" And it's "Can—does the Constitution prohibit it? Has Congress prohibited such action? Has Congress authorized?"

And with respect to waivers, Clean Air Act, Jones Act waivers, these kinds of things, Stafford Act, declarations of emergencies, what waiver authority is there, we have a body of law that kind of—the executive branch runs fairly well. Where it gets diffused and confused is anything there after that.

And so, you can look at all of these different laws, and you don't see anything with respect to, really, energy, critical energy resources, in there. And so, as a consequence, being at the Department, you are kind of brought in at the last minute. In fact, the Department of Defense has a lot of clear, you know, emergency— they weigh in, they can do things. Commerce, even Transportation. MARAD under the Jones Act. The Department of Energy is just like sort of forgotten about. So you are in there, trying to focus on the importance of energy.

Puerto Rico. I mean, the Governor would tell us—they would come to Energy and say, "We can't do anything without energy. Our water doesn't work, our hospitals can't run. We can't do anything without energy. What are you doing for energy?" You don't have those authorities, those clear authorities on the books to be able to have Energy play a key role in solving crises during these emergencies.

Mr. PENCE. So would you agree, as a former Deputy Secretary, that this bill would add a little clarity to that?

Mr. MENEZES. Add clarity——

Mr. PENCE. And help everybody out, both——

Mr. MENEZES. Office of Legal Counsel, believe me, it would——

Mr. PENCE. All right.

Mr. MENEZES [continuing]. Help tremendously.

Mr. PENCE. Thank you. I yield back.

Mr. DUNCAN. OK, the Chair now goes to Ms. Blunt Rochester for 5 minutes.

Ms. BLUNT ROCHESTER. Thank you, Mr. Chairman, and thank you to all of our witnesses today.

Across the country we are already seeing the impacts of the Infrastructure Investment and Jobs Act and the Inflation Reduction Act. We are seeing a commitment to a cleaner economy and cleaner planet, and a commitment to our American manufacturers and workers.

Just in the 6 months since the Inflation Reduction Act became law, companies have announced over 100,000 new domestic clean energy jobs, and billions of dollars in new manufacturing investments. This historic law—or both of these laws have actually created a path to a stronger, more resilient domestic energy system, all while combating the ongoing climate crisis.

But, unfortunately, the bills and resolutions that we see before us today would force us off that path. They neglect to address the global shift toward clean, renewable energy and focus instead on expanding the fossil fuel industry by creating loopholes for important environmental and public health protections. We need legislation that not only protects our domestic energy interests but also protects the environment and health of all Americans.

My questions are for Mr. Garcia. One of the bills being heard today relates to EPA's risk management program rule, also known as the Chemical Disaster Prevention Rule. The proposed legislation would amend the Clean Air Act to exempt refineries that use extremely dangerous hydrofluoric acid from assessing whether they could potentially use safer technologies. I am concerned about any legislation that weakens the Clean Air Act, but I am particularly concerned about a bill that guts a commonsense practice that would save lives and prevent disasters.

Mr. Garcia, can you speak to how clients that your organization represents use foundational laws like the Clean Air Act to protect themselves from dangerous polluters in courts and how creating exemptions for them using the vaguely defined "critical energy resource" would gut those legal protections?

Mr. GARCIA. Yes, absolutely. So what we see is that the way that permitting should work, and the way that—whether it is the Clean Air Act or anything else—it should work in a way that industry along with the communities are equal partners in a way that identifies the best science in order to follow through on whatever the permit would allow, with the health and safety of the communities most impacted. And that is exactly what the what the RNP does, right? It is, in fact, a study. And why would we ever want to stymie innovation in this process?

I believe that, from a lot of folks here today, we keep talking— we keep hearing about the next chemical and the next metal and the next thing that we are going to see. Why aren't we applying the same level of innovation to the standards that would protect our communities in the process?

And so that—I think that is a question that has to be answered. But that is what the process is. And the fact that the bill seeks to exempt it from exploring that innovation is very problematic in all sorts of ways.

Ms. BLUNT ROCHESTER. Another bill that we heard about today authorizes the use of flexible air permitting with respect to certain critical energy resource facilities. In your written testimony, you stated that using flexible air permitting in this manner would take the science out of air permitting decisions. Can you elaborate on that statement and discuss the risks associated with this use of flexible air permitting?

Mr. GARCIA. Yes, absolutely. We have—when we are talking about permitting, there are certain standards that industry or whoever wants the permit needs to comply with. In flexible air permitting, as we have it in the bill, essentially, the administration gets to pick winners or losers, never mind the actual standards.

And so, when you have an administration that is really not looking after the health and safety of the people on the ground, that is really looking for corporate profits and things like that, it really just allows them to pick favorites. And, unfortunately, there is a long history of which industries they pick to be favorites in that process.

Ms. BLUNT ROCHESTER. And just as a followup, can you talk about the public health and environmental implications of that?

Mr. GARCIA. Yes, absolutely. I mean, the idea is that you are leaving out the science, and at the same time you are ramming the most dangerous alternatives out there down communities' throats, right?

And so they are the ones that would have to live with, God forbid, any explosion that happens, the workers themselves. They are the ones who have to deal with the consequences.

Ms. BLUNT ROCHESTER. Thank you so much, and I yield back.

Mr. DUNCAN. The Chair will go now to Dr. Joyce for 5 minutes.

Mr. JOYCE. I want to thank the chairman for holding this critical hearing today.

As we begin the 118th Congress, we continue to hear from our constituents at home that the high costs at the pump and in their electric bills are kitchen-table issues that they deal with every day. The fact is, the only way to bring down prices is to increase supply. We have been blessed in this country to have plentiful reserves of natural resources. In my home State of Pennsylvania, we have significant deposits of coal and natural gas. In the last 20 years alone, new drilling techniques have led to the shale gas revolution, and Pennsylvania is now the second-largest net supplier of energy to all other States. In my conversations with energy producers, they make it clear that the Commonwealth of Pennsylvania has much more that it can give.

What is standing in the way of unleashing the resources under the feet of my constituents? It is the Biden administration's war on American energy.

At every step, this administration has held up and hindered the production of American coal, oil, and natural gas. From creating restrictive and burdensome regulations to attempts to scare away capital investment from fossil fuels, the Biden administration has stifled supply and caused prices to rise.

Mr. Chairman, I ask unanimous consent to submit for the record a report by the National Fire Protection Association, published in July of 2020, which I will submit at the end of my questioning.

Just last month, an administration official raised the possibility of banning gas stoves due to safety concerns. Well, let's look at the science, a report by the NFPA. These claims of unsafe natural gas stoves are incorrect. Gas stoves are incredibly safe to use. In fact, electric ranges are over 200 percent more likely to cause a fire, and over 300 percent more likely to cause a deadly fire, and close to 500 percent more likely to cause fire-related injury than gas ranges.

This misguided attempt to ban gas stoves shows yet another step that the Biden administration is attempting to move the market away from American energy products like natural gas. It is time to stop playing political games and do what is best for our constituents.

No producer single-handedly affects the price, but by giving businesses regulatory certainty and providing confidence to capital markets assures that the domestic production of energy is here to stay. We can lower prices and reclaim American energy dominance.

My first question is for Mr. Eshelman.

How do aggressive and targeted attempts to discourage investment in natural gas affect production?

Mr. ESHELMAN. Well, first off, to hit on your point about the natural gas stoves, I think this is a real personal intrusion, that the Government is trying to tell consumers what kind of choices they can make. So it is what kind of—how you can cook, what kind of cars you can drive, how you can heat your home. So that is a very important point to bring up. I think consumer choice is at risk here with this administration.

Second, business thrives where there is a predictable landscape. So we need State and local governments and the Federal Government all work together to streamline their processes so we can get some permitting done, as well as exploration and production, and particularly in Pennsylvania.

Mr. JOYCE. Thank you. My next question is for Mr. Menezes.

I mentioned the need for regulatory certainty in my earlier remarks. I have a draft bill that would enable critical mineral facilities to work with the EPA in advance of the permitting process so that they can react to surges in the market without needing to restart the permitting process.

Currently, how do regulated critical materials entities deal with large increases in market demand, and what relief would this bill give them?

Mr. MENEZES. You know, currently I believe there are some 300 backlogged applications pending at EPA for the permits that you have requested, and that number is only growing, really, to take advantage of a lot of the provisions that were in the IRA. Capital wants to be expended. This new investment will involve new chemicals, et cetera. They want to get their permits.

A lot of the funding is based on you have to have the permits first. And so frustration is mounting that we can't seem to figure out how to get EPA to really just follow the law, implement the law, get the permits out, and let's try to embrace the new future with these new chemicals and our energy——

Mr. JOYCE. Thank you for addressing this critical issue. My time is expiring. Thank you, and I yield.

Mr. DUNCAN. OK. The Chair is going to go to Ms. Schrier next. And when she finishes her 5 minutes, we are going to take a recess for Members to go vote.

Members are reminded we are going to meet again right after the second vote opens. Vote, come on back. We are going to go to Kelly Armstrong first up when we come back.

So Ms. Schrier is recognized for 5 minutes.

Ms. SCHRIER. Thank you very much, Mr. Chairman. And thank you to our witnesses. This has been a really interesting discussion. You know, last week one of my Republican colleagues acknowledged that there is actually a lot of common ground, that there is bipartisan agreement that we all want to be good stewards of the Earth, that careful use of resources is important, and decreasing greenhouse gas emissions is one element of how we can be good stewards of this earth.

So I was so inspired by these prospects of bipartisanship that the very next day I met with Democratic and Republican House and Senate members at a breakfast discussion about the real practicalities. Like, let's just get down to nuts and bolts about how we are really going to get the changes we need made with clean energy in a timely fashion to actually make the difference that we need to make.

And one important practicality had to do with how delays in permitting and years of litigation could thwart our very best efforts to make the changes that we need to achieve our clean energy goals. And just to be clear on where I stand, I am not suggesting in any way that we gut our bedrock environmental protection laws. But we need to make some pragmatic, necessary reforms, like faster timelines, in order to have a realistic shot of meeting these clean energy and emissions goals.

So I am looking for areas where bipartisanship is possible, and I think we have some opportunity for common ground here.

I also believe and know that our production of renewable energy is only as good as the electric grid that we have. And the broader system and our Federal permitting laws were primarily written for the fossil fuel era. And so it is time to kind of rethink how we can streamline, and it requires some very practical conversations, not trying to speed every answer to yes, but at least getting to a yes or a no quickly so we can move on.

So, Mr. McNamee, I appreciated your testimony. You noted that Congress should consider reforming NEPA and the APA to limit how legal challenges can be made against agency actions. Of course, this is a, as you refer to it, a two-edged sword, and that we want agencies to be accountable, we need public input. And so Congress has to do this balancing act. I was wondering if you could talk a little bit about this.

And I am confining the way I think about this to the speeding of clean energy projects.

Mr. MCNAMEE. Thank you for the question. The—one of the key problems is that the way NEPA litigation—when Congress passed NEPA, there was no cause of action. They thought they just wanted agencies to look at the environmental impacts and make sure they consider it.

The Administrative Procedure Act allows parties to challenge an agency action for being arbitrary and capricious. And so then they say, well, if the agency didn't make the decision properly on NEPA, that is arbitrary and capricious, and so that can be pulled back. An example: FERC in the last few months had a—I think it was the DC Circuit sent back an approval that they made because, even though they addressed the issue, they forgot to cite—make a citation to a reg. And so they had to send it back. That just seems to me to be things that don't need to happen. So there can be ways to streamline that.

In terms of the clean energy, the thing that I find somewhat kind of amusing is that the—that often the same people who didn't like underground natural pipelines suddenly say, "We need to fix permitting for, you know, 400 miles of lattice towers for transmission lines to get wind from the Midwest to the coasts," going through, you know, farms, ranches, et cetera.

And what I think it is, we have to look at what are the real problems. Because I think what the renewable developers have realized is, there is a NEPA problem. There is a problem. And so we need to come up with a solution that fits one size for fitting all, not just picking——

Ms. SCHRIER. And I will give you just some feedback on that. One of those things pollutes, one doesn't. One, in many cases, lines are already there. But we—that is a discussion for another day.

I wanted to just turn at the end to just a really pragmatic example: hydropower. I come from Washington State. Hydropower makes up a third of the renewable energy in the United States. It makes up about half of the electricity we use in Washington State. And with the looming climate crisis, hydropower is going to be critical. You know, it provides baseload, just like nuclear and like natural gas, that wind and solar just can't provide. And so hydropower is critical.

And right now it turns out that just about 3 percent of dams in this country generate hydropower. And this is an area where the chairwoman of this committee and I really agree, that this is exactly the kind of project where, if you consider the net environmental benefit, you consider the economics, you consider the energy benefits, that this sort of project might be ideal for having a speedier pathway to a yes or a no.

So I want to thank you, and I yield back my time.

Mr. DUNCAN. The gentlelady yields back.

So Members are reminded that we are going to come back right after the second vote opens. So vote and come back, and I am going to take them in order. Kelly Armstrong is up next.

Witnesses are asked to remain. If you need to get up, stretch your legs, bathrooms are down the hall. Apparently, there is a lot of people in the lobby out here, so you may not want to go that way.

And we will stand in recess until we get back. [Recess.]

Mr. DUNCAN. All right. We are going to go ahead and get started.

So I will call the subcommittee back in order.

And I will now go to Mr. Armstrong from North Dakota for 5 minutes.

Mr. ARMSTRONG. Thank you, Mr. Chairman. In 2017, when this committee held a hearing on the Promoting Cross-border Energy Infrastructure Act, Ranking Member Pallone remarked at the time, "With President Trump already approving the Keystone XL pipeline, it is unclear to me why Republicans feel it is necessary to strip the President of his authority. Do my colleagues on the other side of the aisle honestly not have confidence in President Trump?" It is not that Republicans didn't have confidence in President Trump, it is that Republicans didn't have confidence in future administrations to make rational decisions regarding energy's infrastructure. We only had to wait a matter of hours after President Trump left office and President Biden—for President Biden to revoke the cross-border permit for the Keystone XL pipeline and prove the exact reason why we need this bill.

Well, it is clear that the FERC and DOE processes are not immune from political influence, which I will get to in a second. They must follow statutory and regulatory guidelines, as opposed to the impulsive decision-making process used by one President. Opponents of carbon energy have routinely exploited the uncertainty of the cross-border process to starve projects of capital investment because they understand that these projects can take decades to complete and are not a short-term investment.

We know that those who are antagonistic to carbon energy oppose every pipeline project. It has nothing to do with the merits or the environmental analysis surrounding a particular piece of infrastructure. To quote a leader of the Keep It In the Ground campaign, the Keystone XL pipeline was never about any single pipeline. It was about establishing a litmus test. Well, the opponents of carbon energy have established that test, and they know that they can rely on domestic—Democratic administrations to undermine our energy infrastructure.

Deputy Secretary Menezes, before we go any further, it is important, I think, to talk about what the Promoting Cross-border Energy Infrastructure Act does not—what it does not do, and we need to be perfectly clear. This bill will have zero effect on NEPA or shortcut environmental reviews. Is that your understanding?

Mr. MENEZES. That is my understanding.

Mr. ARMSTRONG. Thanks. And I am going to switch over to domestic pipelines for a second.

Mr. McNamee, are political considerations playing an outsized role in FERC's permitting decisions?

Mr. MCNAMEE. I am concerned that FERC has not always been focused on the limits of the Natural Gas Act in making its determinations.

Mr. ARMSTRONG. Have there been any policy changes in the last 2 years, or is it simply that the regime has changed?

Mr. MCNAMEE. Two things have happened. One, the regime had changed from majority Republican to majority Democrat Members. And second, there were proposals for new policy statements that would have allowed the Commission to deny natural gas pipelines based on the upstream and downstream natural gas combustion, which, in my opinion, when I was on FERC, FERC did not have the authority to make those decisions under the Natural Gas Act.

Mr. ARMSTRONG. Well, what do you think the practical impact of that would be if they—if prior to putting a pipeline in the ground, a company had to mitigate both the oil well and the SUV?

Mr. MCNAMEE. Well, they may not even have an opportunity to mitigate it. The FERC—if those policies were enacted, the pipeline might not be approved at all because it would be declared too harmful. And so, even though it is in the public interest, it would provide service to customers, and it could lower prices for customers, that may not be approved at all.

Mr. ARMSTRONG. As somebody who grew up in western North Dakota, I would say an oil well is like a five-star steak restaurant and a pipelines is like McDonald's. They both make money, but they make it in very different ways.

Back to Mr. Menezes. In addition to raising costs and limiting growth, roadblocks that delay or prevent the development of energy infrastructure, particularly oil and gas, threaten the overall resilience of our electric grid, as well. Can you touch on why that is important, to support efficient deployment of midstream infrastructure?

Mr. MENEZES. As we have been talking about today, we need to increase supply, but we also need to increase infrastructure to make sure that supply gets delivered to the American people. That will bring downward prices onto the commodity that they are purchasing.

Mr. ARMSTRONG. And there is also global ramifications to an efficient cross-border process in North America. I mean, we import a ton of oil from Canada. They are our closest ally and our neighbor. Maybe I am a little biased, because they are my neighbor directly to the north.

Mr. MENEZES. Well, I don't——

Mr. ARMSTRONG. What are the ramifications——

Mr. MENEZES. I mean, we have about eight refineries that are designed and built to take heavy crude, heavy crude from Canada. It was also for heavy crude from Venezuela until the shale revolution. So we have refineries that are designed to take that oil. It is from Canada, you know, our largest trading partner. So it only makes sense.

Plus, we talked about earlier that would have been—we import about 700,000 barrels per day from Russia. That 830,000 from Keystone would have certainly offset that.

Mr. ARMSTRONG. I always used to say anybody who has ever looked at the process to drill a well in Canada would understand it is significantly more stringent than potentially even drilling on our Federal land. When I would talk to my friends and allies on this issue, I would say, "What can we do to make—change people's minds?"

They would say, "Call it something other than tar sands oil, because it just sounds dirty." It doesn't matter that it is sour crude that we need for marine diesel. I love my Bakken sweet crude, but it doesn't work for everything.

So thank you, Mr. Chairman, and I yield back.

Mr. DUNCAN. I thank the gentleman. So I am assuming I am going to Ms. Kuster for 5 minutes.

Ms. KUSTER. Great. Thank you, Mr. Chairman. Let me get straight to the point. Most of the legislation before us in committee today is merely a handout to the fossil fuel industry, poorly described as an attempt to strengthen our Nation's energy security. So let me be clear. Furthering our Nation's independence— dependance on fossil fuels instead of diversifying our energy sources not only weakens our energy security, but it also harms American families by leaving them vulnerable to global energy price shocks.

In the first legislative hearing of this subcommittee, the Republican majority has put forward partisan legislation which stands no chance of being enacted into law rather than focus our time on meaningful bipartisan efforts to further American energy independence. These bills simply miss the mark. While I disapprove of these misguided proposals, I remain committed to working with my Republican colleagues to find opportunities to actually deliver affordable, reliable, and domestically produced energy to the American people.

One of those opportunities is bolstering our baseload energy resources, such as hydropower and nuclear. In New Hampshire, the Seabrook Nuclear Power Plant provides low-cost baseload energy by operating at full capacity nearly year-round. This past weekend you may have heard about record-breaking cold on top of Mount Washington in my district. Under some of this winter's coldest temperatures, Seabrook continued to deliver reliable baseload energy to the grid.

To protect nuclear energy as a baseload energy resource nationwide, we must also secure the uranium these plants rely upon to operate.

Mr. Menees—Menezes? Thank you.

One of these legislative items before us today would prohibit the import of low-grade uranium produced in Russia. At a time when Putin's regime is using profits from energy exports to fund a gruesome war in Ukraine, I would say cutting off Russian imports is a sound policy. Can you speak to the importance of reducing our dependance on Russian uranium, from an energy security perspective?

Mr. MENEZES. Well, your question touched on the key points. Russia simply is no longer a reliable partner for any critical energy infrastructure resources at all, including the enriched uranium which our civilian nuclear fleet has become dependent on, because it can be cheaply produced in Russia and they export it at cheap prices to ensure that our civilian nuclear fleet that competes in these RTO markets that we talked about, they can't bear any cost increase on the cost of fuel.

The bill not only prohibits it, but it phases it out, and it does provide for some waivers. So it is not as though we are going to be without the fuel. That should give us enough time, because there have been some other provisions of the bill to develop our own abilities for fuel fabrication and, you know, enrichment.

It is not the easiest thing to do, but we need to reclaim our leadership and our own ability, frankly, to not only mine and mill, convert, but also for fuel fabrication. And it is important because, remember, nuclear is our cleanest emission-free source of energy in the United States.

Ms. KUSTER. Thank you. Like many Americans, I remain concerned by the record profits that oil companies continue to rake in while households across the country are struggling to pay their energy bills.

Mr. Slocum, in your testimony you suggested that Congress should press FERC, which is responsible for regulating natural gas markets, to protect American consumers from price manipulation that could increase natural gas prices and America's energy bills. What steps can FERC take to improve price transparency in natural gas markets?

Mr. SLOCUM. Yes. So FERC has jurisdiction over spot natural gas markets if there is evidence of market manipulation. But in the 2005 Energy Policy Act, because of widespread manipulation of the price indices, FERC requested and Congress inserted into the Natural Gas Act a provision that allows FERC to establish its own price transparency natural gas reporting system.

And so, in comments in a rulemaking last year at FERC, I urged the Commission to undertake this never-used statutory authority, and what it would do would shine some badly needed transparency into pricing in natural gas spot markets.

Ms. KUSTER. Thank you. Many of my constituents in New Hampshire rely on natural gas for electricity and home heating, and I am committed to ensuring that Congress and the administration evaluate all available tools to ensure the prices consumers pay are just and reasonable.

To my Republican colleagues across the aisle, please take seriously my offer to work together on commonsense energy legislation and put people over politics.

Thank you, and I yield back.

Mr. DUNCAN. The gentlelady yields back, and the Chair will now go to the gentleman from an energy-producing area—well, actually, nuclear energy—that would be Mr. Allen from Georgia.

Mr. ALLEN. Thank you, Chairman Duncan and Chairman Johnson, for holding this joint subcommittee hearing today to discuss the importance of unleashing our domestic energy production, securing our electric infrastructure and domestic supply chains. This is critical in delivering lower energy costs to Americans and becoming energy dominant.

I can't stress enough the importance of having an all-of-the-above energy strategy, which we have talked about over and over again here today, and ensuring our supply chains here are secure.

Yesterday I hosted my first telephone town hall with thousands of constituents on the line for the 118th Congress. The most questions I got were about, "What are you going to do about energy and the cost of energy? How can we become energy dominant again? I remember those days when we actually set the price of a barrel of oil. That was real power, Mr. Congressman."

My constituents and Americans across the Nation are spending money they don't have on energy, should it be at the pump or trying to heat and cool their homes. The key to affordability and reliability is a diverse energy portfolio and removing regulatory barriers that hinder access to our natural resources right here in the United States. The free market is key in setting the price for energy.

Mr. McNamee, you mentioned in your testimony how subsidized renewables are distorting price formation and regional transmission organizations, which is having a negative impact on important energy sectors like nuclear and coal, and, in turn, the reliability of the grid. Commercial nuclear energy is critical to my district, as it is home to Plant Vogtle, and currently under construction are units 3 and 4, which are scheduled to come online later this year. Nuclear energy is reliable and affordable and emission-free.

Are there ways we can ensure subsidized renewables are not undermining the market so that the coal and nuclear business do not become obsolete?

And how does this degrade reliability?

Mr. MCNAMEE. Thank you, Congressman. The primary challenge in the so-called RTOs, which are in seven parts of the country, is that they tried to use market forces in order to achieve the cheapest energy resource for electricity.

The problem is, once you have subsidized resources like renewables that have no fuel costs, that show up intermittently, they have tax credits, they undermine the way price formation works. And the price formation is for every, basically, 5 minutes of the day during the load curve, generators bid in to see if they can meet the load. And the last generator that is picked sets the price that everybody is paid. So that means if you are wind or solar, you are paid the same price as the natural gas producer or the nuclear producer.

And what has happened is, when you have these intermittent resources coming on and off, they are taking up market share, they are driving a little bit of the marginal price of energy down, but they are undermining the ability of baseload, like nuclear, to be able to make the money to stay operating. And this has become a big crisis in a lot of parts of the country, where you see electricity prices spiking but reliability going down.

In Georgia, one of the things is it still has the traditional vertically regulated, and you mentioned that—planning the resources. That is exactly what utilities do. They do integrated resource plannings, they decide what units are needed to meet the power 24/7, 365 days a year, whether it is cold or whether it is hot.

And they also—the price of energy is averaged. So customers are getting the economic benefits of all the resources.

Mr. ALLEN. Right. And that might explain why we are—have two battery plants that are scheduled to come online, as well in Georgia, one just announced and one is under construction for batteries for the electric vehicles that are going to power the future, because we have the most efficient, most abundant supply of electricity in the country. Thank you for the answer to that question.

Nuclear energy, as I said, is critical to our national security. So I would like to now enter into the record two letters, one from the Nuclear Energy Institute and another from the United States Nuclear Industry Council, which both expressed support for the committee's work to establish a secure nuclear energy fuel supply chain.

Mr. DUNCAN. Without objection, so ordered.

[The information appears at the conclusion of the hearing.]

Mr. ALLEN. And now, Ms. Sweeney—and we are not—we are going to run out of time, but we have talked about the permitting problem in the mineral supply chains. Can you submit in writing to me?

What I am hearing is workforce problems. And maybe you can comment on another question later about the workforce problems we are having throughout the energy sector.

Thank you, and I yield back.

Mr. DUNCAN. The gentleman yields back. The Chair will now go to Ms. Castor for 5 minutes.

Ms. CASTOR. Thank you, Mr. Chairman.

Fossil fuel prices and energy costs were the primary drivers of inflation last year and higher gas prices after Putin's unprovoked invasion of Ukraine. And we had the Big Oil CEOs here at the committee to talk to them about price gouging. And we asked each one of them, were they interested in lowering costs, lowering prices at the pump? We are going through a crisis of a war with an ally. Inflation was hitting consumers hard. And they all said no.

Now, just last week the profits became clear. Mr. Slocum, did you see the profits reported by the Big Oil companies?

Mr. SLOCUM. Yes, ma'am.

Ms. CASTOR. How would you characterize them?

Mr. SLOCUM. Extremely large. And that doesn't even reflect all of their profitability, because they have been plowing billions into share buybacks. Chevron, for example, committing to $75 billion.

Ms. CASTOR. But wait, they could have used that to lower the price at the pump, right?

Mr. SLOCUM. Correct.

Ms. CASTOR. Shell Oil made more in profit than they ever have before. Exxon and Chevron, their $90 billion in profit was called "epic." So this is what profiteering looks like, doesn't it?

Mr. SLOCUM. The proof is in the numbers. Yes, ma'am.

Ms. CASTOR. So I think they have everything they need. Why would we be bending over backwards to provide a wish list for the Big Oil companies when they are making—they are banking the biggest profits ever, and then they are not passing along these savings, they are doing stock buybacks?

Is there anything in this wish list that is on this agenda today that lowers the cost for consumers?

Mr. SLOCUM. I don't see anything in these various bills that are going to reduce cost to consumers or protect consumers from the kind of price gouging that is going on.

Ms. CASTOR. Mr. Garcia, do you see anything?

You have helped shine the light on the litany of this wish list for polluters in Big Oil today. Is there anything in this package that would lower costs for consumers?

Mr. GARCIA. I can't identify a single thing.

Ms. CASTOR. In fact, you have helped us go through some of these bills today. It is like the 1970s have called, and they want their energy policies back. This is 2023, and clean energy is cheaper energy. We, with the help of the Inflation Reduction Act and the Infrastructure Investment and Jobs Act and the CHIPS and Science Act, we are unleashing American innovation to lower costs for consumers, to create jobs, to build healthier, more resilient communities.

Mr. Garcia, is there any reason that you think we should go backwards to the costly policies of the past, and—rather than go forward?

Mr. GARCIA. None. I mean, quite the opposite, right?

I mean, if we do that, there is going to be a lot of damage done, both at a climate level for everything that we are going to face as a country, as a planet, but also in small frontline communities next to these facilities that, unfortunately, bear the burdens of these— largely would bear the burdens of these largely unregulated practices if these loopholes go through.

Ms. CASTOR. I mean, that is one thing that we never really talk about when we have this wish list for polluters in Big Oil, the cost of the climate crisis.

Back home in Florida, because our monopoly electric utilities have kept us—we are the so-called Sunshine State, right? But we are reliant about 75 percent on fracked gas for our electricity. So people are paying higher electric bills as we have warmer days. That is not smart. That is not consumer friendly. We are paying higher property insurance.

There was another story out this morning about displacement because of extreme events. There is a huge pull on the Federal budget because we are outlaying more and more to tackle droughts and floods and these extreme events. And then there is the long-lasting burden of pollution that you highlight.

Again, isn't the future in clean energy? It is cheaper energy. It will help us lower costs across the board.

Mr. GARCIA. Absolutely. I mean, what we are seeing is the attempt by dirty fuels and dirty industries to circumvent laws in order to stay afloat longer, when what we really need to do is make sure that our investments are going to—as you said, the future, making sure that those investments are happening quickly and directly into renewable industries that are also clean industries for our communities.

Ms. CASTOR. Thank you very much. I yield back.

Mr. DUNCAN. I thank gentlelady. The Chair will now go to Mr. Balderson from Ohio.

Mr. BALDERSON. Thank you, Mr. Chairman. Thank you all for being here today. My first question is for Mr. Eshelman.

Thank you for being here, Mr. Eshelman. In your testimony you note that IPAA's member companies are the innovative leaders that broke the code to usher in the shale oil and natural gas revolution in the United States. Chairman Johnson and I can certainly attest to this massive impact the shale revolution had in the Appalachia Ohio region and the economic benefits we have seen in our communities.

As you know, the shale revolution and fracking turned the U.S. into an energy superpower and greatly enhanced our national security. But we must look forward. Natural gas and oil will continue to be essential to our energy portfolio for the foreseeable future.

In addition to prohibiting the President from unilaterally banning fracking, Chairman Duncan's Protecting American Energy Production Act also expresses the sense of Congress that States should maintain primacy for the regulation of fracking for oil and natural gas production on State and private lands.

Mr. Eshelman, why was this aspect of energy policy key to the shale revolution?

Mr. ESHELMAN. Well, hydraulic fracturing is a technology that has been around for a long time but perfected over the past 10 years, which unleashed the shale revolution in the Utica, the Marcellus, and the Permian all around the country and provided jobs and oil and gas to consumers. So it is a very important technology that, if we didn't use it, we would probably lose half of our oil and gas supply tomorrow.

One of the other things I would mention is that there have been efforts in Congress to ban hydraulic fracturing or have it regulated at the Federal level. That doesn't work. A one-size-fits-all regulation does not work for different parts of the country. So the States are doing it well. They are working with communities, and it should remain that way.

Mr. BALDERSON. Thank you. I will follow up with you, sir. You said it best. The IRA's methane tax will jeopardize the operations of many oil and natural gas producers and divert their attention from what they do best: producing the cleanest and safest oil and natural gas in the world.

As you noted, this tax was included in the Inflation Reduction Act despite not being considered at a hearing, not receiving expert testimony, and without an economic analysis. If this isn't repealed, producers will start paying this tax next year.

How will the natural gas tax impact rural producers, rural communities, and, ultimately, how will the tax impact our constituents who rely on natural gas each and every day?

Mr. ESHELMAN. Well, I think you hit on it. There was no congressional hearing on this tax. There was no testimony taken on it, no analysis done on it. So we really don't know the answer to that question. That is the problem with this tax.

Mr. BALDERSON. A followup for Mr. Menezes and Mr. McNamee: Do you think this tax will ultimately hurt rural communities?

Mr. MCNAMEE. Yes.

Mr. MENEZES. We have to keep in mind that those that are benefiting from fracking, they are the small producers. The majors do not frack, OK? They are mom-and-pop operators here. These are the true patriots. These are the true Americans. They have produced the oil to where, for the first time in history, OPEC is forced to negotiate with the United States. That has never before happened since OPEC was in existence.

And so, when you are looking at those provisions, the methane regs, those fees, it is going to be the small producers. The majors, they are not going to be impacted, again, but they don't have the small producing that the IPAA folks have on the methane.

And remember, the methane we talk about is a pollutant. It certainly, in great concentrations, can do great harm. However, it is a product. And so, with the improvement in all the detection devices, what we have seen already is the industry moving to detect, take action, and it is a product. And so, by making the investments, it is more product to actually sell. And so, talk about a circular economy, that is one, and they are taking advantage of it now.

Also, EPA has a supplemental pending rule open for public comment now on methane regulations, which essentially does the same thing without the fee, but you have the full backing of all the penalties in the Clean Air Act if you violate those regulations once they become final.

Mr. BALDERSON. OK. Thank you very much. We are down to 30 seconds. I want to ask one last question. Please be conscious of the time.

Mr. McNamee and Mr. Menezes again, Chairman Johnson's Unlocking Our Domestic LNG Potential Act would streamline the approval process for American companies to export liquefied natural gas.

Do you think—during last week's hearing Under Secretary Paul Dabarr regarding the global environment benefits of the exporting U.S. natural gas to developing nations, many of which are reliant on coal from China.

If you all would respond with just written answers, I would appreciate it.

I yield back, Mr. Chairman.

Mr. MCNAMEE. Exporting natural gas helps both people in foreign countries and domestically.

Mr. BALDERSON. Thank you.

Mr. DUNCAN. All right. I thank the gentleman. The Chair will now go to Ms. Miller-Meeks for 5 minutes.

Mrs. MILLER-MEEKS. Thank you very much, Mr. Chairman. I thank all of our witnesses for having stamina to last through this hearing.

Iowa is a little-known energy State. We have 50 percent of our energy from renewables. Over 50 percent of our electricity is from wind. And we are an exporter, a net exporter of energy that both reduces greenhouse gas emissions and cleaner burning engines as part of our liquid fuel portfolio. And it should be part of a liquid fuel portfolio in all of our energy needs. It offers, you know, affordable energy to allow us, as a country, to compete globally on an affordable level.

Imagine when I went to COP 26 and COP 27 that I found out that energy demand is increasing. It is not decreasing. And I agree with the witness that we need increased electrification, but energy efficiency is only going to be able to give us so much, and it is a very small amount. And to increase electrification we need electricity, which means we need more energy sources and to build new energy sources and resources, not less.

In Europe, as we have seen over the past year—and I saw when I went to both COP 26 and COP 27—demand is going up. Energy efficiency, as tried in Germany and UK, has been reliant upon wind and solar as renewables and have much higher electricity prices than we have in the United States. So it is not bringing down electricity prices.

In Iowa, not only are we a source of energy, we also have the Ames National Laboratory located at Iowa State University just outside of my district, and it currently leads the Critical Materials Institute. CMI is an energy innovation hub of the U.S. Department of Energy. Its focus is innovation to assure supply chains for materials critical to clean energy technologies. CMI carries out scientific and engineering research that facilitates more diverse primary supply chains in addition to mining, which we wholeheartedly agree we need in this country, more efficient manufacturing, re-use and recycling, and development of new materials.

Mr. Menezes, can you speak to the national significance of diversifying supply, developing substitutes, and driving re-use and recovery of critical minerals with respect to securing our Nation's global competitiveness?

Mr. MENEZES. Well, we have talked about the critical minerals, you know, throughout the day. It is important that we figure out a way for us to have our own access to our critical minerals, you know, as we are dependent on 31 of 35, we import 14. So I think that is a—that is one of the lessons learned.

And it was really emphasized after COVID. So Congress in 2020 did take action, you know, to do R&D for critical minerals. And so they were doing it. But after COVID we realized we really cannot depend even on other countries. While we have been talking about China, you know, other countries provide us other critical minerals. So the focus now is really to do this.

And the Biden administration likewise, you know, underscored that. And—but we need to do more. And so that is why today's bill is so important, that we really need to make sure that the Department of Energy has the expertise to do it.

You mentioned Ames. It is great. Ames is the place where you go, where you get to see the actual elements that are on the elemental chart, except for the radioactive ones, of course. But you can just see, and they are all there in these little vials. And it is just an absolute great place to go and visit.

Mrs. MILLER-MEEKS. I wholeheartedly agree. And as you know, in December of 2022 the U.S. Department of Commerce found Chinese solar panel makers had circumvented the U.S. tariffs by doing minor processing in Southeast Asia—and I think that was mentioned earlier—including Cambodia, Malaysia, Thailand, and Vietnam before exporting to the U.S. New tariffs on U.S. imports from these countries, which account for about 80 percent of U.S. panel supplies, do not take effect until June of 2024 because of a 2-year waiver from President Biden.

I have concerns about China taking advantage of the waivers for renewable technologies meant for other countries, especially when the alternative is American-made products. There is a similar issue with Chinese companies operating production facilities for the EV batteries, which we had talked about through the credits with the Inflation Reduction Act.

Can you please speak to the risk that the waivers for critical minerals and renewable technologies in other countries pose to the U.S. national and economic security?

Mr. MENEZES. Right. For the first time, then, if this bill is enacted, the Energy Department will actually have a role in advising the President and other agencies as to what emergency actions can take place to ensure that we can protect those vital sources. And that is why this is needed.

Otherwise, the Department of Energy and its expertise will lose out to other agencies, who will have other equities to argue about and for. And so you want the Energy Department to be at that table to inform the President as to why certain actions need to be taken to protect our critical minerals and resources.

Mrs. MILLER-MEEKS. Thank you so much. I wish I could go through all of the witnesses, but I don't have time. I have some other questions which I will submit to the committee, and then ask you to respond in writing. Thank you very much.

I yield back.

Mr. DUNCAN. I thank the gentlelady, and the Chair will now go to the gentleman from the Republic of Texas, Mr. Pfluger, for 5 minutes.

Mr. PFLUGER. Thank you, Mr. Chair, and I appreciate the opportunity in this hearing. I would like to thank all the witnesses.

You know, I would hope that what has been said by my colleagues on the other side of the aisle is actually true, that we can work together. I am worried about not having electricity.

There is a narrative here that Big Oil is making big profits. Does—Mr. Slocum, I don't know if you know the price of oil on April 20th, 2020. I don't know if anybody on this panel can tell me what that was. I bet there is one or two. Negative 37.63. Do we remember that? And do we have hearings about the profits that we were not making at that time?

Energy security is so important. It is important to our economy. It is important to our livelihoods. It is important to our military. You know, when we look at the concept of making sure we have affordable, reliable energy—Mr. Garcia, can you tell me what the reduction in methane emissions has been over the last 10 to 15 years here in the United States?

Mr. GARCIA. I don't have that data in front of me, but I am happy to provide——

Mr. PFLUGER. Yes, it is 14 percent, 14 percent.

Mr. Slocum, any idea where that ranks in the world?

Mr. SLOCUM. I don't know, but I don't know about that figure. There is——

Mr. PFLUGER. That is from——

Mr. SLOCUM. It depends upon how it is being measured.

Mr. PFLUGER. That is from DOE.

Mr. SLOCUM. Right. Because there has been discrepancies between actual emissions and recorded——

Mr. PFLUGER. OK.

Mr. SLOCUM [continuing]. Methane emissions.

Mr. PFLUGER. The point that I would like to get to here is we lead the world in reduction of emissions, harmful emissions. We have producers that are doing this because it is not only efficient, but it makes sense in a market-based place. And we do leave this Earth better.

And I would invite you to come to my district to see the wind and the solar and the problems that they actually have for the environment, since you have spoken so eloquently on environment, and see the ranchers and the farmers who operate this land for generations—seven in my case, my family—who leave the land better, and use the resources, and put back into it.

Mr. Eshelman, you mentioned—what percentage, can you remind us? What percentage of energy comes from small producers, from independent producers, not from quote/unquote "Big Oil"?

Mr. ESHELMAN. That is 91 percent of the wells.

Mr. PFLUGER. Ninety-one percent. And disproportionately, how is the methane, the natural gas tax, how does it affect those people that don't have economies of scale, that are independent producers, that—one or two people in their businesses?

Mr. ESHELMAN. That is what we are trying to understand. As I mentioned before, there has never been a hearing on this. There has never been a study on this. There has never been testimony on this tax. So we really don't know.

But we are concerned with giving EPA more authority. And once you give an agency authority, they kind of grow with it. So that is——

Mr. PFLUGER. You know, I figured out very quickly when the wind energy folks came to see me and they said, "Mr. Pfluger, we have a problem with the EPA. We can't get a permit to build a windmill in your district."

I said, "Oh, well, welcome to the club."

We have more wind energy, by the way, Mr. Garcia, in my district than the entire State of California. But they can't get a permit to do that.

Mr. Slocum, is wind and solar—do they provide baseload capacity?

Mr. SLOCUM. They do not. But the question is whether or not baseload capacity is always a requirement.

Mr. PFLUGER. Yes.

Mr. SLOCUM. Especially—you can balance off intermittence with——

Mr. PFLUGER. That is fine, thank you. I think the fact is it doesn't, and it is not required in States like California, where they have intermittent electricity, which is unreliable. However, in most of the United States we want reliable electricity.

We have these balls in the air. One of them is cost, one of them is reliability, one of them is geopolitical security. And the climate is certainly always going to be right there. We do it cheaper, more efficiently, and also more environmentally friendly.

Mr. Garcia, you mentioned that—I don't have the exact note, but that there is no boundary on air pollution. Is that—does that summarize a previous comment?

Mr. GARCIA. For the specific provisions that these—and loopholes that these bills would bring about, yes, there would be no——

Mr. PFLUGER. No, you said air knows no borders. I found my note. "Air knows no borders." I just quoted you on——

Mr. GARCIA. Air, yes, the air——

Mr. PFLUGER. Is that true?

Mr. GARCIA. Absolutely, air doesn't know borders.

Mr. PFLUGER. OK, so what are the Chinese doing to curb? Do they have taxes on natural gas in China?

Mr. GARCIA. My understanding is that the jurisdiction of this committee is about what the United States can do.

Mr. PFLUGER. So my understanding of what you said is that air knows no borders. And I want to make the point that, if that is true, then the people in my district are being affected by Chinese air, just like they are in your home, just like they are everywhere throughout the United States or throughout the rest of the world. So the competitive advantage that we are giving them, giving away, the most important resource we have at our fingertips, is also causing the actual harming of our climate. And it is coming from China. And we have to take actions here to compete and to continue to do it better.

I have 15 more minutes of comments. However, the Chair will not let me use that time, and I yield back.

Mr. DUNCAN. I thank the gentleman, and now we will go to Mr. Obernolte from California.

[Pause.]

Mr. DUNCAN. Is your mic not on? Can you slide over one?

Mr. OBERNOLTE. There we go. I swear I pushed the button.

Mr. McNamee, in your testimony you were discussing the need to enhance cybersecurity in our natural gas pipelines. And that really resonated with me because we all lived through the cybersecurity hack of the Colonial Oil pipeline a year and a half ago, which, as I am sure everyone in this room knows, disrupted the supply of gasoline to the entire southeast United States.

So what can be done? What needs to be done to put those protections in place for natural gas pipelines?

Mr. MCNAMEE. Well, the industry has generally been working to try and—to harden their systems, including their electronic systems, but also there is the challenge of physical attacks.

So I think the primary thing that needs to be done is to—like the reporting issues that want to be done for the bulk power system, but also for natural gas. And I think that, you know, we need to encourage natural gas pipelines, oil pipelines, the electric grid to all be focused on trying to find out what is the next thing that is going to happen and harden it.

Because one of the problems with regulations is it tends to look backward at what has happened, and, of course, you need to harden for that. But the bad guys are way ahead, and so they— kind of constantly be working for what is the next thing. And being aware of what the next thing is is going to be helpful, and that is part of what I think that bill will help do.

Mr. OBERNOLTE. Sure. Well, you know, I think that we are having to transition from seeing cybersecurity of privately held companies as a business issue to seeing it as a national security issue, whereas traditionally we just thought of national security being applicable to the Department of Defense and agencies of the Government.

Mr. Slocum, first of all, let me thank you very much for not reading your statement. You know, I think that hearings—we communicate with each other a lot better when we are not just reading at each other. And I just want you to know that we noticed, that we appreciate it, and, you know, I think that is something we should do some more of.

So you were talking in your testimony about objecting to removing the required finding of being in the national interest before liquid natural gas can be exported. And, you know, just—we heard some other testimony from various witnesses, including Mr. McNamee, about the way that government action can create distortions in the markets for energy.

And I am wondering why that wouldn't be the case here, because every time the Government tries to put their finger on the scale to try and protect energy markets and manipulate energy prices, it seems like things can go awry, and particularly in this case, when you are creating, you know, by the restriction of natural gas, of liquefied natural gas exports, you are creating an economic island that then can be vulnerable to things like arbitrage by people with just a profit incentive, and the consumer ends up losing anyway.

So what—why would that not create economic distortions that we need to avoid?

Mr. SLOCUM. Well, I think because natural gas has historically been a regulated commodity.

So when it was enacted in 1938, this provision, it was recognized that natural gas was providing essential services to homes and businesses. And so, as a result, if you were going to export it or import it, there had to be public interest determinations on it because of its essential utility——

Mr. OBERNOLTE. But should it be, is the question. I mean, I think that is the question that the bill that we are we are debating is asking, is should it be? Does that actually have the intended effect?

Because many times, as you know, governmental action does not.

Mr. SLOCUM. Well, I mean, right now, as I pointed out in my written testimony, the Department of Energy is not really performing a meaningful public interest determination. There has never been a rejected application to export natural gas by the Department of Energy, right? So that—so it seems like the legislation is proposing a solution in search of a problem, right, that the Department of Energy is not—there is no backlog of applications. It takes sometimes just months between the submission of the application of the Department of Energy and its approval.

Mr. OBERNOLTE. OK, that is an interesting different kind of line of argument than the one in your testimony. I think both are interesting. I would love to continue this discussion. I only have a few seconds left here.

Mr. Garcia, I was going to ask you about this. Unfortunately, it is going to be a statement, and not an opportunity for you to respond, and I apologize for that.

But, you know, we have been talking about this economic valuation between trying to value human prosperity and ranking that against protecting the environment. And I think everyone here on the dais would agree that we need to do both. But, you know, the problem is that we all represent constituents who are suffering.

In California we pay twice as much for residential electricity than neighboring States, three times for industrial, four times as much for commercial. My constituents pay twice as much for a gallon of gas to put in their car to commute to work. They don't have a choice to do that. And this is the problem. You know, we have to make those economic judgments. And I think that there is a way to effect a win-win.

But I want to thank you all for your testimony. I found it a very interesting hearing.

I yield back, Mr. Chair.

Mr. DUNCAN. I thank the gentleman. The Chair will now go to Mr. Weber from Texas.

Mr. WEBER. Thank you, Mr. Chairman. I am going to make a couple of comments.

First, the—one of the comments on the panel was earlier, before we went to vote—was that bang for energy efficiency—you get more bang for your buck than you do more power plants.

And I would submit this for the record and for the panel that, you know, power plants, I don't know, 100 years ago there was X number. But American population has doubled, tripled, increased. And so we are getting more and more people, and especially with an open border—thank you, President Biden—we are going to have a lot more people, and we are going to need a lot more power.

Now, some of you all know I am from Texas, and we talked about Winter Storm Uri, where we went through that two Februaries ago, and Texas saw record cold. I am born and raised on the Gulf Coast of Texas, 20-mile radius, 69 years. I have never seen it be 18 degrees on the Galveston Island.

So we lost some power. It was a perfect storm. Some of the industries that normally—plants that normally shut down in the wintertime weren't necessarily ready for that kind of cold. The waterlines freeze. South Texas nuclear plant down in Matagorda County was one of them. I represented them as a Texas State rep. And so we lost—wind energy failed us. Wind turbines, they froze up, blades got iced up.

So we are number 1 in wind energy. We are number 2 in solar panels. And you find out very quickly that, in an incident like that, you get a lot of cloudy skies. You don't have as much sun. Solar panels ice up. They got snow on them. And here is the bad news about solar panels: You know the sun goes down at night, and so they are not constantly producing electricity.

We actually met with ERCoT and PUC—the ERCoT, Electric Reliability Council of Texas, and the PUC there in Texas—and they talked about their plan. There is a website called PowertoChoose.org, where you can choose your provider. A lot of the providers—some of the providers advertise 100 percent green energy. And what that was saying is only solar and wind power. They were able to provide—because of the subsidies and whatnot— cheaper rates from those who used natural gas, and some coal, and even nuclear.

And so the PUC told us—four of us Members of Congress met in Austin with them—that they were going to put a rule into place that, if you are a power provider on that website, a retail electric provider—we call them REPs, R–E–Ps—they had to increase—they had to include, rather, a part of a base load that included some reliable, dependable, affordable—and as your discussion with Mr. Pfluger was—that is natural gas.

We have a lot of wind energy in Texas. I didn't know it was— most of it was in his district, but I am mighty proud for that. A lot of gas pipelines, oil pipelines in his district too.

So I want to go back to Obernolte's discussion about the Colonial Pipeline system. The Keystone Pipeline comes into my district in Texas. It is the safest way to move product, period. Not truck, not rail or barge. It is the safest way. The President shut it down.

Colonial Pipeline system, about 2 or 3 months after Winter Storm Uri, was hacked into and was shut down 4 or 5 days. The Colonial Pipeline system feeds the southeastern part of the United States. It carries 3.1 million barrels of product today: gasoline, diesel, and jet fuel. The Keystone pipeline carries 830,000 barrels of product a day. It is literally one-fourth—more than one-fourth the output of the entire—the Colonial Pipeline system that feeds the entire southeastern part of the United States with a leg that goes north.

So it is extremely important that we have a solid baseline—base-load system that is going to be fossil fuels. Renewables are great. We like renewables. Make no mistake about that. I am tired of our friends across the aisle saying somehow we are in the oil companies' back pockets. That is just a mischaracterization. But what else do you expect? Did I say that out loud?

At any rate, renewables can—they can—it can be the supporting actor in this movie. It cannot be the leading actor in this movie. And what we are trying to do today is to make sure that we are building that up.

Mr. Menezes, in your written testimony you outline H.R. 150, Protecting American Energy Production Act, which I am an original cosponsor of, which would prohibit the President from declaring a moratorium on fracking, which has started in my district. You got 29 seconds. Tell us how good it is, fracking.

Mr. MENEZES. What are we talking about, the virtues of hydraulic fracturing?

I mean, certainly for domestic supply, for export to help our friends and allies all over the world get off of Russia, natural gas— they have come to us asking.

You will look at what is going on in the private sector. This is without government interaction. In fact, it is beyond the reach of the Federal Government. Contracts are being entered into by European off-takers and Asian off-takers, our friends and allies, for good old U.S. LNG. If the President thinks that he can declare an emergency and somehow put a moratorium on fracking, he is going to upset the entire geopolitical energy markets.

Mr. WEBER. I will say thank you for that. And Mr. Chairman, I yield back.

Mr. JOHNSON [presiding]. Thank you. The gentleman yields back. The Chair now recognizes the gentleman from Alabama, Mr. Palmer, for 5 minutes.

Mr. PALMER. Thank you, Mr. Chairman. I want to talk a little bit about where some of your funding comes from, where—how it is used, and particularly about the $27 billion slush fund that was put into the Inflation Reduction Act. You stuck in $27 billion that basically is going to go to Wall Street firms to set up a climate bank. But it doesn't help the American people with their utility bills.

It is about 20 million households that are behind on their utility bills. That is about one in six households. It is forcing people in some places to literally choose between keeping their homes adequately warmed and still be able to afford their groceries and their medicine. I think, if Europe is any indication, we are going to find out that there are several thousand people that will be classified as excess winter deaths. We have seen this in the United Kingdom, we have seen it in Europe. It has basically become a conduit for seed money for this new climate bank. And because there was no clarity in it, you have got climate groups fighting over the money.

I am just—Secretary Menezes, will this $27 billion slush fund lower the cost of heating for these American families?

Mr. MENEZES. No, the provision is clear. It is to go to develop zero-emission technologies for States and municipalities——

Mr. PALMER. Will this slush fund provide natural gas for a town like Pembroke Township in Illinois that—85 percent African American? They don't have any means to grow their economy, they are heating their homes with propane and wood. Will it help them?

Mr. MENEZES. Zero-emission technologies, nonprofits, municipalities, and States.

Mr. PALMER. Will it do anything for the people in the northeastern part of the United States that are not using natural gas, they are using heating oil, and the cost has almost doubled in the last couple of years—$5 and something a gallon. Will it help them?

Mr. MENEZES. Not that I am aware of.

Mr. PALMER. Yes, and it has nothing to do with natural gas in that area. It is all about heating oil.

Will it help people who have been—who bought into this stuff like Dharnai, India, where in 2014 Greenpeace went in and convinced them that they could supply power for the village through 100 percent renewables, and 6 years later it is basically an animal shed because they couldn't afford it, they couldn't—it wouldn't power appliances like a refrigerator that most of us take for granted. They couldn't maintain the batteries. So now they are getting their power from fossil fuel. It wouldn't help in a situation like that, would it?

Mr. MENEZES. I was not aware of that, but I would hope that is not the consequence of this fund.

Mr. PALMER. Yes, there is a really good paper on this from the Institute for Energy Research.

My concern is, looking at China's influence and funding various groups that are pushing renewables, that in many respects—and I mean this seriously—it undermines our national security, this mad dash to renewables.

I keep hearing some of my colleagues across the aisle talk about we don't need to be dependent on foreign sources for oil and natural gas. Well, I got news for them. We don't have to be. The only reason that we might be is because we refuse to access what we have.

I hear them basically ignore the fact that natural gas is largely responsible for the tremendous reduction in greenhouse emissions that we have already enjoyed. They ignore the technological advances, yet they want us to be dependent on China for the—for wind turbines and for solar panels. Does that make sense?

Mr. MENEZES. Not to me.

Mr. PALMER. It doesn't to me, either. It does concern me, though, because it becomes a national security issue at that point. China is already, in some critical minerals, holding back on shipping those, and some of those are critical for our national defense, not just our national energy grid.

But having worked for two international companies, having a little bit of an understanding about this—and there is another report from the Electric Power Research Institute—we are not going to be at net zero by 2050. We don't—the engineering doesn't support it. The technology doesn't support it.

We could be a lot more on the renewable side, but if we were really smart about it, we would go to next-gen nuclear and power the world. We would be working in sub-Saharan Africa, South America, Latin America, and the Caribbean, building natural gas facilities so that those people could enjoy the same economic benefits that we do, rather than sitting back and watching China. They have already built 14 coal-fired power plants outside of China, and they built, what, maybe one every two weeks now in China.

So I just—Mr. Chairman, I appreciate the opportunity to raise these issues about the slush fund that I think was put in that Inflation Reduction Act, and I yield back.

Mr. JOHNSON. The gentleman yields back. The Chair now recognizes the gentleman from Idaho, Mr. Fulcher, for 5 minutes.

Mr. FULCHER. Thank you, Mr. Chairman. And coming up on the tail end of this, a lot of the subject has been covered already, so I am going to go—I think I will address this to Mr. Menezes. It has to do with hydro.

In our State, hydropower, 51 percent of our total in-State electricity usage, and it is great, at least for us. It is baseload, it is cheap, it is renewable. Comment on hydro as a source, and if you agree that it is effective and efficient. How do we ensure that that is an ongoing predominant baseload source?

Mr. MENEZES. Excellent question. Well, we should be allowed to continue to use the dams that are in operation to provide the hydropower, first of all. There is a big movement, of course, of dam removal, right? So a lot of communities are facing that. And so that is a real threat to the hydropower that we have today.

The other thing is, it is going to be impossible to build any sizable—a new dam. So you are looking at incremental gain, incremental hydro. And so that is to make improvements in the existing hydro, perhaps some expansion. And on existing dams, perhaps you can put some electricity generating turbines there. But that is really the future on hydro that way.

We do have other potential great technologies. I mean, we have tidal, we have run of river, we have a variety of other technologies in the using of hydropower. So I think the future is still bright on that, but it is going to be very difficult to maintain, the base load that we have now.

The other thing, of course, is pump storage hydro.

Mr. FULCHER. OK.

Mr. MENEZES. That is another good use of hydro power, that——

Mr. FULCHER. I have heard it said that the greatest battery in the world is water behind a dam, storage.

Mr. MENEZES. Yes.

Mr. FULCHER. I am going to shift to Mr. McNamee, if I may, please.

Another asset we have in my State of Idaho is Idaho National Lab. And a lot of research is being done, a lot of work with small modular reactors is being done there. Talk about that a little bit. I hear various experts talk about the benefit of—the greatest benefit being able to decentralize the grid, for example. Of course, the no-carbon emission is another factor. Others say it is great for redundant backup.

What is—in your view, what is the most appropriate use, the most appropriate role for small modular reactors as those come online?

Mr. MCNAMEE. Well, the ones that you have mentioned are absolutely important.

And another one is reliability. SMRs are really, I think, going to transform the way we do the electric grid. And what is interesting is, because, you know, they are between 5 to 300 megawatts, you can build them in a manufacturing facility, you can put them on a rail or on a truck, and put them to where they need to be. The price is going to come down for nuclear. They can run 24/7. They are designed to be much safer than—not that current nuclear isn't, but they are going to be safer, the way they operate.

But what is really going to be amazing is, where we were talking earlier about having to build long transmission lines to get wind, let's say, from the Midwest to the coast, whatever, you can put SMRs on old coal plants and use the transmission. You are going to save billions of dollars in transmission. So SMR technology, it is—we are about to be able to embrace it, and it is going to be a big change.

Mr. FULCHER. Great, thank you. Very helpful for me.

Ms. Sweeney, you are not off the hook. I got a question for you. In your written testimony you talk about, I think, it is 80 percent, nearly 90 percent of global rare earth elements from China.

Ms. SWEENEY. Yes.

Mr. FULCHER. Is that mainly because that is where those earths are, or is that for other reasons?

Ms. SWEENEY. No, the U.S. was the leading rare earths producer throughout, I think, the early 1970s. It is really because China focused on rare earth development, and they started, you know, printing money and cornering the market, and then they were able to bring over anybody who wanted to be an end user of those products. They brought all the processing into China, and they were able to pretty much control that market, and still do.

Mr. FULCHER. And here we are. Right, thank you. That is what I suspected. But one last question for you, and I only have 40 seconds left, but—and you did touch on this, but I would like you to touch on it again, please.

Once again, in your written testimony you talked about the permitting process for mining projects. With the current administration that we have got, what are some of the things that that administration could do to make that permitting process better without going through the statutory process?

Ms. SWEENEY. There are a lot of commonsense solutions out there. I mean, really, it is looking at how to use NEPA the way it was intended to be used.

So essentially, there are—agencies are supposed to try to avoid duplication. They are supposed to tier off of each other's environmental assessments, use the same information. There are a lot of commonsense solutions that don't need any changes statutorily.

Mr. FULCHER. Thank you very much. Mr. Chairman, I yield back.

Mr. JOHNSON. The gentleman yields back. The Chair is—now recognizes the gentlelady from Michigan, Mrs. Dingell, for—oh, I am sorry, Mrs. Dingell. Sorry about that. I got it out of order. I now recognize the gentleman from Texas, Mr. Crenshaw, for 5 minutes.

Mr. CRENSHAW. Thank you, Mr. Chairman. Thank you all for being here.

I want to quote the great Thomas Sowell that the first lesson of economics is scarcity. There is never enough of anything for all the people who want it. And the first lesson of politics is to ignore the first lesson of economics. And that seems to be the rule that I think radical environmentalism plays when it comes to understanding basic supply and demand, and how that affects prices.

Now, we can engage in wishful thinking, and we can wish for a reality that is different than the one we live in, where supply and demand aren't real forces, and that we can just make them up and demand the prices that we feel like having and expect the supply to then be there when we need it. Of course, that is not reality. We can we can go write fiction novels about it, I suppose, if we would like. That might be fun.

So I want to talk about supply and demand on something very specific, and that is our refineries, because this EPA rule has gotten a lot of attention, and our bill, in response to that EPA rule, has gotten a lot of attention today.

It is worth noting, first, that 8 refineries have shut down in the last 5 years. It is one of the great reasons for the bottlenecks in refinery production and, of course, the reason for prices going up and staying up. When demand post-COVID shot up and recovered, the supply could not recover accordingly. Just the basics here.

Now the EPA wants to make that supply harder. The EPA wants to target hydrofluoric acid using refineries. That would affect 41 of our 130 refineries. That is 50 percent of all the product refined.

Mr. Garcia, you talked about this today. I mean, what is really the intent behind this? And you have said it, so I am going to paraphrase you, which—the intent is to get them to change to a sulfuric acid-type refinery. Is that correct?

Mr. GARCIA. No, the intent is for them to actually take the time to explore alternatives that are going to make it safer.

Mr. CRENSHAW. Well, there is only one alternative. So you want them to change, though, right? I mean, you want them to use the different ingredient for the refineries, is that correct?

Mr. GARCIA. We have talked about innovation. And so the analysis actually leads to innovation. And so the idea is that, while there might be one today, there could be more in the future. But you have to do the analysis in order to discover what that innovation is going to look like.

Mr. CRENSHAW. Would a risk analysis discover innovation? I have never heard of an innovation, a new technology, discovered by a risk analysis.

Mr. GARCIA. If you are taking into account—if you are exploring what alternatives are out there, then yes, it will.

Mr. CRENSHAW. How?

Mr. GARCIA. Again, if you are exploring——

Mr. CRENSHAW. Like, you are a——

Mr. GARCIA. If you are looking at new science——

Mr. CRENSHAW. Is there any idea?

Mr. GARCIA. You are looking at new science. I mean, you are looking at new science, new developments in technology that come about. And so, as those new developments in technology come about——

Mr. CRENSHAW. OK, so we can write a fiction novel about it.

Mr. GARCIA [continuing]. Into which you can actually implement them, into the refinery——

Mr. CRENSHAW. I understand, I understand. There is no answer.

Mr. GARCIA [continuing]. That is how you get progress.

Mr. CRENSHAW. There is no answer. Again, we could write a fiction novel about it. That would be fun. But this is reality. So in reality, there is one other alternative, which is the sulfuric acid use of—type of refinery.

Now, if we shut down the 41 refineries to transition to that, which I suppose is deemed safer for some reason, that would take 50 percent of our refining capacity offline.

Mr. Menezes, what would that do to our economy and our gas prices?

Mr. MENEZES. Fifty percent of our refined products offline? Well, we have a demand of about 20 million, 17 million barrels per day, you know, for refined products. If we lost half of that, I can only imagine what it would do on all the economies that depend on trucks, on, you know, transportation, your constituents that need to drive.

I am not even sure our national energy modeling system at DOE can model a 50 percent reduction——

Mr. CRENSHAW. And we have gotten some numbers on this. It would take 2 to 3 years for each facility to change. It would take a minimum of $200 million, or potentially $800 million, depending on the size. So we are talking billions and tens of billions of dollars in costs. For what benefit? I don't even know.

And a guaranteed increase in prices. I mean, we are fighting for the disenfranchised here, for people who can't afford to fill up their tanks, and yet all but guaranteeing that they can't do so for almost no benefit. And that is the theme, that is the general theme of radical environmentalism: great cost for almost no benefit. That is a problem.

And so our bill that has been chastised quite a bit in this hearing is simply in response to that, and says, look, if the plant already exists, if the refining plant already exists or is in construction, it is exempt from this new rule. If it is being planned, then by all means, take into account this particular rule and risk assessment. That is just—that is commonsense environmentalism.

And I urge this committee—because we all want clean energy. Anybody who follows me knows that I am constantly battling for it. But we have to do so in a way that doesn't hurt people more than we want to help them.

And I yield back.

Mr. JOHNSON. The gentleman yields back. And now the Chair is honored to once again recognize the gentlelady from Michigan, Mrs. Dingell, for 5 minutes.

Mrs. DINGELL. Thank you, Mr. Chairman, and thank you to all the witnesses who have to be ready to end this.

And I do have to say I am worried about the number of pieces of this legislation, because I care deeply about making sure that we are taking care of everybody and that they can afford the energy of the future. And yet it is many of our children that are suffering from asthma and many other things, and that what we are talking about here is going to give handouts to oil and gas and undermine our Nation's environmental laws and actually rescind programs that are doing something about greenhouse gas.

But I want to get to two very specific issues today. I want to start with the draft legislation to repeal the Greenhouse Gas Reduction Fund and close by focusing on our critical minerals supply. The Greenhouse Gas Reduction Fund, which was established in the Inflation Reduction Act and based on original legislation I authored—so yes, I care about it—will invest $27 billion to develop clean energy projects aimed at reducing greenhouse gas emissions. Over 40 percent of the funding will also target disadvantaged communities, communities that for far too long have carried the brunt of environmental pollution.

For years I have been a champion of a clean energy accelerator similar to this fund because of its potential to accelerate the clean energy transition. Therefore, I am disappointed to see my colleagues on the other side propose repealing this historic program, which the EPA has yet to even fully implement, and the benefits which have yet to be fulfilled.

I am going to start with Mr. Garcia. My first question is simple.

How will repealing the Greenhouse Gas Reduction Fund affect our ability to meet our climate goals?

Mr. GARCIA. Well, in a lot of the calculations that we are talking about in terms of economic investment across the country for energy production, no one is accounting the actual harm that is coming our way because we keep investing in dirty fuels.

And so that is one thing that I have noticed about the proponents of these bills, and it is a larger narrative, is that they want to sweep all of these consequences under the table.

You know, if you have hydrofluoric acid in a refinery, yes, it can blow up. I think that might hurt the economy, right?

So the same thing happens when we are talking about investment. It is making sure that we are making investments for the right long-term outcomes. And so that is what the Greenhouse Gas Reduction Fund really does.

Now, it is not a slush fund. People keep calling it a slush fund. It is really not. Fifty-five percent of the overall program funding will be dedicated to projects in low-income and disadvantaged communities. Eight billion of the 20 billion is earmarked for low-income and disadvantaged communities, plus a separate 7 billion program. And so that is something that we have to keep in mind. For—since the last industrial revolution, communities of color and low income have bore the burden that all of these industries put on their shoulders. And so this is the least that we could do at this point, and it is something that we need to see fulfilled, and we need—and it is something that won't cure the sins of the past but will certainly help get us to better consequences in the long-term.

Mrs. DINGELL. Thank you. I had a couple more questions, but I am down to almost a minute.

So this fund is specifically designed to provide funding for projects where investment is lacking, effectively filling in a funding gap rather than duplicating Federal programs.

Mr. Chairman, I am going to ask if I can submit some further questions on this, because I think it is really going to hurt frontline communities.

And I would also like to request unanimous consent to submit a letter for the record from the environmental community expressing strong opposition to the legislation repealing the—

Mr. JOHNSON. Without objection, so ordered.

[The information appears at the conclusion of the hearing.]

Mrs. DINGELL. Thank you. My remaining time is on critical minerals.

Strengthening supply chains for electric vehicles and batteries is a top priority for me. We cannot and will not be dependent on China. And I want to keep my own State of Michigan as a leader in this sector.

The Bipartisan Infrastructure Law made historic investments in battery manufacturing and recycling facilities, battery reprocessing, and critical minerals mining and recycling research. These investments are a critical down payment, but we all know that more work needs to be done to meet the demand for these critical minerals.

Mr. Garcia, what policies should this committee be exploring to develop and strengthen our critical mineral supply chains in an equitable, sustainable way?

And you may need to provide more on that for the record.

Mr. GARCIA. Absolutely. I will say quickly that it has to look to strengthen and enforce the laws that are in the books right now in order to make sure that that extraction and that recycling happens in a way that protects communities first but still yields the adequate production that is needed in order to get new kinds of transportation methods, clean types of transportation methods on the roads.

Mrs. DINGELL. Thank you.

Mr. Chairman, I will yield back, and I do want to work with you in a bipartisan way, but I have some real concerns. Thank you.

Mr. JOHNSON. The gentlelady yields back. And now, seeing there are no further Members wishing to ask questions, I would like to thank all of our witnesses once again for being here with us today.

I ask unanimous consent to insert in the record the documents included on the staff hearing documents list.

Without objection, that will be the order.

[The information appears at the conclusion of the hearing.]

Mr. JOHNSON. And pursuant to committee rules, I remind Members that they have 10 business days to submit additional questions for the record, and I ask that witnesses submit their response within 10 business days upon receipt of those questions.

Without objection, the subcommittee stands adjourned. [Whereupon, at 3:21 p.m., the subcommittees were adjourned.] [Material submitted for inclusion in the record follows:[123]]

* * *

February 6, 2023

American Petroleum Institute

The Honorable Cathy McMorris Rodgers
Chair
House Committee on Energy & Commerce
2322 Rayburn House Office Building
Washington, DC 20515

Dear Chair McMorris Rodgers:

The American Petroleum Institute (API) writes regarding the legislation to be considered during the upcoming *Joint Energy, Climate, & Grid Security Subcommittee and Environment, Manufacturing, & Critical Materials Subcommittee Legislative Hearing: "Unleashing American Energy, Lowering Energy Costs, and Strengthening Supply Chains"* on Tuesday, February 7, 2023.

API is committed to meeting the challenge of providing affordable and reliable energy while continuing to reduce emissions. As the leading trade association representing the entire value chain of the U.S. oil and natural gas industry, API supports policies that strengthen our nation's energy security and our economy and that protect our environment. The hearing on February Jth will consider legislative proposals that speak to these objectives, and we commend the Committee for holding this important hearing.

The Committee's renewed focus on restoring American energy leadership is critical to ensuring that American oil and natural gas reserves are prioritized as a long-term strategic asset that will serve as a foundation for economic growth and energy security. Members of the

123 Legislation and some submitted material have been retained in committee files and are available at https://docs.house.gov/Committee/Calendar/ByEvent.aspx?EventID=115306.

Energy and Commerce Committee have already vocalized how America's energy security directly correlates to our national security at the outset of the 118th Congress, and we support the efforts of the Committee, under your leadership, to address these energy challenges.

API remains focused on working in a bipartisan manner to find common-sense, economically-sound solutions to not only maintain our energy independence, but also to assist our allies around the world as they grapple with energy uncertainty resulting from Russia's war in Ukraine. Demand for oil and natural gas is increasing, and we can - and must do more to avoid policies that stymie increased energy production such as development restrictions on federal lands and waters, onerous and duplicative regulations, insufficient infrastructure due to permitting delays and denials, inconsistent standards for environmental reviews, and significant supply chain bottlenecks.

The oil and natural gas industry urges Congress to pass legislation that will enable continued innovation, investment in long-term projects and cleaner energy technologies, and improved infrastructure development. Much of the legislation being offered in this joint legislative hearing is a strong step in this direction. Accordingly, API commends the Committee for considering proposals including:

- H.R. 150, the "Protecting American Energy Production Act."
- H.R. 484, the "Natural Gas Tax Repeal Act."
- H.Con.Res._, Expressing disapproval of the revocation by President Biden of the Presidential permit for the Keystone XL pipeline.
- H.R. 647, the "Unlocking our Domestic LNG Potential Act of 2023."
- H.R. _, To require the Secretary of Energy to direct the National Petroleum Council to issue a report with respect to petrochemical refineries in the United States, and for other purposes.
- H.Con.Res. _, Expressing the sense of Congress that the Federal Government should not impose any restrictions on the export of crude oil or other petroleum products.
- H.R. _, To authorize the Administrator of the Environmental Protection Agency to waive application of certain requirements, sanctions, or fees, with respect to processing or refining of critical energy resources at a critical energy resource facility, and for other purposes.
- H.R. _, To amend the Toxic Substances Control Act with respect to critical energy resources, and for other purposes.
- H.R. _, To amend the Solid Waste Disposal Act to treat the owner or operator of a critical energy resource facility as having been issued an interim permit for the treatment, storage, or disposal of ha_zardous waste, and for other purposes.
- H.R. _, To require the Administrator of the Environmental Protection Agency to authorize the use of flexible air permitting with respect to certain critical energy resource facilities, and for other purposes.

Further, other proposals being considered by the Committee that may, with additional modifications, also support our energy and national security objectives include:

- H.R._, the "Promoting Cross-border Energy Infrastructure Act."
- H.R._, the "Promoting Interagency Coordination for Review of Natural Gas Pipelines Act."
- H.R._, To amend the Clean Air Act to prohibit the phase out of gasoline and prevent higher prices for consumers and for other purposes.

As Congress continues its work on these and other important energy legislation, API and our members stand ready to support policies that strengthen America's energy security and promote economic development and environmental stewardship. We look forward to continuing to work with Members of the Committee and their colleagues in the House and Senate to support legislation that will indeed, "pave the way to unleashing American energy sources and technologies of all kinds."

Sincerely,
Amanda E. Eversole

CC: The Honorable Frank Pallone, Ranking Member, House Energy and Commerce Committee

The Honorable Bill Johnson, Chair, House Subcommittee on Environment, Manufacturing, and Critical Minerals

The Honorable Jeff Duncan, Chair, House Subcommittee on Energy, Climate, and Grid Security

HF – What Is It?

- Hydrogen Fluoride (HF) is a colorless gas or liquid chemical compound. HF is the principal industrial source of fluorine and, when dissolved in water, is known as hydrofluoric acid
  - HF acid is mainly used for industrial purposes (e.g. glass etching, metal cleaning, electronics manufacturing)
  - HF acid is used in refineries as a catalyst in the alkylation process to form alkylate, which is a blending component in high-octane gasoline, jet fuel, and marine diesel fuel

API American Petroleum Institute

HF – A Brief History

- HF has been safely used in refineries since World War II, when alkylate was first required in aviation fuels
  - Only 2% of HF acid in U.S. is used in refinery alkylation
  - Of the 120+ operating refineries in U.S., nearly all have some form of alkylation unit – 42 of which use HF acid
- There have been few HF releases with off-site impacts; there has never been an off-site fatality in the U.S. due to an HF release
- Other types of commercially-proven refinery alkylation technologies include: modified HF (MHF), sulfuric acid alkylation, and ISOALKY

API American Petroleum Institute

HF – Why It Is Important

- Alkylate is the ideal blend stock for gasoline to meet clean fuel regulations
  - Alkylate has high-octane (92-97 RON) and low Reid Vapor Pressure
  - Alkylate helps reduce emissions from automobiles due to its low sulfur content and its very low benzene content
- HF alkylation is not interchangeable with other refinery alkylation process technologies
  - HF alkylation does not require refrigeration
  - Regeneration of HF does not produce SO_2/SO_3 emissions
  - Revamp to other commercially-proven technology is prohibitively expensive and would require at least a 2-6 month shutdown

API American Petroleum Institute

What Are API Members Doing?

- Refineries adhere to strict industry standards and undergo safety assessments to mitigate risk and assure safe operations
  - Refineries employ operational standards and mitigation systems, including those covered by API RP 751, *Safe Operation of Hydrofluoric Acid Alkylation Units*
  - The 5th edition of RP 751 was published in August 2021 and significantly expanded the Materials, Fabrication, and Inspection and Maintenance Practices section
  - PSSAP Program

American Petroleum Institute

Representative Sampling

- Industry is highly-regulated and facilities have multiple RMP covered processes. Therefore, it is not necessary to audit each covered process if it includes a representative sample
  - Employing sampling as part of the audit process is a robust, scientifically proven, method of demonstrating that all covered process and all RMP elements are compliant
  - The refining industry is extraordinarily complex and highly involved. Auditing *all* covered processes and all RMP elements requires not only significant resources but is very disruptive to operations
  - Most facilities have only one process, but certain industries, such as chemical manufacturing and petroleum refining, often have more than one regulated process; about 100 facilities have more than 10 regulated processes with an average cost of $36,500 per process

American Petroleum Institute

QUESTIONS OR COMMENTS?

American Petroleum Institute

Coalition for Chemical Innovations (CCI)

EXECUTIVE SUMMARY

COMMERCIALIZING INNOVATIVE CHEMICAL PRODUCTS

October 2022

Coalition for Chemical Innovations

Who We Are:

The Coalition for Chemical Innovations (CCI) is a group of diverse stakeholders within the chemical industry, including chemical manufacturers, processors, distributors, and users.

Our Mission:

The U.S. Environmental Protection Agency's (EPA) current interpretation and implementation of the Toxic Substances Control Act (TSCA) in the review of new chemicals under Section 5 is stifling chemical innovation at a considerable cost to sustainability, the economy, and improved human and environmental health.

CCI's mission is to educate decision makers, including Congress, EPA, other federal agencies, and related stakeholders on the true costs of these policies and to outline common-sense policy changes to prevent the further stifling of chemical innovation by EPA's implementation of TSCA Section 5.

The Problems We Seek to Fix

EPA's current practice under TSCA Section 5 is to assume that all chemicals for which it identifies a hazard, other than "low hazard" for health and "ecotoxicity," may be used in a way that leads to an exceedance of the hazard threshold. In other words, EPA assumes that any use is "reasonably foreseen," inevitably leading to the regulation of new chemicals and the application of risk mitigation measures even when submitters provide data disproving EPA's assumptions.

This hazard-based approach is inconsistent with TSCA's risk-based framework and reflects a misapplication of TSCA Section 5. It renders meaningless the TSCA statute's requirement to evaluate chemicals in the context of "reasonably foreseen" conditions of use, as every theorized condition of use then becomes foreseen.

In addition, because existing, incumbent substances, even those that pose *the same or higher* hazard, do not face the same requirements (now or potentially for decades into the future), the requirements on new chemicals pose a significant commercial disadvantage and discourage innovation.

CCI commissioned FTI Consulting, Inc. (FTI) to conduct an analysis of new chemical submissions to EPA, both Premanufacture Notice (PMN) and Low Volume Exemption (LVE) applications from 2011 to the present. FTI assessed the nationwide economic losses associated with implementation of TSCA since it was amended in 2016. The analysis accounts for the "upstream" losses to CCI members and their suppliers from lost sales opportunities and the "downstream" losses to CCI members' customers from lost sales and higher costs, along with indirect costs.

The results of the FTI analysis, obtained in August 2022, confirm:

- **Final EPA determinations on PMNs submitted by CCI members have plummeted since the passage of the Act:**

2011

Seventy (70) chemicals **received a final determination from EPA**

2019

Six (6) chemicals **received a final determination from EPA**

- **EPA's average review time for PMNs has increased by 260% since the passage of the Act:**

Prior to 2016

In the five years before TSCA was amended, average PMN review time was 158 days

Since 2016

Since 2016, the average review time of a PMN has increased to 413 days

Under TSCA, PMN reviews are intended to be complete within <u>90 days</u>.

- **EPA's average LVE review time has increased by 443% since passage of the Act:**

Prior to 2016

In the five years before the 2016 amendments, average LVE review time was 44 days

Since 2016

Since 2016, the average review time has been 87 days. In 2021, the average review time increased to 195 days

Under TSCA, LVE reviews are intended to be complete within <u>30 days</u>.

CCI

More Results of the FTI Analysis

80% Drop in R&D

CCI members have historically spent billions on research and development (R&D) in the United States, providing employment for U.S. workers and helping expand U.S. productivity. Current implementation of TSCA Section 5 has stifled R&D:

- U.S. R&D expenditures by CCI members **peaked at $324 million in 2018**; and
- U.S. R&D expenditures began to decrease in 2019 -- dropping from $324 million in 2018 to $187 million in 2019 and then **falling to just $65 million in 2021** -- an 80 percent drop.

Upstream and Downstream Losses

For CCI members alone -- just a subset of affected industry -- forecasted upstream and downstream losses, both direct and indirect, and associated CCI member losses from 2022 to 2026 include:

- An average of **6,600 job losses** each year across the U.S. workforce;
- **$10 billion in lost economic output** and $4.2 billion in lost gross domestic product (GDP);
- Over **$500 million in reduced federal tax revenue**s and nearly **$300 million in reduced state and local tax revenues**; and
- Extrapolated **GDP loss** to the broader U.S. economy would be **$6.2 billion on average** from 2022 to 2026.

The Solution

EPA frequently reminds the regulated community that it "has inherent authority to reconsider previous decisions and to revise, replace, or repeal a decision to the extent permitted by law and supported by reasoned explanation." 87 Fed. Reg. 39511 at 39512 (July 1, 2022). EPA must use its inherent authority here to make three policy changes to avoid further stifling chemical innovation through its implementation of TSCA Section 5. EPA needs to:

- **Explicitly consider risk reduction in comparison to existing chemicals when assessing new chemicals;**

- **Rely on actual, not modeled, data in making TSCA Section 5 decisions; and**

- **Interpret "reasonably foreseen" as Congress intended in enacting the 2016 TSCA amendment.**

CCI has prepared the attached briefing documents on each of these policy changes.

CCI wishes to work with EPA to effectuate them.

For more information, e-mail info@chemicalinnovations.org

Scan QR code for explainer videos:

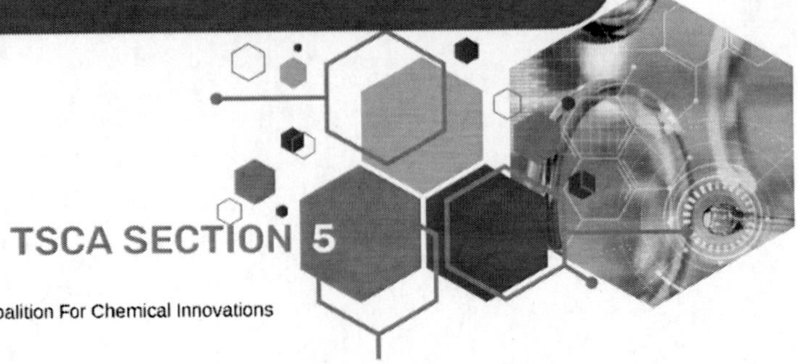

Chet Thompson
President and CEO

American
Fuel & Petrochemical
Manufacturers

1800 M Street, NW
Suite 900 North
Washington, DC
20036

202.844.5505 direct
202.457.0480 office
cthompson@afpm.org
afpm.org

February 7, 2023

The Honorable Bill Johnson
Committee on Energy & Commerce
Chairman
Subcommittee on Environment,
Manufacturing and Critical Materials
U.S. House of Representatives
Washington, D.C. 20515

The Honorable Paul Tonko
Committee on Energy & Commerce
Ranking Member
Subcommittee on Environment,
Manufacturing and Critical Materials
U.S. House of Representatives
Washington, D.C. 20515

The Honorable Jeff Duncan
Committee on Energy & Commerce
Chairman
Subcommittee on Energy,
Climate and Grid Security
U.S. House of Representatives
Washington, D.C. 20515

The Honorable Diana DeGette
Committee on Energy & Commerce
Ranking Member
Subcommittee on Energy,
Climate and Grid Security
U.S. House of Representatives
Washington, D.C. 20515

Dear Chairmen Johnson and Duncan and Ranking Members Tonko and DeGette:

The American Fuel & Petrochemical Manufacturers (AFPM) appreciates the opportunity to provide its perspectives on legislation under consideration at the Joint Energy, Climate, & Grid Security Subcommittee and Environment, Manufacturing, & Critical Materials Subcommittee Legislative Hearing, *Unleashing American Energy, Lowering Energy Costs, and Strengthening Supply Chains.*

AFPM is a national trade association representing the U.S. refining and petrochemical manufacturing industries. AFPM members support close to three million jobs, contribute to our economic and national security, and enable the production of thousands of vital products used by families and businesses throughout the U.S. Our members produce and deliver the gasoline, diesel, and jet fuel that keep us moving, as well as supplying the petrochemicals that are used as building blocks in thousands of products from cell phones to automobiles to medical devices, communications systems, and materials critical for producing renewable energy.

The past year was tremendously challenging for the global energy industry. Russia's invasion of Ukraine and the resulting sanctions and other restrictions have reshuffled global crude oil, natural gas, and trade of gasoline and diesel, made even more challenging by re-opening economies following COVID. U.S. refiners and petrochemical manufacturers demonstrated extraordinary resilience and adaptivity, running near maximum utilization for most of the year to meet U.S. and global demand for our products. At the same time, in the U.S., policymakers of both parties have called for increased refining capacity and reversals of the ongoing rationalization of U.S. refining capacity, which contracted by more than 1 million barrels per day since the beginning of 2020. Some closures were due to market factors, others were policy driven. In fact, more than half of capacity reductions are the result of conversions to renewable diesel production, and AFPM's members are responsible for 80 percent of recent announcements in this drop-in renewable fuel.

AFPM appreciates the Committee's promotion of sound regulations, strong energy production, and enhanced process safety. As you consider legislation to meet these goals, AFPM offers several initial observations:

I. AFPM welcomes the Committee's legislative efforts to provide clear standards for when the Environmental Protection Agency (EPA) may require a safer technology and alternatives analysis (STAA).

Nothing is more important to refiners and petrochemical manufacturers than the safety of our employees, contractors, neighbors, and the communities in which they operate.[1] AFPM members invest significant resources to continually improve our safety programs and practices, both as individual companies and as an industry. As it relates to the bill under consideration, the refining industry uses hydrofluoric acid (HF) as a critical catalyst to produce alkylate. Alkylate is an irreplaceable blending component for today's cleaner-burning motor gasoline.

The refining industry takes management of HF very seriously. Since 1992, AFPM members have followed API Recommended Practice 751 (RP 751), the most rigorous and comprehensive document pertaining to the safe operation of HF alkylation units. RP 751 is the collaborative product of an industry working group that includes nearly 100 of the top global leaders in HF alkylation science and process safety. The most recent edition of RP 751—its 5th—was released in August 2021 and reflects the newest data, industry learnings, and technologies. Under RP 751, HF alkylation operations are getting safer every year.

A STAA is an engineering concept that companies utilize when evaluating technology and design alternative for new process units. Early in the technology selection process, a STAA can be a helpful tool and AFPM supports its use. Once a facility is up and running on technology selected, procured, and built, however, most alternative technologies are no longer feasible to implement, particularly where refinery fuel alkylation is concerned, without completely reconstructing individual process units and potentially reconfiguring entire facilities. Disrupting this one process would have impacts across an entire facility, potentially resulting in a range of other costly problems such as more expensive gasoline; gasoline and other fuel shortages; higher dependence of fuel imports; potential refinery shutdowns; and job losses.

By providing clear standards this legislation would promote both safety and regulatory certainty. AFPM recommends that the Committee consider providing the same certainty to all covered industries and processes and looks forward to engaging to further clarify legislative language.

---

[1] *https://www.afpm.org/issues/safety-health*

II. AFPM agrees with the Committee that banning energy exports would harm U.S. consumers, the economy, and national security.

The U.S. is home to the largest and most complex refining industry in the world, made possible by our skilled workforce, the investments we have made, and our access to reliable and competitive feedstocks globally. In fact, U.S. refiners invested more than $100 billion in the last fifteen years to make our refining sector the most competitive in the world. We upgraded the complexity of our refining kit so that more of our facilities would be optimized to process the toughest-to-refine types of crude oil—feedstocks primarily available from the global market that most other countries' refineries cannot process. The ability to purchase and refine these types of crude oil has been a tremendous advantage to the United States and our standing as a global energy leader.

As a result, the U.S. is the largest global exporter of crude oil and refined petroleum products. Our largest export destinations for refined products include Latin America and Europe— important markets not only for U.S. economic security, but also for U.S. foreign policy and national security.

Banning exports would not only place U.S. interests at risk, but it would also likely place further upward pressure on consumer fuel prices as the U.S. retreats from participating in global markets. U.S. participation in the global market for energy is a strength.

III. The U.S. needs infrastructure and smart regulations if we are going to meet growing global energy demand while addressing climate change.

The U.S. is a global leader in refining and petrochemical production. With a growing global population and millions of people entering the middle class each year demand for more energy and mobility will increase, not to mention the need for infrastructure for clean water, healthcare, fresh food, and countless other aspects of modern life that AFPM's members enable.

To remain a global leader, U.S. regulatory policy needs to strike the right balance between environmental protection and business certainty. Too often, projects become uneconomic because of the length of permitting reviews or regulatory approvals. This is true for both conventional and renewable energy, but also applies to areas like the new chemicals program under the Toxic Substances Control Act and the ability to build and scale carbon capture, hydrogen, and other systems to reduce the carbon intensity of the energy sector. The Committee is considering multiple bills that would make targeted changes in areas such as cross-border pipeline approvals, flexible air permitting, and streamlining reviews under multiple statutes including NEPA, SWDA, the Clean Air Act, TSCA, and others.

Making regulations work better does not mean sacrificing protection for the environment or consumers. We look forward to working with the Committee to build on the legislation under consideration this week to better promote infrastructure development and regulatory certainty.

IV. AFPM supports legislation requiring the National Petroleum Council to produce a report on the role of petrochemical refineries in the United States and the significant contributions these refineries have made to our nation's energy security.

The U.S. petrochemical industry has a crucial and enduring role to play in meeting the needs of a growing world population while simultaneously fulfilling the imperative to produce petrochemicals in a sustainable and clean manner. Petrochemicals are the building blocks for products that improve health, safety, and quality of life for people around the world. The U.S. economy depends on petrochemicals and they are essential for improving and modernizing myriad other industries. Petrochemicals are critical in modern medicine, food safety, efficient construction, advanced electronics, mobility, and transportation as well as our energy infrastructure.

* * *

If policymakers wish to facilitate the long-term health of the refining and petrochemical sectors, we urge a continued focus on developing a regulatory environment that promotes investment and certainty. We are confident that given the opportunity to fairly compete in the market for both consumers and emissions reductions, our industries will continue to lead the world in both.

AFPM appreciates the Committee's attention to these issues of great importance to our members and American consumers.

Sincerely,

Chet Thompson
President and CEO
American Fuel & Petrochemical Manufacturers

February 6, 2023

The Honorable Jeff Duncan
Chairman
Subcommittee on Energy, Climate & Grid Security
U.S. House of Representatives
Washington D.C. 20515

The Honorable Diana DeGette
Ranking Member
Subcommittee on Energy, Climate & Grid Security
U.S. House of Representatives
Washington D.C. 20515

Dear Chairman Duncan and Ranking Member DeGette:

On behalf of the Center for LNG (CLNG), I am writing to provide comments to the U.S. House of Representatives Subcommittee on Energy, Climate, & Grid Security's hearing "Unleashing American Energy, Lowering Energy Costs, and Strengthening Supply Chains."

Commodity prices have risen around the globe, but because the United States is the world's largest natural gas producer, U.S. natural gas prices have increased less than other countries. Due to this advantage, our consumers and businesses have been more sheltered from higher natural gas prices than the rest of the global economy.

Natural gas prices are driven by multiple factors that affect both supply and demand, including weather, natural gas storage levels, changes in industrial demand, and the impact of global events on LNG exports and the U.S. economy. While the European energy crisis has created high demand for U.S. LNG, keep in mind that LNG export volumes represent only about 10% of the total winter market demand for natural gas domestically. Not only are LNG exports one of the smallest natural gas customer sectors compared to residential, electric or industrial demand, the volume of LNG exports is a fairly predictable number that increases very gradually since there are just over a handful of U.S. LNG export facilities.

LNG export terminals take 6-10 years from start to in-service, which gives natural gas producers plenty of time to prepare for the increased demand. The future supply of natural gas is abundant, and estimates are brightened with successive assessments of the resource base. In fact, current LNG exports represent less than 1% of U.S. proven reserves of natural gas and about 0.001% of the total estimated U.S. natural gas resource base. Further, LNG exports provide multiple benefits to domestic consumers, such as their key role in incentivizing production across the country and increasing household purchasing power.

In response to domestic demand for natural gas, producers responded with record levels of natural gas production in November and December of 2022, a 4% increase winter-over-winter. These increases, along with a mild winter, have already helped to ease market conditions as more supply flows into the market. The Energy Information Administration (EIA) recently projected that 2023 natural gas prices will average 24% less than 2022. This is a classic example of supply and demand leveling the playing field without government intervention.

LNG exports will continue to stabilize the domestic market and incentivize U.S. natural gas production, which in turn generates billions of dollars in new investments that benefit local communities, produce tax revenues and support jobs, including those in manufacturing. These jobs number in the thousands during construction and in the tens of thousands throughout the supply chain during operations. Furthermore, our exports help to fund national and international efforts to reduce emissions, with members investing billions in low-carbon and no-carbon solutions.

Globally, these same exports support our allies' efforts to reach net-zero emissions, allowing them to generate electricity with natural gas, rather than higher-emitting energy sources. Most importantly, U.S. LNG is necessary for our allies in Europe to replace Russian natural gas, as well as to help all our trading partners get closer to their COP27 and Paris Accord climate goals.

We are world leaders in natural gas and LNG exports. They support our economy and emission-reductions goals, while enhancing the national security of the United States and that of our allies. Hyperbole and fear have no place in energy policy discussions and limiting future LNG exports will only have a negative impact on the United States and our trade partners.

CLNG appreciates your leadership as well as that of many of your colleagues who continue to recognize the role natural gas and LNG play in reducing emissions across the globe, while also realizing the economic benefits natural gas creates here at home. We particularly want to commend Representative Johnson for his continued leadership in promoting LNG exports and the many benefits they bring.

Sincerely,
Charlie Riedl
Executive Director, Center for Liquefied Natural Gas
900 17th St., NW, Suite 500 Washington, DC 20006
charlie.riedl@ngsa.org

The Center for Liquefied Natural Gas (CLNG) advocates for public policies that advance the use of liquefied natural gas (LNG) in the United States, and its export internationally. A committee of the Natural Gas Supply Association (NGSA), CLNG represents the full LNG value chain, including LNG producers, shippers, terminal operators and developers, providing it with unique insight into the ways in which the vast potential of this abundant and versatile fuel can be fully realized. For more information, please visit www.lngfacts.org.

* * *

February 6, 2023
Chair Cathy McMorris Rodgers
Energy and Commerce Committee
The House of Representatives
Washington, DC 20515

Ranking Member Frank Pallone
Energy and Commerce Committee
The House or Representatives
Washington, DC 20515

Dear Chair McMorris Rodgers and Ranking Member Pallone

ClearPath Action would like to thank the Committee for its leadership on U.S. nuclear energy and for considering the draft bill, Prohibiting Russian Uranium Imports Act, during the joint legislative hearing on "Unleashing American Energy, Lowering Energy Costs, and Strengthening Supply Chains." A secure and robust nuclear fuel supply chain is critical to ensuring American families receive clean, affordable, and reliable energy from our nation's nuclear power plants.

Reducing reliance on Russian fuel provides the certainty required to incentivize domestic industry, build new capacity, and support our allies. This need for secure nuclear fuel supply chains extends beyond traditional, large light-water reactors. The two flagship advanced nuclear reactor demonstration projects underway through the Department of Energy require high-assay, low-enriched uranium fuel (HALEU), which today is only available from Russia. According to modeling by the International Energy Agency in 2021, to meet our global climate goals, around 500 gigawatts of new nuclear power must be brought online by 2050. The United States has the potential to supply both the technology and the fuel for this massive global market.

While the Prohibiting Russian Uranium Imports Act would create certainty for the low-enriched uranium (LEU) supply market, it could also address the HALEU supply market. This bill has the potential to enable the United States to lead globally across both the LEU and HALEU supply chains. The Committee could allow the Department of Energy to identify the remaining need for the Advanced Nuclear Fuel Availability program, established under section 2001 of the Energy Act of 2020, and reappropriate that amount in addition to the funds made available for the American Assured Fuel Supply program.

This comprehensive approach would complete the intent of this legislation by providing an American alternative to Russian nuclear fuel dominance that addresses both near-term and long-term supply chain security. Thank you again for your interest and leadership in furthering American clean energy leadership in the nuclear power industry.

Sincerely,

Rich Powell
Chief Executive Officer
ClearPath Action

* * *

February 3, 2023

The Honorable Cathy McMorris Rodgers
Chair
Committee on Energy and Commerce
U.S. House of Representatives
Washington, D.C. 20515

The Honorable Frank Pallone
Ranking Member
Committee on Energy and Commerce

U.S. House of Representatives
Washington, D.C. 20515

Dear Chair Rodgers and Ranking Member Pallone,

I write to commend the House Energy and Commerce Committee for taking steps to enhance the domestic nuclear fuel cycle and alleviate the national security issues associated with dependence on Russian nuclear fuel imports.

The U.S. commercial nuclear industry is committed to eliminating the import of uranium and related conversion and enrichment services from the Russian Federation. Our industry will continue to work with the U.S. government to establish a secure supply of nuclear fuel conversion and enrichment capabilities so we can move away from Russian fuel imports as expeditiously as possible without placing the U.S. nuclear fleet and new-reactor deployment at risk.

In addition to several proposals in the draft legislation, we encourage Congress and the Administration to explore the feasibility of redirecting a portion of the unexpended funds authorized through the Civil Nuclear Credit Program to support domestic conversion and enrichment expansion capabilities.

Nuclear energy is an essential element of U.S. infrastructure, providing nearly 20 percent of the U.S. electricity generation and half of our carbon-free generation. If U.S. operators are not able to acquire fuel, there will be significant negative impacts on affordability, reliability, and air quality for millions of Americans. It will also significantly impede our ability to reach our carbon-reduction goals.

Chair Rodgers and Ranking Member Pallone
February 3, 2023

Establishment of a secure fuel supply will allow the U.S. and our allies to make continued progress toward climate and energy security goals through the maintenance of existing reactors and the build-out of new reactors using U.S. technology.

We look forward to continuing to work with Congress on a workable solution to developing a competitive domestic fuel supply to strengthen our energy security.

Thank you very much.

Yours very sincerely,

Maria Korsnick

c: The Honorable Jeff Duncan, U.S. House of Representatives
The Honorable Bill Johnson, U.S. House of Representatives

GEOFF MOODY
SENIOR VICE PRESIDENT, GOVERNMENT AND POLICY,
AMERICAN FUEL & PETROCHEMICAL MANUFACTURERS (AFPM)

For years, the US Environmental Protection Agency (EPA) has overseen a regulation known as the Risk Management Plan (RMP). The RMP is a safety performance standard that applies to fuel refineries and a litany of other US manufacturers that use high volumes of certain chemicals. RMP sites are expected to make continuous investments and improvements in safety performance, which is a good thing. Refiners fall under RMP jurisdiction for several chemicals, two of which are catalysts that facilities use to produce alkylate – an irreplaceable component in the cleanest US gasoline and high-octane aviation gas. It is because of alkylate and the EPA's upcoming proposal to change the RMP that the Biden administration might soon find itself making an unforced energy error.

The two primary and commercially-proven catalysts that refiners use to produce alkylate are hydrofluoric acid (HF) and sulfuric acid. Both are on the RMP list, but HF tends to draw more attention and is the focus of major refinery safety investments, training, and risk mitigation technology.

EPA regulations must reflect how thoroughly HF is managed by US refiners. Policy that incorrectly presumes the opposite could put in jeopardy significant US fuel manufacturing.

Refiners account for less than 2% of global HF consumption. HF is near ubiquitous in large-scale manufacturing. It is used in agriculture and also to produce computer chips, refrigerants, hydrogen fuel cells, pharmaceuticals, branded aluminium cans, and even drinking water. Even still, HF in fuel manufacturing draws a disproportionate share of regulatory interest.

The EPA's RMP proposal is expected to single out HF alkylation at fuel refineries, and saddle facilities with an expensive, unconstructive new paperwork burden. Regulators know full well that requiring alternative alkylation technology assessments at refineries that are already up and running will consume valuable man hours, impose significant costs, and yield nothing in terms of actionable results or safety improvements. The point is simply to put a barrier in the way of HF alkylation. This is precisely what some anti-refining groups are lobbying for.

A loss or major reduction in HF alkylation would have devastating consequences for US and global fuel supplies and for affordable energy advocates everywhere. Most policymakers have no idea that those are the stakes with the RMP.

Nearly half of the alkylate produced in the US is made with HF catalyst. Refineries with HF alkylation units account for nearly 40% of US fuel manufacturing. Facilities cannot seamlessly transition from one alkylation technology to another. It is a massive undertaking that could approach US$1 billion – a cost so extreme that it could push some refineries to close. Even if a transition was feasible, safety would not be better served. Alkylation risk would simply be shifted to other parts of the supply chain.

Those who oppose HF alkylation wrongly believe that allowing use of this technology means compromising on safety. Refiners do not accept that. We have done more than any other industry to formalise and evolve HF safety guidelines, such as through API Recommended Practice 751. Our safety procedures are consistently reviewed and enhanced every few years as we gather intel from real-world experiences and capitalise on advancements made in risk-reducing technologies.

Because of all the steps that refiners take to keep employees and community neighbours safe, HF alkylation units pose less life-threatening risk to their local public than vehicle collisions, lightning strikes and sharp objects, just to name a few. When refiners say safety is our priority, we mean it.

EPA regulations must reflect how thoroughly HF is managed by US refiners. Policy that incorrectly presumes the opposite could put in jeopardy significant US fuel manufacturing, and our ability to produce the cleanest possible gasoline and aviation fuels in the US. The effect for consumers would be tighter fuel supplies and potentially higher prices. There is no way that could be considered a win by the Biden team.

April 6, 2022
The Honorable Michael Regan
Administrator
U.S. Environmental Protection Agency
1200 Pennsylvania Avenue N.W.
Washington, DC 20460

Dear Administrator Regan:

We are writing to you today to request the timely review of two Pre-Manufacturing Notice (PMN) applications (case numbers P-21-0216 and P-21-0217) submitted to the EPA under section 5 of the Toxic Substances Control Act (TSCA). These PMN applications are critical to addressing supply chain issues and impact our country's efforts to reduce our carbon footprint and increase job opportunities.

As you may be aware, General Motors and LG Energy Solution, one of the leading manufacturers of EV batteries, created a joint venture, Ultium Cells LLC, to manufacture battery cells for electric vehicles at three new facilities in the United States, including in Lansing, Michigan. The joint venture is expected to invest $2.6 billion and create approximately 1,700 jobs in Michigan alone.

The new EV battery cells that the Ultium facilities will manufacture use a material called carbon nanotubes (CNT), which is produced and supplied by LG Chem. Quality CNTs dramatically improve the batteries' performance, capacity, life span, and charging speed, all of which are central to making EVs a more attractive option for consumers. However, before the CNTs can be used for EV battery production, the EPA must review and approve the Pre-Manufacture Notice (PMN) that LG Chem submitted in September of 2021.

We understand that EPA has a significant backlog of more than 300 TSCA section 5 applications awaiting review and that the agency has taken a "first-in, first-out" approach to review these applications. As it now stands, the PMN applications have been pending for more than five months, although the TSCA statute mandates a 90-day review. Our concern is that if EPA does not complete its review of the PMNs by May then the launch of the three EV battery plants will be delayed.

Given the importance of this project to Michigan and to our country's EV supply chain, we ask EPA to complete the reviews of the PMNs so that these important EV battery projects are not delayed.

Please feel free to reach out to us if you have any questions. In the meantime, we look forward to your response and finding a way we can work together to move this very critical project forward.

Sincerely,

Elissa Slotkin
United States Representative

Gary Peters
United States Senator

Debbie Stabenow
United States Senator

Tim Walberg
United States Representative

* * *

February 6, 2023

The Honorable Cathy McMorris Rodgers
Chair, House Energy and Commerce Committee
United States House of Representatives
Washington, DC 20515

The Honorable Frank Pallone
Ranking Member, House Energy and Commerce Committee
United States House of Representatives
Washington, DC 20515

Dear Chairwoman McMorris Rodgers and Ranking Member Pallone,

On behalf of the Uranium Producers of America (UPA), I write to express our support for the "Prohibiting Russian Uranium Imports Act," which your committee is set to consider during a legislative hearing on February 7. This legislation will help stop the funding of Russia's war in Ukraine with U.S. dollar purchases of nuclear fuel, protect our national and energy security, and take critical steps towards reestablishing American nuclear fuel cycle capabilities.

It is long past time to ban uranium imports from Russia, just as the United States has already done with Russian oil, gas, and coal. For too long we have been funding Ukraine's defense with one hand, and Russian aggression through our nuclear fuel purchases with the other. Russia's state nuclear power company, Rosatom, benefits from hundreds of millions of U.S. dollars in nuclear fuel purchases annually. By continuing to rely on Russian nuclear fuel, we are directly subsidizing the Putin regime and the continued devastation in eastern Europe.

In addition to cutting off another U.S. financial spigot for Russia, banning Russian uranium imports will help create the long-term market conditions necessary for U.S. uranium producers to make investments across the nuclear fuel cycle. The legislation's redirection of unspent and unneeded Civil Nuclear Credit Program dollars towards the nuclear fuel supply chain is a common-sense measure to start securing our nuclear power needs now and well into the future. U.S. policymakers must recognize and address the fact that our global competitors, namely China and Russia, treat their nuclear fuel cycle capabilities as strategic assets and deploy them to achieve geopolitical objectives. Restoring America's global nuclear energy leadership is crucial for U.S. security.

We urge the Committee to move with urgency in considering the "Prohibiting Russian Uranium Imports Act." Thank you for considering this important and timely legislation and we look forward to working with you as the bill makes its way through the legislative process.

Sincerely,
Jon Indall
Counsel, Uranium Producers of America
Date: February 6, 2023

To: The Honorable Cathy McMorris Rodgers
Chairwoman
U.S. House Committee on Energy & Commerce
2125 Rayburn House Office Building
Washington, DC 20515

The Honorable Frank Pallone, Jr.
Ranking Member
U.S. House Committee on Energy & Commerce
2125 Rayburn House Office Building
Washington, DC 20515

RE: USNIC Letter to House Energy & Commerce Committee on Russian Uranium Imports

Dear Chairwoman McMorris Rodgers & Ranking Member Pallone:

The U.S. Nuclear Industry Council (USNIC) and its members are grateful for the Committee's continued support for the U.S. nuclear industry. A robust U.S. nuclear fuels supply chain is crucial to our industry's health. As we progress steadily toward deployment of advanced reactors, the attention paid by the Committee to the importance of supporting nuclear as a clean energy solution is highly valued.

That is why USNIC welcomes this Committee's strong focus on resilient supply chains to support the Nation's critical infrastructures, including the vital importance to strengthen our domestic capability to enrich uranium to support our own energy infrastructure as well as that of partners and allies. In that connection, USNIC urges the Committee to ensure that there is adequate flexibility to apply funds already appropriated under the Infrastructure Investment and Jobs Act (IIJA) for High-Assay Low-Enriched Uranium (HALEU) and low-enriched uranium (LEU) in the event the Secretary of Energy determines it is necessary or appropriate.

We are also appreciative of proposed legislation "To prohibit the importation into the United States of unirradiated low-enriched uranium that is produced in the Russian Federation, and for other purposes." USNIC and our members have reviewed the proposed legislation carefully and are convinced it can be an important element for continued successful development of our vital national industry. We support the Committee's intent with this legislation that clearly, we cannot, and should not, rely on Russia for the enriched uranium to meet current industry needs, and definitely not to meet the needs for HALEU that is vital to U.S. objectives for advanced reactors.

We also note that the proposed legislation provides for a thoughtful process that will allow for the domestic fuel supply chain to ramp up over the next six years, as well as the criteria provided for potential waiver of restrictions if the Secretary of Energy determines that there is no alternative viable source available, or it is otherwise in the national interest. Given the need to protect American interests as we build domestic capacity, we recommend that the Committee include language to ensure that existing contracts signed prior to the invasion are grandfathered during the transition period to ensure adequate supply and avoid rewarding Russia with an opportunity, to sell at high prices in third markets, enrichment that they are now committed to sell at pre-invasion prices. These provisions should alleviate any issues for those who may be concerned that Russian enriched uranium might at some future time be needed to meet U.S. requirements.

Thank you very much for the support you consistently provide for our industry, an industry that is crucial for U.S. energy security, economic well-being, and national security.

Sincerely,

Todd Abrajano President & CEO
U.S. Nuclear Industry Council (USNIC)
Mobile: 913-620-0700; Todd.abrajano@usnic.org

February 5, 2023

The Honorable Cathy McMorris Rodgers
Chairman
Energy and Commerce Committee
United States House of Representatives
Washington, DC 20515

The Honorable Frank Pallone
Ranking Member
Energy and Commerce Committee
United States House of Representatives
Washington, DC 20515

The Honorable Jeff Duncan
Chairman
Energy, Climate, & Grid Security Subcommittee
United States House of Representatives
Washington, DC 20515

The Honorable Diana DeGette
Ranking Member
Energy, Climate, & Grid Security Subcommittee
United States House of Representatives
Washington, DC 20515

The Honorable Bill Johnson
Chairman
Environment, Manufacturing, & Critical Minerals Subcommittee
United States House of Representatives
Washington, DC 20515

The Honorable Paul Tonko
Ranking Member
Environment, Manufacturing, & Critical Minerals Subcommittee
United States House of Representatives
Washington, DC 20515

Dear Chairman Rodgers and Ranking Member Pallone:

Our organizations write to offer our strong opposition to the proposed legislation that would amend the Clean Air Act to exempt refineries that use extremely dangerous hydrofluoric acid (HF) from assessing whether they could potentially use safer technologies. We are united to prevent chemical disasters,[124] and we urge committee members not to advance this

---

[124] Comment submitted by Coalition to Prevent Chemical Disasters, Accidental Release Prevention Requirements: Risk Management Programs Under the Clean Air Act; Safer Communities by Chemical Accident Prevention

legislation. This bill would undermine crucial protections for workers, environmental justice communities, nearby schools, hospitals, and the public, which are under consideration by the Environmental Protection Agency in their rulemaking on the Risk Management Program (RMP).

We stand opposed to this legislation as aligned with the Louisville Charter for Safer Chemicals (louisvillecharter.org), which calls for the prevention of disproportionate exposures and hazards, and reduction of cumulative impacts on Environmental Justice communities; requiring safer substitutes and solutions for a non-toxic economy; to act with foresight to protect health and prevent pollution, particularly when credible evidence shows that a substance or class of substances is potentially hazardous and/or harmful; and to ensure that the public and workers fully have the right-to-know, participate and decide in the decisions that impact their health because of the potential harm from toxic chemicals.

This legislation would exempt facilities that use HF in alkylation units (e.g., HF refineries) from modest requirements to assess whether they could possibly convert to safer chemicals or processes. Safer technologies alternatives assessments encourage innovation. Alternatives identified in such assessments can prevent chemical disasters, such as the one that occurred in 2019 at the HF alkylation unit at the Philadelphia Energy Solutions Refinery in Philadelphia, Pennsylvania.[125] This devastating series of explosions injured five workers and a first responder, led to the evacuation of 4,000 people, and culminated in over 1,000 workers being laid off with no severance and almost no notice when the plant shut down.[126] During the incident, over 5,000 pounds of highly toxic HF were released and the estimated property damage loss was $750 million.[2]

The U.S. Chemical Safety and Hazard Investigation Board (CSB), the U.S. Government's independent agency charged with investigating chemical disasters across the country, identifying their root causes, and making recommendations to government and industry on best practices to prevent future occurrences did not mince words in their recommendations to EPA following the PES disaster. CSB concluded in their final report on this disaster that EPA should:

> "require new and existing petroleum refineries with HF alkylation units to conduct a safer technology and alternatives analysis (STAA) and to evaluate the practicability of any inherently safer technology (IST) identified."

As the climate changes the risk and severity of chemical disasters and releases can be exacerbated by extreme weather. When industrial facilities, including HF refineries located in these areas fail to adequately prepare for extreme storms, wildfires, earthquakes, heat waves, floods, rising sea levels, and other natural disasters this can lead to a cascading series of harms, including toxic chemical exposures, on top of the effects of the weather event itself.[127] The

---

(Docket Number EPA-HQ-OLEM2022-0174). (2022, Nov 4). https://www.regulations.gov/comment/EPA-HQ-OLEM-2022-0174-0269.

125 CSB, Fire and Explosions at Philadelphia Energy Solutions Refinery Hydrofluoric Acid Alkylation Unit at 80 (June 21, 2019), No. 2019-04-I-PA (published Oct. 11, 2022) (CSB PES Report), https://www.csb.gov/philadelphia-energy-solutions-pes-refinery-fire-and-explosions-/.

126 Maykuth, Andrew. (1 July, 2019.) Philadelphia refinery workers sue over abrupt closure, layoffs. The Philadelphia Inquirer. https://www.inquirer.com/business/energy/philadelphia-refinery-fire-workers-sue-pes-closure-job-layoffs-20190701.html.

127 Center for Progressive Reform, Earthjustice, and the Union of Concerned Scientists. (2021, July). Preventing "Double Disasters." https://www.ucsusa.org/sites/default/files/2021-07/preventing-double-disasters%20 FINAL.pdf.

Government Accountability Office (GAO) released a report in 2022 showing that approximately 31% of RMP facilities with at least one program 2 or 3 process, including some that are HF refineries, "are located in areas with one or more selected natural hazards that may be exacerbated by climate change."[128]

This bill would require regulators to defer to the American Petroleum Institute (API) rather than the Government's own data and the recommendations of the very federal agencies responsible for chemical safety and government accountability. The legislation refers to API Recommended Practice 751, Safe Operation of Hydrofluoric Acid Alkylation Units, a voluntary industry guideline. Regulators cannot enforce voluntary guidelines.[129,130] CSB reports of deadly incidents have shown the weaknesses of API guidelines, including the HF guideline.[131] Some refineries that use HF alkylation are not even members of API; they may be members of other trade associations, such as American Fuel and Petrochemical Manufacturers, and likely do not even follow API voluntary guidelines, increasing the potential for "natech" ("natural hazards triggering technological accidents" as defined by GAO) disasters.

Additionally, hazard assessments are crucial for determining if hazards are present, or are likely to be present, in the workplace, and to help determine whether such hazards require the use of additional personal equipment to protect workers. Exempting such requirements in this bill would put workers at risk of death or injury, and by extension, neighboring fenceline communities. Fully informed and protected workers are key to preventing disasters and keeping neighbors safe, as evidenced by a number of recent "near miss" incidents that could have been much more disastrous had it not been for the quick action of highly trained union workers, such as in the case of the PES incident.[132,133] Workers have a right to know the hazards in their workplace, and robust hazard assessments are essential to ensure their full knowledge and protection.

Hazard assessment are also important for determining whether and what hazards are present, or likely to be present, for the purpose of protecting and assisting first responders. In a number of chemical disasters the quick and informed action of first responders prevented far greater catastrophe.[134] Exempting facilities storing and processing extremely hazardous chemicals like HF from commonsense hazard assessments can cause costly delays or miscalculations in emergency response due to the absence of full and clear understanding of the potential hazards.

---

[128] U.S. GAO, Chemical Accident Prevention: EPA Should Ensure Regulated Facilities Consider Risks from Climate Change. (2022, March 2) https://www.gao.gov/products/gao-22-104494.

[129] For example, BP Products North America, Inc., and BP-Husky Refining, LLC, v. Department of Labor, OSHRC Docket No. 10-0637, 2018.

[130] Comments of the United Steelworkers Union on the U.S. Department of Labor, Occupational Safety and Health Administration Process Safety Management Stakeholder Solicitation of Public Comments [Docket No. OSHA–2013– 0020] (pp. 6-8 and p. 17).

[131] See among others CSB investigation reports: Tesoro Anacortes Refinery, Catastrophic Rupture of Exchanger, May 2014; Chevron Richmond Refinery Pipe Rupture and Fire, January 2015; Fire and Explosions at Philadelphia Energy Solutions Refinery Hydrofluoric Acid Alkylation Unit, October 2022.

[132] Bicameral Chemical Disaster Rule Letter to EPA April 2022. https://www.booker.senate.gov/imo/media/doc/final_bicameral_chemical_disaster_rule_letter_to_epa.pdf.

[133] PES Workers' Response to Fire Saves Community From Disaster; Company Announces Shut Down. https://m.usw.org/news/media-center/articles/2019/pes-workers-response-to-fire-saves-community-from-disaster-company-announces-shut-down.

[134] Yellin, D. (2022, January 24). Keeping Passaic fire from chlorine prevented 'one of the biggest disasters in the country'. North Jersey Media Group. https://www.northjersey.com/story/news/passaic/passaic-city/2022/01/16/passaic-chemical-fire-nj-qualco-majestic-industries/6548160001/.

The dangers of HF are well established and the proximity of several HF refineries to residents puts millions of Americans at risk. As the CSB explains in the report referenced above on the 2019 PES Refinery explosion:

> Because HF vaporizes upon release to the air, a large release of HF has the potential to travel off site and expose people, animals, and vegetation to harmful concentrations of the chemical. A 2005 study by the U.S. Public Interest Research Group found that HF releases from refineries could have significant off-site consequences, stating "[s]even petroleum refineries using hydrofluoric acid reported toxic release 'worst-case' scenarios in which more than one million people could be affected." It also found that "15 refineries could place more than 500,000 people in harm's way, and 28 refineries could endanger more than 100,000 people in the event of a worst-case hydrofluoric acid release."[135]

Not all refineries use HF alkylation. As of October 2022 of the 155 U.S. petroleum refineries currently in operation, 46 operated HF alkylation units.[2] Safer alternatives to HF in oil refining exist and are already being used in some refineries. For instance, in April 2021, Chevron and Honeywell announced the start up of an HF-free process for Chevron's Salt Lake City refinery. Big West Oil in Salt Lake City has announced plans to make the same conversion as Chevron SLC (from hydrofluoric acid to ionic liquid catalyst).[136] Honeywell is offering this process to existing and new refineries.[137] Additional facilities are in the process of considering or transitioning to other alternative processes, such as the CVR Wynnewood Refinery in Oklahoma, which has announced plans to convert from hydrofluoric acid catalyst to a solid acid catalyst.[138]

This bill could undermine such transitions and provide an unwarranted exemption from merely considering safer processes that will provide crucial protections for the communities that live around these facilities and the workers inside the facilities,[139] who are disproportionately burdened by chemical disasters. Decades of research and evidence,[140,141]

---

135 U.S. PIRG Education Fund, "Needless Risk - Oil Refineries and Hazard Reduction," August 2005. [Online]. Available: https://uspirg.org/sites/pirg/files/reports/Needless_Risk_USPIRG.pdf. [Accessed 10 September 2019].

136 Big Oil west Proceeds with Honeywell to Revamp Alklation Unit to ISOALKY™ Technology. (2021, November 11.) https://pmt.honeywell.com/us/en/about-pmt/newsroom/press-release/2021/11/big-west-oil-proceeds-with-honeywell-to-revamp-alkylation-unit-to-isoalky-technology.

137 chevron and honeywell announce start-up of world's first commercial isoalky™ ionic liquids alkylation unit. (2021, April 13.) https://www.chevron.com/newsroom/2021/q2/chevron-and-honeywell-announce-start-up-of-isoalky-ionic-liquids-alkylation-unit.

138 CVR Energy Proceeds with KBR on Second Phase Scope for Alkylation Revamp Project. (2021, February 4.) https://www.kbr.com/en/insights-news/press-release/cvr-energy-proceeds-kbr-second-phase-scope-alkylation-revamp-project.

139 A Risk Too Great: Hydrofluoric Acid in U.S. Refineries. (2013.) Tony Mazzocchi Center for Health, Safety and Environmental Education, and the United Steelworkers. https://www.usw.org/workplaces/oil/oil-reports/A-Risk-Too-Great.pdf.

140 Coming Clean and the Environmental Justice Health Alliance for Chemical Policy Reform, Who's in Danger: Race, Poverty and Chemical Disasters, (2014). https://comingcleaninc.org/assets/media/images/Reports/Who%27s%20in%20Danger%20Report%20FINAL.pdf.

141 Coming Clean and the Environmental Justice Health Alliance for Chemical Policy Reform. Life at the Fenceline: Understanding Cumulative Health Hazards in Environmental Justice Communities (2015). https://ej4all.org/life-at-the-fenceline.

including testimonies of the impacted people themselves have documented this historical burden on workers[142] and communities of color and low-income communities.[143,144,145,146]

We strongly oppose this legislation. Protections proposed under EPA's RMP rule that would be undermined by this piece of legislation are long overdue for workers and communities disproportionately impacted by chemical disasters. We urge the Committee, and any other committees with jurisdiction over this legislation, to eliminate catastrophic hazards and injustices and address this legacy of harm, by promptly rejecting this bill.

Respectfully submitted by:

Coming Clean
Environmental Justice Health Alliance for Chemical Policy Reform
Earthjustice
Center for Environmental Health
Union of Concerned Scientists

* * *

February 2, 2023

The Honorable Jeff Duncan
Chairman
Subcommittee on Energy, Climate & Grid Security
U.S. House of Representatives
Washington, DC 20515

The Honorable Diana DeGette
Ranking Member
Subcommittee on Energy, Climate & Grid Security
U.S. House of Representatives
Washington, DC 20515

Re: Manufacturers Strongly Oppose H.R. 647, the "Unlocking our Domestic LNG Potential Act of 2023"

---

[142] Public comments submitted to the EPA docket EPA-HQ-OLEM-2022-0174 by the United Steelworkers, 2022 https://www.regulations.gov/comment/EPA-HQ-OLEM-2022-0174-0216.

[143] Virtual Public Hearings on the Risk Management Program Safer Communities by Chemical Accident Prevention Proposed Rule, September 26, 2022. https://www.regulations.gov/document/EPA-HQ-OLEM-2022-0174-0157.

[144] Virtual Public Hearings on the Risk Management Program Safer Communities by Chemical Accident Prevention Proposed Rule, September 27, 2022. https://www.regulations.gov/document/EPA-HQ-OLEM-2022-0174-0158.

[145] Virtual Public Hearings on the Risk Management Program Safer Communities by Chemical Accident Prevention Proposed Rule, September 28, 2022. https://www.regulations.gov/document/EPA-HQ-OLEM-2022-0174-0160.

[146] Accidental Release Prevention Requirements: Risk Management Programs Under the Clean Air Act, Section 112(r)(7); Rule Retrospection Under Executive Order 13990; Virtual Public Listening Sessions; Request for Public Comment.86 FR 28828. https://www.regulations.gov/document/EPA-HQ-OLEM-2021-0312-0001.

Dear Chairman Duncan and Ranking Member DeGette:

On behalf of the Industrial Energy Consumers of America's (IECA) manufacturing companies, with over 12,000 locations nationwide, we strongly oppose this legislation which strips vital consumer protections from the Natural Gas Act (NGA). Excessive LNG exports can negatively impact U.S. natural gas and electricity reliability and costs. The removed NGA provisions are essential to allow the U.S. Department of Energy (DOE) to act to protect consumers in the event of unforeseen circumstances that impact reliability and price and these provisions have had no impact on slowing down approvals of LNG exports. In fact, approved LNG capacity is equal to 55 percent of U.S. 2021 natural gas demand.

The first responsibility of the U.S. Congress is to assure reliability and reasonable prices in the U.S. market, not to accelerate LNG exports. Increased LNG exports are inflationary. For example, in May 2022, national inventories fell 17 percent below five-year average levels, resulting in Henry Hub prices of over $9.00/MMBtu, a level that is three times higher than in recent history. Increased LNG exports have a direct impact according to both the Federal Energy Regulatory Commission (FERC) and the Energy Information Administration (EIA).[147]

U.S. 2021 natural gas consumption equaled 75.7 billion cubic feet per day (Bcf/d) and now a total of 41.9 Bcf/d of LNG export capacity is either operating, under construction, or approved and not under construction, this equals a volume of exports equal to 55 percent of U.S. demand. The pending applications and projects in pre-filing are equal to another 10.1 Bcf/d.[148,149]

It is important to remember that global LNG customers, which include Chinese state-owned enterprises and utilities of foreign governments, have options to buy from multiple production markets globally, U.S. consumers do not. We are captive and that is why it is paramount to maintain the NGA protections for U.S. consumers. It is predictable that as LNG exports increase, U.S. natural gas reliability and costs could be negatively impacted. Increased demand puts upward pressure on prices.

Increased LNG exports accelerates the threat of price inflation. On October 5, 2022, the chairman of the LNG company Tellurian Inc, Charif Souki, stated that "Getting (U.S.) gas in the water for $4-$5 is something of the past; if you really want to justify an investment...you have to think of $10-$12.[150] On September 28, 2022, the Federal Reserve Bank of Dallas released the results of its Dallas Federal Energy Survey. Sixty-nine percent of the executives surveyed expect the age of inexpensive U.S. natural gas to end by year-end 2025.[151] The 2025 timing is aligned with the next tranche of LNG export capacity coming online.

---

[147] "FERC Says LNG Export Growth Key Factor in Higher Summer Natural Gas, Electric Prices," Natural Gas Intelligence, May 20, 2022, https://www.naturalgasintel.com/ferc-says-lng-export-growth-key-factor-in-higher-summer-natural-gas-electric-prices/.

[148] Natural Gas, U.S. Energy Information Administration (EIA), https://www.eia.gov/naturalgas/.

[149] North American LNG Export Terminals – Existing, Approved not Yet Built, and Proposed, Federal Energy Regulatory Commission, Last Updated on January 24, 2023, https://www.ferc.gov/natural-gas/lng.

[150] U.S. gas at $4-$5 is a thing of the past, says Tellurian chairman; https://www.msn.com/en-us/money/markets/u-s-gas-at-4-5-is-a-thing-of-the-past-says-tellurian-chairman/ar-AA12Axho?ocid=msedgdhp&pc=U531&cvid=2984610aa1c84b43afc051f3c4a55e99.

[151] Federal Reserve Bank of Dallas, https://www.dallasfed.org/research/surveys/des/2022/2203.aspx#tab-forecastcharts.

U.S. LNG Export Terminal Status

Status	Terminal	Location	Capacity (Bcf/d)
Existing			
	Cheniere/Sabine Pass LNG: Trains 1-6	Sabine, LA	4.55
	Dominion-Cove Point LNG	Cove Point, MD	0.82
	Cheniere-Corpus Christi LNG: Trains 1-3	Corpus Christi, TX	2.40
	Sempra-Cameron LNG: Trains 1-3	Hackberry, LA	2.15
	Southern LNG Company: Units 1-10	Elba Island, GA	0.35
	Freeport LNG Dev/Freeport LNG Expansion/FLNG Liquefaction: Trains 1-3	Freeport, TX	2.38
	Venture Global Calcasieu Pass: Units 1-6	Cameron Parish, LA	1.11
Subtotal			*13.76*
Approved, Under Construction			
	Venture Global Calcasieu Pass: Units 7-9	Cameron Parish, LA	0.55
	ExxonMobil-Golden Pass	Sabine Pass, TX	2.57
	Venture Global Plaquemines	Plaquemines Parish, LA	3.40
	Driftwood LNG	Calcasieu Parish, LA	3.81
	Cheniere Corpus Christi: Stage III	Corpus Christi, TX	1.58
Subtotal			*11.91*
Approved, Not Under Construction			
	Lake Charles LNG	Lake Charles, LA	2.27
	Magnolia LNG	Lake Charles, LA	1.19
	Sempra-Cameron LNG: Trains 4 & 5	Hackberry, LA	1.41
	Sempra-Port Arthur LNG: Trains 1 & 2	Port Arthur, TX	1.86
	Freeport LNG Dev: Train 4	Freeport, TX	0.74
	Gulf LNG Liquefaction	Pascagoula, MS	1.50
	Eagle LNG Partners	Jacksonville, FL	0.13
	Texas LNG Brownsville	Brownsville, TX	0.55
	Rio Grande LNG-NextDecade	Brownsville, TX	3.60
	Commonwealth LNG	Cameron Parish, LA	1.21
	MARAD/USCG	Gulf of Mexico	1.80
Subtotal			*16.26*
Pending Applications			
	Sempra-Port Arthur LNG: Trains 3 & 4	Port Arthur, TX	1.86
	Venture Global CP2: Blocks 1-9	Cameron Parish, LA	3.96
	Venture Global Calcasieu Pass	Cameron Parish, LA	0.06
	Sempra-Cameron LNG Vacate T5 & modify T4	Hackberry, LA	-0.45
	Venture Global Plaquemines	Plaquemines Parish, LA	0.45
	Sempra – EnergíaCosta Azul: Phase 1	Baja California, MX	0.40
Subtotal			*6.28*
Projects in Pre-Filing			
	Port Fourchon LNG	LaFourche Parish, LA	0.65
	Delta LNG-Venture Global	Plaquemines Parish, LA	2.76
	Cheniere Corpus Christi Midscale: Trains 8-9	Corpus Christi, TX	0.45
Subtotal			*3.86*
Grand Total			52.07 Bcf/d

We urge you to oppose this legislation. Thank you for your support.

Sincerely,

Paul N. Cicio

Paul N. Cicio

President & CEO

cc: House Committee on Energy and Commerce
The Honorable Jennifer Granholm
Federal Energy Regulatory Commission

The Industrial Energy Consumers of America is a nonpartisan association of leading manufacturing companies with $1.1 trillion in annual sales, over 12,000 facilities nationwide, and with more than 1.8 million employees worldwide. It is an organization created to promote the interests of manufacturing companies through advocacy and collaboration for which the availability, use and cost of energy, power or feedstock play a significant role in their ability to compete in domestic and world markets. IECA membership represents a diverse set of industries including: chemicals, plastics, steel, iron ore, aluminum, paper, food processing, fertilizer, insulation, glass, industrial gases, pharmaceutical, building products, automotive, independent oil refining, and cement.

* * *

Torrance Refinery Action Alliance,
4733 Torrance Blvd. #200, Torrance, Ca 90503
February 6, 2022

Attention: Members, House Energy and Commerce Committee
Chair Cathy McMorris Rodgers, House Energy and Commerce Committee

Representative Jeff Duncan, Chair of the Energy, Climate, & Grid Security Subcommittee

Representative Bill Johnson, Chair of the Environment, Manufacturing, & Critical Materials Subcommittee
Ranking Member Frank Pallone Jr.

Representative Diana DeGette,
Ranking Member of the Energy, Climate, & Grid Security Subcommittee

Representative Paul Tonko,
Ranking Member of the Environment, Manufacturing, & Critical Materials Subcommittee

I speak to you from a non-partisan community organization that is dedicated to protecting the community and refinery workers at Hydrogen Fluoride (HF/MHF) refineries by bringing about conversion to a vastly safer alternative technology in the alkylation process.

The Torrance Refinery Action Alliance (TRAA) was formed after a "near miss" for a mass release of HF/MHF in Torrance California Feb 18, 2015. A multiton piece of equipment flew through the air and landed about six feet away from a large tank of HF. A major release could have resulted in mass casualties in numbers similar to what occurred in Bhopal India in 1984 (the release of a different chemical – but with similar properties resulted in 15,000 dead and over 100,000 people injured.) In all communications with the public and with members of the House of Representatives, TRAA endeavors to only use statements made or confirmed by industry or government.

A proposed bill under consideration by your committee Section 112(r)(7)(B) of the Clean Air Act (42 U.S.C. 7412(r)(7)(B)) states "... The owner or operator of a stationary source described in subclause (II) shall not be required by the regulations under this subparagraph to include in any hazard assessment under clause (ii) an assessment of safer technology and alternative risk management measures applicable to eliminating or reducing risks from the use of hydrofluoric acid in an alkylation unit."

There are at least six commercially proven alternative technologies available for conversion. These alternatives are being actively marketed by Chevron/Honeywell, KBR, Well Resources, and Lummus Technologies.

HF is an exceptionally dangerous chemical used in massive quantities in refineries that inherently have the potential for explosions and are also vulnerable to natural disasters and terrorist attacks. Given the risk associated with its use, it is unfair to communities to allow refineries to continue its use when there are alternatives available. There are 40 such refineries with an estimated 14 million people living within the circles of risk.

It also seems unfair to the companies that have, at great expense, developed and installed alternatives that are vastly safer. Under this bill, the HF refineries would not even have to assess whether these products would save them money and save the community worry and perhaps death and economic ruin.

Vendors of the alternatives state:

1) Alternative methods are inherently safer
2) Alternative methods are more energy efficient
   They produce more alkylate than HF does with the same feedstock, and have greater feedstock flexibility.
3) Alkylation will be increasing in market demand in the coming decade,
4) Alternatives reduce cost by eliminating construction, maintaining safety mitigations, reporting on Risk Management Plan, and the cost of training staff in the danger and preventative processes.
5) Of course it saves the cost of liability (generally self-insured) from a corporate apocalypse resulting from the unlikely but extremely consequential result of a release.

(There have been at least six major near misses in US refineries and at least two foreign releases resulting in significant fatalities.)

Although the three Committee Chairs; Rodgers, Duncan, and Johnson do not have HF refineries in their District, there are about 10 members of the committee who do. The American Petroleum Institute states there are "Commercially Proven Alternatives" - API also says that converting would be prohibitively expensive. However since other refineries have converted, apparently it is not prohibitively expensive. Vendors of Alternatives say that down time for a conversion would be comparable to a normal turnaround. It seems that this is a good business decision. It seems that upgrading is a good business plan by these refineries and would likely reduce gas prices and improve profitability. The committee should draft legislation that would drive conversion with maximum speed.

For the 10 Congressmembers who have HF in one of the refineries in your district, your communities as well as the oil industry as a whole would be greatly served by assessing the alternatives and Incentivizing HF users to convert to one of the commercially proven alternatives with all possible haste. (You may have other refineries in your district, and it is very likely that they do not use HF but an alternative to HF for Alkylation),

As the Fukushima event resulted in the sudden closing of numerous nuclear plants, a major release of HF with mass casualties would likely disrupt the continued use of HF in many if not all of the HF refineries in the US. The risk of this disruption to the US Energy Supply is unacceptable and can only be removed by assessing and requiring conversion from HF to one of the major corporate alternative technologies.

To exclude hazard assessment would be to the detriment of the corporations who are marketing the alternatives. Of course, the public safety and the benefits of transparency for concerned community members is served best by assessing dangers and alternatives but as well, the energy dominance sought by this amendment is also well served by an industry wide upgrade from this antiquated alkylation technology.

We urge your committees to correct this grave error, and not eliminate Hazard and Safer Technology Alternatives Assessment but add to the bill incentives that will result in HF refineries upgrading from this exceptionally hazardous and out of date technology.

We have attached material which supports this argument, we are available for further communication at info@traa.website, WWW.TRAA.Website

Sincerely,
Steven Goldsmith,
President, Torrance Refinery Action Alliance www.TRAA.website
(310) 542-6782 –office (preferred) (310) 227-3111 – cell sgoldsmith84@gmail.com
San Pedro Rotary Club, PP Hawthorne 2006-07
HF release video at now closed Philly refinery [Link]
Attached or included

1) This letter as an attachment
2) A brief summary statement on HF/MHF by TRAA
   HF is an exceptionally hazardous chemical capable of causing thousands of casualties, is used in large quantities in only 27% of US refineries. Refineries are vulnerable to accidents (many have occurred) natural disasters (near misses) and terrorist acts (unknown). There are multiple vastly safer commercially proven alternatives. Not only should HF not be exempt from Hazard Analysis but this bill should carry a bi-partisan amendment requiring a 3rd party assessment of alternatives and subsequent conversion be prioritized over a short period of time."
3) Fact sheet summary with links to key government/industry videos about the exceptional danger.
4) Additional Risks to National Security Concerns - from experts
5) Description of the alternatives - from industry vendors
6) Presentation by API to OMB (stating "commercially proven alternative" - but does have some misstatements)
7) A power point for a quick review of HF, its dangers and the cost savings alternatives
   For a link to in depth review - Science Panel Blog WWW.TRAA.blog
   www.TRAA.website

Chevron and Honeywell Announce Start-up of World's First Commercial Isoalky™ Ionic Liquids Alkylation Unit

SAN RAMON, Calif. and DES PLAINES, III., April 13, 2021 – Chevron Corporation (NYSE: CVX) and Honeywell (NYSE: HON) today announced the commissioning and start-up of the world's first commercial-scale ISOALKY™ process unit that utilizes ionic liquids to produce alkylate. The ISOALKY™ technology represents a major innovation in alkylation technology.

First used in Chevron's Salt Lake City refinery, the ISOALKY™ technology is designed to meet the refining industry's needs for a cost-effective alternative to conventional liquid acid systems that offers process safety advantages. Using a non-aqueous liquid salt, or ionic liquid, the revolutionary new catalytic process is handled with standard personal protective equipment and produces a valuable high-octane blending component that helps lower the environmental impact of gasoline. Pioneered by Chevron U.S.A., Inc., a subsidiary of Chevron, and licensed to Honeywell UOP, the technology is offered to the entire industry under the ISOALKY™ brand name.

"The ISOALKY™ plant is an exciting achievement for Chevron and the Salt Lake Refinery, and it's poised to be a game changer for the refining industry," said Mike Coyle, president of Chevron Manufacturing. "We are proud of the talent and teamwork demonstrated by our people and Honeywell UOP to bring this project to fruition."

"ISOALKY™ is a groundbreaking new technology for refiners, and a lower-risk and economical solution compared to conventional liquid acid technologies that produce alkylate," said Bryan Glover, president and CEO of Honeywell UOP. "Ionic liquids have strong acid properties that enable them to produce alkylate without the volatility of conventional acids, allowing for simpler handling procedures. Together with Chevron, Honeywell UOP has commercialized a solution that meets the rising global demand for cleaner-burning fuels at a lower cost while simplifying complex handling requirements."

ISOALKY™ technology can be used in new refineries as well as in existing facilities undergoing capital expansion or retrofit applications. ISOALKY™ technology has wider and improved feed flexibility relative to conventional alkylation technologies. Ionic liquids are regenerated on-site, eliminating the need for road or marine transportation for offsite regeneration and polymer byproduct handling. More information about ISOALKY™ technology can be found here.

Honeywell UOP (www.uop.com) is a leading international supplier and licensor of process technology, catalysts, adsorbents, equipment, and consulting services to the petroleum refining, petrochemical, and gas processing industries. Honeywell UOP is part of Honeywell's Performance Materials and Technologies strategic business group, which also includes Honeywell Process Solutions (www.honeywellprocess.com), a pioneer in automation control, instrumentation and services for the oil and gas, refining, petrochemical, chemical and other industries.

About Chevron

Chevron U.S.A. Inc. is a subsidiary of Chevron Corporation, one of the world's leading integrated energy companies. Through its subsidiaries that conduct business worldwide,

Chevron Corporation is involved in virtually every facet of the energy industry. Chevron explores for, produces and transports crude oil and natural gas; refines, markets and distributes transportation fuels and lubricants; manufactures and sells petrochemicals and additives; generates power; and develops and deploys technologies that enhance business value in every aspect of the company's operations. Chevron is based in San Ramon, CA. More information about Chevron is available at www.chevron.com.

About Honeywell

Honeywell (www.honeywell.com) is a Fortune 100 technology company that delivers industry specific solutions that include aerospace products and services; control technologies for buildings and industry; and performance materials globally. Our technologies help everything from aircraft, buildings, manufacturing plants, supply chains, and workers become more connected to make our world smarter, safer, and more sustainable. For more news and information on Honeywell, please visit www.honeywell.com/newsroom.

Chevron media contact:
Tyler Kruzich (925) 549-8686
tkruzich@chevron.com

Honeywell media contact:
Tehani Manochio (973) 216-0684
tehani.manochio@honeywell.com

This release contains certain statements that may be deemed "forward-looking statements" within the meaning of Section 21E of the Securities Exchange Act of 1934. All statements, other than statements of historical fact, that address activities, events or developments that we or our management intends, expects, projects, believes or anticipates will or may occur in the future are forward-looking statements. Such statements are based upon certain assumptions and assessments made by our management in light of their experience and their perception of historical trends, current economic and industry conditions, expected future developments and other factors they believe to be appropriate. The forward-looking statements included in this release are also subject to a number of material risks and uncertainties, including but not limited to economic, competitive, governmental, and technological factors affecting our operations, markets, products, services and prices. Such forward-looking statements are not guarantees of future performance, and actual results, developments and business decisions may differ from those envisaged by such forward-looking statements. We identify the principal risks and uncertainties that affect our performance in our Form 10-K and other filings with the Securities and Exchange Commission.

Natural Gas Plummets as Freeport Delays Facility Restart Following Explosion

PUBLISHED TUE, JUN 14 2022; 11:57 AM EDT
UPDATED WED, JUN 15 2022; 7:32 AM EDT
Pippa Stevens; @PIPPASTEVENS1 3

Natural gas prices plunged on Tuesday, after Freeport LNG said its facility that had a fire last week likely won't be back up and running soon.

"[C]ompletion of all necessary repairs and a return to full plant operations is not expected until late 2022," the company said Tuesday in a statement. The facility, located in Quintana Island, Texas, had an explosion last Wednesday.

"Given the relatively contained area of the facility physically impacted by the incident, a resumption of partial operations is targeted to be achieved in approximately 90 days," Freeport LNG said.

"The U.S. natural gas market will now be temporarily oversupplied as 2 bcf/d or a little over 2% of demand for U.S. natural gas has been abruptly eliminated," said Rob Thummel, managing director at Tortoise Capital.

"U.S. natural gas supply will likely remain at current levels as producers won't reduce production by 2 bcf/d. The result is an oversupplied U.S. natural gas market," he added.

Freeport's operation is roughly 17% of the U.S.' LNG processing capacity.

Natural Gas Plummets

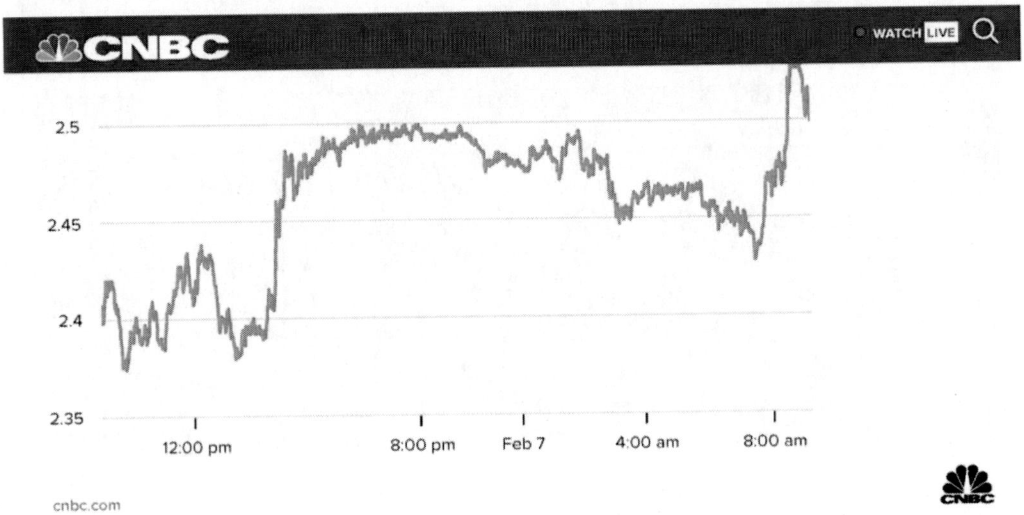

Despite Tuesday's drop, natural gas prices are still up 93% since the start of the year. Demand has rebounded as worldwide economies emerge from the pandemic, while supply has remained constrained.

Russia's invasion of Ukraine upended a market that was already tight. As Europe looks to move away from Russian energy, record amounts of U.S. LNG are now heading to the continent.

Surging prices are adding to inflationary pressures across the economy. Drivers are already grappling with record prices at the pump with the national average for a gallon of gas topping $5 over the weekend, and now utility bills are also set to rise.

Natural gas prices surged above $9 per MMBtu in May, hitting the highest level since August 2008.

After the explosion at Freeport's facility last week, the company initially said the plant would be shut for several weeks.

"The incident occurred in pipe racks that support the transfer of LNG from the facility's LNG storage tank area to the terminal's dock facilities," the company said Tuesday. "None of the liquefaction trains, LNG storage tanks, dock facilities, or LNG process areas were impacted," the company added.

February 7, 2023

The Honorable Cathy McMorris Rodgers
Chair
Committee on Energy and Commerce
2125 Rayburn House Office Building
Washington, D.C. 20515

The Honorable Bill Johnson
Chair
Subcommittee on Environment,
Manufacturing, and Critical Materials
2125 Rayburn House Office Building
Washington, D.C. 20515

The Honorable Jeff Duncan
Chair
Subcommittee on Energy, Climate,
and Grid Security
2125 Rayburn House Office Building Washington, D.C. 20515

The Honorable Frank Pallone, Jr.
Ranking Member
Committee on Energy and Commerce
2125 Rayburn House Office Building
Washington, D.C. 20515

The Honorable Paul Tonko
Ranking Member
Subcommittee on Environment,
Manufacturing, and Critical Materials
2125 Rayburn House Office Building
Washington, D.C. 20515

The Honorable Diana DeGette
Ranking Member
Subcommittee on Energy, Climate,
and Grid Security
2125 Rayburn House Office Building
Washington, D.C. 20515

Dear Chairs Rodgers, Johnson, and Duncan, and Ranking Members Pallone, Tonko, and DeGette:

On behalf of our millions of members and supporters, the 71 undersigned organizations write to oppose the following pieces of legislation that will be considered in the upcoming Energy and Commerce Committee legislative hearing on Tuesday, February 7th. This proposed legislation would exacerbate the climate crisis, perpetuate environmental injustices, and undermine US economic and national security by prolonging reliance on risky and volatile energy sources.

These bills would encourage new fossil fuel production and infrastructure, despite the scientific consensus that there is no room for investment in new fossil fuel production if we are to reach net zero by 2050, keeping the world on a 1.5°C compatible pathway. The bills would also undermine bedrock environmental laws, including the National Environmental Policy Act (NEPA) by short-circuiting permitting processes and limiting public input. NEPA is a critical environmental law and an important tool for frontline and environmental justice communities to influence federal infrastructure projects that will impact them the most.

In addition, by promoting the expansion of fossil fuel production and exports, these bills would delay the transition to a clean, secure, and affordable energy grid that would bring true energy independence and security. The fossil fuel industry is already raking in record profits at the expense of consumers and future generations, yet their supporters in Congress are putting forward these bills to lock us into increased extraction, high and volatile energy prices, and environmental degradation in exchange for even higher profits for oil and gas companies.

We urge opposition to the following bills:

- H.R. 150, the Protecting American Energy Production Act: OPPOSE
    - Prohibits the President from issuing a moratorium on fracking unless authorized by Congress. Fracking releases massive amounts of methane, a potent greenhouse gas that has more than 80 times the power of carbon dioxide over a 20-year period, driving approximately one quarter of the warming our planet has experienced to date. Fracking also harms local communities and ecosystems by releasing air pollutants and contaminating water sources.
- H.R. 484, the Natural Gas Tax Repeal Act: OPPOSE
    - Repeals the Methane Emissions Reduction Program created by the Inflation Reduction Act (IRA). This critical program supports efforts to reduce methane emissions from the oil and gas sector, improve methane monitoring, fund environmental restoration, and help communities reduce the health impacts of pollution.
- H.R. __, the Promoting Cross-border Energy Infrastructure Act: OPPOSE
    - Requires FERC to approve gas pipelines to Canada or Mexico within 30 days, significantly limits requirement for certificates of crossing for modifications to existing pipelines, and limits environmental review and public input. Decreasing scrutiny for cross-border pipelines would lead to increased emissions and deny affected communities sufficient input opportunities.
- H.Con.Res. __, Expressing disapproval of the revocation by President Biden of the Presidential permit for the Keystone XL pipeline: OPPOSE
    - Resolution of disapproval of President Biden revoking the Presidential Permit for Keystone XL pipeline. If built, Keystone XL would have carried 830,000 gallons per day of the dirtiest oil on the planet, threatening our climate, farmland, critical water resources, and wildlife habitat along the pipeline's path.
- H.R. __, the Promoting Interagency Coordination for Review of Natural Gas Pipelines Act: OPPOSE

- This bill would undercut public transparency and input from communities by limiting the time for environmental reviews. The bill alters the approval process for gas pipelines by requiring all other federal and state agencies to defer to FERC.
- H.R. 647, the Unlocking our Domestic LNG Potential Act of 2023: OPPOSE
    - This bill would strip away the federal government's responsibility to examine the full impacts of LNG expansion on US energy markets, the environment, and local communities. It would make it easier to approve LNG exports by removing the first 3 sections of the Natural Gas Act, which require a public interest determination for LNG exports to non-FTA countries. LNG exports negatively impact Americans by exacerbating climate change, raising domestic energy prices, and perpetuating environmental injustices, and these factors need to be taken into account when deciding whether to approve additional LNG export terminals.
- H.Con.Res. __ , Expressing the sense of Congress that the Federal Government should not impose any restrictions on the export of crude oil or other petroleum products: OPPOSE
    - This resolution expresses the sense of Congress that the Federal Government should not restrict the export of crude oil or other petroleum products. Increased oil drilling and exports have enormous climate repercussions and pollute communities and ecosystems. The Federal Government must ensure that these exports do not compromise US climate and environmental justice goals or undermine our global climate leadership.

Sincerely,

350 New Orleans
A Community Voice
Accelerate Neighborhood Climate Action
Businesses for a Livable Climate
Call to Action Colorado
Carrizo Comecrudo Tribe of Texas
CatholicNetwork US
Center for Biological Diversity
Center for Oil and Gas Organizing
Chispa Texas
Citizen's Alliance for a Sustainable Englewood
Clean Air & Water Better Brazoria/Freeport Tx
Climate Reality Project New Orleans
CO Businesses for a Livable Climate
Coalition Against Death Alley
Community for Sustainable Energy
Concerned Citizens of St. John the Baptist Parish
Dayenu: A Jewish Call to Climate Action
Earthjustice
Earthworks
Endangered Species Coalition

Food and Water Watch
For a Better Bayou
Friends of the Earth
Greater New Orleans Housing Alliance
Greater New Orleans Interfaith Climate Coalition
GreenFaith
Greenpeace USA
Healthy Gulf
Hip Hop Caucus
Honor the Earth
I-70 Citizens Advisory Group
Inclusive Louisiana
Indivisible Ambassadors
Larimer Alliance for Health, Safety and Environment
League of Conservation Voters
Littleton Business Alliance
Louisiana Bucket Brigade
Louisiana Environmental Action Network (LEAN)
Louisiana League of Conscious Voters
Mayfair Park Neighborhood Association Board
Mental Health & Inclusion Ministries
Moms Clean Air Force
Montbello Neighborhood Improvement Association
Natural Resources Defense Council
North Range Concerned Citizens
Oil and Gas Action Network
Public Citizen
RapidShift Network
RESTORE
Save EPA (former employees)
Save RGV
Sierra Club
Southern Utah Wilderness Alliance
Southwest Organization for Sustainability
Spirit of the Sun, Inc.
Sunflower Alliance
System Change Not Climate Change
Texas Campaign for the Environment
The Climate Center
The Green House Connection Center
The Vessel Project of Louisiana
The Wilderness Society
Turtle Island Restoration Network
Unite North Metro Denver

Voices for Progress
Wall of Women
WE ACT for Environmental Justice
Western Slope Businesses for a Livable Climate
Womxn from the Mountain
Working for Racial Equity

Inflation Reduction Act's Proposed Methane Fee Would Have Negligible Impact on Natural Gas Prices, Brian C. Prest August 5, 2022

In a 2021 RFF Issue Brief, "Methane Fees' Effects on Natural Gas Prices and Methane Leakage," I modeled the potential effects of a proposed fee on methane emissions from oil and gas facilities, such as those recently considered in Congress (e.g., S.645 and H.R.4084). That modeling found that the impacts of those fees at the proposed levels of up to $1,500 per ton of methane would increase natural gas prices by a very small amount, with a focal estimate of $0.15 per MMBtu in potential price increase. By comparison, as reported in that Issue Brief, residential consumer prices in May 2021 were about $13.50 per MMBtu, indicating a price impact of around 1%.

The Inflation Reduction Act of 2022 (IRA) contains a new version of this fee with a number of modifications, including a one-year delay in implementation and a narrowing of the scope of oil and gas facilities that would be required to pay the fee. Importantly, the new version exempts from the fee facilities that are in compliance with forthcoming Environmental Protection Agency (EPA) regulations on methane. This provision is likely to dramatically reduce the number of companies required to pay the fee, implying that the impact of the methane fee as proposed in the IRA would result in a substantially smaller estimated price impact than the previous estimates of $0.15 per MMBtu and 1% of residential consumer prices.

While it is difficult to model this new provision in detail due to uncertainties around the timing and stringency of EPA's future regulations and compliance decisions by oil and gas operators, the previous estimates for the broader policy were already quite small, indicating the impact on natural gas prices of the methane fee in the IRA would be negligible, and well under 1% of residential consumer prices.

Resources for the Future (RFF) is an independent, nonprofit research institution in Washington, DC. Our mission is to improve environmental, energy, and natural resource decisions through impartial economic research and policy engagement. RFF is committed to being the most widely trusted source of research insights and policy solutions leading to a healthy environment and a thriving economy.
Unless otherwise stated, the views expressed here are those of the individual authors and may differ from those of other RFF experts, its officers, or its directors. RFF does not take positions on specific legislative proposals.

In broader context, Princeton University's REPEAT project recently drew on other work of mine to estimate that the IRA as a whole is likely to reduce natural gas prices by 10-20% in the medium term.[152] I have not independently evaluated those estimates, but that degree of price reduction would offset—many times over—the negligible price impacts of the methane fee.

CLEAN JOBS AMERICA

NEARLY 3.3 MILLION CLEAN ENERGY JOBS[1]

In every region and every state in America, clean energy is creating jobs and careers. Nationwide, more than 110,000 net new clean energy jobs were created in 2018, bringing the total number of Americans who work in clean energy to 3.26 million.

While jobs in solar declined in part because of tariffs on steel and solar panels, wind energy jobs grew by nearly 4 percent and now competes with fossil fuels in many markets.[2]

Energy efficiency continues to lead the clean energy sector in total number of jobs, growing 3.4 percent to 2.3 million jobs.

But the big story in 2018 was around clean vehicles and storage.

Driven by growing consumer demand, the number of jobs in clean vehicles manufacturing increased by 16 percent. About 254,000 Americans now work at companies building hybrid, electric and other clean vehicles, while another 486,000 Americans work in companies that manufacture parts that make vehicles more efficient.

Energy storage saw a 14 percent increase in jobs as utilities, businesses and consumers deployed more batteries in EVs and with solar and wind installations, while grid modernization jobs grew by 3.3 percent.

Smart state policies continue to drive much of the growth in clean energy and the jobs and investments that come with it. But with a new Congress comes new opportunities to pass meaningful legislation on a federal level to keep these jobs growing nationwide. See sidebar for more.

TOP 10 STATES FOR CLEAN ENERGY JOBS

RANK	STATE	TOTAL*	SOLAR	WIND	ENERGY EFFICIENCY	CLEAN VEHICLES
1	California	512,934	126,507	5,785	318,542	22,389
2	Texas	233,447	11,433	25,386	162,816	17,800
3	Florida	158,652	10,528	4,461	118,412	9,360
4	New York	156,059	11,603	3,491	123,292	8,624
5	Michigan	126,081	5,419	4,783	85,061	25,304
6	Illinois	123,247	5,341	8,706	89,469	10,417
7	Massachusetts	116,491	16,527	1,983	86,473	3,184
8	Ohio	112,486	8,108	1,080	81,676	16,646
9	North Carolina	110,913	8,912	908	86,559	7,280
10	Virginia	95,158	4,241	1,628	78,670	5,436

* Total includes renewable energy, energy efficiency, clean vehicles, battery storage, advanced biofuels, low-impact hydro and other sectors.

CLEAN ENERGY GROWTH IN PERSPECTIVE

110,000

Clean energy jobs grew 3.6 percent in 2018, adding jobs in nearly every state and combining to add over 110,000 net new clean energy jobs nationally.

12

Number of states that have or are considering policies that get 100 percent of their electricity from clean energy sources.

3X

Clean energy jobs outnumbered fossil fuel jobs nearly 3 to 1 in 2018.

10

The number of states that generate more than 20 percent of their electricity from wind and solar–Kansas, Iowa, Oklahoma, North Dakota, South Dakota, Vermont, California, Maine, Colorado, and Minnesota.[3]

156 GW

Combined capacity of installed solar and wind surpassed 150GW in 2018. Wind energy is the largest source of renewable generating capacity. A new solar project is installed in America every two minutes.[4,5]

E2

WWW.E2.ORG
WWW.CLEANJOBSCOUNT.ORG
@E2ORG
#CLEANJOBSAMERICA
#CLEANJOBSCOUNT

CLEAN JOBS COUNT

MARCH 2019
E2FS: 19-03-D

[152] See page 10 of https://repeatproject.org/docs/REPEAT_IRA_Prelminary_Report_2022-08-04.pdf.

CLEAN JOBS AMERICA

WANTED: FEDERAL ACTION ON CLEAN ENERGY

To continue creating tens of thousands of new jobs for Americans across the country, Congress should:

- Make sure any infrastructure bill includes policies to modernize the grid and expand electric vehicle charging infrastructure to keep grid and storage jobs growing.
- Stop rollbacks to fuel economy (CAFE) standards that are saving business and consumers money with every visit to the pump and also are driving jobs and American innovation in the clean vehicles sector.
- Upgrade and extend expired energy efficiency tax credits for commercial and residential buildings; clarify the Investment Tax Credit (ITC) to apply to energy storage and offshore wind and lift the cap on the electric vehicle tax credit to create more jobs in these sectors.
- Properly fund R&D investments in clean energy innovation and efficiency at the U.S DOE's Office of Energy Efficiency and Renewable Energy, the Loan Programs Office, and programs such as ARPA-E, and the Advanced Technology Vehicles Manufacturing (ATVM) program.

INDUSTRY BREAKDOWN: JOBS

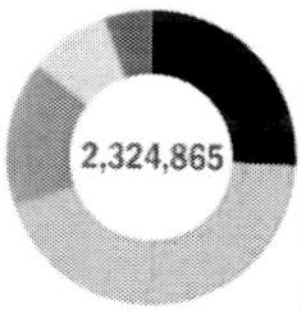

ENERGY EFFICIENCY:
- ENERGY STAR appliances, including High-Efficiency HVAC: **595,331**
- Traditional HVAC: **582,108**
- Advanced Building Materials: **440,188**
- LED, CFL, and Efficient Lighting: **370,562**
- Other: **207,780**
- Renewable H&C: **128,896**

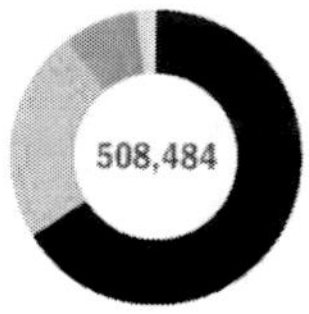

RENEWABLE ENERGY:
- Solar: **334,992**
- Wind: **111,166**
- Geothermal: **8,526**
- Bioenergy/CHP: **42,222**
- Low-Impact Hydro: **11,578**

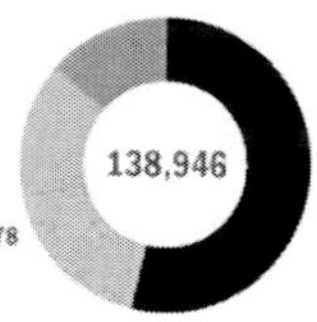

GRID & STORAGE:
- Clean Storage: **74,569***
- Smart Grid: **25,000**
- Micro-Grid: **20,125**
- Other Grid Modernization: **19,252**

* Clean storage includes pumped hydro storage, battery storage, thermal storage, and mechanical storage detailed technologies.

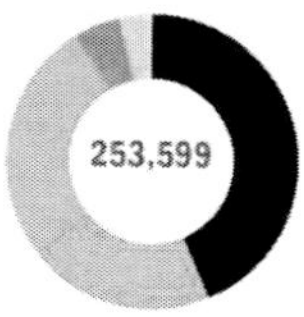

CLEAN VEHICLES:
- Hybrid Electric Vehicles: **110,340**
- Plug-in Hybrid Vehicles: **52,931**
- Electric Vehicles: **67,973**
- Natural Gas Vehicles: **12,545**
- Hydrogen & Fuel Cell: **9,809**

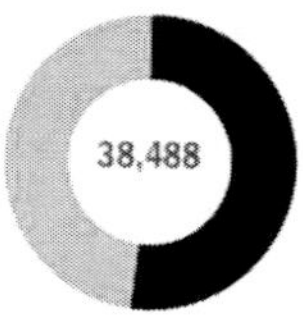

FUELS:
- Other Ethanol/ Non-Woody Biomass: **20,074**
- Other Biofuels: **18,414**

CLEAN JOBS BREAKDOWN BY VALUE CHAIN

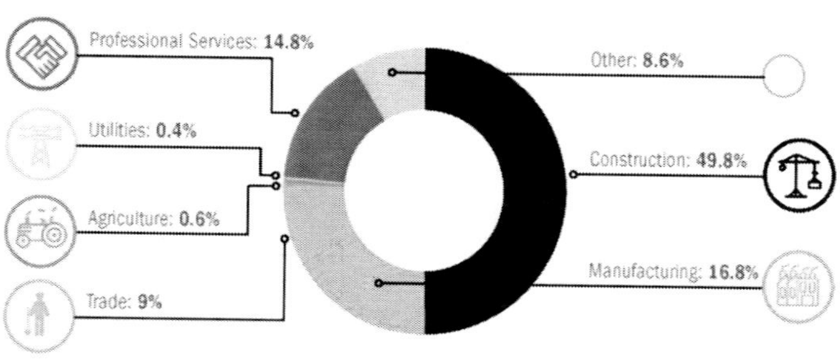

CLEAN JOBS AMERICA

CLEAN VEHICLES **255,000 JOBS**

- Clean vehicles now account for about **13 percent** of all the jobs in the motor vehicles industry
- **15.4 percent** nationwide job growth in 2018
- **42 states** saw double-digit job growth in 2018, with 15 seeing above 20 percent growth
- **486,000** additional employees in the motor vehicle industry work with parts making vehicles more fuel efficient

GRID & CLEAN STORAGE **139,000 JOBS**

- **9 percent** job growth in 2018, with employers projecting growth of 4.4 percent in 2019.
- **Grid modernization** industries added over 2,000 jobs, a 3.3 percent growth rate.
- California and Texas lead the U.S. in Grid & Clean Storage jobs again, but Nevada saw biggest rise in rankings—jumping from 25th to 4th thanks to a **380 percent job surge** in 2018.

SOLAR + WIND **446,000 JOBS**

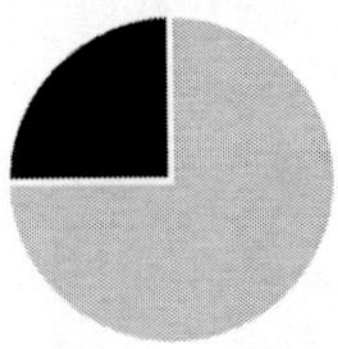

335k SOLAR JOBS

- Solar provides the largest share of electric power generation in the U.S., **100,000** more than the next two sectors combined.
- Solar jobs fell 4.2 percent in 2018—after nearly a decade of growth—but employers are projecting job growth of **over 8 percent** in 2019.
- 89 percent of solar job losses were in maturing in states of MA and CA. In 18, solar jobs increased.
- **66 percent** of solar jobs (224k) are involved in manufacturing and construction.

111k WIND JOBS

- Wind provides the **3rd largest** share of electric power generation jobs, about 1,500 fewer employees than natural gas.
- Jobs grew **3.5 percent** last year, with employers projecting job growth of nearly 5 percent for 2018
- All but **five states** (and D.C.) saw job growth in wind energy in 2018
- **56 percent** of wind jobs (63k) are involved in manufacturing and construction.

ENERGY EFFICIENCY **2.3 MILLION JOBS**

76,000
The number of new jobs added by energy efficiency companies in 2018—accounting for over half of all new energy jobs

7.4%
Expected job growth in 2019 by employers, the highest across the entire energy sector

1.4 M
The number of energy-efficiency jobs in construction

321,000
The number of manufacturing jobs in energy efficiency, an increase of 2 percent

1 in 3
More than one out of every three employees working in the energy sector (from traditional energy to motor vehicles) is involved in energy efficiency

1 in 4
In 2018, ENERGY STAR appliances employed more than 1 out of every 4 energy efficiency workers, largest category of employment

CLEAN JOBS AMERICA

3.26 MILLION CLEAN ENERGY JOBS: HOW DO THEY STACK UP?

3.04 M

That's the number of teachers in elementary, middle & secondary schools[6]

2.25 M

That's how many waiters & waitresses work in America's bars and restaurants[7]

211,000

The number of jobs in coal mines, gas fields and oil patches combined[8]

1.15 M

The number of jobs in the entire U.S. fossil fuel industry

GROWTH ACROSS THE COUNTRY

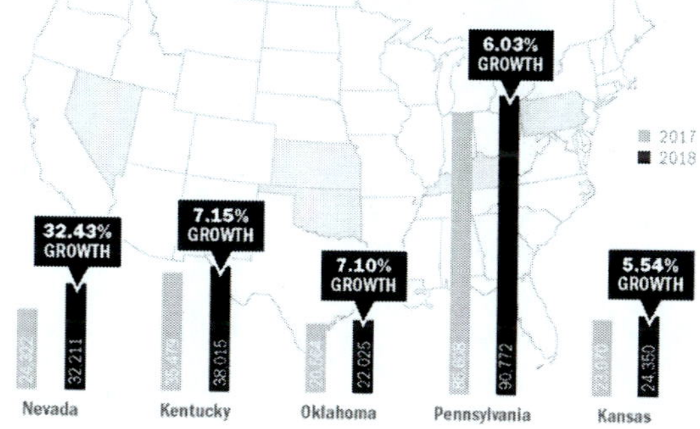

AMERICANS OVERWHELMINGLY WANT MORE CLEAN ENERGY

- More than **85 percent** of Americans—over two-thirds of Republicans (71%)—support requiring utilities in their state to produce 100 percent of electricity from clean energy sources by 2050.[9]
- Between 2013 and 2018, support for renewable energy research increased from 73 percent to **88 percent** among registered voters, including a 30 percent shift among conservative Republicans.
- A majority of Americans (58%) say they think policies to transition from fossil fuels to clean energy will improve economic growth and create new jobs.

WHY SUCH STRONG SUPPORT? CLEAN ENERGY = JOBS

- Wind technicians and solar installers are predicted to be the top two fastest-growing jobs over the next seven years.[10]
- The fastest-growing jobs in 12 states were in renewable energy in 2018.[11]
- 66 percent of new power capacity in 2019 is projected to come from solar, wind, other renewables, and battery storage.[12]

CLOSER LOOK: CLEAN STORAGE[13]

Clean storage added almost 9,500 new jobs for a 14 percent growth rate in 2018.

Solar + storage broke records last year, in 2019...

- The record is projected to be broken yet again because of falling solar and battery prices and continuation of the federal ITC for solar.[14]

The U.S. is projected to regain its spot as the world's largest storage market...

- thanks to residential solar-storage, state incentives, and increased utility-scale storage due to increased market certainty.

Led by increase EV adoption and home installations, lithium-ion battery installations are projected to ...

- increase by 800 percent from 2018-2022.[15]

HIGHWAYS

- More than 1 million EVs are on the road in America, with nearly 50,000 sold nationwide in December 2018 alone.[16]

HOMES

- 10x: Residential energy storage deployments led the way in 2018, growing tenfold from 2017 to 2018.[17]

MANUFACTURING JOBS

- Tesla's Gigafactory in Storey County near Reno, Nevada brought more than 7,000 jobs to the state in 2018 alone. That made Storey the top county in the nation for clean jobs per capita with more than 1,950 jobs per 1,000 residents, attracting workers and new residents from outside the county just to fill all the new jobs.[18]

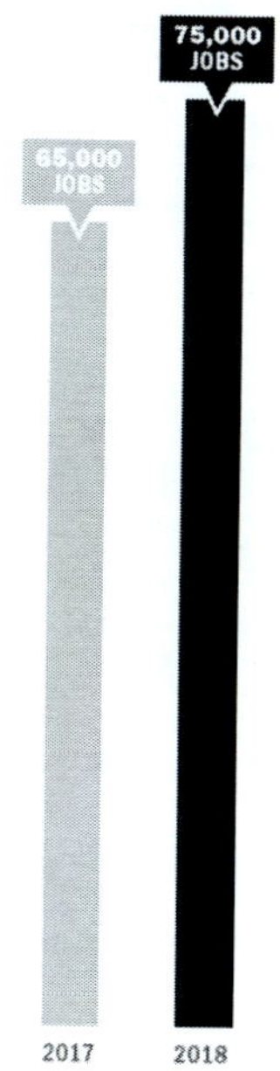

CLEAN JOBS AMERICA

MORE CLEAN ENERGY = MORE CONSTRUCTION JOBS

More than 1 out of 6 of all U.S. construction jobs are in energy efficiency (17%)[19]

38.6%

Wind and solar account for nearly 2 out of every 5 construction jobs in the electric generation sector

Nearly 6 out of 10 of energy efficiency's 2.3 million employees work in construction (1.29 million)

2019 GROWTH

Expected construction of renewable energy projects accounts for the majority of the electric power generation sector's projected 7.1 percent job growth for 2019

CLEAN ENERGY INVESTING A TRILLION DOLLAR MARKET WAITING TO BE UNLEASHED

With the right support and policies, U.S. investors could turn clean energy into an economic powerhouse and unlock a market with near unlimited potential.

A TURNING POINT OPPORTUNITY:

FASTEST-GROWING

Renewable energy resources are now expected to be the fastest-growing source of U.S. electricity generation for at least the next two years, with growth forecasts of 10 percent in 2019 and 17 percent in 2020.[20]

18%

Roughly 18 percent of U.S. energy generation is now supplied by renewable sources, up from 11 percent in 2009.[21]

2035

By the year 2035, renewable energy is expected to become the world's ominant power source. And by 2050, renewables are expected to supply 75 percent of the world's energy.[22]

A GROWING MARKET:

$1 TRILLION

Financial institutions in the U.S. (including banks, asset managers, and private-equity firms) are expected to double planned investments in renewable energy, with the potential to mobilize $1 trillion in cumulative private capital by 2030.[23]

13.4 GW

Corporate contracts for clean energy technology more than doubled in 2018 to 13.4 GW (up from 6.1GW in 2017), totaling now more than 32 GW since 2008– the generation capacity of the Netherlands.[24]

60%

The U.S. is beginning to dominate a new global market, accounting for 60 percent of global corporate clean energy purchases in 2018.

GOOD JOBS FOR VETERANS:

Clean energy employs a greater percentage of veterans than most industries–including oil and gas. Percentage of veterans in the workforce of:

10%
WIND

9%
SOLAR PV

10%
ENERGY EFFICIENCY

8%
CONCENTRATED SOLAR POWER

6%
NATIONAL AVERAGE

6 to 9%
FOSSIL FUEL AND NUCLEAR

CLEAN JOBS AMERICA

ENDNOTES

1 Unless otherwise stated, the data and analyses presented in Clean Jobs America E2 (Environmental Entrepreneurs) are based on data collected for the 2019 U.S. Energy Employment Report (2019 USEER), produced by the Energy Futures Initiative (EFI) in partnership with the National Association of State Energy Officials (NASEO) and collected and analyzed by BW Research Partnership (BWRP). See Pages 9-13 for methodology questions. For more questions regarding methodology, visit https://www.e2.org/clean-jobs-america-faq.

2 https://www.energy.gov/eere/wind/downloads/2016-wind-technologies-market-report

3 https://www.eia.gov/todayinenergy/detail.php?id=37233

4 https://www.awea.org/wind-101/basics-of-wind-energy/wind-facts-at-a-glance

5 https://www.seia.org/solar-industry-research-data

6 https://www.bls.gov/iag/tgs/iag61.htm

7 https://www.bls.gov/iag/tgs/iag722.htm

8 By sector, fossil fuel jobs from electric power production are: coal (86,200), natural gas (43,500), advanced natural gas (69,100) and oil and petroleum (12,500).

9 http://climatecommunication.yale.edu/publications/energy-in-the-american-mind-december-2018/2/

10 https://www.bls.gov/emp/tables/fastest-growing-occupations.htm

11 http://www.projectionscentral.com/Projections/ShortTerm

12 https://www.eia.gov/todayinenergy/detail.php?id=37952

13 Clean Jobs America, 2017 USEER, 2018 USEER, 2019 USEER

14 https://www.greentechmedia.com/articles/read/five-predictions-for-the-global-energy-storage-market-in-2019#gs.FUHXVNmT

15 https://www.woodmac.com/reports/power-markets-the-future-of-lithium-ion-batteries-demand-technologies-and-investments-29646

16 http://www.edisonfoundation.net/iei/publications/Documents/IEI_EEI%20EV%20Forecast%20Report_Nov2018.pdf

17 https://www.woodmac.com/research/products/power-and-renewables/us-energy-storage-monitor

18 http://www.diversifynevada.com/wp-content/uploads/2018/12/2018-Tesla-Economic-Impact-Study.pdf

19 Share based on the Bureau of Labor Statistic's preliminary estimate for construction sector employment in December 2018 https://www.bls.gov/iag/tgs/iag23.htm

20 https://www.eia.gov/todayinenergy/detail.php?id=38053

21 http://www.bcse.org/factbook/#

22 https://www.mckinsey.com/industries/oil-and-gas/our-insights/global-energy-perspective-2019

23 https://www.greentechmedia.com/articles/read/global-renewable-energy-investment#gs.UtbXYdUa

24 https://about.bnef.com/blog/corporate-clean-energy-buying-surged-new-record-2018

PRESENTED BY:

E2 is a national, nonpartisan group of business leaders, investors and others who advocate for smart policies that are good for the environment and good for the economy.

Clean Jobs Count is a campaign to raise awareness of the economic importance of the clean economy. Visit www.cleanjobscount.org to join thousands of business leaders, workers and others to tell lawmakers and policymakers that clean jobs count.

THANKS TO SUPPORT FROM:

E2 wishes to express its appreciation to the **National Association of State Energy Officials** (NASEO), the **Energy Futures Initiative** (EFI) and **BW Research Partnership** ("BWRP") who made this report possible by producing the USEER and its underlying data.

USEER | 2019 U.S. Energy and Employment Report

NASEO National Association of State Energy Officials

ENERGY FUTURES — INITIATIVE —

[bw] RESEARCH PARTNERSHIP

WWW.E2.ORG
WWW.CLEANJOBSCOUNT.ORG
@E2ORG
#CLEANJOBSAMERICA
#CLEANJOBSCOUNT

Good for the Economy.
Good for the Environment.

* * *

February 6, 2023

Re: Legislative Hearing Titled: "Unleashing American Energy, Lowering Energy Costs, and Strengthening Supply Chains."

Dear Representative:

The undersigned organizations write in opposition to five pieces of legislation. These draft bills are among other pieces of legislation being heard in the the Energy and Commerce Committee legislative hearing on Tuesday, February 7th. These bills are being labeled "critical minerals" bills, but the definition of the term "critical energy resources" means these bills could apply to virtually anything related to the energy sector, whether that is oil and gas, coal, petrochemicals or nuclear production, processing and refining.

H.R. __ , To authorize the Administrator of the Environmental Protection Agency to waive application of certain requirements, sanctions, or fees, with respect to processing or refining of critical energy resources at a critical energy resource facility, and for other purposes: OPPOSE

The undersigned oppose the draft legislation authorizing the Environmental Protection Agency to waive the Clean Air Act (CAA) and Solid Waste Disposal Act (SWDA) requirements for waste produced by certain energy facilities. The CAA and SWDA protect human health and the environment from the hazards of industrial waste. The draft legislation would exempt certain energy facilities, potentially including everything from fracking wastewater to mine processing facilities and tailing sites to nuclear facilities, from provisions of the SWDA that are "necessary to protect human health and the environment" from hazardous waste. Exempting energy waste from these laws threatens the health of people in frontline communities, as well as our air and water. The waste from energy production are some of the most threatening products and sites, and often they exist for hundreds of years, even in perpetuity, which is part of the reason why the Superfund program is overwhelmed.

H.R. __ , To amend the Toxic Substances Control Act with respect to critical energy resources, and for other purposes: OPPOSE

The undersigned oppose the draft legislation amending and undermining the Toxic Substances Control Act. Our organizations support clean energy that does not jeopardize human health and the environment. The draft legislation would short circuit the review and approval process for new chemicals used in the energy sector, whether that is for fracking, petrochemicals, mining or dozens of other products. This rushed and weak assessment, which would lead to default approvals, would result in the blind rubberstamping of chemicals for use in energy that have deleterious impacts on human health and the environment. Virtually any chemical that plays a role in the production, refining, distribution, and use of energy could be designated as "critical" by the Department of Energy.

H.R. __ , To amend the Solid Waste Disposal Act to treat the owner or operator of a critical energy resource facility as having been issued an interim permit for the treatment, storage, or disposal of hazardous waste, and for other purposes: OPPOSE

The undersigned oppose the draft legislation regarding interim permitting for certain energy facilities. The draft legislation would exempt certain energy facilities from requirements to secure an interim permit before operating, instead allowing the facilities to operate before securing such a permit. The result could be the release of harmful pollutants into our air and water, threatening the environment and health of people in frontline communities. The facilities that could receive a permit without an accurate assessment of their impact include everything from radioactive waste to petrochemicals to fertilizer to mining waste, all extremely toxic industries.

H.R. __ , To require the Administrator of the Environmental Protection Agency to authorize the use of flexible air permitting with respect to certain critical energy resource facilities, and for other purposes: OPPOSE

The undersigned oppose the draft legislation regarding flexible permitting for certain energy facilities, allowing the Environmental Protection Agency (EPA) Administrator to circumvent the scientific process of approving or denying flexible air permitting at the agency. Doing so could potentially allow the EPA Administrator to increase air pollution from so-called "critical energy resource facilities," subsequently harming environmental and public health. As with the other pieces of legislation we oppose in this letter,a broad spectrum of facilities that emit toxic air pollution could evade scrutiny for health impacts, including processing and refining of oil and gas and coal products, minerals and fertilizers.

H.R. __ , the "Securing America's Critical Minerals Supply Act.": OPPOSE

The undersigned oppose this draft legislation modifying the organization of the Department of Energy. The draft legislation is about far more than critical minerals, as it uses the definition of critical energy resource, which could apply to virtually anything pertaining to energy. It takes the authority on many issues and processes that are vital for the protection of communities, air, lands and water away from those who have the expertise in understanding the potential impacts of extraction and production, whether that is the Department of Interior (DOI) or Environmental Protection Agency (EPA). In doing so it makes the only metric for consideration economic, which would mean that communities, lands and waters would be sacrificed.

Sincerely,
Center for Biological Diversity
Earthjustice
Earthworks
Grand Canyon Trust
League of Conservation Voters
Oxfam America
Southern Utah Wilderness Alliance
The Wilderness Society

Statement for the Hearing "Unleashing American Energy, Lowering Energy Costs, and Strengthening Supply Chains" before the Subcommittee on Energy, Climate & Grid Security and Subcommittee on Environment, Manufacturing & Critical Materials, February 7, 2023

Dear Chairwoman McMorris Rogers and Ranking Member Pallone,

On behalf of the Natural Resources Defense Council, Toxic Free Future, Earthjustice, Environmental Defense Fund, League of Conservation Voters, Environmental Working Group, Center for Environmental Health, Breast Cancer Prevention Partners, Defend Our Health and Clean Water Action, we are writing to express our strong opposition to the draft bill, currently unnumbered and unsponsored, "To amend the Toxic Substances Control Act for critical energy resources, and for other purposes" which is one of the bills covered by today's legislative hearing. In just three short pages, the draft bill would reverse and eviscerate several of the core reforms to the Toxic Substances Control Act (TSCA) that passed the House and Senate with overwhelming bipartisan support just a few years ago.

The draft legislation would make it virtually impossible for EPA to meaningfully review the safety of new chemicals that are classified as "critical energy resources," regardless of their health risks. The bill promotes cursory assessments, followed by default approvals, of any new chemical that is deemed necessary for a "critical energy resource," no matter how toxic, how persistent, or how mobile in the environment.

The revisions would sacrifice the health and safety of the public – including children, workers, the elderly and fenceline communities – to expedite production of any potentially toxic chemical that the industry can persuade the Department of Energy, which is not charged with reviewing the health and safety of chemicals, to deem a 'critical energy resource.' We already know the limitless scope of what the chemical industry is likely to claim as "critical" based on their previous insistence that some of the most toxic chemicals in existence are "critical" for renewable energy or energy security, including PFAS,[153] asbestos,[154] and lead.[155]

Congress acknowledged TSCA's failure to address a host of dangerous chemicals including asbestos, TCE, methylene chloride and PFAS; and strengthened the law seven years ago with near unanimous support; this draft bill would roll back those protections.

The draft bill would:

- Mandate that EPA's risk evaluation of chemicals, rather than continuing to focus on their potential health risks, must also include the consideration of all cost and other "non-risk factors" when evaluating whether the chemical substance poses an unreasonable risk (as opposed to basing safety determinations solely on risks to health or the environment). The prioritization of economic considerations over public health protection was the major flaw that stymied progress under the old TSCA, and Congress' deliberate shift to risk-based evaluations and decision-making was the fundamental reform that brought the law back to life after being rendered ineffective and badly in need of reform.
- Allow new chemicals to begin production before EPA has completed its determination whether they pose an unreasonable risk to human health or the environment. Because Congress wanted EPA to make an affirmative determination of safety for all new chemicals, TSCA explicitly provides that no new chemical can enter production until that determination has been made. The bill would completely reverse this policy.
- In addition to the newly added consideration of costs and any other non-risk factors to EPA's analysis, which will lengthen the time necessary for review, the bill simultaneously prevents EPA from extending the review period for chemicals designated "critical energy resources." The inevitable result will be rushed and superficial reviews that fail to identify risks to health and the environment or incomplete reviews that result in default approvals of unsafe chemicals. As we have seen over and over, when a toxic chemical begins manufacture without a thorough review by EPA, it is almost impossible to end its production, or retrospectively

---

153 ACC, *PFAS: Critical to Renewable Energy*, https://www.americanchemistry.com/chemistry-in-america/chemistries/fluorotechnology-per-and-polyfluoroalkyl-substances-pfas/pfas-critical-to-renewable-energy.

154 ACC, *ACC Urges EPA to Reconsider its Flawed Chlor-alkali Proposal*, https://www.americanchemistry.com/chemistry-in-america/news-trends/press-release/2022/acc-urges-epa-to-reconsider-its-flawed-chlor-alkali-proposal.

155 International Lead Association, *Using Lead Responsibly is Critical to Achieving a Sustainable and Low Carbon Future*, https://ila-lead.org/sustainability/).

establish sufficient protections from the chemical to protect the public.

- Create a limitless loophole from TSCA's chemical assessment and health protection requirements. "Critical energy resource" is an open-ended and undefined concept that could apply to virtually any chemical that plays a role in the production, refining, distribution, and use of energy and is designated as "critical" by the Department of Energy. Once a substance is deemed to be a "critical energy resource" and therefore fast tracked through the PMN process, there is no limit on how the substance can then be used beyond its ostensible "critical energy resource" use and no constraint on non-energy applications that could be harmful to health and the environment.

The bill would establish a precedent for enacting further loopholes to gut the health protective provisions of the Act. If it is acceptable to gut health reviews of chemicals for "critical energy resources," what is the principle that will prevent other broad categories or uses of toxic chemicals from also getting special treatment under Section 5 of TSCA?

Notably missing from the draft bill are any findings demonstrating the need for the bill. In fact, there is no evidence that the public must sacrifice health protections from toxic chemicals in exchange for clean energy. We can develop and deploy new energy technologies without waiving chemical review requirements or placing the communities burdened by PFAS and other toxic chemicals at risk. The draft bill's rejection of that clean and health-protective energy future sells American innovation short.

Overwhelmingly, the public needs and wants more, not less, protection from toxic chemicals.[156] Yet the draft bill would roll back critical public health protections and weaken a core environmental law.

We look forward to speaking to all Members about the pernicious and dangerous nature of this legislation and working to prevent it from becoming law.

Daniel Rosenberg
Director of Federal Toxics Policy
Natural Resources Defense Council (NRDC)

Liz Hitchcock
Safer Chemicals Healthy Families Director
Toxic Free Future

Daniel Savery
Senior Legislative Representative
Earthjustice

Maria J. Doa, Ph.D
Senior Director, Chemicals Policy
Environmental Defense Fund

[156] University of California San Francisco Program on Reproductive Health and the Environment, *Public Opinion on Chemicals,* https://prhe.ucsf.edu/public-opinion-chemicals.

Madeleine Foote
Deputy Legislative Director
League of Conservation Voters

Scott Faber
Senior VP, Government Affairs
Environmental Working Group

Arthur Bowman III, Ph.D
Policy Director
Center for Environmental Health

Nancy Buermeyer
Director of Program and Policy
Breast Cancer Prevention Partners

Sarah Woodbury
Director of Advocacy
Defend Our Health

Lynn Thorp
National Campaigns Director
Clean Water Action

* * *

January 25, 2023

The Honorable Michael S. Regan
Administrator
U.S. Environmental Protection Agency
1200 Pennsylvania Avenue NW
Washington, D.C. 20460

Dear Administrator Regan:

We write to express our appreciation and support for your agency's work to protect Americans from chemical disasters by issuing the Safer Communities by Chemical Accident Prevention proposed rule. As the Environmental Protection Agency (EPA) works to finalize the rule, we write to reiterate our earlier call for you to deliver the strongest possible protections for workers in Risk Management Program (RMP) facilities, first responders, and environmental justice communities who live near these facilities and are most vulnerable to the cumulative impacts of chemical exposure. As you know, the RMP, as conceived in Section 112 of the Clean Air Act, was intended as a tool to prevent chemical facility disasters but to date the program has been largely reactive. After the prior Administration rolled back safety measures that ignored both the statute's core prevention objectives and the facts showing a strong need for action to prevent chemical disasters, EPA is making the right decision to change course and follow the science and the law. With this new rulemaking, EPA has the opportunity to finally fulfill the promise of the Act by making the proposed improvements, and by further strengthening the rule to prevent chemical disasters from happening in the first place.

Recent chemical disasters have highlighted shortcomings in federal regulations that fail to sufficiently protect workers and communities living near hazardous chemical facilities. From 2004-2020, EPA found that industry reported a total of 3,425 incidents, and acknowledged both reporting delays and under-reporting.[157] This does not account for near-misses, and some of these events could have been much worse if not for highly trained workers acting to prevent a more disastrous outcome. For example, just earlier this year, a chemical fire at a facility in Passaic, New Jersey nearly reached a warehouse storing three million pounds of chemicals, including chlorine pellets.[158] Catastrophe was largely averted due to the efforts of roughly 200 firefighters who worked for three days to keep the fire contained while nearby residents sheltered in place.

As climate change continues to drive more frequent and intense severe weather, EPA must also do more to protect communities from the "double disasters" that result when chemical disasters coincide with earthquakes and extreme weather events like hurricanes, floods, and wildfires. A February 2022 Government Accountability Office (GAO) report highlighted this need, finding that roughly a third of RMP facilities are at increased risk from climate impacts and that the current RMP rule does not adequately protect against these climate risks.[159]

We are encouraged by the steps that EPA has taken with this proposed rule toward protecting communities from the danger of chemical disasters, and we urge the agency to further strengthen the rule in several key ways. As many of us wrote to you in April of last year, the updated RMP rule should prioritize hazard reduction and prevention measures, including transitioning to inherently safer chemicals and processes and requiring third-party audits to verify compliance. However, the proposed rule excludes the vast majority (95%) of RMP facilities from the requirement to conduct a Safer Technologies and Alternatives Analysis (STAA), and in most cases only requires a third-party audit after two incidents have occurred. We encourage EPA to broadly require the transition to inherently safer chemicals and processes at RMP facilities, and to lower the threshold that prompts third-party compliance audits. In addition, given EPA's mandate to prioritize environmental justice established in E.O. 13990, "Protecting Public Health and the Environment and Restoring Science to Tackle the Climate Crisis," we urge the agency to require cumulative impact assessments as part of stationary source siting evaluations in the final rule.

To further ensure adequate protections for the millions of Americans that live near RMP facilities, the final rule should improve requirements for outreach to inform the public about RMP facility hazards and emergency response plans before and during incidental releases, and require that this information be made available in multiple languages. To foster information

---

157 U.S. Environmental Protection Agency. (19 April, 2022). Technical Background Document for Notice of Proposed Rulemaking: Risk Management Programs Under the Clean Air Act, Section 1 12(r)(7) Safer Communities by Chemical Accident Prevention. Docket# EPA-HQ-OLEM-2022-0174. https://www.regulations.gov/document/EPA-HQ-OLEM-2022-0174-0066.

158 Coming Clean and Environmental Justice Health Alliance for Chemical Policy Reform. (2022.) Preventing Disaster: Three chemical incidents within two weeks show urgent need for stronger federal safety requirements. https://comingcleaninc.org/assets/media/images/Reports/Preventing%20Disaster%20final.pdf.

159 United States Government Accountability Office. (2022, February). Chemical Accident Prevention: EPA Should Ensure Regulated Facilities Consider Risks from Climate Change. https://www.gao.gov/products/gao-22-104494.

access and transparency, EPA should maintain a publicly accessible RMP database and commit to delivering that database on the fastest possible timeline.

To adequately recognize and protect chemical facility workers and their unions as key partners in incident prevention, the final rule should: expand the conditions in which workers and their representatives are granted stop work authority; ensure that authority extends to workers in all RMP facilities regardless of program level; clarify and require safety reporting that better protects workers' anonymity; allow workers and their representatives to be meaningfully involved in all elements of the rule; require employers to provide workers information and training about the rule and to develop a written program to ensure there is no retaliation against employees for using their rights to prevent a chemical disaster.

Lastly, we commend the EPA for taking steps to recognize climate change as a threat multiplier and proposing important requirements to assess and plan for natural hazards and power loss. We encourage EPA to strengthen these provisions by also requiring implementation of mitigation measures to prevent these climate-related "double disasters," as emphasized in the GAO's February 2022 report. Additionally, the proposed rule does not require real-time air fenceline monitoring, leak detection, nor full facility back-up power, and would be strengthened by each of these requirements. We also note concerns about air monitoring and control equipment being removed from service before extreme weather events, as occurred during Hurricane Harvey, which leaves community members and regulators in the dark as to the full extent of air pollution and chemical disasters that may be exacerbated by extreme weather and/or power loss. The final rule can be strengthened by requiring penalties for intentionally removing air monitoring and control equipment from service, including before extreme weather events.

Thank you for your ongoing work on this and other environmental justice issues. We look forward to continuing to work with you to ensure that the communities we represent, and those across the country, are protected from the danger of chemical disasters by a truly preventative Risk Management Program.

Sincerely,

Cory A. Booker
United States Senator

Nanette Diaz Barragan
Member of Congress

Thomas R. Carper
United States Senator

Lisa Blunt Rochester
Member of Congress

Tammy Baldwin
United States Senator

Suzanne Bonamici
Member of Congress

Benjamin L. Cardin
United States Senator

Tony Cardenas
Member of Congress

Tammy Duckworth
United States Senator

Andre Carson
Member of Congress

Richard J. Durbin
United States Senator

Steve Cohen
Member of Congress

Dianne Feinstein
United States Senator

Mark DeSaulnier
Member of Congress

Edward J. Markey
United States Senator

Robert Menendez
United States Senator

Debbie Dingell
Member of Congress

Adriano Espaillat
Member of Congress

Jeffrey A. Merkley
United States Senator

John Garamendi
Member of Congress

Alex Padilla
United States Senator

Robert Garcia
Member of Congress

Bernard Sanders
United States Senator

Raul M. Grijalva
Member of Congress

Chris Van Hollen
United States Senator

Val Hoyle
Member of Congress

Elizabeth Warren
United States Senator

Jared Huffman
Member of Congress

Sheldon Whitehouse
United States Senator

Ro Khanna
Member of Congress

Ron Wyden
United States Senator

Barbara Lee
Member of Congress

Doris Matsui
Member of Congress

Betty McCollum
Member of Congress

Jerrold Nadler
Member of Congress

Grace F. Napolitano
Member of Congress

Eleanor Holmes Norton
Member of Congress

MarkPocan
Member of Congress

Jamie Raskin
Member of Congress

Mary Gay Scanlon
Member of Congress

Adam Smith
Member of Congress

Melanie Stansbury
Member of Congress

Haley M. Stevens
Member of Congress

Dina Titus
Member of Congress

Rashida Tlaib
Member of Congress

Jill Tokuda
Member of Congress

Paul D. Tonko
Member of Congress

Maxine Waters
Member of Congress

Bonnie Watson Coleman
Member of Congress

* * *

February 7, 2023

The Honorable Cathy McMorris Rodgers
Chair
Committee on Energy and Commerce
2125 Rayburn House Office Building
Washington, D.C. 20515

The Honorable Bill Johnson
Chair
Subcommittee on Environment,
Manufacturing, and Critical Materials
2125 Rayburn House Office Building
Washington, D.C. 20515

The Honorable Jeff Duncan
Chair
Subcommittee on Energy, Climate,
and Grid Security and Grid Security
2125 Rayburn House Office Building
Washington, D.C. 20515

The Honorable Frank Pallone, Jr.
Ranking Member
Committee on Energy and Commerce
2125 Rayburn House Office Building
Washington, D.C. 20515

The Honorable Paul Tonko
Ranking Member
Subcommittee on Environment,
Manufacturing, and Critical Materials
2125 Rayburn House Office Building
Washington, D.C. 20515

The Honorable Diana DeGette
Ranking Member
Subcommittee on Energy, Climate,
and Grid Security and Grid Security
2125 Rayburn House Office Building
Washington, D.C. 20515

Dear Chairs Rodgers, Johnson, and Duncan, and Ranking Members Pallone, Tonko, and DeGette:

We, the undersigned, write to express our strong opposition to legislation that would repeal Section 134 of the Clean Air Act, otherwise known as the Greenhouse Gas Reduction Fund

(GHGRF), and to urge Members of the Committee on Energy and Commerce to join us in opposing such legislation.

The Inflation Reduction Act of 2022 amended the Clean Air Act to establish the GHGRF, a $27 billion first-of-its-kind program to support the rapid deployment of low- and zero-emission technologies. The program is split between a $20 billion green finance program, through which nonprofit entities will leverage private capital to fill gaps in access to green finance that have long persisted, and a $7 billion program to assist states, local governments, and Tribes in deploying zero-emission technologies in low-income and disadvantaged communities. Together, these two funding streams will ensure that all communities – particularly the most underserved and marginalized communities – can benefit from the transition to a low-carbon economy.

The GHGRF is a critical tool for combatting the climate crisis, lowering energy costs, and advancing environmental justice. This program has unprecedented potential to improve lives by investing in high-impact, community-based projects that reduce climate pollution and improve public health. Importantly, at least $15 billion of GHGRF investments – more than 55 percent of overall program funding – will be dedicated to projects in low-income and disadvantaged communities that have long been overlooked by financial markets. These communities contribute the least to climate change, yet they are hit hardest by the economic and public health impacts of a warming world. The GHGRF will not only help these communities deploy clean energy technologies and enhance their resilience to climate change, but it will also spur local and regional job creation, as well as enable wealth-building and community ownership of clean energy projects.

The GHGRF promises to play a unique and vital role filling gaps in access to green capital. By statute, the GHGRF's $20 billion green finance program is required to support projects that otherwise lack access to capital. In other words, it will not duplicate investments currently being made elsewhere – by the public or private sector – but will instead be complementary and spur investment where it is needed most.

Moreover, the GHGRF does not seek to reinvent the wheel. Rather, its green finance program builds on the successful model and track record of green banks and community-based lenders like community development financial institutions (CDFIs) and credit unions that have expanded access to green capital in states, cities, and regions – and, in particular, in environmental justice communities. Similarly, the GHGRF's $7 billion program to deploy zero-emission technologies through states, local governments, and Tribes will ensure that clean energy solutions are tailored to a given jurisdiction's geographic, market, and regulatory needs.

With a one-time infusion of public dollars, the GHGRF will supercharge the deployment of green capital and accelerate the transition to an equitable clean energy future. All told, the GHGRF is expected to catalyze far more than $27 billion in pollution-reducing investments, making it a key part of our efforts to build a more inclusive and more affordable low-carbon economy. We urge Members of Congress and this Committee to recognize the GHGRF's vast potential to deliver benefits for everyone and to oppose efforts to undermine this program before it even has the chance to yield results. Thank you for your consideration.

Sincerely,
Earthjustice
Environmental Defense Fund
Evergreen Action

League of Conservation Voters
Natural Resources Defense Council
Sierra Club
WE ACT for Environmental Justice

House Committee on Energy and Commerce, February 7, 2023, Joint Hearing of the Subcommittee on Energy, Climate, and Grid Security and the Subcommittee on Environment, Manufacturing, and Critical Materials, "Unleashing American Energy, Lowering Energy Costs, and Strengthening Supply Chains," Responses to Submitted Questions for the Record Mark W. Menezes, Former Deputy Secretary, Department of Energy, Former Chief Counsel, Energy and Environment, House Committee on Energy and Commerce

Griffith question:

Is distributing grants to lending institutions a core purpose of the Clean Air Act?

Answer: The "core purpose" of the Clean Air Act is to protect our Nation's air quality through the development, implementation and enforcement of emission prevention and control standards, not distributing grants to lending institutions. In fact, in the Clean Air Act, Congress made clear that implementation of the Act is the primary responsibility of the States and local governments, not EPA. To help the States and local governments achieve the goal of air pollution prevention and control, Congress authorized the EPA to provide both technical and financial assistance to air pollution control agencies to develop and execute their air pollution prevention and control programs. To my knowledge, EPA has not made these financial grants available to lending institutions.

Allen question:

Can you give me a list of actions taken by the Biden Administration that has made the U.S. more reliant on foreign energy and has decreased U.S. domestic energy production?

Answer: The list of actions taken to curtail domestic energy production has been widely reported to include cancelling or placing moratoria on onshore and offshore leases, taking federal lands out of production, and the like. A few actions that might not be as obvious are delaying the issuance of permits on necessary energy infrastructure projects, issuing GHG guidance to federal agencies to use the social cost of carbon under NEPA, minimizing the use of categorical exceptions under NEPA, withdrawing federal support of interstate pipelines to move shale oil and gas to New England to replace foreign imports, and asking OPEC countries to increase production for America rather than ask American producers to fill the gap caused by Putin's invasion of Ukraine.

Miller-Meeks question:

As we think about global competitiveness in the context of critical materials, and as we discuss legislation related to this topic how can we continue to build off of exiting efforts to support our nation's energy supply, while also recycling and reusing minerals that have been discarded in technologies such as unused hard drives electronic devices, and others?

Answer: Today, the US finds itself dependent on other countries, especially China, for many critical minerals and materials necessary for the technologies necessary in the daily lives of all Americans and in the component parts of our energy infrastructure, particularly in electric vehicles and renewable energy. The US is not a global leader or even a global competitor in producing or supplying these critical minerals or materials. Since it will take some time to develop our own sources or discover or create replacements, recycling and reuse of discarded devices is an important component in our pathway to competitiveness. As you mentioned, the Ames Laboratory leads our nation's Critical Materials Institute, which I had the distinct honor to visit when I was then Under Secretary of Energy. The CMI is doing innovative, groundbreaking work on recovering lithium, cobalt, platinum—a growing number of critical minerals and materials—through a growing number of processes.

Any legislation to address this need to recycle and reuse should consider enlarging the Department of Energy's role in ensuring and adequate and reliable supply of critical energy resources that are essential to the energy security of the US under the DOE Organization Act. That statutory priority will ensure the necessary resources and the continued leadership of the CMI's role to accelerate the discovery of new materials and more cost-effective and efficient ways to recycle and reuse discarded materials. Additionally, legislation could include inducements to encourage Americans to collect, transport, process and recycle, including rebates, tax incentives, and educational support.

Burgess question:

Can you speak to the beneficial environmental impact of pipelines as compared to other forms of energy transportation?

Answer: Pipelines are the most efficient transport of oil, refined products, natural gas, and industrial gases from production to processing to end-users. Unlike other forms of transportation with congestion and bottleneck logistics and being prone to frequent accidents, pipelines provide a safe, uninterrupted supply of gases and liquids necessary for our vibrant economy. The environmental impacts of pipelines are several including fewer spills or discharges, small above-ground footprint, and lower amounts of energy to transport large volumes of energy and supplies.

What role can Congress play in creating a resilient and diverse baseload generation ecosystem?

Answer: Congress can play an important role in creating such an ecosystem in several ways. Congress can pass your legislation to improve the permitting process for natural gas pipelines. Today, most new generation being placed in service is natural gas and renewables. Both complement one another as natural gas plants ramp up and down to accommodate the intermittency of wind and solar. However, NERC is warning that we need to have more access to natural gas in order to have natural gas generation available to provide replacement power.

In addition, in my view, Congress should mandate FERC to develop market rules that send adequate price signals for the development of fuel-secure, low-emitting generation sources like nuclear, natural gas, clean coal, and pumped storage hydro. Until we develop breakthrough technologies for reliable battery storage, we will continue to need all sources of generation to make our electric system reliable, affordable and resilient.

Congress should also provide parity incentives to natural gas, clean coal and other low-emitting energy sources as Congress provided to other fuel resources in the Inflation Reduction Act. The world needs to reduce emissions, not become vulnerable to energy outages due to an over dependency on few intermittent sources of generation.

House Committee on Energy and Commerce, February 7, 2023, Joint Hearing of the Subcommittee on Energy, Climate, and Grid Security and the Subcommittee on Environment, Manufacturing, and Critical Materials, "Unleashing American Energy, Lowering Energy Costs, and Strengthening Supply Chains," Responses to Submitted Questions for the Record, Katie Sweeney, Executive Vice President and General Counsel, National Mining Association

The Honorable Rick W. Allen

1) Permitting is a primary hindrance to securing our mineral supply chains. Can you describe another key hinderance – a qualified and robust mining workforce?
    a) How can we support our workforce to ensure we have the tools and the women and men necessary to achieve our energy and security objectives?

Sweeney Response: A strong domestic mining workforce is critical to the success of American economic competitiveness, national security, and emissions reductions for our current and future generations. Unfortunately, nearly all mining company have highlighted worker shortages and trouble with recruiting enough qualified and skilled workers as an issue affecting their business. This is in both skilled trades and also professional occupations like engineers, geoscientists, and metallurgists. In addition, the industry is experiencing increasing labor shortages due to retirements, with increasing numbers of workers nearing retirement age.

Modern mining is a people first industry that supports the communities where we live and work. Mining depends on a 21st century workforce to drive innovation while continuing to prioritize health and safety and working to reduce environmental impacts. To do this, we must recruit, educate, train, and develop tomorrow's workforce today, giving them the tools they will need to meet the challenge head on.

Coordinated federal, state, local, and industry efforts are needed to address this issue and invest in the mining workforce. The National Mining Association strongly supports federal actions to support the recruitment and training of a robust domestic mining workforce, specifically:

- The bipartisan and bicameral Mining School Act, which directs the Department of Energy, in coordination with the Department of the Interior, to establish a competitive grant program for mining schools to support recruitment and education opportunities for engineers and other qualified professionals in the field of mining.
- The Consolidated Appropriations Act of 2021 authorized DOE under Section 7002(K) to promote a secure and robust critical minerals supply chain by requiring the Secretary of Labor and the Director of the National Science Foundation to develop curriculum and an interdisciplinary program for institutions of higher education to strengthen the workforce for exploration and development of critical minerals and critical minerals manufacturing. The program will be designed to support the critical mineral supply chain and improve the ability of the United States to increase domestic, critical mineral exploration, development, production, manufacturing, research, including fundamental research into alternatives, and recycling.
- To date, no funding has been appropriated for these activities. NMA encourages Congress to support these activities and fully funding the authorized amount under Sec. 7002(K) at $50 million annually through FY 2029.

The Honorable Mariannette Miller-Meeks, M.D.

The draft bill "to prohibit the importation into the United States of unirradiated low-enriched uranium that is produced in the Russian Federation, and for other purposes" currently includes waiver authority until January 2028, meaning that the Secretary of Energy can authorize imports to sustain U.S. nuclear reactors or operations, or if it is in the national interest. In January 2028, imports of low-enriched uranium from Russia will no longer be allowed under any circumstances. I agree that importing uranium from Russia even in select circumstances needs to be limited, but Mrs. Sweeney, I am interested to get your thoughts on 2028 as the cutoff point.

1) Would you say waiver authority until January 2028 is reasonable, or does this timeline need to be adjusted?
   Sweeney Response: Our uranium import dependence is a case study in how our vital domestic minerals supply chains have atrophied to levels that result in a dire national security risk. The U.S. is home to the world's largest fleet of nuclear power plants, significant uranium reserves, and yet we import virtually all of the uranium we use – half of which comes from Russia, Kazakhstan and Uzbekistan. Aggressive action must be taken to address this vulnerability and immediately reinvest in American-sourced, essential mined materials produced under world-leading environmental and labor standards.
   As we continue to rely on Russia's artificially low-priced uranium to supply U.S. commercial reactors, domestic uranium production and employment are at levels not seen since the dawn of the nuclear industry in the 1940s. The Prohibiting Russian Uranium Imports Act (H.R. 1042) is a lifeline providing needed certainty to supply chains in the future, however, given the precarious state of the front-end of America's nuclear fuel cycle, it is urgent that Congress and the Department of Energy take

immediate action to support and fully fund the Strategic Uranium Reserve. This reserve will ensure the availability of nuclear fuel in the very likely event of a market disruption. Further, it will preserve our uranium production capabilities and provide a means to supply the U.S.-origin uranium required for defense programs, which currently rely on stockpiles that are finite and diminishing.

2) What actions would Congress need to take to prepare for an end to all imports of Russian LEU by January 2028?
Sweeney Response: Congress must fully fund the Strategic Uranium Reserve, as recommended by the U.S. Nuclear Fuel Working Group and originally proposed by the Department of Energy's Office of Nuclear Energy to address our nation's overreliance on imported uranium that has undermined energy security and impacted fuel supply capabilities. The time to act is now while we still have the infrastructure and licensed capacity in place to begin supplying the uranium reserve, capabilities, and human resources that would take decades and enormous financial commitments to rebuild if lost.

3) What means at the federal level would be needed for the U.S. to withstand a ban on all imports of Russian LEU *even earlier* than January 2028?
Sweeney Response: Continued funding of $150,000,000 annually to support the Strategic Uranium Reserve will support the uranium mining and conversion industries, further ensuring that energy and national security supply chains will not be impacted by the loss of Russian uranium imports.

* * *

Bernard L. McNamee

March 15, 2023
Kaitlyn Peterson
Legislative Clerk
Committee on Energy and Commerce,
2125 Raybmn House Office Building
Washington, D.C. 20515
Kaitlyn.Peterson@mail.house.gov.

Re: Responses of Bernard L. McNamee to Question for the Record made on March 2, 2023 in relation to appearance on February 7, 2023

Dear Ms. Peterson:

Thank you to Chair Rodgers and Ranking Member Pallone, Chair Duncan and Ranking member DeGette, Chair Johnson and Ranking Member Tonko, the Members of the Committee, and Committee Staff for providing me an opportunity to appear before the Subcommittee on Environment, Manufactming, and Critical Materials and Subcommittee on Energy, Climate, and Grid Security on Tuesday, Febrnaiy 7, 2023, and to testify at the joint legislative heai·ing entitled "Unleashing American Energy, Lowering Energy Costs, and Strengthening Supply Chains."

Below are my responses to the QFRs posed by Representatives Griffith and Miller-Meeks.

The Honorable H. Morgan Griffith

1. I am certainly supportive of energy infrastructure permitting efficiencies such as the interagency pipeline review bill this Committee is considering. It is my understanding that, under the National Environmental Policy Act, the Natural Gas Act, and specifically Title 18 of the Code of Federal Regulations Sec. 380.15, FERC should consider the siting of projects in existing right-of- way. In your experience as a FERC commissioner, does the Commission give due consideration to existing rights-of-way on these applications and in pre- filing discussions?

McNamee Response Morgan Q1: It has been over two years since I served on the Federal Energy Regulatory Commission ("FERC" or "Commission") and I will not comment on any specific proceeding in which I participated. Furthermore, I cannot testify with personal knowledge as to how FERC reviews have taking place since my term ended. However, I am able to provide a general description of how applications for natural gas pipeline certificates are reviewed by FERC.

There are multiple steps and checks on the Commission when considering an application for a natural gas pipeline. The process for approving a natural gas pipeline under Section 7 of the Natural Gas Act usually begins with the pre-filing process in which an applicant will engage with Commission Staff (primarily the Office of Energy Projects) regarding the proposed pipeline. This provides an opportunity for the applicant and staff to discuss routing issues, including potential use of existing rights-of-way. With the development of either an Environmental Assessment or an Environmental Impact Statement impacts of the propose pipeline on the route are examined. The EA and EIS are initially published as drafts and subject to public comment and are often revised in consideration of those comments. Such comments can also raise issues of rights-of-way. When FERC considers whether to approve a pipeline under section 7 of the NGA, it considers the environmental impacts of the proposed pipeline. Parties can file comments on the proposed pipeline application, including about the rights-of-way. The Commission is supposed to address all of the comments filed in relation to the proposed application for the pipeline, including if rights-of-way issues are raised. The Commission's decisions are ultimately subject to rehearing by the Commission and review by the courts. A simple flow chart of the FERC process is presented on FERC's website.[160]

As a general matter, I believe there are there are likely times where the evidence supports and does not support using the existing rights of way. Those decisions should be driven by the particular facts of the case. But the process should promote the proper consideration of rights-of-way.

2. How does FERC independently assess and verify multiple projects from different applicants, make use of single rights-of-way? Even if only for a short portion of the project route.

McNamee Response Morgan Q2: As discussed above, it has been over two years since I served on the Federal Energy Regulatory Commission ("FERC" or "Commission") and I will

160 FERC, Natural Gas Certificate Process, https://www ferc.gov/sites/default/files/2020-04/FERCNaturalGasCertificateProcess.pdf (accessed March 13, 2023).

not comment on any specific proceeding in which I participated. Furthermore, I cannot testify with personal knowledge as to how FERC reviews have taking place since my term ended. However, I surmise that the process described in McNamee Response Morgan Q1 would help ensure that such use of rights of way by multiple projects was utilized as appropriate.

The Honorable Mariannette Mill-Meeks, M.D.

While I do not represent a district on the east coast, I was struck last May by a top U.S. fuel pipeline operator, Colonial Pipeline, shutting its entire network after a cyber attack that involved ransomware. Colonial is the source of nearly half of the U.S. East Coast's fuel supply, and the incident is a prime example of how vulnerable U.S. energy infrastructure is to hackers. I am fully aware of the consequences of a cyberattack t energy infrastructure in my district. [citation omitted]

1. What sort of actions can Congress take to reduce the vulnerability of domestic pipelines and electric infrastructure to cyber-attacks?

McNamee Response Mill-Meeks Q1: The threats from foreign adversaries and non-state actors on American energy infrastructure are real and continue to grow.[161]

Background on Current Efforts to Protect Energy Infrastructure from Cyberattacks

The federal government has multiple agencies and initiatives working on assisting the private sector protect energy infrastructure. These agencies include, the Department of Energy's Office of Cybersecurity, Energy Security, and Emergency Response (CESER) and the Department of Homeland Security.

As discussed on the Department of Energy's CESER website these initiatives include:

> Industry partners include the Electricity Subsector Coordinating Council (ESCC), the Electricity Information Sharing and Analysis Center (E-ISAC), the Oil and Natural Gas Subsector Coordinating Council, and industry-led research partnerships. Federal partners include DHS via the Industrial Control Systems Cyber Emergency Response Team (ICS-CERT), Science & Technology, and the National Cybersecurity and Communications Integration Center (NCCIC); NIST Smart Grid Interoperability Panel (SGIP); DARPA; DOD and others. In particular, CESER is a member of the Networking and Information Technology Research and Development (NITRD) program that provides a forum for inter-agency coordination of networking and information technology research activities.[162]

In addition, "[t]he Cybersecurity and Infrastructure Security Agency (CISA), through the National Risk Management Center (NRMC), is working with government and industry partners

---

[161] U.S. Department of Energy, "Securing America's Energy Infrastructure from Cyber Threats," July 26, 2021 https://www.energy.gov/articles/securing-americas-energy-infrastructure-cyber-threats (accessed March 11, 2023).

[162] U.S. Department of Energy, Office of Cybersecurity, Energy Security, and Emergency Response, Cybersecurity, https://www.energy.gov/ceser/cybersecurity (accessed March 11, 2023).

to identify cybersecurity risks and develop strategies to strengthen the security and resilience of the Nation's pipeline infrastructure."[163]

DHS's Transportation Security Administration (TSA) revised and reissued its Security Directive regarding oil and natural gas pipeline cybersecurity in July 2022.[164]

Under the Energy Policy Act of 2005, FERC oversees the reliability of the bulk power system (a/k/a "the electric grid"). Under this authority FERC has approved mandatory cybersecurity reliability standards. These standards are developed with the assistance of the North American Electric Reliability Corporation (NERC). These standards are referred to as Critical Infrastructure Protection (CIP) cyber security reliability standards.[165]

My Thoughts and Recommendations

The issue with protecting America's energy infrastructure from cybersecurity threats is not merely one of needing more government regulation or standards on the private sector (though such standards can help). Securing our energy infrastructure requires: 1) ongoing vigilance and investment by the private sector to protect assets; 2) government and the private sector sharing threat information with each other on a continual and timely basis; 3) not sourcing critical components (such as chips, switches, and transformers) from hostile nations, like China, and instead obtain such critical technology from American companies and our allies; and 4) establishing a new national security policy (with the resources and personnel to support it) that will make it clear to adversaries that the U.S. government/military will impose economic sanctions and military retaliation for cyber or physical attacks on American energy infrastructure as the situation merits.

Once again, thank you for allowing me to participate in the Committee's important work on behalf of the American people.

Sincerely,
/s/
Bernard L. McNamee

---

[163] Cybersecurity and Infrastructure Security Agency, Pipeline Cybersecurity, https://www.cisa.gov/sites/default/files/publications/fact sheet pci 508.pdf (accessed March 11, 2023).

[164] Transportation Safety Administration, TSA revises and reissues cybersecurity requirements for pipeline owners and operators, July 21, 2022, https://www.tsa.gov/news/press/releases/2022/07/21/tsa-revises-and-reissues-cybersecurity-requirements-pipeline-owners (accessed March 11, 2023).

[165] FERC, Cyber and Grid Security, https://www.ferc.gov/industries-data/electric/industry-activities/cyber-and-grid-security (accessed March 11, 2023).

Index

N

O

P

R

S

T

U

V

W

Y

Z